Amphibians and Reptiles of Georgia

Amphibians and

The
University
of Georgia
Press

Athens and
London

Reptiles of Georgia

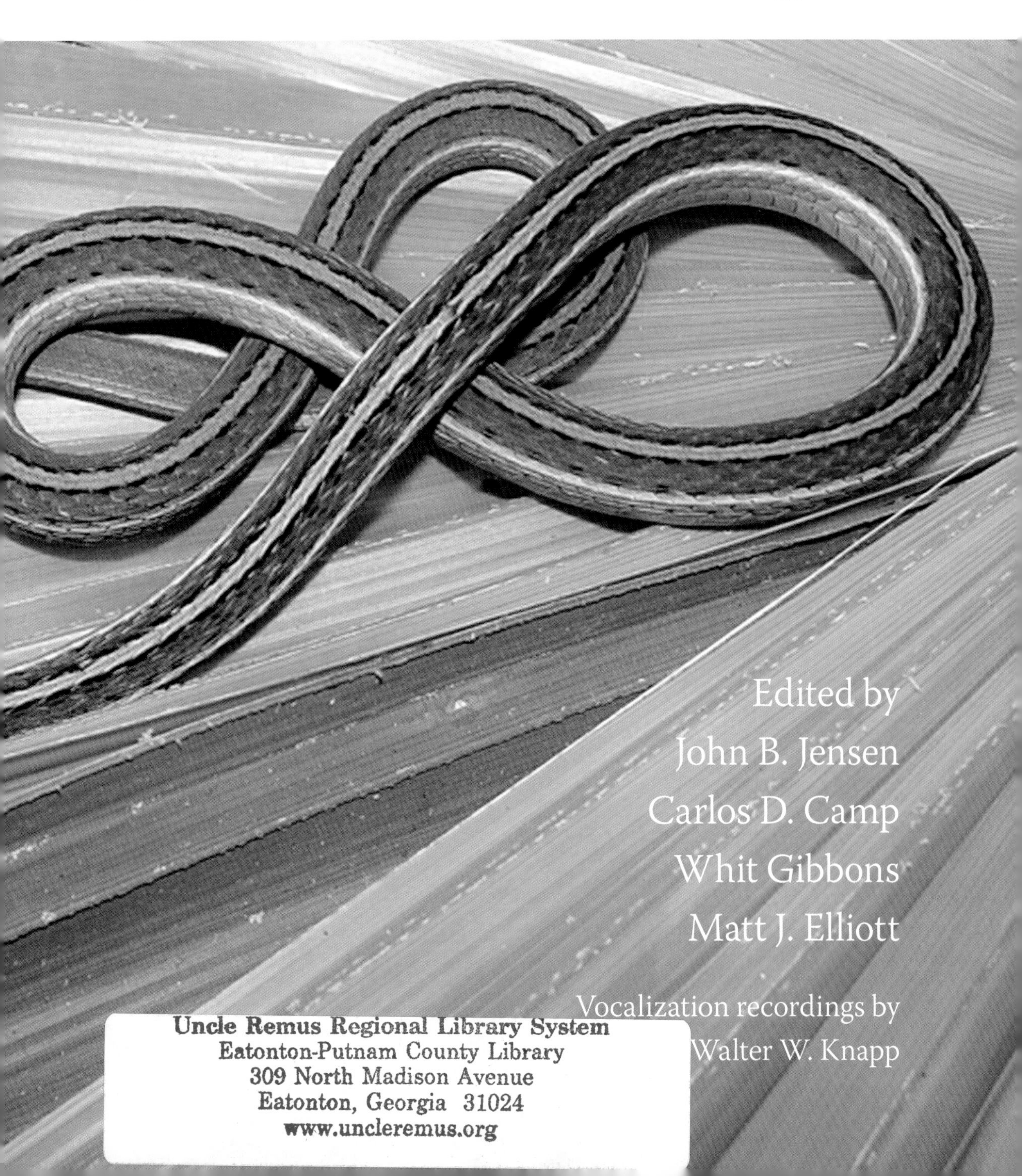

Edited by
John B. Jensen
Carlos D. Camp
Whit Gibbons
Matt J. Elliott

Vocalization recordings by
Walter W. Knapp

Designed by April Leidig-Higgins
Line art by Bricelyn H. Strauch
Set in Warnock Light by
Graphic Composition, Inc.; Bogart, GA.
Printed and bound by Everbest Printing Company
for Four Colour Imports

The paper in this book meets the guidelines for permanence and durability of the Committee on Production Guidelines for Book Longevity of the Council on Library Resources.

Printed in China
12 11 10 09 08 P 5 4 3 2 1

Library of Congress Cataloging-in-Publication Data

Amphibians and reptiles of Georgia / edited by
John B. Jensen . . . [et al.].
p. cm.
Includes bibliographical references and index.
ISBN-13: 978-0-8203-3111-9 (pbk. : alk. paper)
ISBN-10 0-8203-3111-2 (pbk. : alk. paper)
1. Amphibians—Georgia. 2. Reptiles—Georgia.
3. Amphibians—Georgia—Identification.
4. Reptiles—Georgia—Identification.
I. Jensen, John B., 1966–
QL653.G4.A46 2008
597.909758—dc22 2007034022

British Library Cataloging-in-Publication Data available

This book has been sponsored in part by

Contributing Authors

Thomas S. B. Akre, Longwood University
Kimberly M. Andrews, Savannah River Ecology Laboratory
Mark A. Bailey, Conservation Southeast Inc.
Roger Birkhead, Auburn University
Berkeley W. Boone, Savannah River Ecology Laboratory
Kurt A. Buhlmann, Savannah River Ecology Laboratory
Carlos D. Camp, Piedmont College
Todd S. Campbell, University of Tampa
W. Ben Cash, Maryville College
Dean A. Croshaw, University of New Orleans
Mark S. Davis, North Georgia College and State University
Michael E. Dorcas, Davidson College
Matthew J. Elliott, Georgia Department of Natural Resources
Kenneth M. Fahey, Brenau University
Thomas M. Floyd, Georgia Department of Natural Resources
Whit Gibbons, Savannah River Ecology Laboratory
Xavier Glaudas, University of Nevada–Las Vegas
James C. Godwin, Alabama Natural Heritage Program
Gabrielle J. Graeter, Savannah River Ecology Laboratory
Judith L. Greene, Savannah River Ecology Laboratory
Gregory C. Greer, International Expeditions
Bob Herrington, Georgia Southwestern State University
Pierson Hill, Davidson College
Jeff Humphries, North Carolina Wildlife Resources Commission
Natalie L. Hyslop, University of Georgia
John B. Jensen, Georgia Department of Natural Resources
Kenneth L. Krysko, Florida Museum of Natural History
Trip Lamb, East Carolina University
Emily C. Moriarty Lemmon, University of Texas
W. H. Martin, Harpers Ferry, W.Va.
Peri A. Mason, University of Georgia
D. Bruce Means, Coastal Plains Institute
Brian S. Metts, Savannah River Ecology Laboratory
Mark S. Mills, University of Southern Missouri
Tony Mills, Savannah River Ecology Laboratory
Paul E. Moler, Florida Fish and Wildlife Conservation Commission
Robert A. Moulis, Savannah, Ga.
Sean M. Poppy, Savannah River Ecology Laboratory
Steven J. Price, Davidson College
Robert N. Reed, Southern Utah University
James I. Richardson, University of Georgia
Betsie B. Rothermel, Austin Peay State University
David E. Scott, Savannah River Ecology Laboratory
Lora L. Smith, Joseph W. Jones Ecological Research Center
Stacy N. Smith, Athens, Ga.
Kristina Sorensen, United States Fish and Wildlife Service
Philip B. Spivey, Georgia Department of Natural Resources
Dirk J. Stevenson, Fort Stewart Military Installation
Brian D. Todd, Savannah River Ecology Laboratory
Tracey D. Tuberville, Savannah River Ecology Laboratory
D. Gregory Waters, Georgia Department of Natural Resources
John D. Willson, Savannah River Ecology Laboratory
Christopher T. Winne, Savannah River Ecology Laboratory
Cameron A. Young, Center for Snake Conservation

Contents

Amphibian and Reptile Vocalizations

Available for download at www.ugapress.org/AmphibsandReptiles.html
(recordings by Walter W. Knapp except where noted; narration by John B. Jensen)

01_American_toad.mp3
Bufo americanus, Franklin Co., Ga.
02_Fowler's_toad.mp3
Bufo fowleri, Whitfield Co., Ga.; Madison Co., Ga.
03_Oak_toad.mp3
Bufo quercicus, Long Co., Ga.
04_Southern_toad.mp3
Bufo terrestris, Coffee Co., Ga.
05_Northern_cricket_frog.mp3
Acris crepitans, Lamar Co., Ga.
06_Southern_cricket_frog.mp3
Acris gryllus, Ben Hill Co., Ga.; Polk Co., Ga.
07_Bird-voiced_treefrog.mp3
Hyla avivoca, Burke Co., Ga.; Hancock Co., Ga.
08_Cope's_gray_treefrog.mp3
Hyla chrysoscelis, White Co., Ga.; Terrell Co., Ga.; Fayette Co., Ga.
09_Green_treefrog.mp3
Hyla cinerea, Pulaski Co., Ga.; Ben Hill Co., Ga.; Ben Hill Co., Ga.; Toombs Co., Ga.
10_Pine_woods_treefrog.mp3
Hyla femoralis, Clinch Co., Ga.
11_Barking_treefrog.mp3
Hyla gratiosa, Dooly Co., Ga.; Oglethorpe Co., Ga.
12_Squirrel_treefrog.mp3
Hyla squirella, Burke Co., Ga.
13_Mountain_chorus_frog.mp3
Pseudacris brachyphona, Dawson Co., Ga.
14_Brimley's_chorus_frog.mp3
Pseudacris brimleyi, Screven Co., Ga.
15_Spring_peeper.mp3
Pseudacris crucifer, Jasper Co., Ga.
16_Upland_chorus_frog.mp3
Pseudacris feriarum, Walton Co., Ga.
17_Southern_chorus_frog.mp3
Pseudacris nigrita, Taylor Co., Ga.

18_Little_grass_frog.mp3
Pseudacris ocularis, Jenkins Co., Ga.
19_Ornate_chorus_frog.mp3
Pseudacris ornata, Taylor Co., Ga.
20_Greenhouse_frog.mp3
Eleutherodactylus planirostris, Glynn Co., Ga.
21_Eastern_narrow-mouthed_toad.mp3
Gastrophryne carolinensis, Dade Co., Ga.; Dougherty Co., Ga.
22_Eastern_spadefoot.mp3
Scaphiopus holbrookii, Jenkins Co., Ga.
23_Gopher_frog.mp3
Rana capito, Taylor Co., Ga.
24_Bullfrog.mp3
Rana catesbeiana, Lowndes Co., Ga.; Appling Co., Ga.; Walker Co., Ga.; Wilkes Co., Ga.
25_Green_frog.mp3
Rana clamitans, McDuffie Co., Ga.; Hart Co., Ga.; Walton Co., Ga.
26_Pig_frog.mp3
Rana grylio, Ware Co., Ga.; Seminole Co., Ga.; Terrell Co., Ga.; Ben Hill Co., Ga.
27_River_frog.mp3
Rana heckscheri, Tattnall Co., Ga.
28_Pickerel_frog.mp3
Rana palustris, Hall Co., Ga.
29_Southern_leopard_frog.mp3
Rana sphenocephala, Jackson Co., Ga.
30_Wood_frog.mp3
Rana sylvatica, White Co., Ga.
31_Carpenter_frog.mp3
Rana virgatipes, Aiken Co., S.C.
32_American_alligator_hatchling.mp3
Alligator mississippiensis (hatchlings), Levy Co., Fla.
33_American_alligator_adult_male.mp3
Alligator mississippiensis (adult males), St. Johns Co., Fla. (Recording by Kent A. Vliet)

Species of Possible Occurrence in Georgia

34_Pine_barrens_treefrog.mp3
Hyla andersonii, Escambia Co., Ala.
35_Gray_treefrog.mp3
Hyla versicolor, Stone Co., Mo.
36_Cuban_treefrog.mp3
Osteopilus septentrionalis, DeSoto Co., Fla.

Acknowledgments

MANY CONTRIBUTING authors and photographers, who are credited along with their work, were instrumental in making this volume possible. Countless other people also were critical to the completion and quality of this book. Here we acknowledge those who contributed their time, expertise, knowledge, and support. We hope those whose names we inadvertently omitted will forgive us and buy the book anyway.

Jim Flynn deserves special thanks for developing, maintaining, and managing a Web site for soliciting photographs and for posting and evaluating submitted photographs. Roger Birkhead, John Friel, Steve Gotte, Bob Herrington, Kenney Krysko, Lance McBrayer, Elizabeth McGhee, Paul Moler, Ronald Nussbaum, Dennis Parmley, Greg Schneider, and George Zug examined or loaned us specimens from various museums, ensuring that our range maps and distributional information are as accurate as possible. Terry Schwaner provided genetic identification of questionable salamander larvae. Sarah Berckman and Thomas Floyd provided critical specimen data from the Georgia Natural Heritage database and obscure literature records, respectively. Todd Schneider developed the database that organized the entire Georgia amphibian and reptile records; Jeanne Bohannon, Thomas Floyd, Katrina Lyons, and Phil Spivey entered much of that information. Melissa Hayes assisted greatly with the organization of the manuscript and range maps. Thomas Floyd, Sean Graham, Delton Hilliard, Grace Hilliard, Terry Johnson, Walter Knapp, Ron Lee, Erin McGuire, and Cameron Young secured live specimens of amphibians and reptiles desperately needed for photographing. Audrey Owens provided many helpful suggestions for illustrations used in the book. Liz Kramer, Jason Lee, Kevin Samples, and Lisa Anderson at the University of Georgia's Natural Resources Spatial Analysis Laboratory assisted with several of the figures. Rick Lavender graciously helped organize the index.

A great deal of the amphibian and reptile occurrence data was obtained by volunteers of the Georgia Herp Atlas (GHA) project. GHA supervolunteer Walter Knapp provided nearly half of the records obtained during this project. We thank the following other GHA volunteers for their time and effort: Sandy Abbott, Audrey Albright, Gary Albright, Judy Albright, Logan Albright, Bill Alley, Jim Allison, Fred Alvery, Jerry Amerson, Marie Amerson, Ann Amis, Brandon Anderson, Mary Andrew, Lee Andrews, Mike Armstrong, Anselm Atkins, Brent Auhlenbach, Holly Bailey, Keith Bailey, Kerry Bailey, Mark Bailey, Terry Bailey, Barbara Bales, Jim Bales, Katie Bales, Anne Barilla, Becky Beaton, Giff Beaton, Bern Bechtel, Amanda Beck, Arch Beckelheimer, Mark Beebe, Carolyn Belcher, Robert Belcher, Jodie Bell, Jeremy Bennett, Sarah Berckman, Chris Bergquist, Desi Bergquist, Katherine Bergquist, Michael Bergquist, John Biagi, Paul Binionm, Bill Birkhead, Roger Birkhead, Ronald Bjorkland, Mary Kay Blalock, Pamela Blockey-O'Brien, Pillon Bonds, Catherine Bowers, John Bowers, Ken Boyd, Paul Bradshaw, Sara Broome, Brian Brown, Joann Brown, Laura Bryant, John Buford, Kurt Buhlmann, David Burlingame, Liz Burlingame, Toby Burnham, Beverly Burns, Amy Byrom, Shan Cammack, Shannon Camp, Sharon Camp, Miles Carella, Kris Cargile, David Carlock, Linda Carroll, Todd Carroll, Tom Carroll, Kathleen Casses, Joe Caudell, Sally Cersosimo, C. Chambley, Holt Champman, Hazel Childers, Matt Chipman, Arlene Clark, Chad Clark,

Ken Clark, Erin Clarke, Charlie Cockran, Don Cohrs, Doris Cohrs, Mitzi Cole, Scott Connelly, Alex Cook, Connie Cook, Kevin Cook, McKenzie Cook, Cris Cooke, J. D. Cooper, Rebecca Crader, Leigh Creech, Alan Cressler, Brenda Dailey, Mike Dailey, Shaun Dailey, Rhonda Davis, Dean Demarest, Judy Demersman, Sarah Demersman, Andy Denmon, Marci DeSart, Bob Dix, Mark Dodd, Jac Dorminy, Bill Dunn, Trey Dunn, Jimmy Dutton, Jason Dyer, Barbara Edwards, Josh Eldridge, Mary Elfner, Dan Emsweller, Katie Emsweller, Michelle Emsweller, Chris Ericksen, Margaret Ericksen, Byron Feimster, Justin Feimster, Patti Feimster, Jim Ferrari, Danise Fields, David Fields, Bill Fletcher, W. H. Fletcher, W. O. Fletcher, Cathy Flynn, Jacob Flynn, James Flynn, Scott Flynn, Sylvie Ford, Jane Frazier, Scott Frazier, Dot Freeman, Michael Frick, Atrice Frye, David Funderburk, Joe Gametta, Z. Ganaway, Greg Garner, Rusty Garrison, Jennifer Gibson, Jon Gilbert, Laine Giovanetto, Robin Goodloe, Minter Goodson, J. Goodwin, Michael Gowen, John Graham, Mike Graham, Sean Graham, Stephanie Graham, Ben Granitz, Fred Granitz, Mallory Granitz, Georgia Graves, Ted Gregory, Russell Grey, Sue Grigalunas, Gary Grosserman, Dan Guynn, Pam Guynn, Ghislane Guyot, Laura Hackenbrock, Andrew Halfinger, Bruce Hallet, Allison Hamilton, Joe Hamilton, Jordan Hammond, Sharma Hammond, Matt Harlfinger, Theresa Harper, Christi Harrell, David Harrell, Ashley Harrington, Bess Harris, Deanna Harris, Duane Harris, Suki Harris, Walton Harris, James Harrison, Joyce Harrison, B. J. Hatch, Harris Hatcher, Kim Hatcher, Fred Hay, Wayne Helfrich, Rosanna Henderson, Evan Hernandez, Jay Herrington, Johnny Hester, Roy Hester, John Hiers, Phyllis Hiers, Phillip Higgins, Delton Hilliard, Grace Hilliard, David Hinson, Clay Hodge, Donna Hodge, Hanson Hodge, Larson Hodge, Malcolm Hodges, Sheila Hoffman, Ann Hogue, Cherie Hogue, Margaret Hogue, Michael Hogue, Paul Hoinouski, Ronny Holcomb, Noel Holcomb, Eddie Holsey, Earl Horn, Caroline Howard, Mark Hughes, Amanda Hyde, Gary Hyde, Karen Hyde, Kevin Hyde, Kelly Irwin, Ty Ivey, Andrew Jackson, Dale Jackson, Rhett Jackson, Becky James, Paul Jastram, Ronny Johnson, Sara Johnson, Steve Johnson, Terry Johnson, Jeff Jones, Jessica Jones, William Jones, Kent Kammermeyer, Margaret Kavanaugh, Tim Kendrick, Eda Kenney, Claire Kersey, Mark Kersey, Burnie Kessner, Kim Kilgore, Hoover Kincaid, Benjamin Kirk, Matthew Kirk, Nancy Jo Kirk, Joan Knapp, Walter Knapp, Brad Koelin, Andrew Koen, Adaire Krementz, John Krider, C. Krissman, Cortnie Krissman, Nichole Krissman, Michael Krogh, Katy Kyle, Carol Lambert, Telia Landers, Patti Lanford, Lou Laux, Angela Leverett, Bill Leverett, Robin Liles, John Lillis, Anne Lindsay, Kelli Littlefield, Carol Lowe, Adam MacKinnon, Charles Maley, Mike Maley, Sean Maley, Sean Manny, Barry Mansell, Nicole Marlin, John Martin, Monroe Matherly, Tim Matthews, Linda May, Bryan McCartney, Ken McClung, Jim McCurdy, Georgette McElfresh, Lee McElfresh, Petra McElfresh, Joshua McKinney, Phelan McKinney, Michelle McLaurin, Andrew Metcalfe, Francis Michael, Brent Minder, Penelope Minder, Carla McGowan, Emmy Minor, Chad Minter, Brian Mitchell, Paul Moler, Tom Moltz, Steve Mooneyham, Stan Moore, Marcus Moran, Don Moreley, Maria Morena, Catherine Morlen, Michael Morrill, Robert Moulis, Jaclyn Mullen, John Murphy, Tom Nalcott, Homer Nelson, John Newhouse, Anthony Nolen, Sherie Nolen, Dexter Norris, Terry Norton, Susan Nunnily, Mark Oberle, Danny O'Dendal, T. R. Oever, Eric O'Neill, Leonard O'Neill, Jim Ozier, Elaine Pack, Laura Pallas, John Palis, Jep Palmer, Stephanie Palmer, Jeff Parish, Dianne Parrish, Tom Patrick, Mark Patterson, Tommy Patterson, Buck Payne, Jerry Payne, Rose Payne, Susan Pearson, Michael Petelle, Laura Phillips, David Phlegar, C. Pinilla, Theresa Pinilla, Season Platt, Shannon Poe, Caroline Pope, Hillary Pope, Howard Pope, Earl Possardt, Donna Prince, W. F. Prince, Phillip Purser, H. Quillen, Charles Rabolli, C. Randall, Greag Raney, Jessica Raney, Robert Redmond, Will Reeves, John Reid, Kathy Richterkessing, David Richterkessing, Ken Riddleberger, Bob

Ringer, Mike Robinson, Beverly Rollins, Betsie Rothermel, Patricia Rovadi, Shannon Rowe, Carol Ruckdeschel, Steve Ruckel, Aaron Ruther, Gary Ruther, Phyllis Ruther, Jack Sandow, Bob Sargent, Lynn Schlup, Georgann Schmalz, Eric Schneider, Todd Schneider, Ann Scott, Seburn Scott, Sandra Sellers, Jeff Sewell, Chad Sexton, Pete Shaffer, Brian Shaner, Bob Shoop, Jimmy Shurley, Chris Skelton, Michelle Smith, Stacy Smith, Kristina Sorensen, Seth Sorrells, Blair Spearman, William Spearman, Ricky Spears, Rebekah Sperman, Kris Spikes, Kitty Spivey, Phil Spivey, Jon Stallings, Renee Stanfield, Scott Stanfield, Peter Stangel, Dirk Stevenson, Ben Stewart, Dan Stewart, Donna Stewart, Janice Stewart, Mark Stewart, Mike Stewart, Bobby Strange, Johnathan Streich, John Strong, Shirley Strong, Jim Sweeney, Misty Szabo, Terry Tatum, Andrew Tavolacci, Joanna Taylor, Jonathan Taylor, Vince Taylor, Reggie Thackston, Josh Thomason, Shawn Tidwell, Michael Tortorella, Randy Trammel, Tommy Truman, Rusty Trump, Tyler Trump, Lisa Turner, Eric VandeGenachte, James Vandiver, Vic Vansant, Beth Vaughn, and Jack Vaughn.

Finally, the editors would like to thank the agencies and organizations that helped fund the production of this book and/or allowed us to work on it as part of our job—the Georgia Department of Natural Resources, Piedmont College, The Environmental Resources Network, and the University of Georgia—Savannah River Ecology Laboratory.

Counties of Georgia

Amphibians and Reptiles of Georgia

Introduction

AMPHIBIANS AND REPTILES (collectively called herpetofauna) are remarkably abundant and diverse in the southeastern United States. Amphibians (frogs, toads, and salamanders) typically have a smooth, moist body; lack scales and claws; and spend at least a portion of their life cycle in water. Reptiles, which traditionally include turtles, snakes, lizards, and crocodilians, characteristically have scales and claws, and either lay shelled eggs or bear live young on land. Exceptions exist, but these general biological traits readily separate these two major groups of vertebrates (animals having backbones) from one another in most instances.

Anyone interested in Georgia's amphibians and reptiles is the beneficiary of a rich store of knowledge that has been compiled and conveyed by countless herpetologists, natural historians, and museum curators dedicated to observing and cataloguing the distribution and abundance of Georgia's natural flora and fauna. *Amphibians and Reptiles of Georgia* (University of Georgia Press, 1956) by the late Bernard S. Martof was the first published overview of Georgia's herpetofauna. At that time, 142 species of amphibians (63) and reptiles (79) were recognized as being in the state, although 2 (American crocodile, *Crocodylus acutus;* and spectacled caiman, *Caiman crocodilus*) were said to be introduced species not native to Georgia. A half-century later, 30 additional species were known from the state, reaching totals of at least 85 species for both amphibians and reptiles. Taxonomic revisions such as the elevation of subspecies to the status of full species, descriptions of newly discovered species, and the discovery of already described species not formerly known to be present in the state account for the difference.

The Georgia Herp Atlas and the collection of numerous locality records that it involved (see **Georgia Herp Atlas** section for more information) contributed enormously to our understanding of the distribution patterns of the state's herpetofauna. The greatest and most significant contributions to the compilation of herpetological records in the state began with the pioneering efforts in field research, museum curation, and scientific publication by Gerry Williamson, Robert Moulis, Win Seyle, and others associated with the former Savannah Science Museum. Curators of the preserved herpetological collections at the Georgia Natural History Museum at the University of Georgia in Athens, especially the late Joshua Laerm, the former director, as well as Elizabeth McGhee and Laurie Vitt, also provided statewide records and a better understanding of Georgia's amphibians and reptiles. The founding of Partners in Amphibian and Reptile Conservation (PARC) in Atlanta in 1999, representing more than 170 organizations from federal and state governments, private industries such as the pet trade and timber industry, conservation groups, zoos, and academia, is one of Georgia's most recent significant contributions to herpetofauna conservation. Many participants were herpetologists, and all had a deep concern for the conservation status of amphibians and reptiles in the Southeast and the country. PARC's mission is "to conserve amphibians, reptiles, and their habitats as integral parts of our ecosystem and culture through proactive and coordinated public/private partnerships."

The state of Georgia, with its well-deserved reputation for herpetofaunal diversity in natural habitats from the coast to the mountains, was indeed the most appropriate region to launch such an organization.

Nonetheless, grounds for concern about the environmental welfare of herpetofauna remain—and not just in the United States. Declines in the numbers of individuals as well as of local and regional populations have been documented for amphibian and reptile species from every continent and almost every country in which they occur. Some Georgia species have also declined, and the decline of others seems imminent.

Georgia's assets in regional biodiversity of native plants and animals, especially the herpetofauna, are widely recognized not only among herpetologists, but also among all field ecologists and conservation biologists familiar with the southeastern United States. The state's hidden biodiversity, particularly of most amphibians and reptiles, can be experienced firsthand only by those willing to undertake outdoor adventure. Wary river cooters (*Pseudemys concinna*) and brown watersnakes (*Nerodia taxispilota*) still bask by the hundreds on logs and limbs along many of Georgia's larger rivers, although few people except those with boats are likely to see them. Highly vocal southern (*Bufo terrestris*), American (*B. americanus*), and Fowler's (*B. fowleri*) toads still come by the thousands to woodland pools, man-made ditches, and even tire ruts to breed during warm spring rains, but only people willing to leave their vehicle or home on rainy nights are likely to see them. Likewise, night rains in late autumn and winter bring tens of thousands of ornately patterned tiger (*Ambystoma tigrinum*), marbled (*A. opacum*), and spotted (*A. maculatum*) salamanders to isolated wetlands, where they mate and lay eggs before returning to their terrestrial, typically subterranean, homes in pine or hardwood forests. Yet once again, the average person is seldom aware that such animals are underfoot. Thus, most herpetofauna qualify as hidden biodiversity because most species go unseen even by people who live around them. And that is unfortunate because amphibians and reptiles exhibit an array of appearances and behaviors that are fascinating to almost anyone who has the opportunity to experience them. Our purpose in this book is to reveal this hidden biological world to the people of Georgia in the belief that this remarkable component of our natural heritage is underappreciated only because so many know so little about these intriguing animals.

Our audience for the book includes the ever-growing array of Georgians interested in nature, biodiversity, and the environment. The book will also serve as an authoritative reference source for amateur, student, and professional herpetologists in Georgia, the Southeast, and beyond. Land managers and others responsible for maintaining or altering private and public properties who need or want to know more about the status of amphibian and reptile species in their part of the state will find the book a key reference. Information on the animals themselves will be valuable for everyone interested in amphibians and reptiles. In addition, the book will introduce regional experts on particular species, in essence serving as a directory of herpetologists for the state of Georgia.

Physiography, Climate, and Habitats of Georgia

Georgia owes its rich herpetofauna largely to the state's size and its high geologic, climatic, and habitat diversity. The largest state east of the Mississippi River, Georgia spans nearly 15 million ha (58,000 sq mi) and hosts many different freshwater, saltwater, terrestrial, and subterranean habitats. The distribution patterns and local abundances of native amphibian and reptile species are a product of the number, types, and extent of these habitats in a region. Habitats themselves vary in response to climate and across particular topographic patterns. The type of habitat and its geographic location within the state dictate the presence or absence of certain

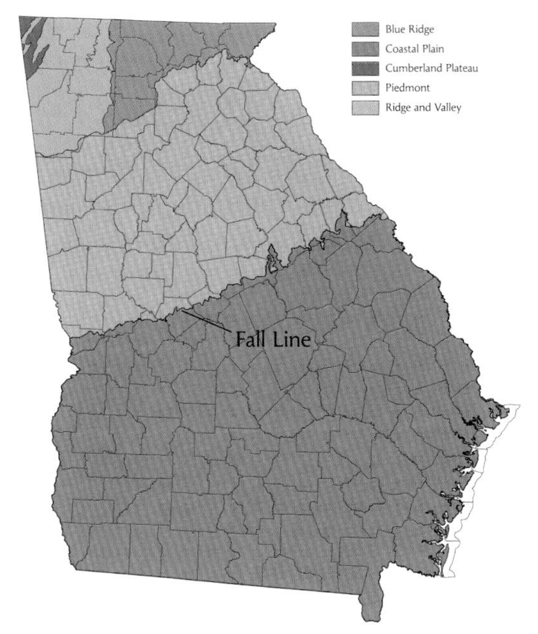

Physiographic Provinces (derived from G. E. Griffith, J. M. Omernik, J. A. Comstock, S. Lawrence, G. Martin, A. Goddard, V. J. Hulcher, and T. Foster, 2001, Ecoregions of Alabama and Georgia; Reston, Va.: U.S. Geological Survey)

species. Thus, a southern hognose snake (*Heterodon simus*) is most likely to be found in a sandhill habitat in the Coastal Plain, whereas a green salamander (*Aneides aeneus*) would be expected in rocky hardwood forests in the northern part of the state.

Georgia is typically recognized as having five physiographic provinces. Patterns of topography, soil and vegetation types, and geologic history make each distinctive. We have not included an exhaustive summary of the habitat types found in these provinces but instead highlight those that support unique herpetofaunal arrays. Table 1 (p. 5) shows the species diversity within each province.

The **Blue Ridge Province** in northeastern Georgia is the southern extent of the Appalachian Mountains. The highest elevations in the state are found here, including 30 summits 1,220 m (4,000 ft) or more above sea level. This region is also characterized by extreme topographic variability, ranging from 300 m (~1,000 ft) at the bottom of deep gorges to above 1,370 m (~4,500 ft)

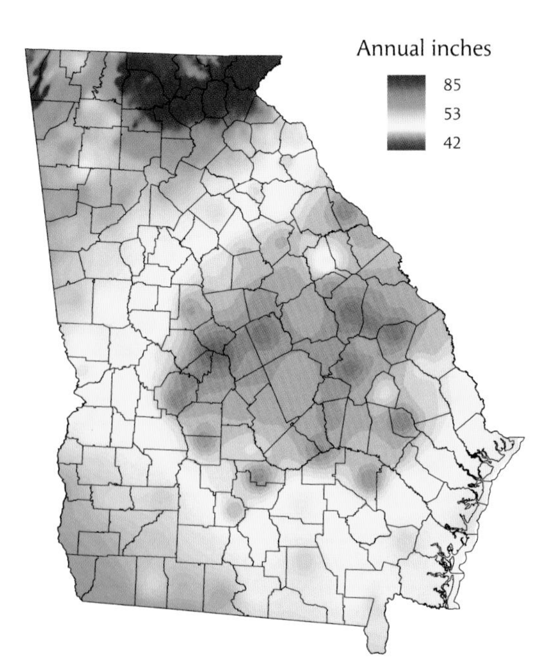

Mean Annual Precipitation, 1971–2000 (PRISM Group, Oregon State University, Corvallis; http://www.prismclimate.org)

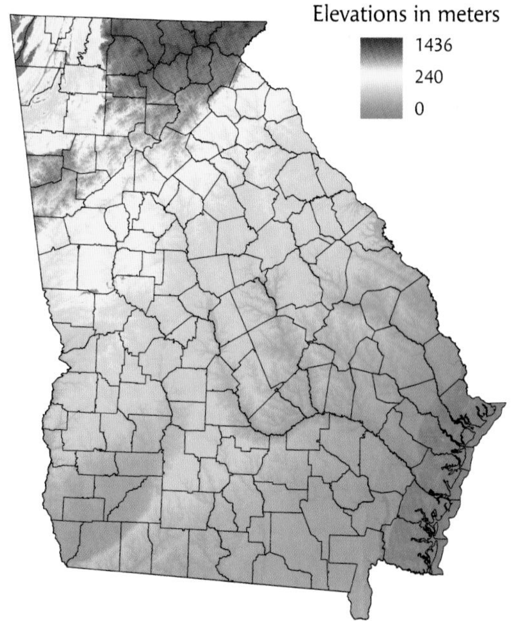

Topography of Georgia (U.S. Geological Survey, 1999, National Elevation Database)

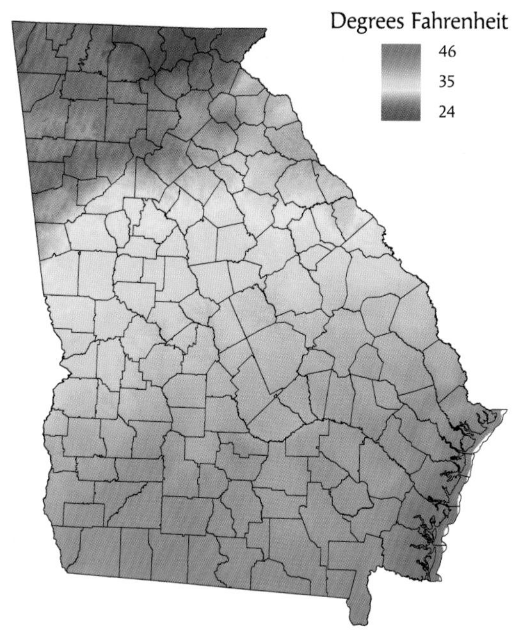

Average Daily Low Temperatures, January 1971–2000 (PRISM Group, Oregon State University, Corvallis; http://www.prismclimate.org)

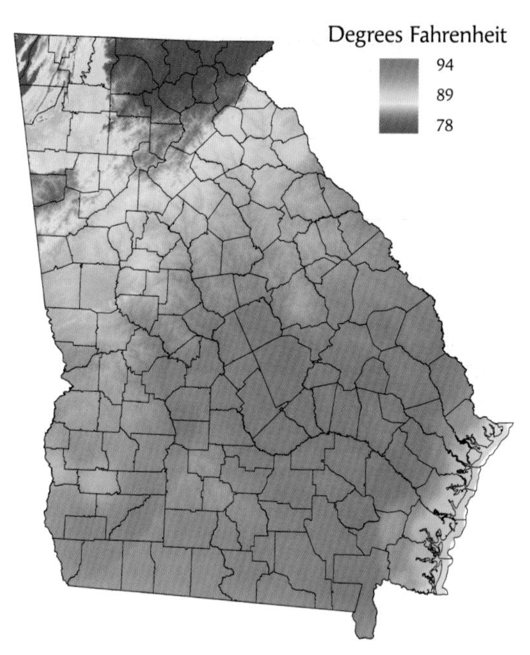

Average Daily High Temperatures, July 1971–2000 (PRISM Group, Oregon State University, Corvallis; http://www.prismclimate.org)

on the highest peaks. Not surprisingly, this province experiences the state's harshest winters and mildest summers. In general, reptiles, while still numerous, are less well represented here than they are farther south, but amphibians, especially salamanders, are both abundant and diverse. In fact, the high annual precipitation, which may exceed 200 cm (80 in) in some areas, combines with unique habitat features to make the Blue Ridge of Georgia and surrounding states home to the most diverse salamander fauna in the world.

Much of the landscape is covered with deciduous hardwood forests that provide a closed canopy, a dense leaf-litter zone, and moist conditions. Woodland salamanders of the genus *Plethodon* are particularly abundant. Seepages and small, rocky streams harbor a different suite of plethodontid salamanders, most notably dusky salamanders (genus *Desmognathus*). Curiously, the latter form assemblages of three to six species organized so that the largest species are the most aquatic and the smallest ones occur

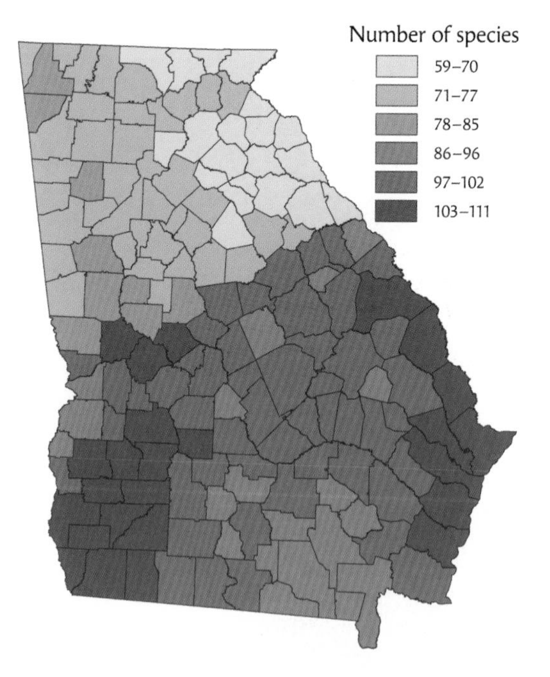

Amphibian and Reptile Species Richness by County, Based on Predicted Ranges

TABLE 1. Species Diversity within Physiographic Provinces

	Number of Species in					
	Georgia	Cumberland Plateau	Ridge and Valley	Blue Ridge	Piedmont	Coastal Plain
All species	170	61	80	68	98	140
Amphibians	85	35	35	34	42	58
Frogs	31	16	17	12	18	28
Salamanders	54	19	18	22	24	30
Reptiles	85	26	45	34	56	82
Crocodilians	1	0	0	0	1	1
Worm Lizards	1	0	0	0	0	1
Lizards	15	5	8	8	12	15
Snakes	41	16	26	21	31	41
Turtles	27	5	11	5	12	24

Blue Ridge Mountains in the Chattahoochee National Forest, White County (Dirk J. Stevenson)

Deciduous forest on Pigeon Mountain, Walker County (John B. Jensen)

farthest from water. The eastern hellbender (*Cryptobranchus alleganiensis*), the most massive salamander in the United States, occurs in larger streams of the Tennessee River drainage. Mountain bogs in the low valleys provide critical habitat for the rare bog turtle (*Clemmys muhlenbergii*).

If the Blue Ridge provides the harshest winters and mildest summers in Georgia, the reverse is true of the **Coastal Plain Province.** Spanning the entire lower half of the state, this province has been covered by successive ocean inundations that left behind deposits of sand and clay when they receded. With few exceptions, the Coastal Plain lacks significant surface exposures of rock. Despite the relatively gentle and low topography, this region has many unique habitats that contribute significantly to its status as the most herpetologically diverse in the state.

The endangered longleaf pine ecosystem of the Coastal Plain consists of a variety of habitat types, including sandhills and pine flatwoods, and is home to many endemic species. Sandhills have the highest elevations in the Coastal Plain; deep, well-drained, sandy soils; and scrubby vegetation. The gopher tortoise (*Gopherus polyphemus*) is an especially important member of the sandhills fauna because it builds and maintains

Small stream on Pigeon Mountain, Walker County (John B. Jensen)

Mountain bog in the Chattahoochee National Forest, Rabun County (Thomas M. Floyd)

Sandhill on Fort Stewart Military Reservation, Long County (Dirk J. Stevenson)

Pine flatwoods on Fort Stewart Military Reservation, Long County (Dirk J. Stevenson)

long, deep burrows in which many other animals, including other reptiles and amphibians, escape hot summer days, cold winters, naturally occurring fires, and predators. Pine flatwoods are located in low areas with poorly drained soils, typically have a fairly dense ground cover of wiregrass and other bunchgrasses, and are inhabited by numerous amphibians and reptiles, the flatwoods salamander (*Ambystoma cingulatum*) being among the most habitat dependent.

The Okefenokee Swamp, Georgia's largest freshwater wetland, deserves special acknowledgment as an important herpetological habitat, even though much smaller versions of it are scattered about the lower Coastal Plain. The Okefenokee is actually a mosaic of wet prairies, bogs, cypress and gum swamps, and interspersed upland habitats on islands. The American alligator (*Alligator mississippiensis*) is arguably the most characteristic species, but species such as the Florida red-bellied cooter (*Pseudemys nelsoni*) and striped crayfish snake (*Regina alleni*) exist at very few sites elsewhere in Georgia.

Coastal Georgia's barrier islands, also part of

Panoramic view of the Okefenokee Swamp, Ware County (John B. Jensen)

Salt marsh and tidal creek, McIntosh County (Dirk J. Stevenson)

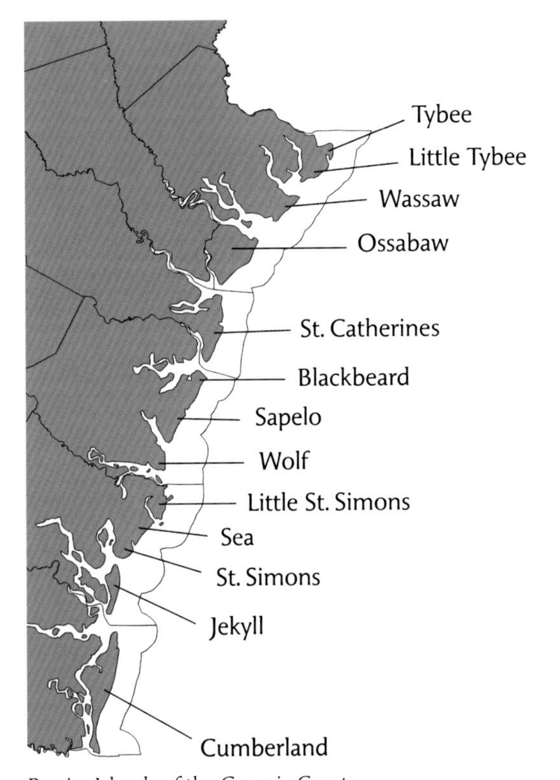

Barrier Islands of the Georgia Coast

the Coastal Plain, are home to numerous species of amphibians and reptiles, and their beaches, estuaries, and the ocean itself are where Georgia's sea turtles can be found. The brackish waters and estuarine habitats of the salt marsh have no resident amphibians but do have one characteristic reptile, the diamondback terrapin (*Malaclemys terrapin*).

The **Piedmont Province** covers the second-largest area of the state and, from landscape, habitat, and faunal standpoints, is functionally transitional between the mountains to the north and the Coastal Plain to the south. The topography ranges from flat to nearly mountainous, but "hilly" would best characterize most of the region. The soil is strongly clay based, and both igneous and metamorphic rock outcroppings are common. Hardwood, pine, and mixed forests all occur here, as do a variety of aquatic and wetland habitats, but no habitats that are especially important to amphibians and reptiles are unique to the Piedmont. Webster's salamander (*Plethodon websteri*) is the only amphibian or reptile restricted primarily to this province. The Piedmont is not lacking in reptile and amphibian diversity, however; it contains a combination of species that range statewide (or nearly so), some with more northern and mountainous affinities, and some more typically associated with the Coastal Plain. Fewer Coastal Plain species overlap into the Piedmont than mountain species because of the more abrupt change at the Fall Line—the boundary between the crystalline bedrock of the Piedmont and the sedimentary conditions of the Coastal Plain. As streams flow south across this boundary, they more readily erode the Coastal Plain side, creating a waterfall or cascade at the transition. This transition coincides with the ancient shoreline of Mesozoic seas and also marks the inland-most point of stream navigability. Not coincidentally, three of

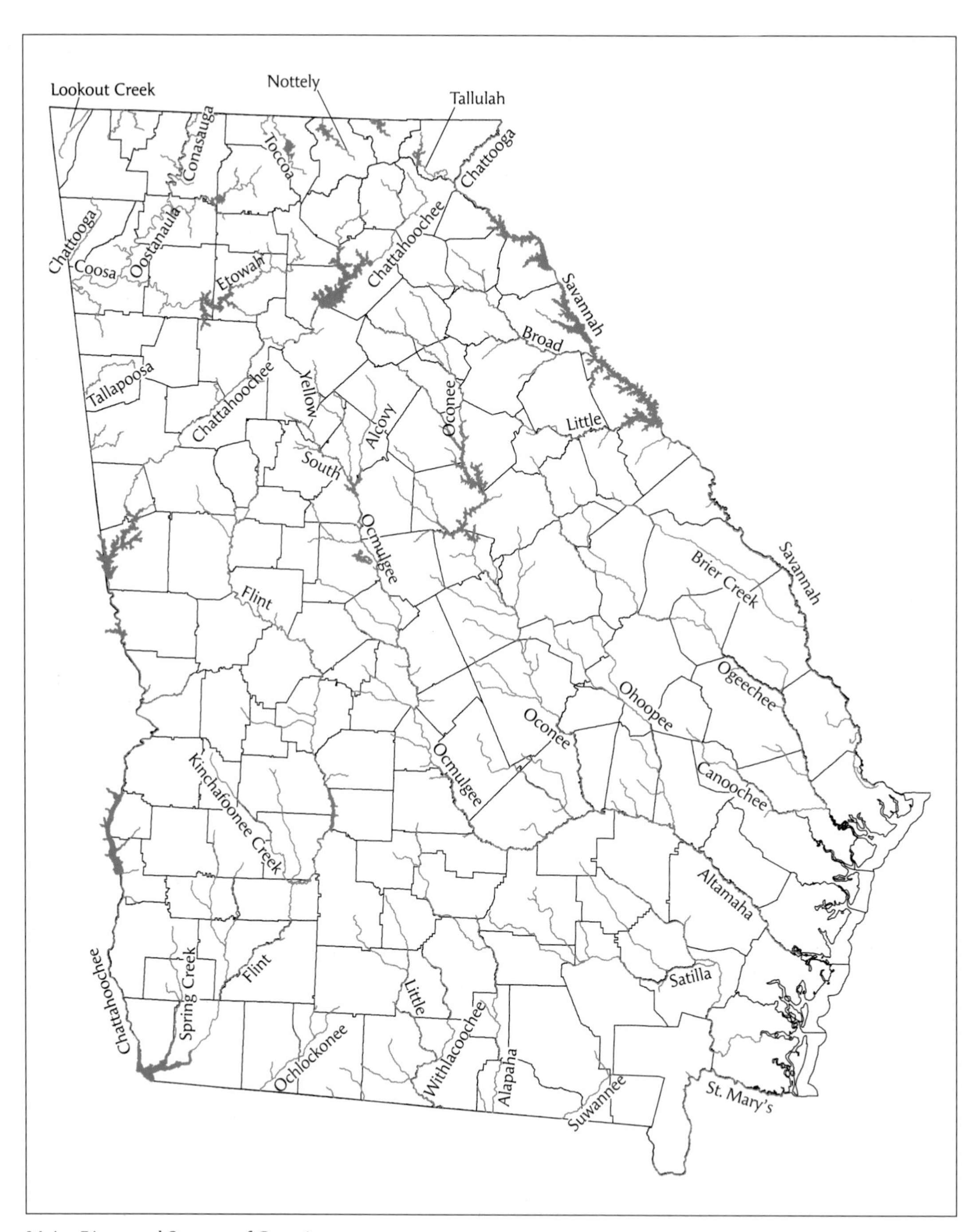

Major Rivers and Streams of Georgia

Georgia's historic port cities—Augusta, Macon, and Columbus—are located at the Fall Line. Because many Coastal Plain amphibians and reptiles require loose, sandy soils for various aspects of their life history, their ranges often do not extend across the Fall Line. The ranges of two closely related species, the American toad (*Bufo americanus*) and the southern toad (*Bufo terrestris*), meet at the Fall Line but do not overlap.

The topographically similar **Cumberland Plateau** and **Ridge and Valley** provinces are found in the northwestern portion of the state and are discussed as one here. Like the Piedmont, the Cumberland–Ridge and Valley region combines features of otherwise widely divergent provinces, but in an interdigitating rather than a transitional way. Roughly parallel ridges and valleys generally trend northeast–southwest, a feature that is readily apparent on a road map of the eastern United States because most major roads follow the valleys. The ridges have habitat features most similar to the mountains of the Blue Ridge, while the wide, intervening valleys often resemble the typical Coastal Plain landscape. Cottonmouths (*Agkistrodon piscivorus*), barking treefrogs (*Hyla gratiosa*), and southern cricket frogs (*Acris gryllus*), each more characteristic of the Coastal Plain, can be found here.

The ridges are formed primarily of sandstone and limestone, and in many areas the latter is riddled with caves. The relatively stable temperature and humidity of caves are attractive to several species of salamanders. The Tennessee cave salamander (*Gyrinophilus palleucus*) is an aquatic obligate to these caves, while the cave (*Eurycea lucifuga*), long-tailed (*Eurycea longicauda*), and Pigeon Mountain (*Plethodon petraeus*) salamanders use them frequently but not exclusively. Some of these same species seasonally occupy the crevices of rock outcroppings, where they may encounter other species such as the green salamander (*Aneides aeneus*).

Other important reptile and amphibian habitats are not as specific to certain physiographic provinces. Large streams and rivers are essential to several species of turtles and a few snakes. The map turtles (genus *Graptemys*) and alligator snapping turtle (*Macrochelys temminckii*) are decidedly riverine and rarely if ever disperse across land. Thus, like many species of fish, these turtles are strongly tied to specific drainages. Rainbow snakes (*Farancia erytrogramma*) and brown watersnakes (*Nerodia taxispilota*) primarily use riverine habitats, where they hunt for their preferred prey, American eels and bullhead catfish, respectively. The floodplain forests, or bottomlands, in low-lying areas adjacent to streams boast their own array of amphibians and reptiles. The three-lined salamander (*Eurycea guttolineata*), bird-voiced treefrog (*Hyla avivoca*), and striped mud turtle (*Kinosternon*

Cumberland Plateau, Walker County—view from Pigeon Mountain toward Lookout Mountain overlooking McLemore Cove (John B. Jensen)

View from cave mouth on Lookout Mountain, Dade County (John B. Jensen)

Rock outcrop on Lookout Mountain, Dade County (John B. Jensen)

Spring Creek, an example of a large stream in the coastal plain, Decatur County (John B. Jensen)

Floodplain forest in Bond Swamp National Wildlife Refuge, Bibb County (John B. Jensen)

Partially filled isolated wetland, Early County (John B. Jensen)

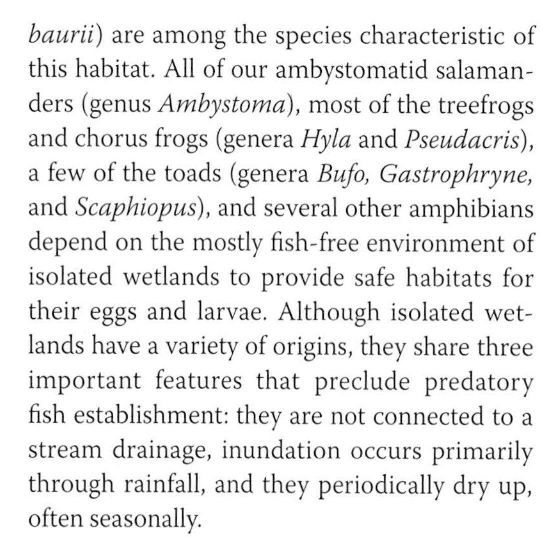

baurii) are among the species characteristic of this habitat. All of our ambystomatid salamanders (genus *Ambystoma*), most of the treefrogs and chorus frogs (genera *Hyla* and *Pseudacris*), a few of the toads (genera *Bufo, Gastrophryne,* and *Scaphiopus*), and several other amphibians depend on the mostly fish-free environment of isolated wetlands to provide safe habitats for their eggs and larvae. Although isolated wetlands have a variety of origins, they share three important features that preclude predatory fish establishment: they are not connected to a stream drainage, inundation occurs primarily through rainfall, and they periodically dry up, often seasonally.

FOR THOSE SEEKING additional information on physiography, climate, geology, and habitats in Georgia, *The Natural Environments of Georgia* by Charles H. Wharton is a must read.

Georgia Herp Atlas

The need for a repository of the distribution information obtained through the Georgia Herp Atlas (GHA) project was a significant impetus for this book. Initiated by the Georgia Department of Natural Resources in 1996, the volunteer-supported atlas effort provided more than 6,500 herpetofaunal records that substantially enhanced our understanding of the distributions of these species and improved the precision of the range maps found in the following species accounts.

During the 5-year GHA period, 492 volunteers submitted verifiable records of amphibians and reptiles they encountered, either through dedicated searches or incidental to other activities. Although volunteers were not assigned different regions, records were reported from 158 of the 159 counties in the state (Towns County was not represented by any records). Verification came mainly in the form of photographs and, in the case of frogs and alligators, audio recordings. Videos, preserved specimens, and shed snake skins were also sufficient vouchers. Not all records were acceptable—some lacked verification or documentation, and many photos were not recognizable—but 89 percent of the 7,452 submitted records were accepted.

Each of 133 species was represented by one or more records. Table 2 shows the 10 most reported species within each major category. Many biases preclude viewing this list as representing the most abundant species in the state; however, most of these would certainly qualify as being among the more conspicuous members of Georgia's herpetofauna. Frog records far outnumber those of other groups, primarily because their presence is relatively easy to document through audio recordings. One particular volunteer, Walter Knapp, was especially zealous in his pursuit of audio frog records. In fact, most of the recordings available on the University

TABLE 2. Top Ten Most Reported Species of the Georgia Herp Atlas by Major Group

Frogs	Number of Records
Green treefrog	450
Cope's gray treefrog	344
Green frog	310
Spring peeper	283
Bullfrog	269
Northern cricket frog	230
Southern leopard frog	203
Fowler's toad	188
Bird-voiced treefrog	182
Eastern narrow-mouthed toad	170
Total frogs (all species)	3,578

Green treefrog (James Kiser)

Salamanders	Number of Records
Slimy salamander complex	87
Southern two-lined salamander	65
Marbled salamander	46
Red salamander	42
Three-lined salamander	38
Spotted dusky salamander	28
Spotted salamander	22
Eastern newt	22
Southern red-backed salamander	19
Mole salamander	12
Total salamanders (all species)	484

Slimy salamander complex (Dirk J. Stevenson)

(continued)

TABLE 2. *Continued*

Lizards	Number of Records
Green anole	159
Fence lizard	95
Ground skink	38
Six-lined racerunner	27
Five-lined skink	19
Broadhead skink	19
Eastern glass lizard	7
Slender glass lizard	5
Mediterranean gecko	4
Coal skink	2
Total lizards (all species)	377

Green anole (John B. Jensen)

Snakes	Number of Records
Rat snake	192
Common kingsnake	106
Black racer	98
Northern watersnake	86
Timber rattlesnake	85
Copperhead	79
Common garter snake	64
Ringneck snake	61
Brown snake	61
Cottonmouth	52
Total snakes (all species)	1,295

Rat snake (Bradley Johnston)

Turtles	Number of Records
Eastern box turtle	279
Pond slider	147
Common snapping turtle	88
Eastern mud turtle	53
Painted turtle	52
River cooter	50
Gopher tortoise	45
Common musk turtle	43
Loggerhead musk turtle	23
Spiny softshell	16
Total turtles (all species)	856

Eastern box turtle (Bradley Johnston)

of Georgia Press's Web site (www.ugapress.org/AmphibsandReptiles.html) are GHA records made by Knapp. The relative rarity of particular species affected their encounter rates, of course, but cryptic or elusive behaviors and the inhospitable nature of certain habitats resulted in some fairly common species being reported infrequently or not at all. Quite a few species, especially toads, cricket frogs, dusky salamanders, several species of skinks, and mud turtles, were underreported simply because many photographs did not capture the often subtle characteristics needed to differentiate between closely related and similar-looking species. Some species may actually be underreported because of their abundance. For example, a volunteer who saw a dozen spotted dusky salamanders and one red salamander during one observation period would likely turn in only one photo for each species. Considering all of the potential biases, the number of records for a species should probably be viewed as representing the relative ease of documenting species' occurrence rather than as an indication of actual species abundance.

Taxonomy

Among the features of herpetology that perplex the broader lay public is the regular use of Latin names rather than familiar, easier-to-learn English counterparts. The problem is that English names fail to meet two primary goals of taxonomy, the science of naming organisms: (1) to provide a common lexicon for scientists worldwide through the development of standardized names, and (2) to do so in a manner that accurately reflects evolutionary relationships. Common names, although often colorful, fail to do either. For example, in most areas of the country, the name gopher refers to a burrowing mammal. In the Coastal Plain of Georgia and surrounding states, however, gopher is the colloquial name for the gopher tortoise (*Gopherus polyphemus*). Mammalian gophers occur in these areas, too, but they are often locally called salamanders. Likewise, true salamanders are obviously amphibians, not mammals, but in the mountains of northern Georgia salamanders are referred to as lizards, with wood lizards (*Plethodon*) living in forests and spring lizards (*Desmognathus*) in streams. Real lizards are, of course, reptiles, not amphibians.

Attempts have been made to standardize common names, but herpetologists have never been able to agree on universal rules to govern the process. Even if common names were successfully standardized, as has been done for North American birds, the names lose their significance for scientists who do not use English as a primary language. To make matters worse, common names do not reflect evolutionary relationships. Spreading adders are not closely related to European death adders, and waterdogs have no recent kinship to Labrador retrievers. The most workable system at this juncture uses Latin names, which are developed within a universally accepted framework of rules to ensure their standardization.

The fundamental unit of nature requiring a standardized name is the species. Each species has a unique name consisting of two Latin terms, a practice dating from Carolus Linnaeus's 1758 work, *Systems of Nature*. The first term is the name of the genus, the larger group to which the species and any closely related species belong, and the second often reflects a peculiar

Seal (pictured here) and other salamanders of the genus *Desmognathus* are often colloquially named "spring lizards." (Thomas M. Floyd)

characteristic of the species; both terms together comprise the binomial name of the species.

Exactly what a species is, however, remains open to interpretation. For most of the last 100 years, "species" was nearly universally defined in terms of "biological species," a group of potentially freely interbreeding organisms. This definition works well when two different types of organisms live in the same region (in sympatry) without interbreeding. But this notion of a species fails when applied to organisms that do not come into contact because they are separated either by geography (a condition known as allopatry) or by time, one or both of them being extinct. Furthermore, forms whose life cycle lacks sexual reproduction—such as certain mole salamanders (genus *Ambystoma*) in the midwestern United States and racerunners (genus *Cnemidophorus*) in the desert Southwest—cannot be distinguished as biological species because they produce offspring through a type of cloning called parthenogenesis.

An alternative idea currently gaining popularity among taxonomists is that of the evolutionary species, defined as a lineage of populations that is on a unique "evolutionary trajectory," meaning that it is different from all other lineages and will continue to be so. Reproduction with other groups is acceptable under this concept as long as it is limited and does not swamp developing differences. Therefore, different subgroups of the slimy salamander (*Plethodon glutinosus*) have been named as species because they are genetically distinct, and if they interbreed with their neighbors, do so on a limited basis along zones of contact. The evolutionary species concept also solves the problem created when geographical barriers separate populations. If the populations are different, then they must have evolved their differences, and because they do not come into contact, interbreeding cannot slow their continuing evolution as independent entities. Such was the justification for recognizing the geographically isolated Ozark zigzag salamander (*Plethodon angusticlavius*) as a distinct species.

Regardless of the species concept invoked, taxonomists must be able to determine differences in order to define species. Historically, species descriptions were based on physical differences, and in many cases they still are. The discovery of cryptic species, however, groups so similar as to be very difficult to distinguish physically, demanded better tools with which to recognize them. Research on the genetic role of DNA made such tools available. One of the first techniques to be widely used in herpetology was electrophoresis. DNA contains specific codes for specific proteins, and electrophoresis measures the differences between similar proteins (called allozymes). More recent techniques involve the direct determination of the genetic code from DNA itself. The use of genetic techniques has accelerated the discovery and description of new species. For instance, the number of recognized species of dusky salamanders (genus *Desmognathus*) has doubled since 1958, and the number of recognized species of woodland salamanders (genus *Plethodon*) in the eastern United States has grown nearly fourfold during that time. Practically speaking, however, the increased reliance on genetic information to determine species has made it difficult, if not impossible, to verify species identifications in the field and for nonpractitioners to identify the species at all.

Once a species is determined, it must be

The slimy salamander complex in Georgia consists of eight genetically distinct but morphologically similar species, including the northern slimy salamander, pictured here. (John B. Jensen)

classified in a genus, whose name becomes the first part of the species' binomial name. Species are grouped into genera to reflect evolutionary relationships. All the members of a genus must have descended from a common ancestor, and all of the species descended from that ancestor must be included in the genus. For example, the shovel-nosed salamander (*Desmognathus marmoratus*) was until recently placed in its own genus, *Leurognathus,* because it has certain unique physical features. When genetic data showed its close relation to the black-bellied salamander (*Desmognathus quadramaculatus*), however, it was moved into the genus *Desmognathus,* which now includes all the descendants of the presumed common ancestor.

Additionally, a species may be subdivided into subspecies. The idea of subspecies is directly rooted in the biological species concept. Its purpose is to recognize distinct races that, although able to interbreed with other races, have unique characteristics and occupy definable geographic regions. The justification for subspecies is based on the presumption that populations living in different regions evolve differences as adaptations to different environments, and that such differential evolution leads to the development of definable races. A subspecies is named using a trinomial, with a third term tacked on to its species name (e.g., the Florida pine snake, *Pituophis melanoleucus mugitus*). The result can be confusing to nonprofessionals, who sometimes mistakenly equate subspecies with species because each has a name and a description. The operative word in the concept of subspecies is "race," however; subspecies are merely variants of species and are not unique evolutionary units in their own right. In fact, recent studies have demonstrated that subspecies often do not accurately reflect evolutionary patterns within species. In addition, a group of populations that legitimately fits the criteria for a subspecies also fits neatly into the definition of an evolutionary species. Its very existence as a geographic race indicates that it is evolving independently. Closer scrutiny of subspecies has combined with the increasing use of the evolutionary species concept to produce a steady decline in the number of subspecies, and the concept may eventually fade from use. The first edition of Roger Conant's *A Field Guide to Reptiles and Amphibians,* published in 1958, listed 22 subspecies of woodland salamanders in the eastern United States. Most taxonomists now recognize none.

THOSE SEEKING MORE information on general taxonomic issues should see **Selected References for Further Study** at the end of this book for suggested reading material.

Conservation

Amphibians and reptiles are integral parts of properly functioning ecosystems; in many Georgia habitats they compose the highest vertebrate biomass. Because of their abundance, amphibians and reptiles serve as critical prey for other animals, play important roles in predation and the control of prolific species, or, often, perform both functions. The existence of some "keystone" species is important to the existence of others that share their habitat, even in the absence of predator-prey relationships; the gopher tortoise (*Gopherus polyphemus*) and American alligator (*Alligator mississippiensis*) are excellent examples. Gopher tortoises dig long, deep burrows in harsh sandhill habitats that are subject to extreme summer heat and frequent natural fires. A host of vertebrates and invertebrates depend on these burrows to escape fires, extreme weather conditions, and pursuing predators. Because they are vegetarians, gopher tortoises also assist plants through seed dispersal. Burrow excavations return leached nutrients to the surface. What gopher tortoises do for terrestrial species, alligators do for those that inhabit swamps and other wetlands. Alligators create deep "gator holes" that are often the only significant refuge for fish and fully aquatic amphibians during times of drought. Many wading-bird rookeries are strategically placed above alligator-rich wetlands to prevent nest predators such as raccoons from reaching nest trees. Some turtle species, most notably the Florida red-bellied cooter (*Pseudemys nelsoni*), deposit their eggs in alligator nests. While the turtles show no parental care of the eggs, mother alligators do, and inadvertently protect the turtle eggs from predators while protecting their own. In actuality, though, we still have a great deal to learn about the beneficial impacts amphibians and reptiles have on other animals and plants and their environments in general.

Unfortunately, many people value wildlife only if they see utilitarian benefits to keeping it around. While the authors and editors of this book believe that amphibians and reptiles should be respected and valued simply because they deserve the same opportunity to participate in life as humans do, we also note that conserving amphibians and reptiles has direct benefits for humans. Virtually all of the active compounds used in pharmaceuticals were originally discovered in or on animals and plants. Indeed, a new weapon in the battle against HIV, the virus that causes AIDS, may come from the skin secretions of certain frogs.

Amphibians and reptiles are also very important bioindicators and biomonitors. The decline or disappearance of certain species in certain habitats may signal problems affecting the overall health of the ecosystem. For example, declining or disappearing frog, watersnake, or freshwater turtle populations in a reservoir used for municipal drinking water or in a lake where fish are commonly harvested for the dinner table can be a warning sign that water quality, and thus human health, may be compromised.

Additionally, when sustainably harvested or collected through regulated means, some amphibians and reptiles provide reliable food, other useful products, and enjoyable pets. Children, especially those who live in urban areas devoid of many forms of wildlife, often learn to respect wildlife and wild places by taking care of pet amphibians and reptiles and studying what these animals need to survive in terms of food, water, shelter, and space.

Despite the legitimate reasons to conserve amphibians and reptiles, many Georgia species are declining, some rapidly. Threats are many, varied, and often additive, but they can be lumped into seven main categories: habitat loss and degradation, unsustainable use and incidental catch, introduced invasive species, disease, environmental pollution, global climate change, and malicious killing.

Habitat Loss and Degradation

If the major threats were ranked in order of their impact on Georgia's amphibians and reptiles, habitat loss and degradation would easily top the list. Most amphibians and reptiles in Georgia have specific habitat needs and cannot persist in areas that have undergone significant habitat alteration, degradation, or outright destruction. Further, their relative immobility often makes amphibians and reptiles more vulnerable to habitat disturbance than other vertebrate groups. Many birds and mammals, for example, can simply leave if their habitat is disturbed, but slow-moving, underground, or dormant amphibians and reptiles are usually destroyed along with the habitat.

Some important amphibian and reptile habitats are especially vulnerable. For example, the longleaf pine community (comprising a variety of habitat types) harbors many endemic amphibians and reptiles, but today no more than 3 percent of its former extent remains. Not surprisingly, many species endemic to this community, including the federally Threatened flatwoods salamander (*Ambystoma cingulatum*) and eastern indigo snake (*Drymarchon couperi*), have also declined tremendously. Much of the loss of longleaf pine forests can be attributed to forest clearing and conversion for agriculture, silviculture (tree farming), and real estate development. Ironically, a less apparent but similarly destructive activity is the suppression of lightning-ignited fires and the natural disturbance they provide. Without these fires, longleaf pine forests give way to fire-intolerant trees and shrubs, which eventually shade out the ground cover and render the forest unsuitable for much of the biota of the longleaf pine forest.

The wetland habitats that many amphibians and reptiles require are also increasingly being destroyed or degraded. From the 1950s to the 1970s, the southeastern United States had a net loss of 154,000 ha (386,000 acres) of wetlands per year, more than any other region of the country. An additional 20 percent—31,200 ha (78,000 acres)—of the state's wetlands were lost from the mid-1970s to the mid-1980s.

Unsustainable Use and Incidental Catch

Some of Georgia's amphibian and reptile species are recreationally or commercially collected for use as pets or fishing bait, slaughtered for their meat and/or hides, or captured and killed during the harvest of other animals. While the collection of some species (e.g., dusky salamanders [genus *Desmognathus*], which are caught and sold as "spring lizards" for fishing bait) is currently limited in scope and unlikely to affect natural populations significantly, other species, especially turtles, are vulnerable to overcollection. Turtles mature more slowly and have naturally lower levels of recruitment into breeding populations than most other vertebrates do, putting a premium on adult survivorship. During the 1970s and early 1980s, the turtle soup industry targeted adult alligator snapping turtles (*Macrochelys temminckii*) for harvest. In less than 20 years the adult population in the Flint River declined by more than 80 percent, prompting the state to protect the species from further harvest.

Diamondback terrapins (*Malaclemys terrapin*) were also harvested for human consumption with a similar result. This species made a startling comeback but is now facing a new threat: drowning in traps set for blue crabs.

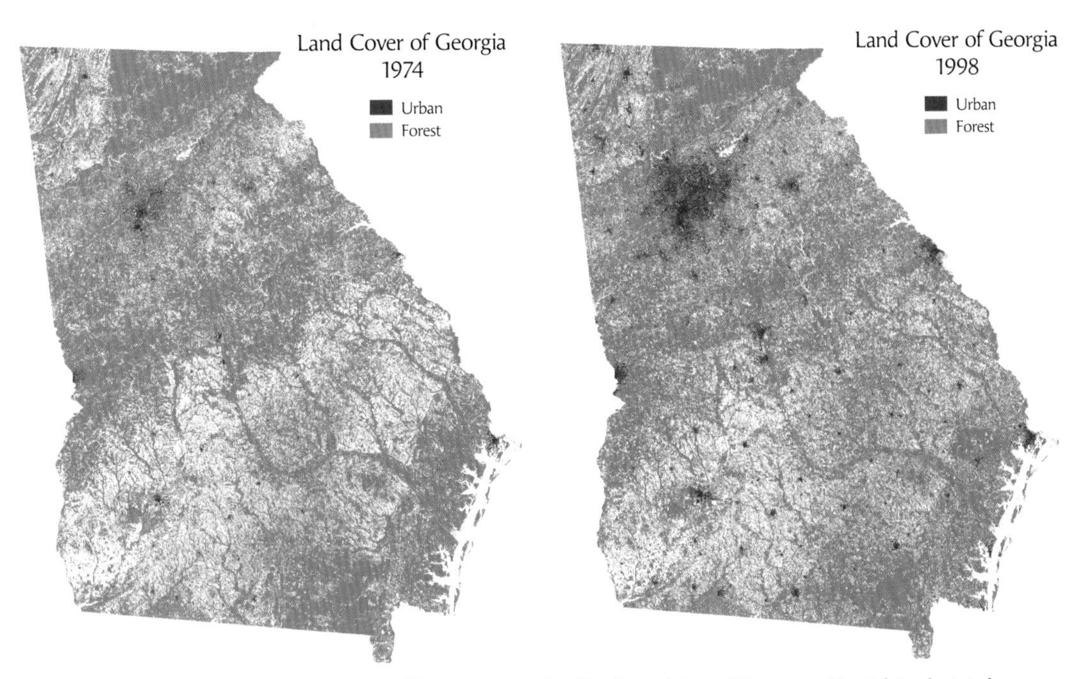

Land Cover Change, 1974–1998, as Mapped from LANDSAT Satellite Data (Natural Resources Spatial Analysis Laboratory, Institute of Ecology, University of Georgia). 1974: This Georgia Land Use Trends (GLUT) land cover map shows that in 1974, Georgia was 3 percent urban and 67 percent forested (including deciduous forest, evergreen forest, mixed forest, forested wetlands, and also clear-cuts). 1998: The 1998 GLUT land cover map shows that Georgia was 5 percent urban and 68 percent forested in 1998. This year largely marks the end of a long-term trend in Georgia from declining agriculture toward reforestation. Not shown in this map are major shifts within forest types since 1974, including large decreases in deciduous forest and forested wetlands and increases in evergreen forest, mixed forest, and clear-cuts.

Destruction and conversion of habitat is the greatest threat to Georgia's gopher tortoise and many other species. (Brad Winn)

Overharvest of alligator snapping turtles for turtle soup led to significant declines in many Georgia streams. (John B. Jensen)

Similarly, Georgia's sea turtle species are often incidentally captured and drowned in commercial shrimping gear. The growing demand for many other turtle species, collected both legally and illegally for meat and the pet trade, is now causing concern. A single bog turtle (*Clemmys muhlenbergii*), for example, may fetch a thousand dollars in the black-market pet trade.

Introduced Invasive Species

Nonnative plants and animals brought to Georgia for a variety of reasons or hidden in cargo from distant lands often have deleterious affects on native amphibians and reptiles. Plants such as kudzu, Chinese privet, and Chinese tallow tree can quickly spread, dominate certain landscapes, and alter the habitat so that it is no longer capable of supporting viable populations of resident amphibians and reptiles. Exotic animals can have more direct impacts on our herpetofauna. For instance, nonnative fire ants, which quickly and efficiently envenomate and consume defenseless prey and eggs, are suspected of playing a role in the decline of many ground-nesting reptiles in Georgia, including the southern hognose snake (*Heterodon simus*) and common kingsnake (*Lampropeltis getula*). Other exotic animals may outcompete native species. The tropical brown anole (*Anolis sagrei*), for example, is far more prolific than the native green anole (*A. carolinensis*), and populations of the former can quickly dominate shared habitats. Some of Georgia's native species, including bullfrogs (*Rana catesbeiana*) and pond sliders (*Trachemys scripta*), have been introduced in foreign lands and have caused similar problems to native species there.

Disease

Herpetologists around the world are increasingly concerned about the rapid decline and disappearance of amphibians from areas affected by a type of chytrid fungus. Although chytrid fungi occur naturally and may play important roles in their ecosystems under natural conditions, a chytrid fungus pathogenic to vertebrates, especially frogs, was recently documented. The particular chytrid species (*Batrachochytrium dendrobatidis*) that causes amphibian chytrid fungus disease attacks the beaks of tadpoles and the skin of adults found primarily in aquatic

Many herpetologists suspect that nonnative fire ants are at least partially to blame for declines and disappearances of southern hognose snake populations. (Gabriel J. Miller)

habitats, usually those in cool climates. Much remains to be learned about the transmission and spread of this fungus. Some researchers have suggested that it is a natural parasite of African clawed frogs (*Xenopus laevis*) and that the pandemic nature of this disease is the result of the worldwide sale of African clawed frogs for research and the aquarium trade. So far, no evidence indicates that this fungus is affecting populations of Georgia's amphibians, but discoveries in other parts of the United States warrant consideration of this pathogen as a realistic future threat. A different fungus (*Saprolegnia*) has been implicated in the destruction of amphibian embryos that have previously suffered from the damaging effects of ultraviolet (UV) radiation from the sun. A type of pathogenic virus called an iridovirus is yet another cause of a potentially far-reaching disease that has affected American amphibians and may be a future cause of concern in Georgia.

Although no widespread diseases are currently affecting entire groups of reptiles, specific diseases are now affecting two of Georgia's turtles. Upper respiratory tract disease (URTD), caused by the bacterium *Mycoplasma,* has caused some gopher tortoise populations in Florida to decline. Die-offs have not been documented in Georgia, but the disease has been confirmed from several localities. Like the chytrid fungus in amphibians, URTD is caused by a relatively recently discovered pathogen, and more research and monitoring are necessary to understand its impact and potential spread. A virus that causes a type of skin tumor called a fibropapilloma affects many green sea turtles. These grotesque tumors can severely impair an infected turtle's vision, locomotion, and feeding ability.

Environmental Pollution

Wetlands and aquatic habitats are often the final resting spot for a cocktail of chemicals—including fertilizers, pesticides, herbicides, heavy metals, and fuels—and animals that inhabit these environments are vulnerable to contamination. Their porous skin makes amphibians especially susceptible to chemical poisoning. Because they live many years and are carnivores, waterdogs (genus *Necturus*) can accumulate such chemical pollutants as polychlorinated biphenyls (PCBs) in their fat tissues. These poisons may be incorporated into the liver and transferred to eggs, where they have been found in high levels.

Few, if any, streams in Georgia have escaped damage caused by adjacent or upstream forest clearing, which can exacerbate erosion and lead to sedimentation. Many naïve Georgians believe that streams are supposed to be thickly stained by suspended red-clay sediments after every

Eastern newts are among several amphibians in Georgia that have tested positive for the potentially deadly fungus *Batrachochytrium dendrobatidis,* but whether they are susceptible to the disease it causes is not known at this time. (Giff Beaton)

rain; in fact, this is not natural and is instead very destructive to aquatic life. The fully aquatic hellbender (*Cryptobranchus alleganiensis*) requires space under large, flat rocks in mountain streams for shelter and nesting spots, but the rocky bottoms of many streams in which the species occurred historically have been completely smothered by fine sediments. All three map turtles (genus *Graptemys*) found in Georgia eat a diet that consists largely of mussels and aquatic snails. These mollusks suffocate under heavy sediment accumulations, and their documented declines from Georgia's waterways are almost certainly one of the reasons map turtles are also in apparent decline.

Litter is clearly an aesthetic problem in Georgia, but some forms of it can be extremely destructive to wildlife as well. For example, improperly or carelessly discarded plastic shopping bags make their way into waterways and eventually into the open ocean, where they drift and look very much like the jellyfish that leatherback sea turtles (*Dermochelys coriacea*) eat. If ingested, these bags can cause intestinal blockage resulting in death.

Global Climate Change

Although the causes for the current rate of global climate change are still being debated, its consequences for amphibians, reptiles, and their habitats could be significant. Predicting what those consequences may be is difficult, however, because of the complex interaction of temperature, evaporation rates, rainfall, and oceanic currents. The most obvious effect of a warming climate is the "migration" of habitats to more northern or more elevated areas, much as habitats shifted when temperatures naturally warmed and continental glaciers receded following the Pleistocene Ice Age. Their limited dispersal abilities may make amphibians and reptiles more vulnerable to shifting habitats than birds and many mammals are.

Changing temperatures may have other consequences as well. The sex of alligators and many of Georgia's turtle species is determined by the incubation temperature of the eggs. Eggs incubated at temperatures above a species-specific threshold will be all or mostly one sex, and those incubated at cooler temperatures will be all or mostly the other sex. The potential effects of a rapid rise in environmental temperatures on sex ratios are unknown.

Many stream-dwelling amphibians and reptiles are significantly affected by land-clearing activities and poor erosion control, which lead to heavy sediment loads in waterways after substantial rains. (John B. Jensen)

Malicious Killing

Ignorance and irrational human fears have led to the destruction of many a Georgia snake. In most cases malicious killing alone is not widespread enough to affect snake populations significantly, but added to the stress of other threats such as habitat loss, it may exacerbate declines. The destruction of eastern diamondback rattlesnakes (*Crotalus adamanteus*) caused by the two remaining Georgia rattlesnake roundups is perhaps an example of this. Although their organizers tout "safety" reasons for reducing rattlesnake numbers, many of the snakes are taken from natural areas where they posed no threat to humans.

The so-called sport of "plinking" affects some riverine turtle populations. Some Georgians apparently enjoy floating down a lazy river and shooting basking turtles off logs and rocks. The degree to which this affects populations is unknown, but plinking should not be

The sex of painted turtles, as with many other turtles, is determined by the egg incubation temperature, so global warming could skew sex ratios. (John B. Jensen)

That's a big snake!

and killed this enormous eight foot black snake in their yard on Friday. They live near Guysie. Alma Times 9-23-99

Even federally protected species, such as this eastern indigo snake, from 1999, are not immune from malicious killing. (*Alma Times*)

dismissed as inconsequential for the turtles—or for other people boating on the river or walking on nearby land.

THE DECLINE OR vulnerability to decline of many of Georgia's amphibians and reptiles has led to regulatory and legal actions intended to stop or, ideally, reverse these trends. Placement on state or federal protected species lists, and the recovery efforts that go along with it, is the ultimate legal protection declining animals can receive. Federal and state protected species lists include Endangered and Threatened species. State and federal laws define Endangered species as those "that are in danger of extinction in the foreseeable future throughout all or part of their ranges." Threatened species are those "that are likely to become endangered in the foreseeable future throughout all or part of their ranges." Two other protected statuses apply at the state level in Georgia. A Rare species is one that "may not be endangered or threatened but which should be protected because it is so scarce." An Unusual species has "special or unique features that make its survival important." Although the classification includes no legal protection, the state of Georgia also categorizes some species as Species of Concern. Generally these are organisms that have sufficiently few documented occurrences in the state to warrant attention. Some are arguably deserving of state or federal protection (see Table 3, pp. 22–23), but their status has not been formally reviewed. Other laws and regulations offer protection to certain additional nonlisted species, but because these laws are subject to frequent change, the Georgia Department of Natural Resources should be consulted for updated information.

THOSE SEEKING MORE information on general amphibian and reptile conservation issues should see **Selected References for Further Study** for suggested reading.

Herpetology: The Study of Amphibians and Reptiles

Reptiles and amphibians, although not particularly closely related to each other, have historically been lumped together by those interested in one group or the other. Perhaps it is because they frequently share common habitats and hiding places, and people looking for a member of one group are often just as likely to find one of the other. The word "herpetology" comes from the Greek word *herpeton,* which means "creeping" or "crawling thing." Certainly, many reptiles and amphibians do creep or crawl, and this characteristic has contributed to their combination as a single focus for intellectual pursuit. Amphibians and reptiles are often collectively referred to as herpetofauna, or more affectionately

TABLE 3. Protected and Rare Amphibians and Reptiles of Georgia

	Federal Status	State Status
Frogs		
Pseudacris brachyphona, mountain chorus frog		SC
Pseudacris brimleyi, Brimley's chorus frog		SC
Rana (Lithobates) capito, gopher frog		R
Rana (Lithobates) virgatipes, carpenter frog		SC
Salamanders		
Ambystoma cingulatum, flatwoods salamander	T	T
Ambystoma tigrinum, tiger salamander		SC
Amphiuma pholeter, one-toed amphiuma		R
Aneides aeneus, green salamander		R
Cryptobranchus alleganensis, hellbender		T
Desmognathus aeneus, seepage salamander		SC
Desmognathus apalachicolae, Apalachicola dusky salamander		SC
Desmognathus folkertsi, dwarf black-bellied salamander		SC
Desmognathus marmoratus, shovel-nosed salamander		SC
Eurycea chamberlaini, Chamberlain's dwarf salamander		SC
Gyrinophilus palleucus, Tennessee cave salamander		T
Haideotriton wallacei, Georgia blind salamander		T
Hemidactylium scutatum, four-toed salamander		SC
Necturus cf. *beyeri,* Alabama waterdog		SC
Necturus maculosus, mudpuppy		SC
Necturus punctatus, dwarf waterdog		SC
Notophthalmus perstriatus, striped newt		T
Plethodon metcalfi, southern gray-cheeked salamander		SC
Plethodon teyahalee, southern Appalachian salamander		SC
Plethodon petraeus, Pigeon Mountain salamander		R
Plethodon shermani, red-legged salamander		SC
Plethodon websteri, Webster's salamander		SC
Pseudobranchus striatus, northern dwarf siren		SC
Stereochilus marginatus, many-lined salamander		SC
Crocodilians		
Alligator mississippiensis, American alligator	T(S/A)	
Worm Lizards		
Rhineura floridana, Florida worm lizard		SC
Lizards		
Eumeces (Plestiodon) anthracinus, coal skink		SC
Eumeces (Plestiodon) egregius, mole skink		SC
Ophisaurus attenuatus, slender glass lizard		SC
Ophisaurus compressus, island glass lizard		SC
Ophisaurus mimicus, mimic glass lizard		R

TABLE 3. *Continued*

Snakes		
Crotalus adamanteus, eastern diamondback rattlesnake		SC
Drymarchon couperi, eastern indigo snake	T	T
Farancia erytrogramma, rainbow snake		SC
Heterodon simus, southern hognose snake		T
Lampropeltis triangulum triangulum, eastern milk snake		SC
Micrurus fulvius, eastern coral snake		SC
Nerodia floridana, eastern green watersnake		SC
Pituophis melanoleucus, pine snake		SC
Regina alleni, striped crayfish snake		SC
Rhadinaea flavilata, pine woods snake		SC
Seminatrix pygaea, black swamp snake		SC
Tantilla relicta, Florida crowned snake		SC
Turtles		
Caretta caretta, loggerhead sea turtle	T	E
Chelonia mydas, green sea turtle	T	T
Clemmys guttata, spotted turtle		U
Clemmys (Glyptemys) muhlenbergii, bog turtle	T(S/A)	E
Dermochelys coriacea, leatherback sea turtle	E	E
Eretmochelys imbricata, hawksbill sea turtle	E	E
Gopherus polyphemus, gopher tortoise		T
Graptemys barbouri, Barbour's map turtle		T
Graptemys geographica, common map turtle		R
Graptemys pulchra, Alabama map turtle		R
Lepidochelys kempii, Kemp's ridley sea turtle	E	E
Macrochelys temminckii, alligator snapping turtle		T
Malaclemys terrapin, diamondback terrapin		U
Pseudemys nelsoni, Florida red-bellied cooter		SC

E = Endangered
T = Threatened
T(S/A) = Threatened Due to Similarity of Appearance
R = Rare
U = Unusual
SC = Species of Concern (no legal protection)

as herps, and those interested in them, whether scientists (herpetologists) or hobbyists (herpetoculturists), are called herpers. Searching for amphibians and reptiles in the field is known as herping.

Broadly defined, herpetology is any scientific study involving amphibians or reptiles. It includes virtually any biological topic; anatomy, physiology, ecology, evolution, and behavior are just a few of the possibilities. Herpetological studies encompass topics as different as the description of sperm storage structures in salamanders and the aggressive interactions of territorial male alligators. The results of herpetological research can be published in broad scientific journals (e.g., *Nature* and *Science*), journals devoted to specific biological concepts (e.g., *Ecology, Molecular Systematics, Journal of Comparative Physiology*) or geographic regions (e.g., *Georgia Journal of Science, Southeastern Naturalist*), or journals that focus specifically on the biology of amphibians and reptiles. Several popular magazines and a host of internet Web pages offer information on less technical aspects of amphibian and reptile biology.

U.S. herpetologists have organized themselves into three large national scientific societies: the American Society of Ichthyologists and Herpetologists (ASIH), in which those who study amphibians and reptiles have allied themselves with those who study fish; the Herpetologists' League (HL); and the Society for the Study of Amphibians and Reptiles (SSAR). These societies publish the major national journals devoted to herpetology, specifically *Copeia* (ASIH), named after Edward Drinker Cope, the acknowledged father of North American herpetology; *Herpetologica* and *Herpetological Monographs* (HL), and the *Journal of Herpetology* (SSAR). SSAR also publishes *Herpetological Review*, a combination journal and newsletter, and the *Catalogue of American Amphibians and Reptiles.* In addition, local and state herpetological societies often include amateur hobbyists as well as scientists.

The experience of enjoying amphibians and reptiles is open to anyone. Many pet stores cater to the needs of amphibian and reptile hobbyists. Many wild species, particularly amphibians and aquatic turtles, may be safely kept in a terrarium, although always in strict accord with the laws governing their collection and captivity. It is illegal, for instance, to collect or keep harmless native snakes or lizards in Georgia.

Most research herpetologists have considerable academic training. Although there are notable exceptions, the general track involves obtaining an undergraduate college degree in biology, zoology, ecology, or a related area and one or more graduate degrees (typically a master of science [MS] or doctor of philosophy [PhD]) that require an independent research project (thesis or dissertation) under the guidance of a practicing research herpetologist. Employment opportunities for professional herpetologists include teaching and/or research at both the public school and college levels, conservation positions with government (e.g., Georgia Department of Natural Resources, U.S. Fish and Wildlife Service) and private (e.g., The Nature Conservancy) agencies, and work for private environmental consulting firms.

Explanation of Species Accounts

Unlike most state guides to amphibians and reptiles, the species accounts included here were not written by one or a few authors. They were provided by a multitude of biologists who, through their field experience, are personally familiar with the species they were assigned to cover. In many cases, these account authors are arguably the leading authorities on their species within the southeastern United States or Georgia.

Overviews of each major group (frogs, salamanders, crocodilians, worm lizards, lizards, snakes, and turtles) and each family precede the accounts. Species accounts within each major group are alphabetically arranged by scientific name within their families, which are also arranged alphabetically.

The term "species account" is used loosely here. We lumped cryptic species—that is, those that are closely related and exhibit little physical differentiation—together into single accounts for their respective species complexes. Any differences in life history, ecology, and distribution among cryptic species are addressed within the account. In contrast, the eastern milk snake (*Lampropeltis triangulum triangulum*) and the scarlet kingsnake (*L. t. elapsoides*) are treated in individual accounts for reasons described therein.

The information presented in the species accounts derives from a variety of sources, including research and observations by the author(s) and/or the available scientific literature. This information often combines the results of research and observations on particular species from throughout their studied ranges. Descriptions, behaviors, habitats, and so on based strictly on Georgia studies are so indicated. For sources of additional information on particular species, please refer to **Selected References for Further Study**.

Scientific and Common Names

Scientific names seem to change almost daily as our knowledge of genetic interrelationships is refined, making inconsistencies between names used in various media a given. Although new scientific names have been proposed for some of the reptile and amphibian species found in Georgia, we have chosen in some instances to use the older, more familiar scientific name with the proposed name in parentheses; for example, *Rana catesbeiana* (*Lithobates catesbeianus*) for the bullfrog. Similarly, common names of particular species differ from book to book and person to person. Although herpetologists have developed several standardized lists of common names, we chose to use the colloquial names most familiar to the collective body of both scientists and the general public. For more information on both scientific and common naming of species, please refer to the **Taxonomy** section.

Description

Sizes of adults and juveniles are usually given as either body length or total length, according to which measurements were available in the literature for the various species. Body length, measured from the tip of the snout to the back of the vent, is often the preferred measurement because many amphibians and reptiles lose all or part of their tail after injury or attack. Carapace length is used as the measure of body size for turtles. It is impractical to use measurements of body or total length for turtles because they often retract their head and tail within their shell. Measurements are presented in both metric and English units.

The physical features of each species, possible variations, and characters that distinguish the species from similar-looking species are also provided. Accurate identification of amphibians and reptiles requires knowledge of anatomical features and names that may be unfamiliar to nonherpetologists. The diagrams given within major group overviews and the glossary should help familiarize readers with these important characters.

Taxonomy and Nomenclature

Alternative scientific and common names and any controversies about them are discussed in this section, as are recognized subspecies native to Georgia. Please refer to the **Taxonomy** section for further explanation on the use of species, subspecies, and common names.

Distribution and Habitat

Distribution information that is not easily gleaned from the range maps, such as the physiographic provinces or river drainages in which particular species occur, is given in this section. Major habitats, and in some cases microhabitats, are also indicated.

Reproduction and Development

Details of courtship, mating, nesting, hatching/birth, growth, and age at maturity are provided when these are known. Unfortunately, details of reproductive and developmental biology are poorly known for many amphibians and reptiles.

Habits

This section provides information on ecology and behavior, including daily and seasonal activity patterns, diet, predators, defense mechanisms, and other pertinent habits.

Conservation Status

Legal Georgia and federal listings, if applicable, are given here along with causes of declines and current threats specific to the particular species in question. Please refer to the **Conservation** section for major threats that affect many species in the state. Species that remain relatively easy to find using proper search techniques, and

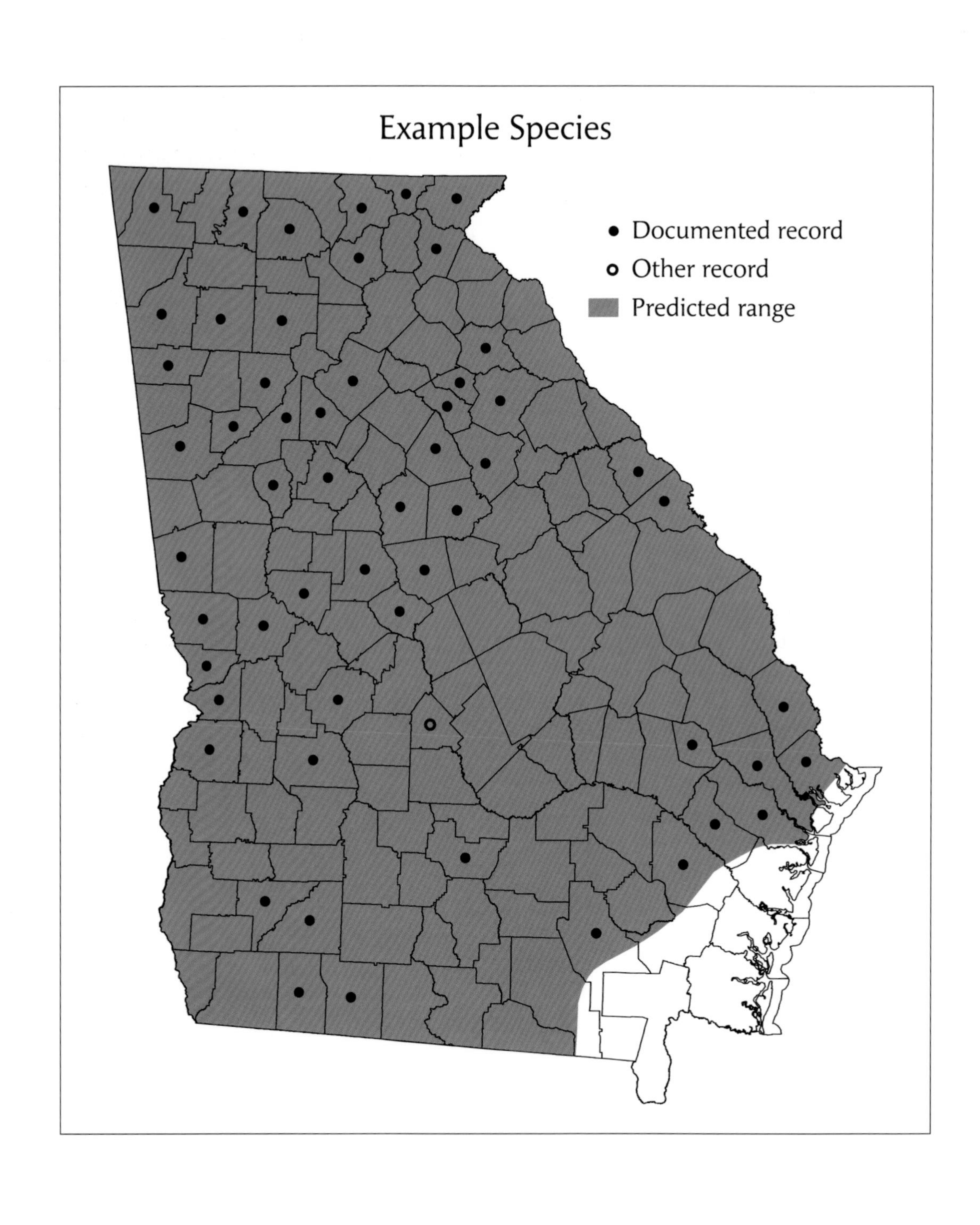
Example Species
Documented record
Other record
Predicted range

for which significant population declines are not known, are referred to as common.

Photographs

Because many species of amphibians and reptiles show significant variation in color, pattern, and other aspects of physical appearance, we tried to include several photographs with each account, although space and photograph availability precluded our representing all the intraspecific variation that exists. We also tried to include photographs of various life stages, especially larval and adult amphibians, of appropriate species. With few exceptions, photographs are of Georgia specimens, with county of occurrence indicated.

Range Maps

Counties where species are known to occur are indicated by dots or open circles in the center of the county. Solid dots were derived from a variety of reliable sources, including scientific literature, technical reports written by herpetologists, Georgia Natural Heritage Program Element Occurrence Records, master's theses and PhD dissertations, specimens in natural history museums, photographs with accompanying locality information, and the Georgia Herp Atlas project. Open circles represent counties where the appropriate account author(s) or the editors personally observed the species but for which no formal documentation exists. The shaded range on the state map represents an estimate made using known locality information and knowledge of the species' preferred habitat, soil type, watershed, and/or topographic limitations. A question mark on the range map indicates that too little is known to even approximate the species' range. An inset map shows the approximate U.S. geographic range.

FINAL NOTE: To ensure accurate identification of a particular amphibian or reptile, readers should pay close attention to the descriptions, distributions, preferred habitats, and especially the range maps, in addition to the photographs. Referring to photographs alone will inevitably lead to misidentifications.

Amphibians

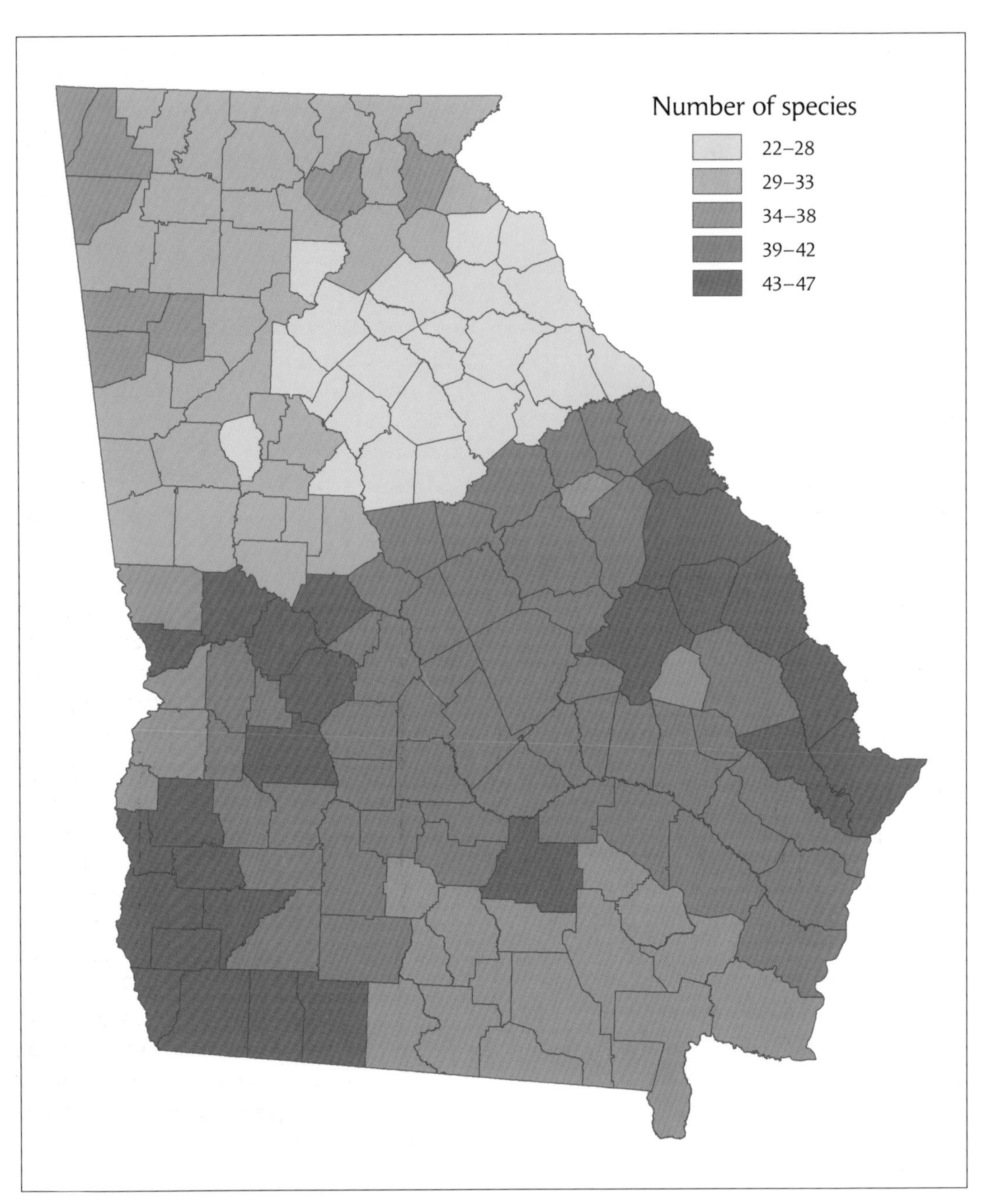

Amphibian Species Richness by County, Based on Predicted Ranges

SOME 380 MILLION YEARS AGO, during the Devonian Age, an adventurous fishlike vertebrate set a finny foot onto land and began the invasion of terrestrial habitats by backboned animals. Although those first four-footed creatures, or tetrapods, have traditionally been classified in the vertebrate class Amphibia, they bore little resemblance to living amphibians. In fact, taxonomists no longer use that term to describe the group that encompasses both early tetrapods and modern amphibians. The earliest tetrapods had heavy, bony skeletons; tail fins supported by rays; scaly bodies; and feet bearing more than five digits. Relative to their early ancestors, contemporary forms have smaller and fewer bones. The skin is smooth, scaleless, and usually wet from mucous-producing glands, and some species have poison-secreting glands there as well. All amphibians can breathe through their skin, at least to some degree. The tail fin, when present, is not supported by bony rays, and the typical pattern for digits is four on the forefoot and five on the hind foot. Current amphibians have many other distinctive characteristics as well, including short, uncurved ribs; a skull mounted on the neck to prevent side-to-side rotation of the head; and, except for caecilians, the ability to raise and lower the eyeballs, giving the face a distinct "pop-eyed" appearance.

One feature that most present-day species presumably share with their early ancestors is a complex life cycle that includes an aquatic larval stage and a terrestrial adult stage. It is this characteristic life cycle that gave rise to the name amphibian, which means literally "double life." The basic pattern is as follows: adults mate; the female lays soft, jelly-protected eggs in water; the embryo hatches and finishes its development as an aquatic larva; the larva metamorphoses into its "adult" form (actually a small, sexually immature version of the adult); this juvenile grows and becomes mature, at which time it completes the cycle by mating and producing offspring. There are many variations to this pattern. Some species lay eggs in humid sites on land rather than in water; some species virtually complete the entire larval stage while still within the egg; and some never metamorphose, living out their lives as overgrown, sexually mature versions of their larvae, a condition known as paedomorphosis.

Modern amphibians are classified into three distinct groups, or orders: frogs, salamanders, and the tropical legless caecilians. Georgia has representatives of the first two groups, with at least 31 frog and 54 salamander species. Although Georgia has fewer frogs than salamanders, frogs are more diverse on a global scale. More than 4,000 species of frogs occur worldwide, but salamanders have only about one-tenth of that number. Frogs are apparently the oldest group, dating back 220 million years in the fossil record, making them contemporaries of dinosaurs.

Frogs differ from salamanders in having enlarged hind legs for jumping. Their larvae, called tadpoles or pollywogs, have internal gills and lack legs until they approach metamorphosis, at which time they also lose their tail. Salamander larvae, on the other hand, have both legs and external gills, and the tailed adults have hind limbs that are about the same size as the forelimbs (except for sirens, genera *Pseudobranchus* and *Siren,* which have no hind limbs).

Frogs

Frogs are familiar members of Georgia's fauna. Whether from bedtime stories ("The Frog Prince"), from the pulpit (the second plague of Egypt), or from an elementary school science project (raising tadpoles to watch them metamorphose), most of us become acquainted with these unique creatures at an early age. Part of our familiarity comes from the fact that frogs are like no other animals. Their compact body, long and powerful hind legs, and drastic metamorphosis make them unlikely to be confused with any other type of animal.

Although Georgia has only 31 species, there are some 4,200 species of frogs worldwide, most of them in the tropics. Georgia's species show a similar distribution pattern; that is, most are found in the southern half of the state. Although all members of this order are frogs, those with relatively short hind limbs—so that they seem to hop rather than jump—are commonly called toads. Georgia's toad species include the common toads of the genus *Bufo,* the eastern spadefoot, *Scaphiopus holbrookii;* and the eastern narrow-mouthed toad, *Gastrophryne carolinensis.*

The typical adult frog is semiaquatic and lives along the edge of a pond or stream. As might be expected of a group with so many living species, however, frogs exploit a diversity of habitats. Some forms are terrestrial or fossorial, and others are arboreal. Georgia's frogs represent all of these lifestyles.

Mating in this group involves the male grasping the female from behind in an embrace called amplexus. Fertilization is external for most species, including all of those found in Georgia; that is, the male's sperm does not fertilize the female's eggs until she has released them into the water. All of Georgia's native species include aquatic larvae or tadpoles that hatch from eggs deposited in water. Tadpoles are very different from the adults they will become, both in physical form and in ecology. They breathe by taking water into the mouth, passing it over internal gills, and then expelling it from the gill chamber through a small opening called a spiracle. The mouth is bizarre, consisting of a keratinized beak as well as rows of tiny teeth, and typically is specialized for grazing on algae. The intestinal tract is long, coiled, and designed—much like a cow's—to digest plant cellulose (dietary fiber). Metamorphosis involves not only the development of legs and eyelids, but radical changes to the mouth and digestive system as well. Paedomorphosis—the retention of larval characters in the adult form—so common in salamanders, never occurs in this group.

An important characteristic of frogs is their reliance on vocalizations as a means of communication. The onset of the amphibian breeding season is signaled throughout the state by a symphony of trills, snores, whistles, and clucks. The number of different calls a single individual may produce is surprising, and males may have calls that are specific for attracting mates, for advertising territorial ownership, or for indicating that the wrong sex has been amplexed during courtship. Males of many species have a vocal sac that appears dark when not expanded. Female frogs rarely call, but some emit catlike screams when caught by a predator. Effective

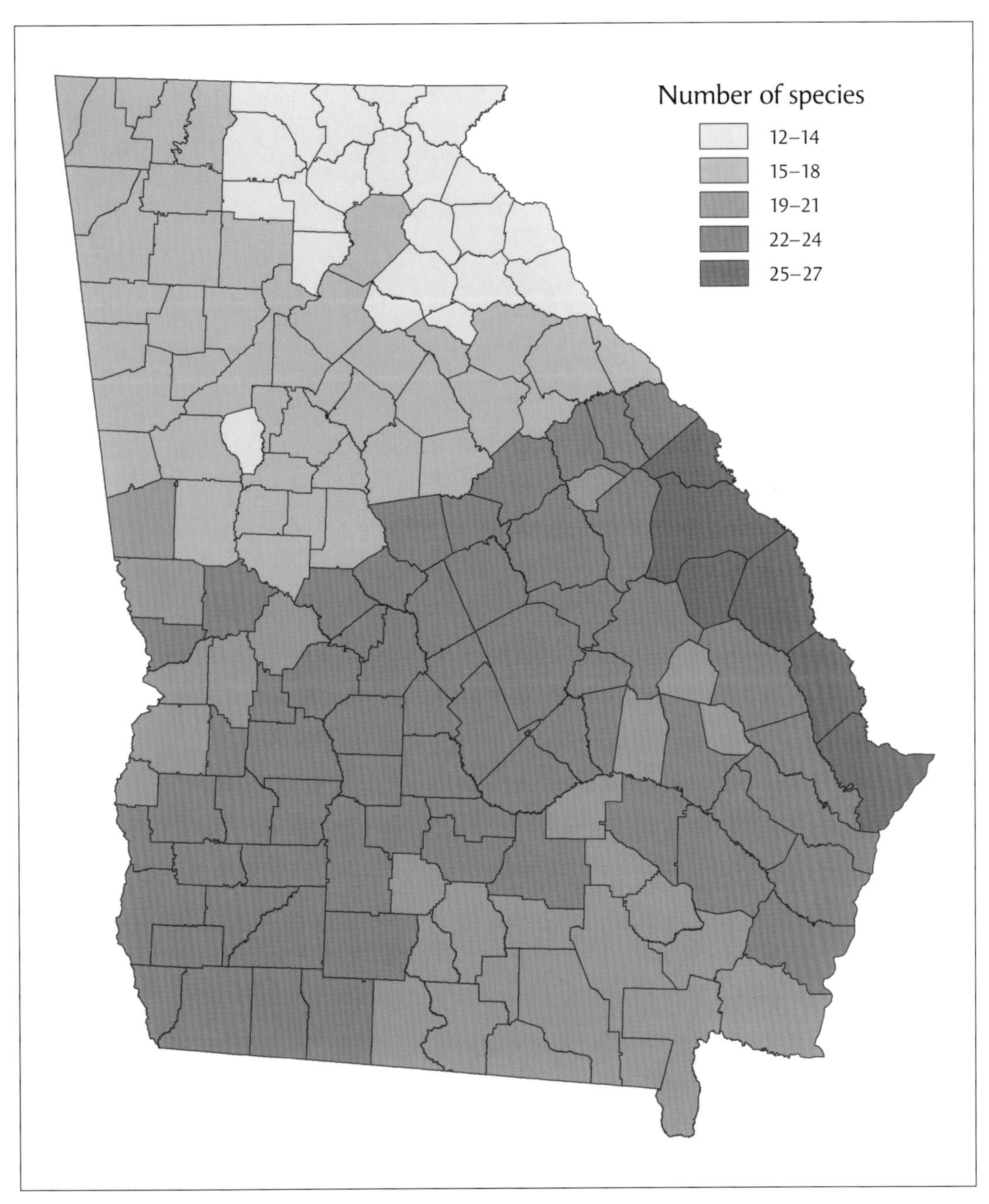

Frog Species Richness by County, Based on Predicted Ranges

vocal communication requires an acute sense of hearing, and a frog's ear is evident in the form of a distinct, round eardrum, or tympanum, located just behind each eye.

Many frogs produce toxic skin secretions as a defense. These range in their effects on predators from simply tasting bad to being lethal. In Georgia, the eastern spadefoot, the eastern narrow-mouthed toad, and the toads of the genus *Bufo* all secrete skin toxins. Although

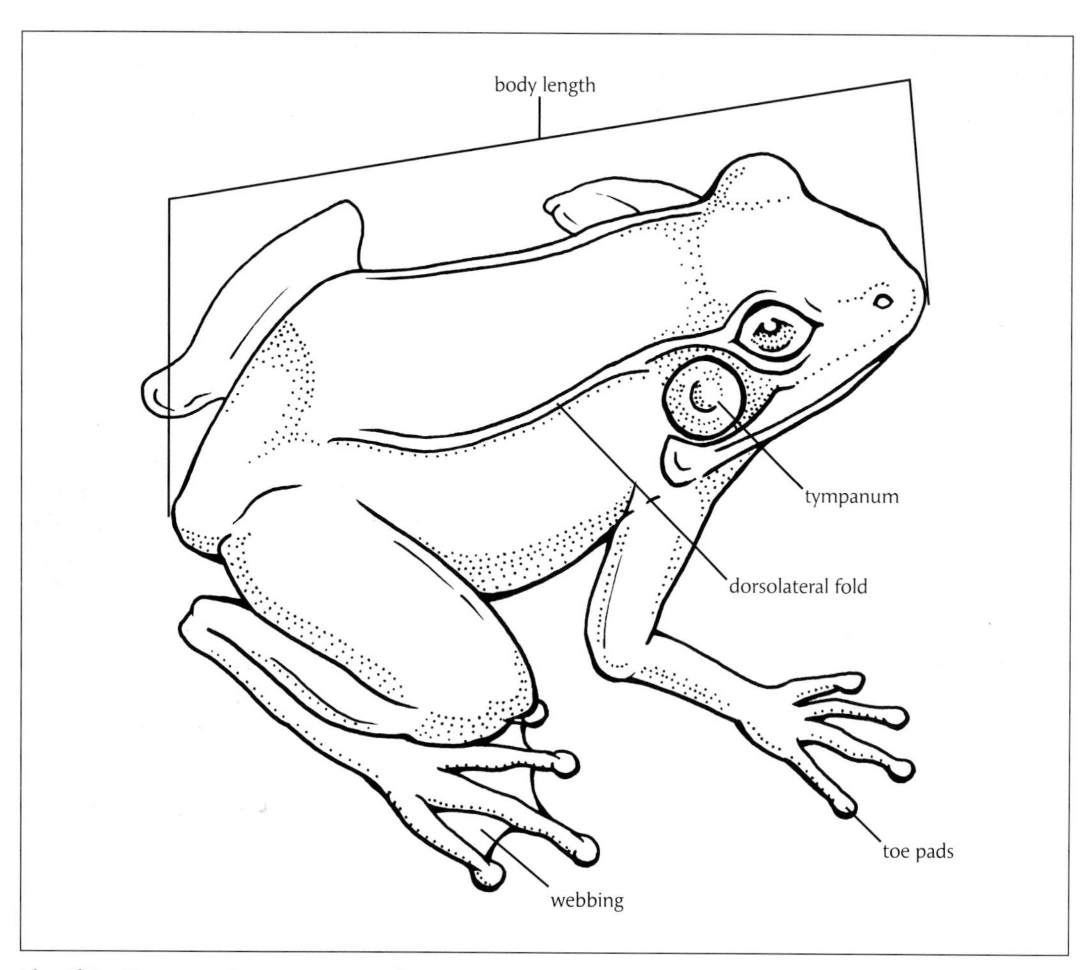

Identifying Features and Measurements of Frogs

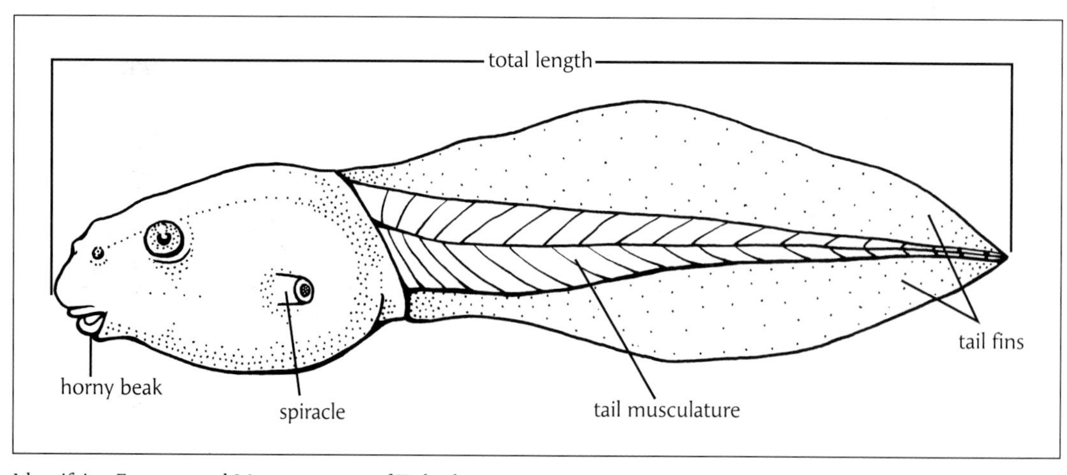

Identifying Features and Measurements of Tadpoles

handling these frogs is not dangerous, touching the moist membranes of one's eyes or nose after doing so may result in unpleasant, allergic-like reactions such as itchy, watery eyes. The brightly colored frogs that have been popularized on T-shirts and coffee mugs and are known for their deadly toxins are found only in the tropics.

True Toads

FAMILY BUFONIDAE

The four species found in Georgia (genus *Bufo*) are relatively small frogs that are 100 mm (4 in) or less in body length. In contrast, the tropical cane toad (*Bufo marinus*) reaches 230 mm (9.2 in). Toads have short hind limbs relative to most other frogs and thick, warty skin. They also have enlarged, horny protuberances on the hind feet for digging and conspicuous poison-producing glands called parotoids behind the eyes. Our toads are terrestrial as adults and exploit a variety of woodland habitats. They breed in both temporary and permanent wetlands, and females lay long strings of eggs in quiet water. Many Georgians are accustomed to seeing rain-filled puddles teeming with small (less than 25 mm, or 1 in, in total length), black tadpoles, which, following metamorphosis, may exit these pools by the thousands.

The family occurs statewide in Georgia. Its larger distribution is essentially worldwide, and its approximately 400 species are native to every continent except Australia and Antarctica.

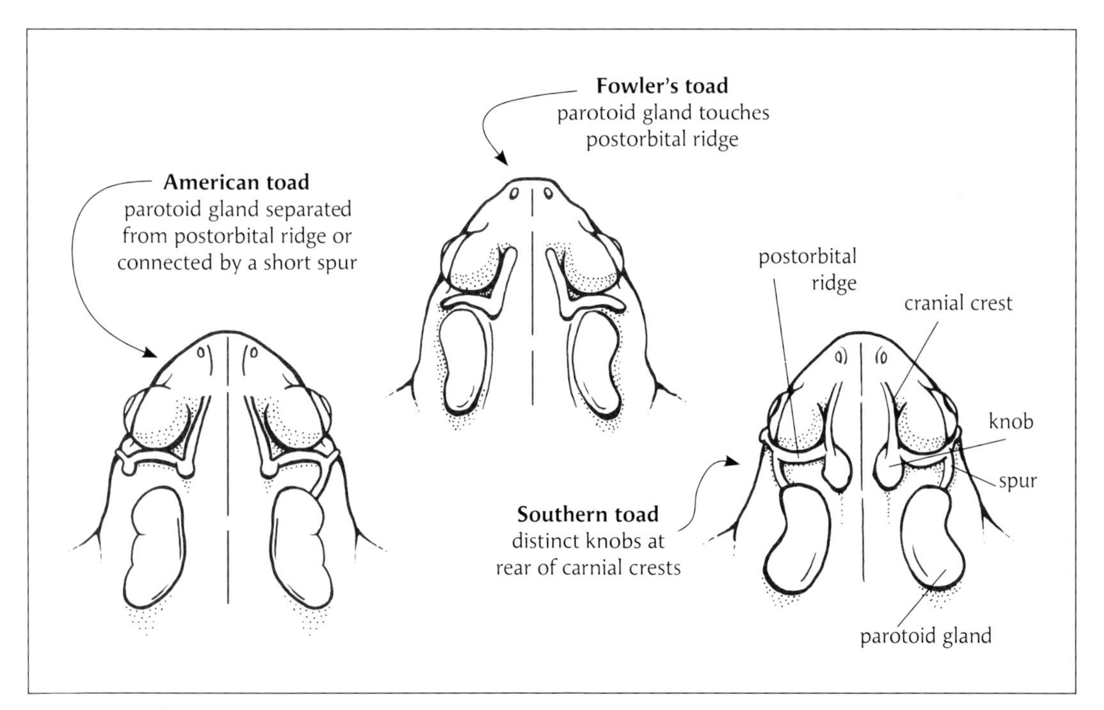

Comparison of Heads of Three Similar Toads

Adult American toad, Banks County (John B. Jensen)

American Toad

Bufo (Anaxyrus) americanus

Description

Adults average 51–90 mm (2.0–3.5 in) in body length. The largest individuals in Georgia are found in the mountains and reach lengths of 107 mm (4.2 in). Males have a dark throat, at least during the breeding season, and are typically smaller than females. The brown or gray base color may be supplemented with lighter patches of yellow or tan with scattered dark spots on the back. A light stripe down the center of the back may be present. Some individuals are almost entirely brick red without dark spots on the back. Older individuals are uniformly drab. The skin is dry and heavily covered with warts. Like other toads of the genus *Bufo,* the American toad has an enlarged parotoid gland behind each eye. The American toad is very similar to Fowler's toad (*B. fowleri*), and the two species are best distinguished by the postorbital ridge behind each eye and its relation to the parotoid. The postorbital ridges of the American toad are separated entirely from the parotoids, although the two are usually connected indirectly by a short spur. The postorbital ridge touches the parotoid in Fowler's toads (see figure, p. 35). American toads usually have only one or two large warts within each of the largest dark spots on the back; Fowler's toads have three or more. The southern toad (*B. terrestris*) is also similar in appearance, but in Georgia its range does not overlap that of the American toad. Tadpoles of the American

American toad tadpole, Walker County (John B. Jensen)

Adult American toad, Banks County (John B. Jensen)

toad may reach 27 mm (1.1 in) in total length and are dark brown to black. Their tail musculature is dark on the top and distinctly lighter underneath. Newly metamorphosed toadlets are 7–12 mm (0.3–0.5 in) long.

Taxonomy and Nomenclature

American and southern (*B. terrestris*) toads were once considered subspecies of a single species, but protein analysis indicated that the American toad is more closely related to Woodhouse's toad (*B. woodhousii*). Two subspecies of the American toad are currently recognized, but only the eastern American toad (*B. a. americanus*) occurs in Georgia. The American toad and Fowler's toad are known to hybridize. Some taxonomists have suggested placing most North American members of the genus *Bufo*, including all of those in Georgia, in the genus *Anaxyrus*.

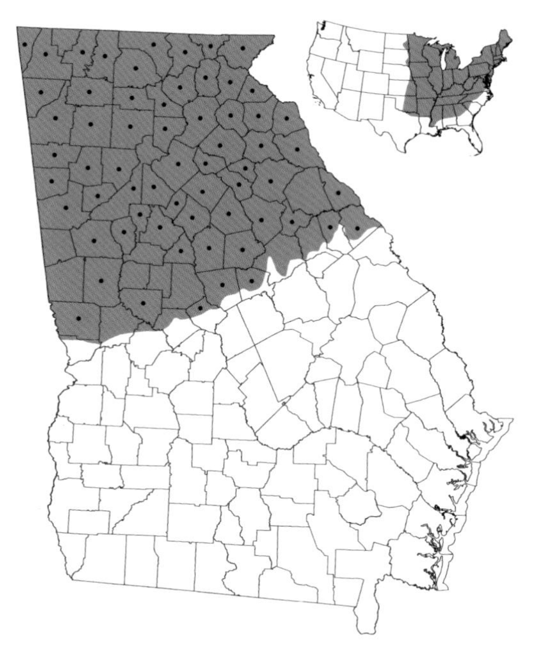

Distribution and Habitat

The American toad is the most widely distributed member of the family Bufonidae in North America. In Georgia, it occurs widely throughout the Piedmont, Blue Ridge, Cumberland Plateau, and Ridge and Valley provinces. Within most of its Georgia range the American toad appears to be outnumbered by Fowler's toad; however, this trend is reversed in mountainous regions, especially at higher elevations. Terrestrial habitats range from moist to dry hardwood and pine-hardwood forests to white pine–eastern hemlock forests, and from forested wilderness to agricultural lands; American toads may even be found in residential areas. They prefer spots with accumulated leaf litter and sandy or loamy soil that makes burrowing relatively easy. Breeding habitats include temporary ponds, isolated pools in floodplains, shallow margins of lakes, roadside ditches, and flooded wheel ruts.

Reproduction and Development

In Georgia, American toads usually begin calling and breeding in late January or February and continue through early April. Males congregate in choruses to attract females to wetlands where breeding takes place. The call of the male, made while floating or while sitting near the water's edge, is a long, high-pitched, musical *bu-r-r-r-r* that is 6–30 sec in duration (01_American_toad .mp3). The call of the southern toad is similar but about an octave higher. Once the female has selected a male, he approaches her from the rear and grasps her behind her forelimbs, inducing her to lay eggs in long double strands while the two toads float on the water. Each female lays 2,000–20,000 eggs, which hatch within 3–12 days. Tadpoles metamorphose after a developmental period of about 2 months. Newly transformed toadlets dispersing to other areas are often found in large numbers near their natal wetlands. Reproductive maturity is reached in 2–3 years.

Habits

American toads are most active just after dusk, usually seeking refuge within burrows or under objects that offer cover during the day. Unlike southern and Fowler's toads, which are common during the summer months, American toads are seldom encountered during this time. Prey for metamorphosed toads includes earthworms, insects, and other ground-dwelling arthropods. The skin and the parotoid glands produce thick, milky white toxic secretions that irritate the moist membranes of many animals. Even so, these toads have several predators, including birds, snakes, and mammals that either are immune to the toxins or have adapted strategies that enable them to tolerate these chemicals. Although toxins are incorporated into the eggs, making them unpalatable, the toxicity is apparently lost or reduced during larval development and then regained after metamorphosis. Raccoons and striped skunks eat American toads but often avoid the more toxic portions of the body. The diet of the eastern hognose snake (*Heterodon platirhinos*) is composed largely of toads, and other snakes, including watersnakes (genus *Nerodia*) and common garter snakes (*Thamnophis sirtalis*), occasionally prey on them. On encountering a predator, an adult toad crouches and remains motionless. If the threat involves a snake, the toad often inflates its body and extends its hind limbs, presumably to make itself larger and more difficult to swallow. Toadlets exhibit similar crouching behavior and also have the ability to lighten or darken their skin color, making detection more difficult.

Conservation Status

American toads are common in Georgia.

Thomas M. Floyd

Adult Fowler's toad, Bibb County (John B. Jensen)

Fowler's Toad

Bufo (Anaxyrus) fowleri

Description

Adults average 51–76 mm (2.0–3.0 in) in body length but can reach 95 mm (3.7 in). Males are typically smaller than females and, at least during the breeding season, have a dark throat. Coloration generally ranges from shades of brown or gray—or more rarely greenish—to brick red. A light stripe usually runs down the middle of the back. Many individuals have six well-defined dark brown or black spots on the back. The skin is dry and heavily covered with warts. The postorbital ridge behind each eye touches the parotoid gland (see figure, p. 35). The postorbital ridges do not directly contact the parotoid in the similar American toad (*B. americanus*). Fowler's toads tend to have three or more warts within the dark spots on the back while American toads usually have only one or two. Fowler's toad is distinguished from the southern toad (*B. terrestris*) by the absence of pronounced knobs extending from the two cranial crests. Tadpoles reach 27 mm (1.1 in) in total length and are dark brown to black, and the body and back are often slightly mottled with lighter markings. In contrast with the American toad tadpole, the tail musculature is usually not distinctly bicolored. Newly metamorphosed toadlets are approximately 8–11 mm (0.3–0.4 in) long.

Taxonomy and Nomenclature

Fowler's toad was long considered a subspecies of Woodhouse's toad (*B. woodhousii*) but is currently recognized as a separate species. Its call is distinct from that of Woodhouse's toad, and

Fowler's toad tadpole, Jasper County (John B. Jensen)

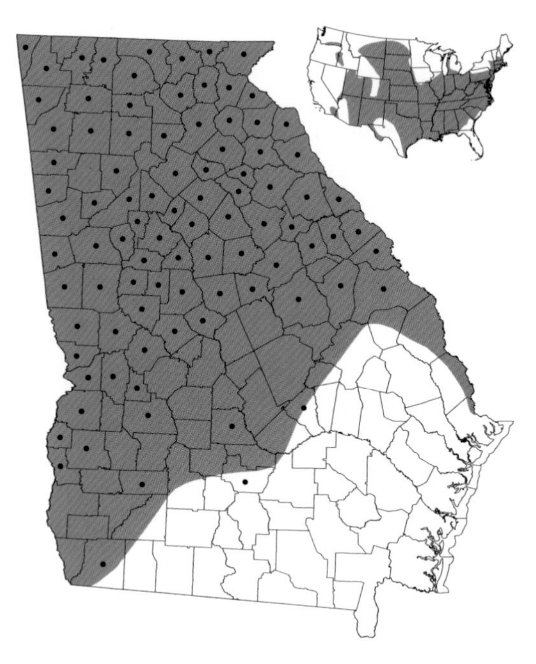

analysis of protein variation has determined that Fowler's toad is actually more closely related to the southern toad. No subspecies of Fowler's toad have been described. Fowler's toad is known to hybridize with both the southern toad and the American toad in Georgia. (See the **American Toad** account concerning the genus name *Anaxyrus.*)

Distribution and Habitat

Fowler's toad is common throughout the Piedmont, Blue Ridge, Cumberland Plateau, and Ridge and Valley provinces but is rather scarce below the Fall Line. Coastal Plain populations are often associated with major river corridors. Except in mountainous habitats, Fowler's toads tend to outnumber American toads where the two species occur together. Fowler's toads favor terrestrial habitats with sandy or loamy soil, which facilitates burrowing. Frequented habitats include moist or dry forests of hardwood, pine, or mixtures of the two. Fowler's toads are also found in agricultural lands and are perhaps the most common amphibian inhabitant of residential areas. Fowler's toads breed in habitats ranging from temporary ponds to the shallow margins of large lakes, but preferred sites are typically more permanent than the wetlands used by either the southern or the American toad.

Reproduction and Development

Breeding typically begins by April and continues through August, and perhaps later at the southern end of the range. Males congregate in large choruses and position themselves near the water's edge to attract females. The call is a short *w-a-a-a-h,* a nasal bleat lasting 1–4 sec (02_Fowler's_toad.mp3). A male accidentally grabbed by another male during mating will give a short chirp as a release call. The American toad usually breeds earlier than Fowler's toad in any given locality, but their breeding periods occasionally overlap, sometimes leading to isolated cases of hybridization. Fowler's toads and southern toads may hybridize as well in areas where their ranges overlap. Each female lays 7,000–10,000 eggs in long, coiled double strands that hatch within about a week. The tadpoles metamorphose and disperse after a developmental period of 40–60 days.

Habits

Fowler's toads are most active at dusk and during the night, but individuals often forage during the day in humid, overcast weather. During most days the toads retreat into burrows or under the cover of objects. Unlike the American toad, this species is frequently encountered at night throughout the summer. Postmetamorphic Fowler's toads prey on a variety of insects and other arthropods. Individuals show site fidelity to their natal wetlands and are able to use solar, lunar, and stellar cues to return to familiar habitats. Although data is lacking for Georgia, individuals within a Connecticut population were observed to keep the same home ranges in consecutive years, moving an average of about 25–30 m (82–98 ft) between successive captures. Throughout their range Fowler's toads congregate in large numbers under outdoor lights that attract insects, often suffering high rates of mortality on heavily traveled streets and parking lots. For protection Fowler's toads depend on toxic skin secretions that irritate the mucus membranes of many would-be predators. Hognose snakes (genus *Heterodon*), whose diet is largely composed of toads, are immune to the skin toxins. Mammals such as raccoons and striped skunks eat only the less toxic parts of the toad.

Conservation Status

Fowler's toad is very common and apparently secure in Georgia.

Thomas M. Floyd

Adult oak toad, Liberty County (Dirk J. Stevenson)

Oak Toad

Bufo (Anaxyrus) quercicus

Description

This diminutive toad is the smallest member of the family Bufonidae in Georgia; adults reach a maximum body length of only 35 mm (1.4 in). Males are smaller than females and, at least during the breeding season, have a darker throat. Oak toads nearly always have a yellow to white stripe down the middle of the back. The base color ranges from light gray to dark brown and at times can appear almost black. The back typically has two to five pairs of dark brown to black blotches, each pair split by the light stripe. Warts cover the entire surface of the back, are relatively uniform in size, and are often reddish brown, bright red, or orange. Each limb has several black bars. Tadpoles reach a maximum total length of 28 mm (1.1 in). The body of the tadpole is mottled with dark or black coloration, and the upper surface of the tail musculature may have light bands. The sides of the tail fins are virtually transparent with scattered specks of pigmentation. Newly transformed frogs are 7–9 mm (0.3–0.4 in) in body length.

Taxonomy and Nomenclature

No subspecies are recognized. (See the **American toad** account concerning the genus name *Anaxyrus.*)

Distribution and Habitat

The oak toad occurs throughout the Coastal Plain of Georgia. Populations inhabit sandy environments such as longleaf pine–wiregrass

Oak toad tadpole, Wakulla County, Florida (Steve A. Johnson)

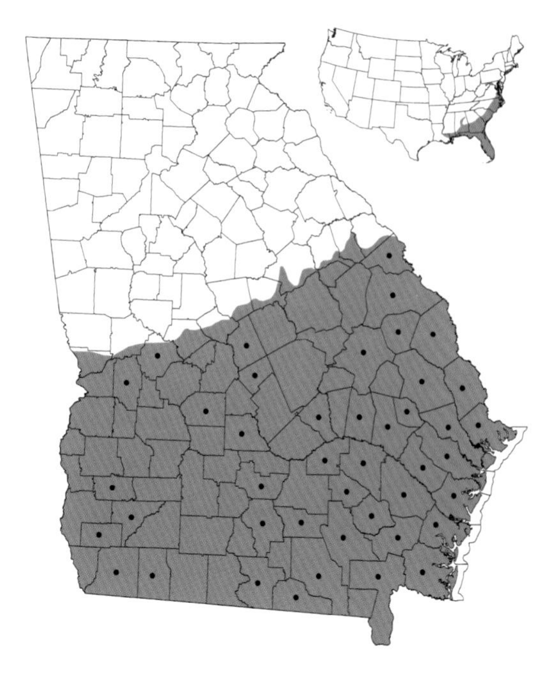

flatwoods and longleaf pine–turkey oak sandhills. Breeding occurs in temporary aquatic habitats such as seasonal ponds, borrow pits, and roadside ditches.

Reproduction and Development

Information on the reproduction of oak toads is limited. They appear to breed during warm rains from April through September. In Georgia, oak toads have been observed active and calling as early as March and as late as November. The call is often compared with the peeping of a hatchling chicken (03_Oak_toad.mp3). This birdlike, high-frequency whistle, usually emitted in bursts of three or four notes, can be heard 200–300 m (0.1–0.2 mi) away. Although large choruses of oak toads are uncommon, they can be very loud, often drowning out other calling frogs. Each female lays up to 500 eggs in strands containing 2–8 eggs each. The eggs are laid on the wetland bottom or attached to flooded grass stems or other debris (e.g., pine needles), often in water as shallow as 25–80 mm (1.0–3.1 in). Long-term research has revealed that oak toads do not breed every year, but the environmental characteristics influencing their pattern of reproduction remain unclear. Hatching occurs in 2 or more days, depending on water temperature. Metamorphosis occurs approximately 1 month later.

Habits

Oak toads are much more frequently encountered during their breeding season than at any other time of the year. Individuals spend most of their time underground or beneath objects (e.g., logs) that offer cover, but appear to be more active in the day than other toad species. They spend much of their time in a "lie-in-wait" posture, from which they ambush small invertebrate prey. Predators include toad-eating snakes such as eastern (*Heterodon platirhinos*) and southern (*H. simus*) hognose snakes as well as watersnakes (genus *Nerodia*).

Conservation Status

Oak toads are fairly common in suitable habitats; however, the drastic loss of their preferred upland habitat has reduced or eliminated many populations.

W. Ben Cash

Adult southern toad, Baker County (Gabriel J. Miller)

Southern Toad

Bufo (Anaxyrus) terrestris

Description

Adults average 41–75 mm (1.6–2.9 in) in body length, but some individuals, especially those found on barrier islands, reach 113 mm (4.4 in). Males are smaller than females and, at least during the breeding season, have a darker throat. The skin is dry and heavily covered with warts. The base color is usually brownish, but some individuals may be dark gray, blackish, or brick red. Darker spots or blotches containing one or more warts are randomly scattered over the back and upper sides. A light line down the middle of the back is usually present but is often obscure. The belly is grayish, and the chest is spotted. The lower and upper surfaces of the hind limbs often have dark blotches that form bars when the limbs are folded in their normal resting position. Like all members of the genus *Bufo*, southern toads have an elongated and swollen parotoid gland behind each eye. The presence of two pronounced knobs extending from the cranial crests between and behind the eyes is the best characteristic for distinguishing this species from Fowler's (*B. fowleri*) and American (*B. americanus*) toads, although the ranges of southern and American toads are not known to overlap in Georgia (see figure, p. 35). Tadpoles are mostly solid black with a diagonal gold line behind the eye and may reach 26 mm (1.0 in) in total length prior to metamorphosis. Newly metamorphosed toads are 7–11 mm (0.3–0.4 in) long.

Southern toad tadpole, Baker County (Anna Liner)

Taxonomy and Nomenclature

Although southern and American toads are quite similar in appearance and vocalizations, genetic analysis indicates that the southern toad is actually more closely related to the less similar appearing and sounding Fowler's toad. (See **American Toad** account concerning the genus name *Anaxyrus*.)

Distribution and Habitat

Southern toads occur throughout the Coastal Plain of Georgia but are replaced by American toads above the Fall Line. When they are not actively breeding, southern toads occupy a wide variety of terrestrial habitats, including agricultural fields, pine woodlands, hardwood hammocks, and residential areas. Favored habitats typically have sandy soil that makes burrowing relatively easy. Southern toads breed in both temporary and permanently aquatic habitats ranging from flooded tire ruts to the shallow margins of large lakes.

Reproduction and Development

The breeding season typically begins in February and may last until October, although most breeding activity ceases by early summer. Breeding aggregations can be enormous, and

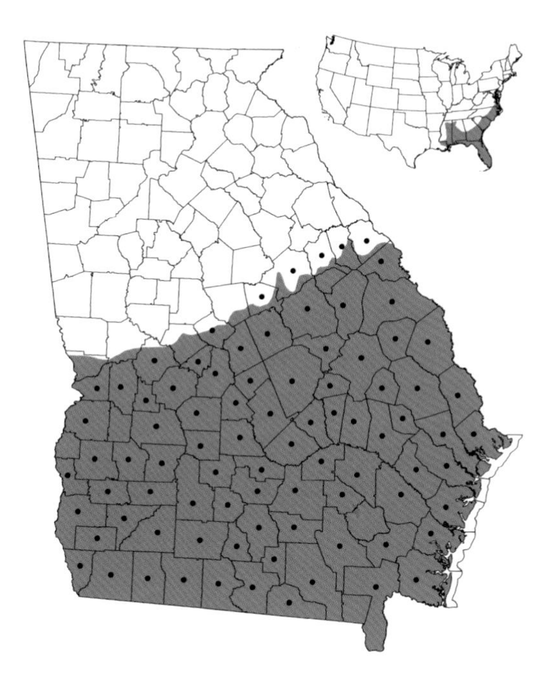

their chorusing can be deafening. The male issues a shrill, musical trill that is 2–8 sec in duration (04_Southern_toad.mp3). The call is harsher and about an octave higher than that of the American toad. Heavy rains outside the breeding season may trigger some males to call. Females lay 2,500–4,000 eggs, each 1.0–1.4 mm (less than 0.1 in) in diameter, in long coiled strings that hatch in 2–4 days. The tadpoles metamorphose and leave the water after a developmental period of 30–55 days. Newly transformed toadlets can often be found in huge numbers near their natal wetlands. Sometimes they are so abundant that it is difficult to take a step without trampling one or more underfoot. Individuals probably reach sexual maturity at about the same age as similar species of toads, 2–3 years.

Habits

Activity is greatest at dusk and during the night; by day these toads usually retreat into burrows or under objects that offer cover. Burrowing, which is done rear-end first, is aided by two keratinized, spadelike protuberances on the back of each hind foot. Adults capture prey with their sticky tongue and target almost anything they can swallow, including beetles, earwigs, ants, cockroaches, crickets, snails, and bees. Southern toads frequently feed on insects attracted to outdoor lights at night. Tadpoles graze on aquatic vegetation and algae but also opportunistically feed on carrion. Southern and other bufonid toads may move around in the open apparently unconcerned about being spotted by predators. This nonchalance is certainly due to the protection offered by the toxic skin secretions, particularly those produced by the parotoid glands. The eggs and tadpoles are also toxic to many, but not all, potential predators. Hognose snakes (genus *Heterodon*) feed heavily on southern toads, and other snakes, including watersnakes (genus *Nerodia*), eastern indigo (*Drymarchon couperi*), common garter (*Thamnophis sirtalis*), and eastern ribbon (*T. sauritus*) snakes, take them occasionally. A southern toad threatened by a snake will inflate its lungs to pump up the body and stretch out its hind legs so that it stands up on all four feet, apparently trying to appear too large for the snake to swallow. Sliding a snakelike object, such as a broom handle or a garden hose, along the ground toward an unwary toad evokes the same response. Unfortunately for toads, the teeth of some snakes, particularly hognose snakes, penetrate and quickly deflate the blown-up toad. Two-toed amphiumas (*Amphiuma means*), lesser sirens (*Siren intermedia*), and dragonfly larvae are known predators of southern toad tadpoles.

Conservation Status

Southern toads are very common and apparently secure.

John B. Jensen

Treefrogs

FAMILY HYLIDAE

Hylids are small to moderate-sized frogs; Georgia's 15 species are all less than 80 mm (3.2 in) in body length. Most members of this family have expanded pads on the ends of their toes. In the true treefrogs of the genus *Hyla* these pads are very large and sticky. An extra joint in each digit gives the toes greater flexibility in grasping small limbs and twigs. These characteristics make the true treefrogs excellent climbers. The toe pads on most of our chorus (genus *Pseudacris*) and cricket (genus *Acris*) frogs are relatively small in comparison, and these species tend to be ground dwellers. Treefrogs exploit a wide variety of terrestrial and wetland habitats. Breeding sites range from permanent wetlands such as beaver ponds to shallow temporary pools in pastures and wheel ruts. The tadpoles are small and have eyes on the sides of the head. Hylids occur statewide, and the family's 700 or so species are distributed on both American continents as well as in Europe, Asia, Australia, and northern Africa.

Adult northern cricket frog, Clarke County (John B. Jensen)

Northern Cricket Frog

Acris crepitans

Description

Adults reach a body length of 19–38 mm (0.7–1.5 in). Males are smaller than females and, at least during the breeding season, have a darker throat. This small, slender frog has warty skin that is variable in color but typically is light brown or gray with patches of light green, black, or yellow. Many adults have a rear-pointing dark triangle on the top of the head. The hind legs have dark stripes running along the back of the thighs. Northern cricket frogs are easily confused with southern cricket frogs (*A. gryllus*) but have a wider body and a nose that is relatively short and blunt. In contrast, southern cricket frogs have a more slender body and a longer and more pointed nose. Northern cricket frogs also have shorter hind legs and more extensive webbing on their feet than southern cricket frogs have; the webbing on their hind feet reaches to the next-to-the-last joint of the longest toe. At least two and one-half joints are free of webbing in southern cricket frogs. The heel of the hind foot does not reach the snout of the animal when the leg is brought forward. Stripes on the thighs of northern cricket frogs have ragged borders rather than the more sharply defined stripes on the thighs of southern cricket frogs, but this is not always a reliable character. Tadpoles of northern cricket frogs often have a distinctive black tail tip and reach a total length of 35–36 mm (1.4 in), although an occasional individual may reach 45 mm (1.8 in). They can be distinguished from southern cricket frog tadpoles

Northern cricket frog tadpole, Monroe County (John B. Jensen)

Northern (*left,* Monroe County) and southern (*right,* Bartow County) cricket frogs. Note the ragged-edged thigh stripe on the northern versus the more smooth-edged stripe on the southern. (John B. Jensen)

by examining the spiracle—the tubular vent that exits the gill chamber; it is much shorter in northern cricket frogs. Individuals typically emerge as tiny frogs with a body length of 13–15 mm (0.5–0.6 in).

Taxonomy and Nomenclature

Although three subspecies occur rangewide, the only subspecies present in Georgia is the northern cricket frog (*A. c. crepitans*), which has the same common name as the species.

Distribution and Habitat

Northern cricket frogs are common above the Fall Line but also occur in the Coastal Plain, especially along major river corridors. They prefer the edges and shores of sunny ponds and wetlands where vegetation is abundant, and they can be found along forested seasonal wetlands and even along slow-moving streams and water-filled ditches. During periods of high humidity, individuals may move several hundred meters or more into surrounding terrestrial habitats.

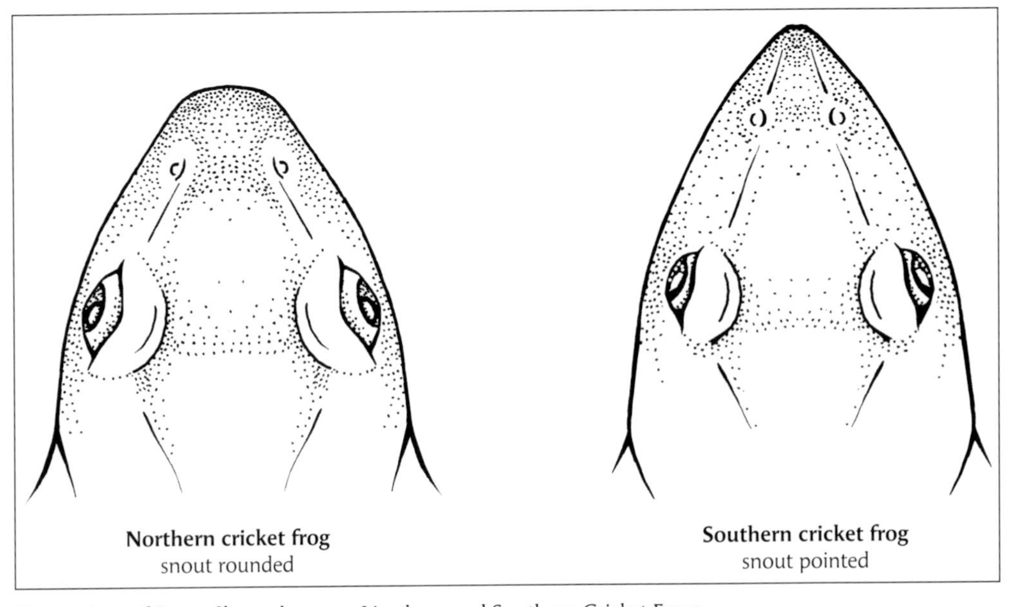

Comparison of Snout Shapes between Northern and Southern Cricket Frogs

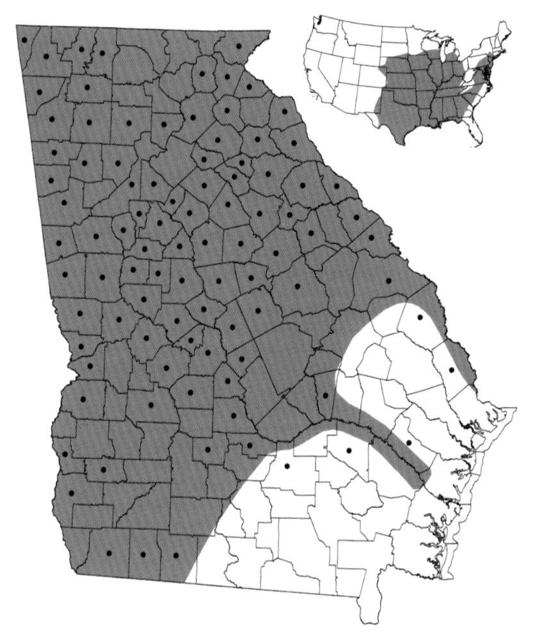

Adult northern cricket frog, Appling County (Dirk J. Stevenson)

Adult northern cricket frog, Paulding County (John B. Jensen)

Reproduction and Development

The reproductive period begins in March and April and lasts until August. Males emit a continuous clicking call; a large group of calling males sounds very much like a chorus of crickets or cicadas (05_Northern_cricket_frog.mp3). Females usually lay eggs singly or in small clusters of 10–20 along pool margins in shallow areas, although large egg masses containing up to 400 eggs have been recorded. Eggs may be laid on the bottom of a pool of water or attached to aquatic vegetation. Females may produce multiple egg masses in one season. Metamorphosis occurs in 1–2 months, and sexual maturity is reached in 1 year.

Habits

Northern cricket frogs in Georgia may be active year-round, temperature permitting, and at all times of day. They tend to remain inactive under leaf litter during cold weather. These frogs can often be seen during daylight in sunlit, grassy areas near bodies of water. Common prey items include small insects and other arthropods. Predators undoubtedly include snakes such as the northern watersnake (*Nerodia sipedon*) and common garter snake (*Thamnophis sirtalis*). Aquatic insects feed on the tadpoles; the black tip on the tail of some tadpoles may misdirect the attacks of dragonfly larvae.

Conservation Status

Populations of the northern cricket frog have been declining in many northern states such as Michigan and New York, and the species is listed as a Species of Special Concern in South Carolina; however, populations in Georgia appear to be stable.

Brian D. Todd

Adult southern cricket frog, Taylor County (John B. Jensen)

Southern Cricket Frog

Acris gryllus

Description

This is a very small frog with slightly warty skin. Adult females are 16–33 mm (0.6–1.3 in) in body length, and the smaller males are 15–29 mm (0.6–1.1 in). The throat of males, at least during the breeding season, is darker than that of females. Overall coloration is quite variable, but most individuals are generally gray, brown, or almost black. Some specimens appear nearly uniform in color; others have very distinctive and colorful patterns. A dark triangle between the eyes on the head is probably the most consistent marking, but some individuals lack even this. A bright green, yellow, or reddish stripe generally runs down the middle of the back, often forking around the triangular blotch on the head. The sides may be variably marked with dark blotches. The legs have blotches that may appear to be bands, but they do not connect beneath the limbs. Most individuals have a light diagonal line below the eye. The presence of one or two bold, dark stripes running across the rear of the thigh is often used to separate this species from the similar northern cricket frog (*A. crepitans*), which typically has a single thigh stripe with ragged edges. Rear-thigh striping is not always a diagnostically reliable character, though, at least in many parts of Georgia where both species occur together. The snout of the southern cricket frog is more pointed (see figure, p. 48), the hind limbs are longer, and the webbing between the hind toes is less extensive (at least two and one-half joints of the longest toe are free of webbing) than in the northern cricket frog. Tadpoles of both species usually

Adult southern cricket frog, Baker County (Gabriel J. Miller)

Southern cricket frog tadpole, Bryan County (Dirk J. Stevenson)

Adult southern cricket frog, Lowndes County (John B. Jensen)

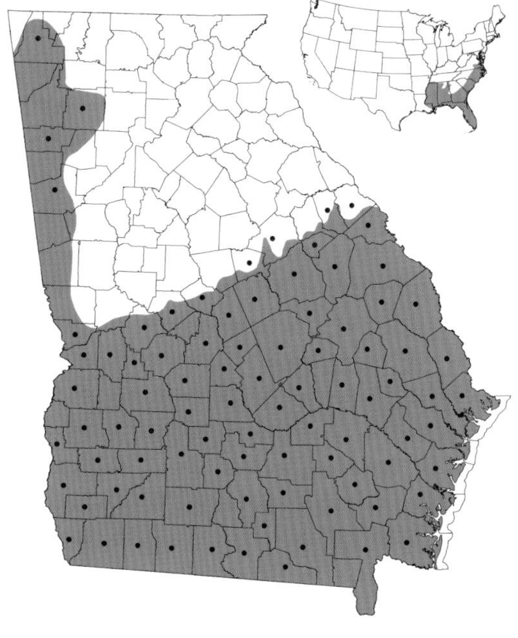

have a distinctive dark tip on the tail. Tadpoles of southern cricket frogs can reach 38 mm (1.5 in) total length and have a much longer spiracular tube than northern cricket frogs have. Newly metamorphosed frogs are 9–15 mm (0.4–0.6 in) long.

Taxonomy and Nomenclature

Two subspecies are recognized: the Coastal Plain cricket frog (*A. g. gryllus*) and the Florida cricket frog (*A. g. dorsalis*); both occur in Georgia. Florida cricket frogs are characterized by two (rather than one) bold stripes on the rear of the thigh, but this feature has been noted in many Georgia populations well outside the previously reported range of that subspecies, especially in populations in Fall Line sandhills. Southern and northern cricket frogs have been reported to hybridize.

Distribution and Habitat

This species is primarily associated with the Coastal Plain, but in Georgia it also occurs in the Ridge and Valley and Cumberland Plateau provinces near the Alabama border. Many Piedmont specimens previously assigned to this species are actually northern cricket frogs. These frogs will use nearly every type of freshwater habitat in their range, both permanent and temporary. As is the case with most native frogs, however, they prefer still-water habitats over moving water. Southern cricket frogs often occur in large numbers along the open margins of well-vegetated ponds, lakes, and swamps. Individuals are often found in terrestrial habitats well away from water.

Reproduction and Development

Breeding may occur from February through October, but activity peaks in April–June. Males call from the ground, on floating leaves such as those of water and cow lilies, or on emergent vegetation just a few inches above the surface. The call is a series of raspy clicks often described as *gick-gick-gick* or the sound created by repeatedly striking two sharp rocks together (06_Southern_cricket_frog.mp3). Males call both day and night. Females attach egg masses to stems of submerged vegetation or drop single eggs directly onto the pond bottom. Each female may deposit up to 250 eggs, usually in clusters of 7–10 eggs each. Hatching may occur within 4 days after deposition, and tadpoles require another 40–90 days to develop and leave the pond as frogs. If southern cricket frogs are like other small frogs in this family, individuals become reproductively mature within a year.

Habits

Adults eat a variety of invertebrates, including springtails, ants, spiders, flies, beetles, and leafhoppers; tadpoles graze on aquatic vegetation and algae. Watersnakes (genus *Nerodia*), common garter snakes (*Thamnophis sirtalis*), other frogs, fish, and birds consume both tadpoles and adults. Predaceous aquatic insects probably feed on the tadpoles as well. Adults attempt to avoid predators by making long, erratic leaps and then swimming to hide within submerged vegetation.

Conservation Status

Southern cricket frogs are very abundant in Georgia and thus are of little conservation concern.

John B. Jensen

Adult bird-voiced treefrog, Walton County (Walter W. Knapp)

Bird-voiced Treefrog

Hyla avivoca

Description

Bird-voiced treefrogs are relatively small frogs; adults are usually 25–51 mm (1.0–2.0 in) in body length. Males are smaller than females and, at least during the breeding season, have a darker throat. The toes have expanded, adhesive disks. The body may be green, gray, or pale yellowish with dark mottling or blotches on the back; a light spot is present under the eye. Individuals often resemble miniature Cope's gray treefrogs (*H. chrysoscelis*), but the hidden surfaces of the bird-voiced treefrog's hind legs are washed with light green or greenish yellow instead of the bright yellow or orange wash found on Cope's gray treefrogs, and their skin is less granular than that of Cope's gray treefrogs. The tadpoles are very distinctive; reddish saddles on the top of the tail musculature alternate between black bands, and the belly is black, making the gut invisible. Tadpoles reach a total length of about 35 mm (1.4 in). Recently metamorphosed frogs are about 13 mm (0.5 in) long.

Taxonomy and Nomenclature

Two subspecies have been described, but their validity has been questioned. Only the eastern bird-voiced treefrog (*H. a. ogechiensis*) is thought to occur in Georgia. Bird-voiced treefrogs and

Calling adult male bird-voiced treefrog, Laurens County (Walter W. Knapp)

Bird-voiced treefrog tadpole, Tattnall County (Dirk J. Stevenson)

Cope's gray treefrogs are known to hybridize in Georgia. The hybrid's call resembles that of the gray treefrog (*H. versicolor*), and the color of the hidden surface of the hind leg is intermediate between orange and green.

Distribution and Habitat

Bird-voiced treefrog populations are distributed throughout central Georgia, primarily across the upper Coastal Plain and middle Piedmont. These frogs frequent cypress, tupelo, and other trees and shrubs in forested wetlands, especially in or near river floodplains. Alder thickets surrounding ponds and small lakes are also favored habitats.

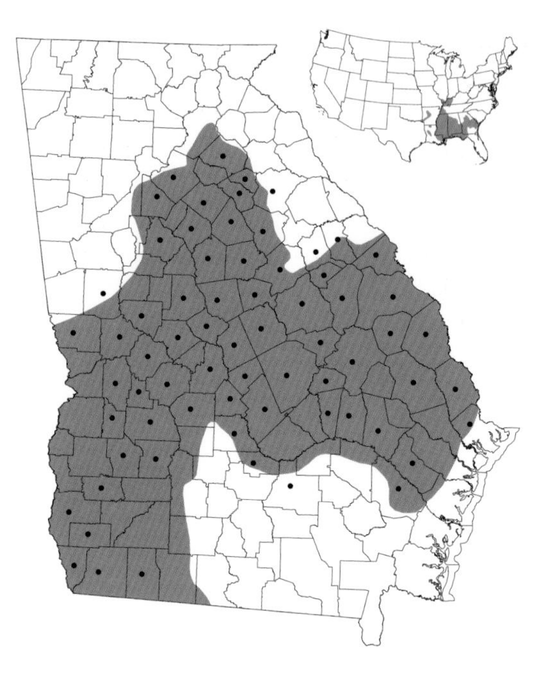

Reproduction and Development

Reproduction may begin in March, but in Georgia it occurs primarily between May and September. Calling males sit 1.0–1.5 m (3.3–4.9 ft) above the ground in bushes and trees and emit a distinctive, high-pitched, birdlike *whit, whit, whit* that is rapidly repeated up to 20 times or more (07_Bird-voiced_treefrog.mp3). Females lay about 650 eggs in clusters of 6–15 in shallow water. Eggs may be deposited on the bottom or on inundated vegetation or other debris. In the warm temperatures of Georgia, the eggs hatch in about 40 hours and the tadpoles metamorphose in about 1 month. Age at maturity has not been reported for Georgia populations, but similar-sized treefrogs mature in 1 or 2 years.

Habits

Bird-voiced treefrogs are nocturnal and are difficult to locate during the day because they often climb high into trees. Outside the breeding season, when males do not call, they are even more difficult to find. Their diet consists primarily of ants and beetles. The fact that the diet does not include several ground-dwelling arthropods eaten by other treefrogs suggests that bird-voiced treefrogs forage largely on tree-dwelling insects. Likely predators include birds and watersnakes (genus *Nerodia*), the latter while adults are more vulnerable at breeding wetlands. Their camouflage and secretive nature probably provide these frogs with considerable protection from predation. Fish, watersnakes, and aquatic insects may eat the tadpoles.

Conservation Status

The species is not given any special protection in Georgia; however, it is state listed as Endangered in Illinois, Threatened in Oklahoma, and is a Species of Special Concern in Kentucky and neighboring South Carolina. Although not as common as some treefrog species, Georgia populations of bird-voiced treefrogs appear stable.

Brian S. Metts

Adult Cope's gray treefrog, Baker County (Gabriel J. Miller)

Cope's Gray Treefrog

Hyla chrysoscelis

Description

This medium-sized treefrog is generally 28–44 mm (1.1– 1.7 in) in body length, with extremely large individuals exceeding 52 mm (2 in). Males are smaller and, at least during the breeding season, have a darker throat than females. Both sexes have granular skin. The toes have expanded, adhesive disks. Individuals may be greenish, but the typical Georgia specimen is light to dark gray with dark, lichenlike blotches on the back and flanks. The blotches frequently connect with similar markings on the exposed surfaces of the hind legs. Identification characteristics include a bright yellow or orange wash on the back and the underside of the hind legs and a small white, almost square blotch under each eye. The similar bird-voiced treefrog (*H. avivoca*) is green or greenish white underneath the legs and has a smaller maximum body size. Newly hatched tadpoles of Cope's gray treefrog are yellowish and about 6–7 mm (0.2–0.3 in) in total length. They gradually become olive, sometimes with a reddish tail, and reach a total length of about 64–65 mm (2.5 in) prior to metamorphosis. The small froglets, approximately 20 mm (0.8 in) in body length, are functional miniatures of the adults but are more frequently green.

Taxonomy and Nomenclature

Cope's gray treefrog and the more northerly distributed gray treefrog (*H. versicolor*) comprise

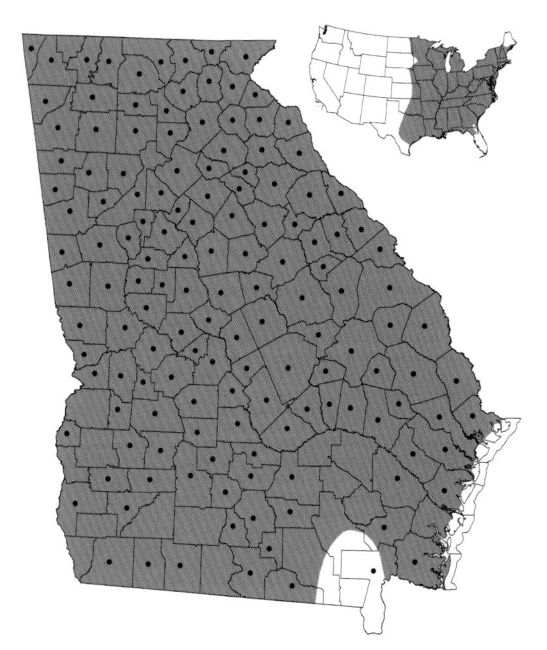

U.S. map is a composite of the Gray Treefrog (*Hyla versicolor*) and Cope's Gray Treefrog (*Hyla chrysoscelis*)

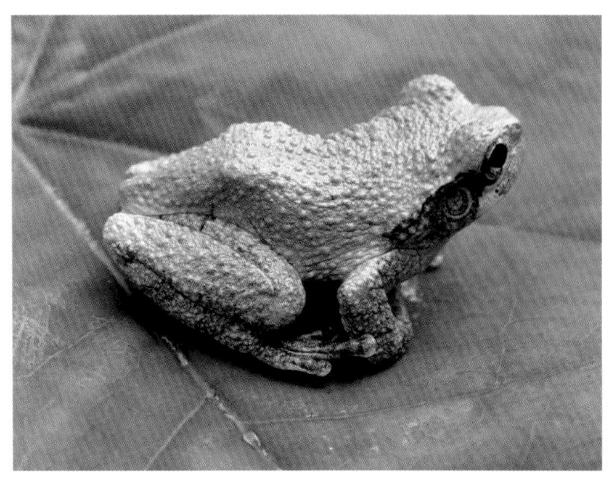

Juvenile Cope's gray treefrog, Floyd County (Bradley Johnston)

Cope's gray treefrog tadpole, Liberty County (Dirk J. Stevenson)

the gray treefrog (*H. versicolor*) species complex. The two are called "cryptic" species because they are virtually identical in appearance. Nevertheless, they are incapable of interbreeding because the gray treefrog's cells have twice the number of chromosomes (a condition known as polyploidy) as the cells of Cope's gray treefrog. Interestingly, Cope's gray treefrogs, with half the number of chromosomes, call at roughly twice the rate of gray treefrogs. Because of the adults' slightly warty skin, this species is sometimes called a tree toad. Hybridization with bird-voiced treefrogs has been confirmed in Georgia.

Distribution and Habitat

Cope's gray treefrogs are found in all physiographic provinces of Georgia in a wide variety of habitats, including urban and residential settings, woodlands, hardwood swamps, bogs, and along the fringes of marshes. Mature hardwood forests on moist slopes and along stream bottoms are favorite locations. Breeding sites include temporary pools; roadside ditches; and heavily vegetated shallow margins of swamps, beaver ponds, and impoundments. When not breeding, Cope's gray treefrogs may be found in a variety of locations, including trees, shrubs and vines, outbuildings, and on windows or under eaves of houses, especially where lights may attract insects.

Reproduction and Development

Adult males begin to arrive at breeding sites during the first relatively warm, rainy nights of the year. Depending on temperature and rainfall conditions, this can be as early as March and as late as August. During drought years breeding may be delayed. Males produce a variety of calls, many of which require sophisticated sound-recording equipment to identify. The primary advertisement call is a loud, lengthy trill that is typically given at breeding sites when females are nearby (08_Cope's_gray_treefrog.mp3). An aggressive *weep, weep, weep* greets a male intruding on another's territory. Females typically deposit their eggs in small masses of 20–40 in a single layer on the water's surface or attached to emergent vegetation. A single female may produce up to 2,000 eggs. Eggs hatch within 2–5 days, and the tadpole stage may last 30–60 days. They become sexually mature in about 2 years.

Habits

These frogs feed primarily on insects. Like several other frog species, Cope's gray treefrogs can survive subfreezing temperatures without any lasting harm because their tissues contain antifreeze-like components. The skin secretions are noxious to many predators and are significant irritants to humans if they come into contact with eyes or other mucous membranes. Their cryptic coloration makes these frogs difficult for predators to detect. Even so, birds or snakes may occasionally eat them; adults are more vulnerable to snakes when they are at breeding sites. Tadpoles are vulnerable to predation by aquatic insects.

Conservation Status

This is a relatively abundant treefrog, even in urban environments, and it is not currently thought to be threatened with broadscale population declines.

Gregory C. Greer

Adult male green treefrog, Richmond County (Thomas Luhring)

Green Treefrog

Hyla cinerea

Description

The green treefrog is a sleek, slender-bodied frog of moderate size. Adult males and females overlap broadly in size and generally fall within the range of 32–64 mm (1.3–2.5 in). The throat of males is gray to pinkish, at least during the breeding season, and that of females is white. The toes have expanded, adhesive disks. This species is noted for the brilliant yellow-green color of its back and the light line resembling a racing stripe extending along its sides and hind legs. The white or yellowish stripes are usually distinct but may be entirely absent or may extend only partway along the body. In contrast, squirrel treefrogs (*H. squirella*), which may also be bright green, have indistinct stripes, and their green color does not extend onto the throat as it does on green treefrogs. Like other treefrogs, green treefrogs change color in response to temperature, stress level, and other factors, and can be pale yellow to dull olive gray rather than bright green. Many individuals have small yellow spots scattered widely across the back. Smooth-skinned green treefrogs are easily distinguished from barking treefrogs (*H. gratiosa*), which have granular skin and a stockier body. Green treefrog tadpoles are greenish yellow with pale lines running from the eyes to the nostrils. The tail musculature is mottled but not striped, and the eyes are on the sides of the head as in other treefrogs. Only about 5 mm (0.2 in) long at

Green treefrog tadpole, Long County (Dirk J. Stevenson)

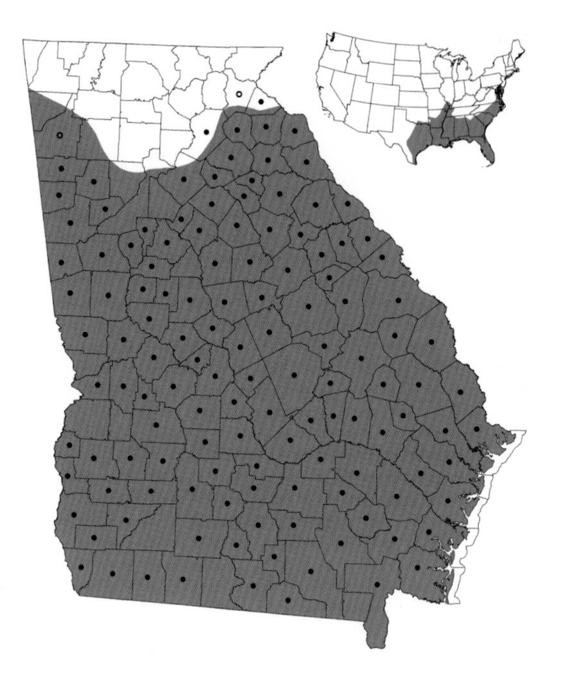

hatching, the tadpoles grow to a maximum total length of 45 mm (1.8 in) prior to metamorphosis. Young froglets are 12–17 mm (0.5–0.7 in) long at metamorphosis.

Taxonomy and Nomenclature

Slight variations in color patterns and body form throughout the range were once the basis for designating subspecies, but these designations are no longer recognized. Hybrids produced by interbreeding between green and barking treefrogs have been found in various places in Georgia and in neighboring states. Habitat alteration or disturbances that cause crowding may lead to the breakdown of the behavioral mechanisms that normally prevent interbreeding. The green treefrog is the official state amphibian of Georgia.

Distribution and Habitat

Green treefrogs reside in wetlands throughout the Coastal Plain and Piedmont. Once restricted to the southern part of Georgia, their range has expanded in recent decades to include many areas north of the Fall Line. Green treefrogs can colonize and thrive in essentially any relatively permanent wetland with abundant emergent vegetation, including lakes, river swamps, Carolina bays, cypress ponds, and farm ponds. Their spread in Georgia may have been helped by the rebound of beaver populations during the last century and the proliferation of farm ponds and small reservoirs. Green treefrogs typically breed in permanent bodies of water but sometimes use temporary ponds that remain flooded for extended periods.

Reproduction and Development

Mating and egg-laying take place between mid-April and mid-August. The *queenk, queenk, queenk* of males begins emanating from swamps and ponds as the sun sets on warm (above 20° C, or 68° F), humid nights. The calls, which have been likened to the sound of quacking ducks or distant bells (09_Green_treefrog.mp3), are repeated about 80 times per minute and can be detected by members of the same species up to 100 m (328 ft) away. The loud chorusing reaches a crescendo shortly after dark and usually tapers off by midnight. Calling males perch on shrubs, cattails, or other emergent vegetation. Some males silently station themselves near a vigorously calling male and await the appearance of a female. These lurking "satellite" males occasionally intercept females attracted to the calling

Adult male green treefrog, Washington County (Walter W. Knapp)

male and amplex them. Females lay an average of 790 eggs (range 275–1,160) in multiple clusters at or just below the surface of the water, often attaching them to vegetation. Eggs hatch in 2 or 3 days. A few females produce a second or even a third clutch within the same season. The larval stage lasts 5–9 weeks depending on temperature and other conditions. Unlike some species that prefer temporary ponds, green treefrog tadpoles appear incapable of speeding up their development in response to pond drying. Most tadpoles metamorphose sometime between July and early August.

Habits

Microhabitat use and activity patterns outside the breeding season are virtually unknown. An "autumnal disappearance" when green treefrogs apparently leave the breeding ponds in late summer has been noticed at several localities in the species' range, including Georgia's Okefenokee Swamp. At a breeding pond near the northernmost extent of the species' range in Illinois, adults disappeared in August, and juveniles moved up to 90 m (295 ft) into the surrounding woodlands during autumn, suggesting that they may overwinter in terrestrial habitats. Green treefrogs prey on a wide variety of invertebrates. A population studied in Florida ate primarily moth and beetle larvae, followed by crickets, stinkbugs, and a variety of adult beetles. Green treefrogs may fall prey to watersnakes (genus *Nerodia*) and other predators. Tadpoles are vulnerable to predatory fish and aquatic invertebrates.

Conservation Status

There is no evidence of widespread population declines or other trends that raise concern for the long-term persistence of this species. Indeed, green treefrogs are abundant in many places and have colonized new areas of the state.

Betsie B. Rothermel

Adult pine woods treefrog, Long County (Dirk J. Stevenson)

Pine Woods Treefrog

Hyla femoralis

Description

Pine woods treefrog adults range in body length from 25 to 45 mm (1.0–1.8 in). Males tend to be slightly smaller than females and, at least during the breeding season, have a darker throat. Each toe has an expanded, adhesive disk. The color of the back ranges from deep russet brown to grayish green, usually with one or two dark, irregularly shaped blotches; another blotch, which may be triangular, is often present between the eyes. Some individuals have a dark stripe or band extending from the snout, through the eye, and across the tympanum to the insertion of the front limb. The belly is usually white or cream with no markings. Orange to silver spots or blotches cover the rear surface of the thighs; these markings are typically sufficient to distinguish the pine woods treefrog from other treefrogs. Small warts usually cover the skin, giving it a finely granular appearance, but some individuals lack warts entirely. Tadpoles are 25–38 mm (1.0–1.5) in total length and are usually brownish red to dark green. The prominent tail fins are translucent and frequently red, and the margins are usually covered with dark blotches. The bottom part of the otherwise dark tail musculature usually has a light stripe running down its length. Unless it has been damaged or shortened by attempted predation the tail has a short, threadlike extension, or flagellum. Tadpoles often have a bright yellowish gold belly. Newly transformed juveniles measure approximately 13 mm (0.5 in) in body length.

Adult pine woods treefrog, Baker County (Anna Liner)

Pine woods treefrog tadpole, Liberty County (Dirk J. Stevenson)

Adult pine woods treefrog, Ware County (Walter W. Knapp)

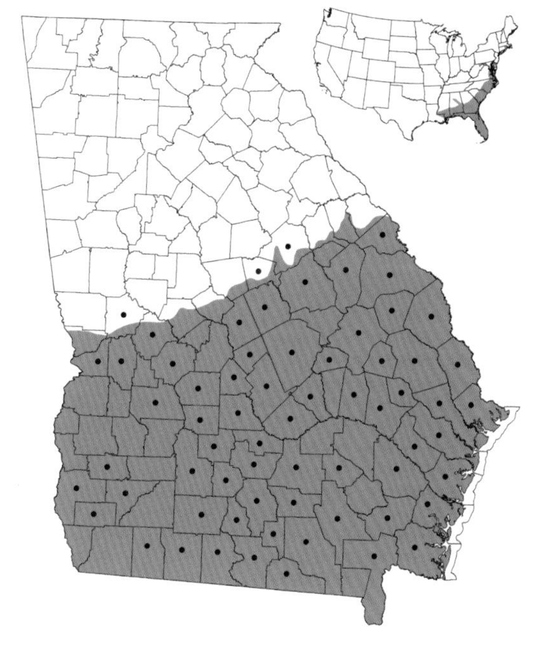

Taxonomy and Nomenclature

No subspecies are recognized. Natural hybridization occasionally occurs between the pine woods treefrog and Cope's gray treefrog (*H. chrysoscelis*).

Distribution and Habitat

Pine woods treefrogs occur throughout the Coastal Plain in longleaf pine–wiregrass flatwoods and longleaf pine–turkey oak sandhills. They spend the nonbreeding season in these terrestrial habitats, typically under bark or in rotting logs, and frequently climb high off the ground in pines or other trees. The adults breed in temporary aquatic habitats such as shallow, open, grassy ponds and isolated wetlands dominated by pond cypress. They also breed in roadside ditches and borrow pits.

Reproduction and Development

Pine woods treefrogs breed from late March through early September in Georgia, primarily in April–June. Breeding is usually associated with rain. Male frogs call from shrubs on the pond's perimeter or from emergent vegetation in the pond proper. The call has been described as resembling Morse code because of its raspy, staccato pattern, and has also been likened to that of a cicada. Large choruses are said to be reminiscent of riveting machines at work (10_Pine_woods_treefrog.mp3). Although it seems to have no discernible repetitive pattern, the call, often heard from as high up as 9 m (29.5 ft) in trees, is audible from a great distance. Eggs are laid in loose, oblong masses 12–18 cm (4.7–7.0 in) long that are attached to grass stems at or near the water's surface. Typical egg masses contain 100–250 eggs, and each female may lay up to 700 or more eggs. The eggs and the masses they form are brown to creamy yellow. They hatch in as little as 3 days, and tadpoles metamorphose in 50–75 days.

Habits

Pine woods treefrogs are adept climbers known for their tendency to call as storm fronts approach. This "rain call" is a slower version of the advertisement call and gives the impression that the frog is having trouble getting the call started. These frogs eat small insects and other invertebrates, but specific foods are unknown. Some researchers believe the frogs' use of high-canopy habitats gives them access to canopy-dwelling invertebrate prey. Common garter snakes (*Thamnophis sirtalis*), eastern ribbon snakes (*T. sauritus*), black racers (*Coluber constrictor*), and rat snakes (*Elaphe obsoleta*) are known predators. The tadpoles undoubtedly fall prey to predaceous aquatic insects.

Conservation Status

Georgia populations appear to be stable, and there are no specific conservation concerns linked with the pine woods treefrog.

W. Ben Cash

Adult barking treefrog, Baker County (Gabriel J. Miller)

Barking Treefrog

Hyla gratiosa

Description

The barking treefrog is the largest native treefrog in the United States; most adults are between 50 and 70 mm (2.0–2.7 in) long. Males are smaller than females and, at least during the breeding season, have a darker throat. The skin is noticeably granular, and the body is generally plumper than that of other treefrogs. Expanded pads on the tips of the toes are obvious. These frogs are usually bright green with numerous dark spots along the back and sides, and yellow flecking on the back. Typically, an irregular white stripe runs along each side of the body from the angle of the jaw to the groin. Like most treefrogs, adults can change their color and pattern quickly, and occasional individuals are uniformly dark tan or gray without any blotches or stripes. Although rarely confused with other treefrogs, the barking treefrog is most similar to the green treefrog (*H. cinerea*); it differs in its slightly larger size, fatter body, spotted pattern, and more granulated skin. Barking treefrog tadpoles, which may reach 50 mm (2 in) in total length, have a translucent tail and body wall. The upper tail fin is tall and inserts farther forward on the body (almost even with the eyes) than on other treefrogs. Small tadpoles often have a distinctive black smudge on the front part of the upper tail fin. As is the case with other hylids, the eyes are on the sides of the head. Newly transformed froglets are approximately 14–20 mm (0.6–0.8 in) long.

Adult barking treefrog, Taylor County (John B. Jensen)

Adult male barking treefrog in calling position, Emanuel County (Trip Lamb)

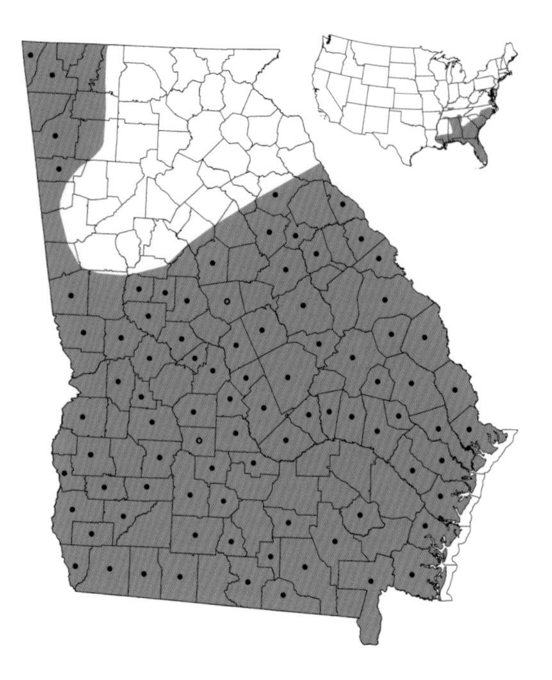

Taxonomy and Nomenclature

No subspecies are recognized. The barking treefrog is known to hybridize with the green treefrog in wetland habitats—including several in Georgia—that have been modified by human activities.

Distribution and Habitat

Barking treefrogs are widely distributed over the Coastal Plain, and populations also occur across the lower Piedmont and in the Cumberland Plateau and Ridge and Valley provinces in northwestern Georgia. Temporary or semipermanent wetlands that lack fish are the preferred breeding sites, but these treefrogs also breed in shallow areas of more permanent bodies of water that contain fish. During nonbreeding periods they occupy a wide range of habitats, including moist or swampy deciduous forests and pine woodlands.

Reproduction and Development

Reproduction is typically triggered by periods of soaking rainfall occurring between March and August. The call, from which the species receives its common name, is an explosive *boonk* or *moonk* repeated every 1 or 2 sec (11_Barking_treefrog.mp3). Choruses are often large, and males call while floating on the surface of the water or from low emergent vegetation. A large chorus makes a noise reminiscent of a distant industrial turbine or other heavy machinery. Eggs are deposited singly, and a single female can lay up to 2,000. Larvae typically metamorphose after approximately 2 months of aquatic life. Age at maturity is unknown for Georgia populations, but Georgia's other large treefrogs mature in 1–2 years.

Early stage barking treefrog tadpole, Liberty County (Dirk J. Stevenson)

Late stage barking treefrog tadpole, Baker County (John B. Jensen)

Habits

Although infrequently encountered when not moving to or from breeding pools, barking treefrogs are widespread and apparently abundant in suitable habitats. During spring rains, these frogs can routinely be seen and heard as breeding choruses assemble. They apparently spend a considerable amount of time above the ground foraging for insects and other small invertebrates and are much less likely to be attracted to outdoor lights than some of our other treefrogs. In contrast to most frogs in general and treefrogs in particular, individuals will often sit quietly on a finger while they are observed and admired. Barking treefrogs rely on their camouflage for defense, but frog-eating predators, especially watersnakes (genus *Nerodia*), eat them when they are more exposed and vulnerable during breeding. Tadpoles are probably prey for a variety of predaceous invertebrates.

Conservation Status

Barking treefrogs are common in Georgia.

Bob Herrington

Adult squirrel treefrog, Taliaferro County (Walter W. Knapp)

Squirrel Treefrog

Hyla squirella

Description

The squirrel treefrog is a somewhat small, delicate, smooth-skinned frog with a short head. Adults are typically 22–41 mm (0.9–1.6 in) long; the maximum body length recorded is 45 mm (1.8 in). Females are larger than males. During the breeding season, at least, adult males have a sienna or yellow throat while females' throat is lighter. Each toe has an expanded, adhesive disk. The skin color ranges from bright green and yellow to dark brown, with many shades and variations in between. Spots or yellow flecks may also be present. An individual's color and the presence of spots can change depending on temperature, activity, stress, and other factors. The belly is whitish, and there is yellow on the groin, thigh, armpit, and lower leg. Some individuals have a light stripe with an indistinct lower edge extending along the side from the top of the mouth to the thigh. In addition, there may be a spot or dark bar between the eyes. The green treefrog (*H. cinerea*) is the species most likely to be confused with the squirrel treefrog, but the former's light side stripe is much bolder and has a more distinct lower border. Tadpoles of the squirrel treefrog are small, averaging 32 mm (1.3 in), and have a greenish body and a yellow belly with a black center. The olive tail has a few dark spots and a dark tip. Newly transformed froglets are 11–13 mm (0.4–0.5 in) long.

Taxonomy and Nomenclature

No subspecies are currently recognized. The squirrel treefrog is sometimes called the rain frog because individuals usually call during and after summer rains.

Calling adult male squirrel treefrog, Morgan County (Walter W. Knapp)

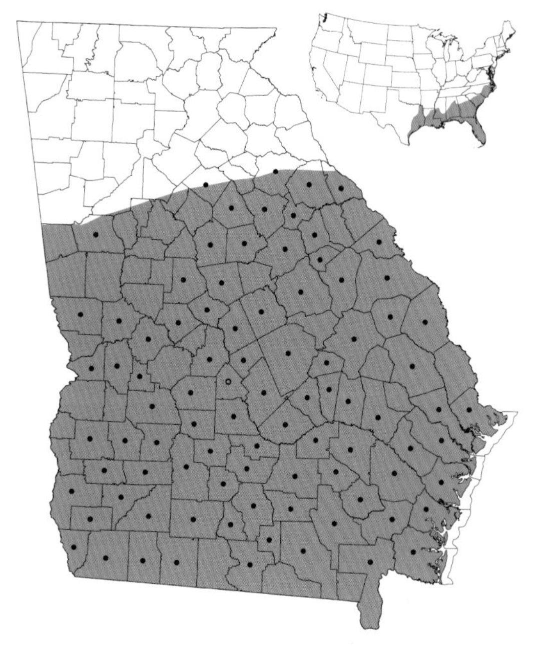

Adult squirrel treefrog, Baker County (John B. Jensen)

Squirrel treefrog tadpole, Bryan County (Dirk J. Stevenson)

Distribution and Habitat

In Georgia, squirrel treefrogs are found primarily below the Fall Line and are common throughout the Coastal Plain. Like the green treefrog, however, this species is apparently expanding its range northward through the Piedmont. The preferred habitat is open, moist forests, but individuals can be found in swamps; in open pine woods; among trees, shrubs, and vines; under leaf litter; and around houses and outbuildings. Adults often seek shelter under loose bark or in tree holes. Open, temporary wetlands or shallow roadside pools and ditches are favored breeding habitats.

Reproduction and Development

Squirrel treefrogs breed from March to late summer. The advertisement call is a harsh, nasal, ducklike *waaaak* or *quank* that is repeated about two times per second during the height of the breeding season but is not very loud (12_Squirrel_treefrog.mp3). Males usually call from a perch on a vertical stem up to about 1 m (3.3 ft) above the water or from the ground. Females lay clutches of 800 or more eggs in shallow standing water with abundant vegetation. The eggs are light brown above and white below, and take about 48 hours to hatch. The tadpole stage lasts about 25–55 days (average = 45 days). Reproductive maturity is reached in 1–2 years.

Habits

Squirrel treefrogs are nocturnal and generally spend the day resting and hiding in moist areas, although in rainy weather they move about during the day. Their rain call, generally used during and after summer showers, is a squirrel-like, scolding rasp that is usually voiced away from water. This species exhibits some activity throughout the year, including the winter during warm spells. Individuals sometimes retreat into insect burrows and rotting logs during cold weather. Adults forage at night and feed on insects in trees and shrubs; juveniles primarily eat small insects. Predators include snakes and turtles. The ability to quickly change colors and the inclination to hide during the day are the squirrel treefrog's major defenses against predators. These are among the most familiar frogs to rural southerners, often showing up in surprising places such as basements and campground bathrooms. They occasionally find their way into people's homes, where they dry out and leave their mummified remains to be discovered behind a piece of furniture or in the corner of a closet.

Conservation Status

This species is common throughout Georgia's Coastal Plain, and populations seem to be stable.

Gabrielle J. Graeter

Adult mountain chorus frog, Cherokee County (John B. Jensen)

Mountain Chorus Frog

Pseudacris brachyphona

Description

Mountain chorus frogs reach a slightly larger size than the physically similar upland chorus frogs (*P. feriarum*). Adult males are 24–32 mm (0.9–1.3 in) long; adult females are 27–34 mm (1.1–1.3 in). At least during the breeding season, adult males have a brown or yellowish vocal sac while females have a white, smooth throat. Color patterns of adults are quite variable in Georgia, although most have a light brown back. Some individuals have dark brown stripes shaped like reversed parentheses running down the back while others are speckled or lack a discernible pattern. Most have a dark brown stripe that runs from the tip of the snout through the eyes (giving the appearance of a mask) and continues down each side. The belly is white and may have a few dark speckles. Tadpoles are uniformly medium brown or brassy and have a lower tail fin than the tadpoles of other chorus frogs. The tail musculature may be darkly mottled. Tadpoles may reach approximately 35 mm (1.4 in) in total length prior to metamorphosis. Newly transformed froglets are about 8 mm (0.3 in) long.

Taxonomy and Nomenclature

DNA studies indicate that this species is most closely related to Brimley's chorus frog

Adult mountain chorus frog, Dawson County (Walter W. Knapp)

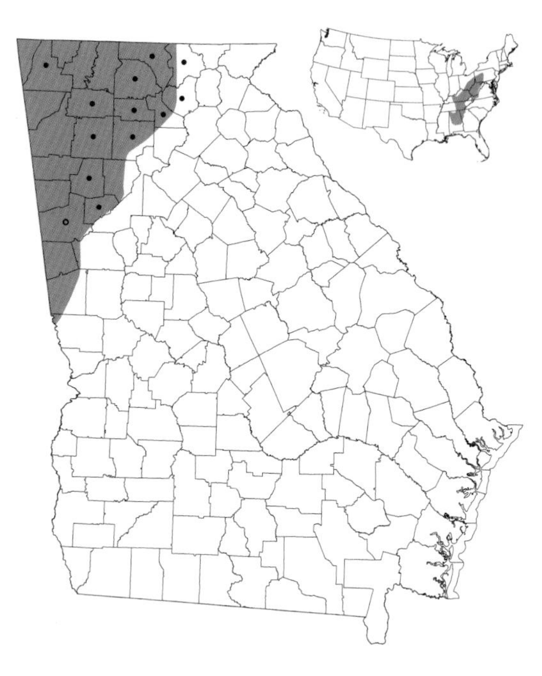

Mountain chorus frog tadpole, Walker County (John B. Jensen)

(*P. brimleyi*). Together, these two species form a genetic group separate from the one containing the southern chorus frog (*P. nigrita*) and the upland chorus frog.

Distribution and Habitat

Mountain chorus frogs are generally found in the Appalachian Mountains and surrounding foothills. Specifically, populations are known from the western half of the Blue Ridge, the Ridge and Valley, and the Cumberland Plateau provinces. This species appears to replace the upland chorus frog ecologically in some of the mountainous areas of the state, although occasionally the two species use the same breeding sites, which are most often grassy pools and ditches. Outside the breeding season, mountain chorus frogs inhabit the surrounding fields and woodlands.

Reproduction and Development

After late winter or early spring rains, at temperatures above approximately 4.5° C (40° F), mountain chorus frogs congregate to breed in temporary pools. Males produce a short trill that sweeps upward in pitch. The call, which contains approximately 20–25 pulses and lasts 0.25 sec, has been compared to the sound of a squeaky wagon wheel (13_Mountain_chorus_frog.mp3). It is very similar to the advertisement call of Brimley's chorus frog but much shorter in duration than the calls of either the southern chorus frog or the upland chorus frog. The call is perhaps the best character for distinguishing the mountain from the upland chorus frog in Georgia. Females deposit up to 1,500 eggs in multiple clusters along grass stems in shallow ditches and pools. Eggs hatch in 1 or 2 weeks, and tadpoles metamorphose 50–60 days later. Most individuals reach sexual maturity the following breeding season.

Habits

Female mountain chorus frogs remain at the breeding sites only long enough to mate, then disperse to nearby woodlands and meadows. Males remain near the pools for 2–3 months and call when weather conditions are optimal, but they also disperse as summer approaches. Mountain chorus frogs hide during the day and forage at night on small arthropods. Adults and/or tadpoles may be taken by a wide variety of predators, including dragonfly larvae, aquatic beetles, fishing spiders, fish, mole salamander larvae (genus *Ambystoma*), eastern newts (*Notophthalmus viridescens*), common garter snakes (*Thamnophis sirtalis*), watersnakes (genus *Nerodia*), and birds. When a predator approaches, mountain chorus frogs initially remain still, relying on their cryptic color pattern to protect them, but on further disturbance they jump or swim away.

Conservation Status

The status of the mountain chorus frog in Georgia has not been well established. This species is currently treated as a Species of Concern by the state.

Emily C. Moriarty Lemmon

Adult Brimley's chorus frog, Screven County (John B. Jensen)

Brimley's Chorus Frog

Pseudacris brimleyi

Description

Males reach a body length of 24–28 mm (0.9–1.1 in), and females are 27–35 mm (1.1–1.4 in) long. At least during the breeding season, males have a darkened throat with loose skin corresponding to the vocal sac area while females' throat is white and smooth. Adult frogs have a light brown background color with three dark brown longitudinal stripes running down the back. The stripes narrow as they extend onto the head and do not expand into a triangle or patch between the eyes as is frequently the case in the upland chorus frog (*P. feriarum*). A bold black or dark brown stripe extends along each side from the snout through the eye to the groin. The leg stripes or spots run down the legs rather than across them as in other chorus frogs. The belly is either white with dark speckles or unmarked. Tadpoles, which reach a total length of about 30 mm (1.2 in) prior to metamorphosis, have a dark brown upper surface with a few brassy flecks. Newly transformed froglets are 9–11 mm (0.4 in) in body length.

Taxonomy and Nomenclature

DNA analyses indicate that Brimley's chorus frog is most closely related to the mountain chorus frog (*P. brachyphona*). Together these species form a genetic group, or clade, distinct from the one containing the southern chorus frog (*P. nigrita*), the upland chorus frog, and several other chorus frog species found outside Georgia.

Brimley's chorus frog tadpole, South Carolina (Ron Altig)

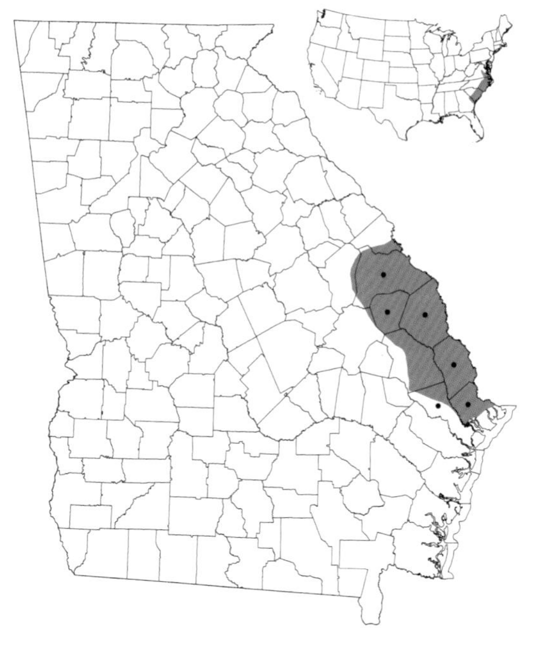

Adult Brimley's chorus frog, Screven County (John B. Jensen)

Distribution and Habitat

Brimley's chorus frogs are restricted to the extreme northeastern Coastal Plain of Georgia. Adults congregate to breed in temporary pools, ditches, and shallow wooded wetlands during late winter and early spring and disperse to surrounding fields and woodlands during the remainder of the year.

Reproduction and Development

Following late winter or early spring rainstorms, when nighttime temperatures rise above approximately 4.5° C (40° F), Brimley's chorus frogs gather to breed in ephemeral pools. Males compete for prime calling locations and attempt to drive off rival males. Their advertisement call is a short trill that contains 15–22 pulses and lasts 0.25 sec (14_Brimley's_chorus_frog.mp3). The call is very similar to that of the mountain chorus frog but is much shorter in duration than the calls of southern and upland chorus frogs. Females deposit up to 290 eggs, usually in multiple clumps, along grass stems and twigs just below the water's surface. Tadpoles hatch within 1–2 weeks and metamorphose after 6–10 weeks. Juvenile frogs emigrate from the breeding pool shortly after absorbing their tail. Most are sexually mature by the following mating season.

Habits

After the breeding season Brimley's chorus frogs scatter throughout the swampy woodlands surrounding the breeding pools. They hide in leaf litter and under logs during the daylight hours and emerge at night to feed on small arthropods. Known predators of chorus frogs include common garter snakes (*Thamnophis sirtalis*), watersnakes (genus *Nerodia*), mole salamander larvae (genus *Ambystoma*), eastern newts (*Notophthalmus viridescens*), fishing spiders, aquatic beetles, dragonfly larvae, various fish species, and birds. These frogs rely mainly on their cryptic coloration and propensity to hide to protect them from predators.

Conservation Status

Recent records of Brimley's chorus frogs in Georgia are very few. A number of historical populations no longer exist because of urban expansion near the coast. This species is treated as a Species of Concern by the state of Georgia.

Emily C. Moriarty Lemmon

Adult spring peeper, Baker County (Gabriel J. Miller)

Spring Peeper

Pseudacris crucifer

Description

Adult male spring peepers are about 18–34 mm (0.7–1.3 in) long; females are 20–35 mm (0.8–1.4 in). During the breeding season, at least, males have a dark brown or yellowish throat with loose skin indicating the vocal sac area, while females have a smooth, light-colored throat. The background color is light tan with a characteristic dark brown X on the back. In some individuals the right and left halves of the X do not connect. A brown stripe runs from the tip of the snout through the eye and along the side. The belly is white and may have a scattering of brown speckles. These frogs are better climbers than other Georgia chorus frogs because they have enlarged toe pads similar to those of true treefrogs (genus *Hyla*). Tadpoles are medium brown and often have gold flecking on the body. The fins are frequently blotched, with a clear area near the tail musculature. Tadpoles reach a total length of 30–36 mm (1.2–1.4 in) before undergoing metamorphosis. Newly transformed frogs are 9–14 mm (0.4–0.6 in) in body length.

Taxonomy and Nomenclature

The spring peeper was once considered a member of the genus *Hyla*, but protein and DNA sequence data support its inclusion in *Pseudacris*. Its closest relative is the little grass frog (*P. ocularis*). Until recently Georgia's spring peepers were subdivided into two subspecies—the southern spring peeper (*P. c. bartramiana*)

Adult spring peeper, Walker County (John B. Jensen)

Spring peeper tadpole, Liberty County (Dirk J. Stevenson)

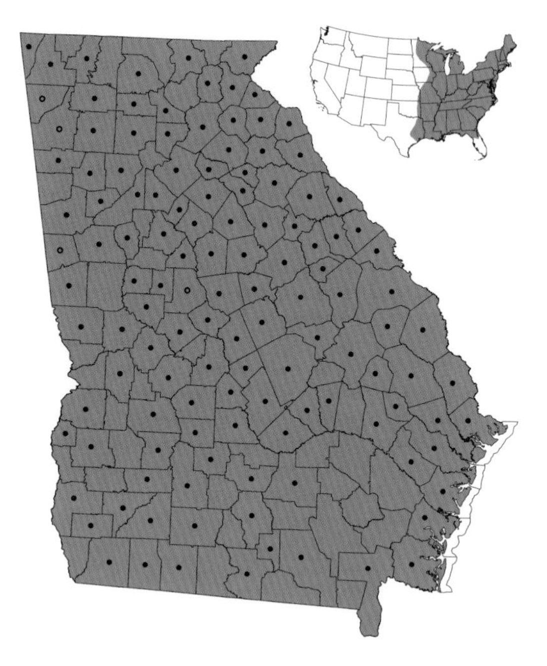

and the northern spring peeper (*P. c. crucifer*)—but genetic research has shown these forms to be invalid.

Distribution and Habitat

The spring peeper is the most abundant and widespread chorus frog in Georgia. During the peak of the breeding season, adults can be found in bodies of water throughout the state, including temporary wetlands. Unlike Georgia's other chorus frogs, spring peepers often breed in relatively permanent or semipermanent pools, particularly in the weedy portions of beaver ponds and older, well-established farm ponds. A few individuals call at semipermanent seeps. In the Coastal Plain, spring peepers are more likely than ornate (*P. ornata*) and southern (*P. nigrita*) chorus frogs to breed in swamps associated with floodplains. After the breeding season, spring peepers disperse to nearby woodlands and fields for the remainder of the year.

Reproduction and Development

Spring peepers call from late fall through spring when nighttime temperatures range between 2 and 21° C (35–70° F). During the peak of the breeding season, they call during the day as well as at night. The advertisement call is a smooth whistle without pulses or trills that ascends in pitch (15_Spring_peeper.mp3). Males switch to a trilled territorial or aggression call when other males are calling too close to them; this aggression call is somewhat similar to the advertisement call of the upland chorus frog (*P. feriarum*). Males compete for the best calling sites and evict neighboring males with their aggression call or, in some cases, with physical force. Females deposit up to 1,000 eggs along submerged twigs or grass stems. Eggs hatch in 1–2 weeks, and tadpoles metamorphose within 2 months. Most individuals reach sexual maturity during their first year.

Habits

Outside the breeding season spring peepers hide in leaf litter or low shrubs during the day and emerge at night to forage on small invertebrates. They sometimes become active during the daylight hours following rainstorms. Tadpoles and adults fall prey to common garter snakes (*Thamnophis sirtalis*), watersnakes (genus *Nerodia*), mole salamander larvae (genus *Ambystoma*), newts (genus *Notophthalmus*), fishing spiders, aquatic beetles, dragonfly larvae, fish, and birds.

Conservation Status

Spring peepers are abundant throughout Georgia.

Emily C. Moriarty Lemmon

Adult male upland chorus frog, Dade County (John B. Jensen)

Upland Chorus Frog

Pseudacris feriarum

Description

The upland chorus frog is about the size of a quarter. Adult males reach a body length of 21–30 mm (0.8–1.2 in); the slightly larger females are about 22–33 mm (0.9–1.3 in) long. During the breeding season, at least, males have loose brown or yellow skin on the throat in the vocal sac area; females have a smooth, white throat. Adult frogs typically have a tan background with up to three darker brown longitudinal stripes or rows of spots running down the back. The stripes usually merge on top of the head to form a triangle or patch between the eyes. Some individuals lack stripes and instead have a few dark speckles on the tan background. The belly is white and may have scattered brown speckles. A distinct white line is present along the upper lip. Tadpoles reach approximately 34–36 mm (1.3–1.4 in) in total length. They are medium brown or black and typically have small black spots on the back; the belly is unpigmented. Body size at metamorphosis is 8–12 mm (0.3–0.5 in).

Taxonomy and Nomenclature

The upland chorus frog was once considered a subspecies of the western chorus frog (*P. triseriata*), but DNA analyses suggest that its closest Georgia relative is the southern chorus frog (*P. nigrita*). Along with several other chorus frog species found outside Georgia these two species form a group distinct from the one containing the mountain chorus frog (*P. brachyphona*) and Brimley's chorus frog

Adult upland chorus frog with aberrant (speckled) pattern, Bibb County (John B. Jensen)

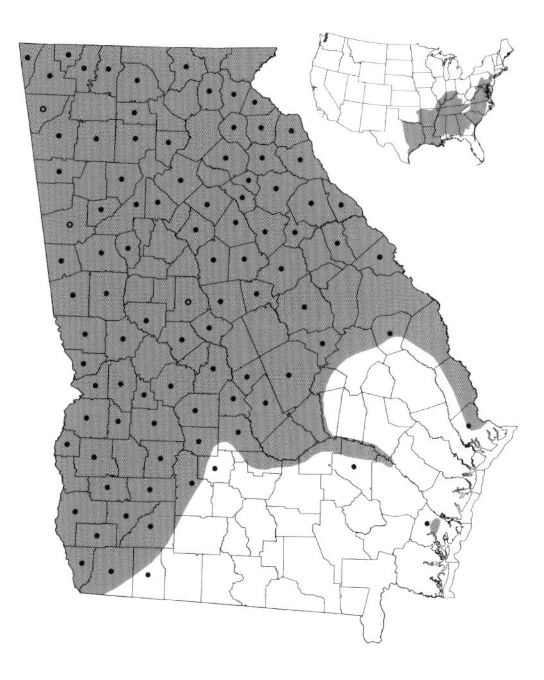

Upland chorus frog tadpole, Banks County (John B. Jensen)

(*P. brimleyi*). Recent genetic data also indicates that no subspecies of the upland chorus frog should be recognized.

Distribution and Habitat

Upland chorus frogs are found throughout most of the northern half of Georgia and extend into the Coastal Plain along several river drainages, including the Flint and Chattahoochee rivers as well as the Altamaha River and its tributaries. North of the Fall Line, these frogs congregate to breed primarily in open habitats such as flooded fields and ditches. In the Coastal Plain, where the range of this species overlaps that of the southern chorus frog, adults typically breed in forested gum or cypress swamp habitats. After mating, frogs disperse from breeding pools to adjacent woodlands and meadows.

Reproduction and Development

In late winter and early spring, when nighttime temperatures rise above 2° C (35° F), upland chorus frogs congregate to breed in ephemeral pools. During the peak of the breeding season, the males may call day and night. Males outnumber females at breeding congregations because females migrate to pools for a single night and return to terrestrial habitats after depositing their eggs. Males remain at the pools for several weeks and compete with other males for calling locations and mates. They produce a trilled advertisement call lasting 0.5–1.25 sec. It sweeps upward in pitch, creating a sound reminiscent of dragging a thumbnail across the teeth of a comb (16_Upland_chorus_frog.mp3). The length of the call depends on the temperature; cold frogs give long calls, and warm frogs give short calls. The call contains 15–23 pulses in the northern half of Georgia but may have as many as 35 pulses in the Coastal Plain, where the range of the upland chorus frog overlaps that of the closely related but slower-calling (5–11 pulses) southern chorus frog. The call of the latter species has individual pulses slow enough to be counted—a useful character for distinguishing between the two species in areas of overlap. This phenomenon, in which closely related species differ more in areas where they live together, is called reproductive character displacement. Females deposit multiple clutches of 15–300 eggs along twigs and grass stems in the water. Eggs hatch within a week or two, and tadpoles metamorphose 48–80 days later, depending on environmental conditions. Juvenile frogs leave the breeding pool and disperse to surrounding fields and woods. Most frogs reach sexual maturity during their first year.

Habits

Following the mating season, adult upland chorus frogs may disperse up to 200 m (656 ft) from their breeding pools. During the summer, they hide under logs, in grass tussocks, or in other damp places during the day and emerge at night to forage on small arthropods. They are occasionally encountered jumping across the forest floor on humid days. Tadpoles and adults may fall prey to common garter snakes (*Thamnophis sirtalis*), watersnakes (genus *Nerodia*), mole salamander larvae (genus *Ambystoma*), newts (genus *Notophthalmus*), fishing spiders, aquatic beetles, dragonfly larvae, fish, and birds. When approached by a predator, frogs remain still and rely on their cryptic coloration for protection; if physically disturbed, they jump away.

Conservation Status

Populations of the upland chorus frog appear stable in most areas of the state. These frogs flourish in low-intensity agricultural areas, particularly those with numerous forested woodlots and pastures where open habitat and ephemeral pools are available.

Emily C. Moriarty Lemmon

Adult southern chorus frog, Liberty County (Dirk J. Stevenson)

Southern Chorus Frog

Pseudacris nigrita

Description

This is a small, slender, dark frog with a somewhat pointed snout. Adults are 19–32 mm (0.7–1.3 in) long. Males are smaller and, at least during the breeding season, have a darker throat than females. Three prominent black stripes that typically break up into rows of large spots are present on the back. A distinct white line along the upper lip is bordered above by a continuous black stripe that extends from the snout to the groin. The base color is often dark brown or black but may be light gray or olive. The skin is granular and has numerous small warts. A close relative, the upland chorus frog (*P. feriarum*), is similar in size and appearance but usually has a dark triangle between the eyes, a slightly stouter body, and a less pointed snout. The southern cricket frog (*Acris gryllus*) is smaller, has a dark triangle between the eyes, and has dark stripes on the rear of each thigh. The small, slender tadpole of the southern chorus frog is brown with a brassy or golden belly. The upper portion of the tail fin inserts at or behind the spiracle; the upper and lower tail fins are clear and marked with numerous small, dark flecks; and the tail musculature is uniformly dark. Tadpoles may reach 35 mm (1.4 in) in total length; newly metamorphosed frogs are 12–16 mm (0.5–0.6 in) long.

Taxonomy and Nomenclature

Two subspecies were formerly recognized, but recent genetic research indicates that their continued recognition is unwarranted. DNA analyses suggest that the southern chorus frog's closest relative in Georgia is the upland chorus frog.

Adult southern chorus frog, Liberty County
(Lang Elliott)

Southern chorus frog tadpole, Liberty County
(Dirk J. Stevenson)

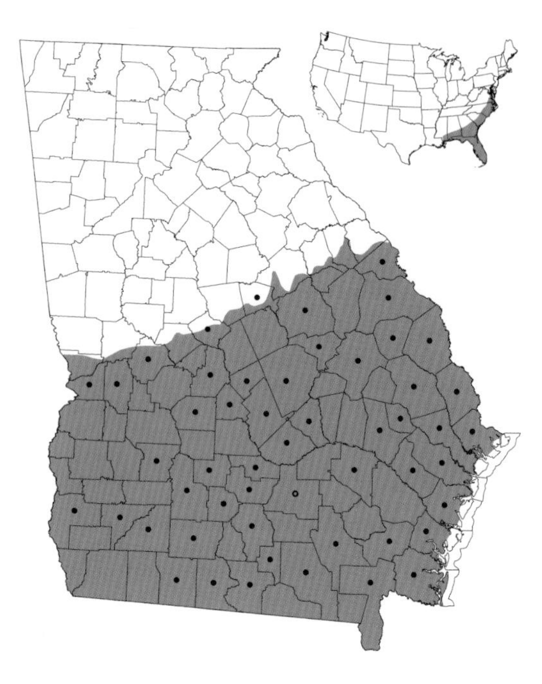

Distribution and Habitat

The southern chorus frog is common throughout the Coastal Plain. It is primarily an upland species, preferring habitats with sandy or loamy soils; it avoids floodplains and other sites on heavy soils. This frog is widespread in moist pine flatwoods, sandhills, mixed pine-oak communities, and disturbed examples of these habitats. Small, rainwater-filled grassy depressions such as the shallow edges of sinkhole ponds, cypress ponds, and Carolina bays are preferred breeding sites. These types of wetlands are often forested with scattered pond cypress, swamp black gum, and slash pine. This tolerant frog thrives in disturbed landscapes and often breeds in man-made habitats, especially roadside ditches, flooded fields, old vehicle tracks, and other small pools and puddles.

Reproduction and Development

Southern chorus frogs breed from December to April. Breeding activity peaks in February and March, and choruses are regularly heard during the day. In southern Georgia this species is a true harbinger of spring, unlike the spring peeper (*P. crucifer*), which begins breeding much earlier (November–January). Choruses are occasionally heard following heavy summer rains, but whether or not actual breeding occurs is unknown. Males call from shallow grassy areas, each frog emitting a slow, metallic, staccato trill consisting of five to eight sharp, ratchetlike notes, best imitated as *ik-ik-ik-ik-ik* (17_Southern_chorus_frog.mp3). Breeding southern chorus frogs avoid permanent ponds, perhaps because of the presence of predatory fish, and are especially attracted to grassy ditches and other wet-weather ponds. Egg clutches containing from several dozen to more than 100 eggs are attached to submerged grasses not far below the water's surface. Clumps of eggs are oval to oblong and about 30 mm (1.2 in) long. A single female may lay 500–1,500 eggs. Eggs hatch within a few days to a week. Transformation to froglets occurs 6–8 weeks later. This frog reaches sexual maturity at approximately 1 year of age.

Habits

Except for some aspects of its breeding ecology, the habits of this frog are not particularly well known. Southern chorus frogs are seldom encountered outside their breeding period. Individuals may spend the warmer months (May–November) underground or sheltered in thick grass or leaf litter. Southern chorus frogs eat insects, spiders, and other small invertebrates. Likely predators include turtles, small snakes, and birds. Aquatic invertebrates such as dragonfly larvae undoubtedly eat the tadpoles.

Conservation Status

The southern chorus frog is very common. Its abilities to adapt to highly disturbed landscapes and to reproduce in ditches and other man-made wetlands have enabled it to thrive throughout its historic range.

Dirk J. Stevenson

Adult little grass frog, Liberty County (Dirk J. Stevenson)

Little Grass Frog

Pseudacris ocularis

Description

This delicate frog is the smallest land vertebrate in North America. Adults range in body size from 11 to 20 mm (0.4–0.8 in); males are slightly smaller than females. At least during the breeding season, the male's throat is darker than its belly. The base color may be gray, tan, brown, greenish, or brick red. The skin is smooth. A bold, dark stripe extends from the pointed snout through each eye and along each side of the body. Faint stripes may be present down the middle and sides of the back. The belly is yellowish white. The hind feet are only slightly webbed, and the toes have slightly expanded tips that are useful in climbing. Recently metamorphosed southern chorus (*P. nigrita*) and southern cricket (*Acris gryllus*) frogs may be confused with little grass frogs, but the former two species have granular skin. Juvenile Brimley's chorus frogs (*P. brimleyi*) are most similar in appearance but usually have a spotted chest. The tail musculature of the tadpoles has a prominent wide, dark stripe above a narrower light stripe. A pale silver to orange stripe may be present between the eye and the base of the tail on each side. The tail fins have abundant diffuse blotches. Tadpoles may reach 23 mm (0.9 in) in length, and newly metamorphosed froglets are 7–9 mm (0.3–0.4 in) in body length.

Little grass frog tadpole, Bryan County (Dirk J. Stevenson)

Adult little grass frogs, Charlton County (John B. Jensen)

Taxonomy and Nomenclature

The taxonomic placement of this species has historically been controversial. Little grass frogs have been placed in seven different genera at one time or another: *Auletris, Calamita, Chorophilus, Cystingnathus, Hyla, Hylodes,* and *Limnaoedus.* Mitochondrial DNA evidence indicates that this species is most closely related to the spring peeper (*P. crucifer*).

Distribution and Habitat

Little grass frogs are restricted to the Coastal Plain. Breeding and nonbreeding habitats are not known to differ significantly and include a variety of shallow, typically fish-free wetlands such as cypress ponds, bogs, wet pine flatwoods,

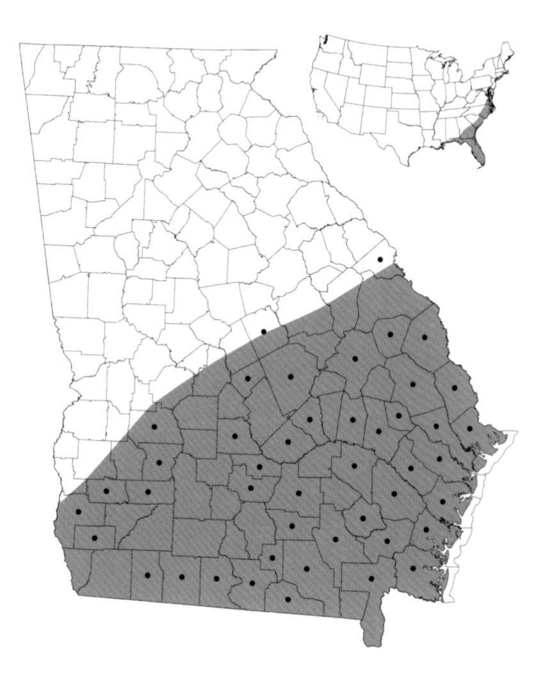

savannas, wet meadows, river swamps, and roadside ditches. These habitats characteristically have abundant emergent grasses, sedges, and/or rushes. Individuals sometimes call from vegetation in brackish ditches.

Reproduction and Development

The breeding season is rather long relative to those of other Georgia frogs, extending from January to September. Males attract females by calling from clumps of vegetation, often several feet above the ground. The call is an insectlike tinkle that is surprisingly loud for such a tiny frog (18_Little_grass_frog.mp3). Its pitch is so high that some people are unable to hear it. Females deposit 100–200 eggs singly or in small clusters of 25 or more on the pond bottom or attached to submerged vegetation. Hatching follows within a few days, and tadpoles metamorphose 45–70 days later. Sexual maturity likely is reached within the first year, as is the case with other small frogs such as chorus, cricket (genus *Acris*), and greenhouse frogs (*Eleutherodactylus planirostris*).

Habits

Adults are active both night and day and feed on a variety of small arthropods associated with leaf litter and soil, including springtails, ants, rove beetles, leafhoppers, and mites. Tadpoles graze on algae and aquatic vegetation. Wolf spiders are known predators, and many other animals probably eat these frogs as well. The dark lateral stripe helps to break up their outline and may protect them from visually oriented predators. If pursued, little grass frogs attempt to escape by leaping horizontal distances up to 0.4 m (1.3 ft).

Conservation Status

This species is common and apparently secure throughout its range in Georgia.

John B. Jensen

Adult ornate chorus frogs, Liberty County (Andy Day)

Ornate Chorus Frog

Pseudacris ornata

Description

Adult males range in body length from 25 to 39 mm (1.0–1.5 in); the slightly larger adult females are 28–40 mm (1.1–1.6 in). During the breeding season, at least, males have a darker throat than females. This well-named frog displays a wide variety of attractive colors and patterns, and it is unusual to find two individuals that appear identical. The background color is usually predominantly gray, brown, reddish brown, pinkish, or green, but some individuals are adorned by a combination of some or all of these. The overall impression is of a small frog made of fine porcelain. Individuals of one particular color dominate some populations, yet other populations are highly variable. Most ornate chorus frogs have a bold black or chocolate brown stripe extending from the nostril on each side through the eye to the shoulder or down to the side of the belly. This stripe may be broken along the side. A dark elongated blotch is usually present on each side above the groin area. The top of the head between the eyes may have a dark triangular blotch, which may be faded. Other dark or faded blotches or stripes may be present on the sides of the back. The exposed surfaces of the fore and hind legs often have blotches that appear to be bands, but they do not connect underneath the limbs. The concealed portions of the groin and thighs are typically bright yellow. Ornate chorus frogs are significantly plumper than other Georgia chorus frogs. Tadpoles can reach 43 mm (1.7 in) in total length and have a very high tail fin that inserts just behind the eyes—a feature shared in Georgia only by

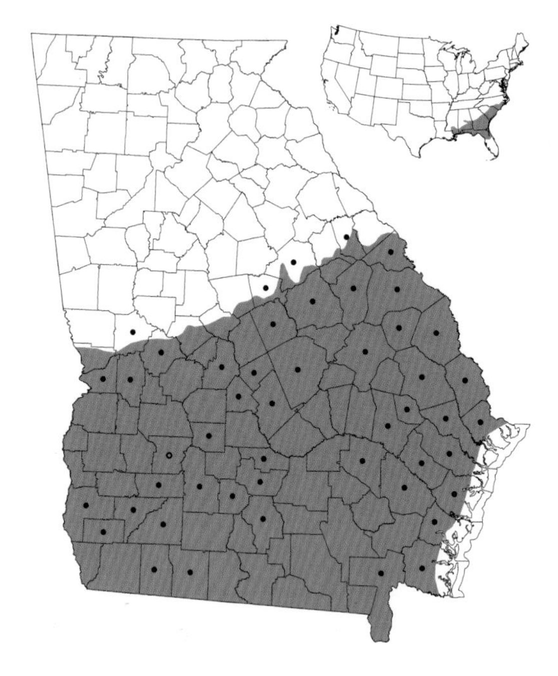

Gravid adult female ornate chorus frog, Baker County. Note the eggs visible through the dark spot on the side. (Gabriel J. Miller)

Ornate chorus frog tadpole, Baker County (John B. Jensen)

barking treefrog (*Hyla gratiosa*) tadpoles. Two gold or brassy stripes are often present on the sides of the back. The tail musculature may be strongly bicolored or mottled. Froglets are typically 14–16 mm (0.6 in) in body length immediately following metamorphosis.

Taxonomy and Nomenclature

No subspecies are recognized. Ornate chorus frogs belong to a genetic lineage, or clade, separate from our other chorus frogs and are both genetically and morphologically most closely related to Strecker's (*P. streckeri*) and Illinois (*P. illinoensis*) chorus frogs of the south-central and midwestern United States, respectively.

Distribution and Habitat

Ornate chorus frogs are restricted to the Coastal Plain, where they typically occupy longleaf pine habitats such as sandhills, pine flatwoods, and upland pine forests. Nonbreeding adults also inhabit fields surrounding such areas, particularly those that are lying fallow and not treated with pesticides. Within these habitats, the frogs burrow into sandy soils, often among the roots of herbaceous vegetation. Temporary, typically fishless, wetlands such as sinkhole ponds, Carolina bays, cypress ponds, and borrow pits serve as breeding and larval habitats. This species is less inclined to breed in ditches, tire ruts, and other shallow man-made depressions than other native chorus frogs. Tadpoles seek sheltered clumps of submerged and emergent herbaceous plants.

Reproduction and Development

During late fall or early winter, adults migrate up to 425 m (1,394 ft) or perhaps more from their upland burrows to breeding wetlands. Males arrive earlier than females and remain in the ponds longer. Breeding may take place from November to March; peak activity in Georgia occurs in January and February. Males, which have a sharp, metallic, peeping call likened to the sound of a hammer hitting a chisel

(19_Ornate_chorus_frog.mp3), vocally advertise themselves to females while floating or perched 2.4–24 cm (1.0–0.9 in) above the water on inundated clumps of aquatic vegetation or floating debris. Eggs are laid in small clusters of typically 20–40, but as many as 106 have been recorded. The eggs are attached to submerged stems of herbaceous plants. Hatching usually occurs within a week. Tadpoles transform and leave the ponds 3–4 months later. Little information is available on age at maturity; other similar-sized chorus frogs mature within the first year of life.

Habits

Very little is known about the habits of this species in its upland nonbreeding habitats. Young grasshoppers and crickets are documented prey of newly transformed individuals, and other insects, earthworms, and roundworms (nematodes) have been suggested as likely food items. Tadpoles graze on algae and aquatic plants. Predatory aquatic invertebrates and salamander larvae (especially those of the mole salamander genus *Ambystoma*) likely consume tadpoles. The southern hognose snake (*Heterodon simus*) is a documented predator of adults, although other vertebrates certainly eat them as well.

Conservation Status

Growing concerns about declines of this species warrant attention. Although ornate chorus frogs still seem to be fairly common in Georgia, certain populations in other states have experienced notable declines and disappearances. Intensive soil disturbance associated with industrial pine farming has reduced or eliminated some populations. Destruction of breeding ponds is also a limiting factor for this and many other associated amphibians.

John B. Jensen

Rain Frogs

FAMILY LEPTODACTYLIDAE

Georgia's single member of this family is less than 25 mm (1 in) in body length, although some tropical species may reach 180 mm (7.2 in). Individuals have less webbing between their toes than most of Georgia's frogs. This family is not native to Georgia, but the greenhouse frog (*Eleutherodactylus planirostris*) has been introduced into several cities in the southern part of the state. At least 850 additional species are distributed in Central and South America and the West Indies.

Adult greenhouse frog, mottled phase, Chatham County (Dirk J. Stevenson)

Greenhouse Frog

Eleutherodactylus (Euhyas) planirostris

Description

This tiny, granular-skinned frog is only 12–30 mm (0.5–1.2 in) in length. The background color is brown, reddish brown, or bronze, and two pattern phases exist: a mottled phase with dark and light coloration, often with a faint chevron-shaped band on the back and an equally faint, irregular band between the eyes; and a striped phase with two light longitudinal stripes on the back. The belly is light gray or white with fine brown stippling. The eyes are often reddish. There is no webbing between the toes. Hatchlings look very similar to adults but have a tiny tail that is soon lost. New hatchlings are 9–11 mm (0.4 in) long.

Taxonomy and Nomenclature

The greenhouse frog is the only Georgia representative of a frog family that is widely distributed in tropical America. Some taxonomists suggest placing the greenhouse frog in the genus *Euhyas.*

Distribution and Habitat

This nonnative species likely found its way from Cuba to Georgia via Florida as a stowaway in tropical plant shipments. The first known Georgia population was discovered in Savannah in 1998. Since then the species has become common on Saint Simons Island and in nearby

Adult greenhouse frog, striped phase, Glynn County (John B. Jensen)

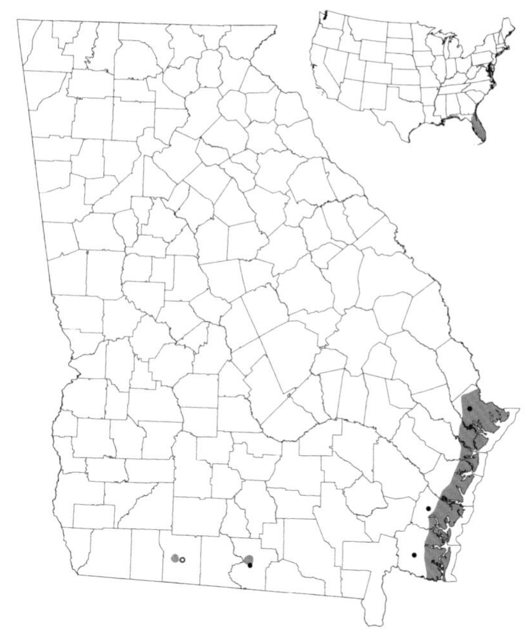

Brunswick. Specimens have also been found in Thomasville and Valdosta. The greenhouse frog's full distribution in Georgia is currently unknown but is probably expanding. It is commonly found in residential areas but may also occur in a variety of natural terrestrial habitats.

Reproduction and Development

Greenhouse frogs breed during spring and summer in damp, but not wet, situations such as clumps of vegetation, under debris, in leaf litter, and even in the soil of potted plants. Each female lays approximately 20 eggs; often she remains to attend them. Eggs hatch directly into tiny froglets after an in-egg development and metamorphosis period of about 2 weeks. The advertisement call, a series of short, bird-like chirps (20_Greenhouse_frog.mp3), is often triggered by rain or sprinklers. Individuals reach sexual maturity within the first year of life.

Habits

This is a terrestrial species that never voluntarily takes to water. Like most amphibians, greenhouse frogs are primarily nocturnal and are most active during or following rains. During the day and during dry weather they seek the shelter of various objects, both natural and artificial. The diet consists of small invertebrates. These frogs are probably eaten by a wide variety of carnivorous animals.

Conservation Status

Relative to many other nonnative species in Georgia, the greenhouse frog seems benign. However, subtle impacts, such as competition with native frogs for prey items and adverse effects on native insects, may be difficult to detect.

John B. Jensen

Narrow-mouthed Toads

FAMILY MICROHYLIDAE

Georgia's single species, *Gastrophryne carolinensis,* is a small toad (body size less than 40 mm, or 1.6 in), although other microhylids range in size from 10 to 85 mm (0.4–3.4 in). As the common name implies, these frogs have a very small mouth and short hind limbs. Like the Leptodactylidae, microhylids have reduced webbing between the toes. They are the only frogs in Georgia that lack a visible tympanum. The tadpoles have fleshy lips rather than the horny beak and small teeth typical of other Georgia tadpoles. Many members of this family are burrowers, and some live in near desertlike conditions. Although some species, including Georgia's sole representative, are temperate in distribution, many of the more than 300 other species occur in tropical regions of North and South America, Africa, Asia, and Australia.

Adult eastern narrow-mouthed toad, Long County (Dirk J. Stevenson)

Eastern Narrow-mouthed Toad

Gastrophryne carolinensis

Description

Adults range from 22 to 35 mm (0.9–1.4 in) in body length; females are slightly larger than males. Adult males can be distinguished from females, at least during the breeding season, by the dark-pigmented vocal sac on their throat. Viewed from above, the body of this plump, smooth-skinned toad has a teardrop shape accentuated by its small head and pointed snout. Body color ranges from dark gray to yellow-brown on the back, and some individuals have an indistinct broad, light line running along each side that can be flecked with gold. The belly is dark with light mottling. The tympanum is concealed, and a transverse fold of skin runs across the head just behind the eyes. The rear feet have horny protuberances that are used in digging. No webbing is present between the toes of any of the feet. Tadpoles are black on the back with obscure light mottling on the belly, and they lack the beaklike mandibles and specialized labial teeth associated with the lips of other tadpoles. They are also unique in that the top of the head and the body are relatively flattened. Some individuals have a light lateral tail stripe. Total length ranges from 15 to 30 mm (0.6–1.2 in). Newly transformed frogs are 7–12 mm (0.3–0.5 in) in body length.

Taxonomy and Nomenclature

No subspecies are currently recognized. The Great Plains narrow-mouthed toad (*G. olivacea*)

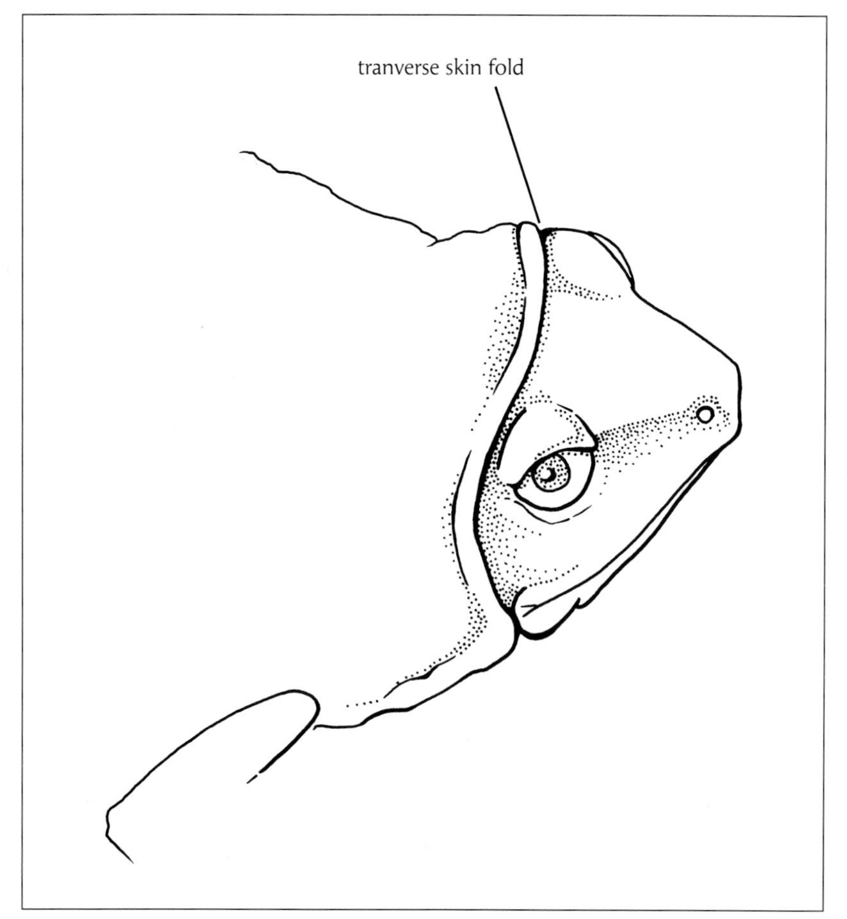

Head of Eastern Narrow-Mouthed Toad

was once considered a subspecies of the eastern narrow-mouthed toad.

Distribution and Habitat

Eastern narrow-mouthed toads have been recorded from all of Georgia's physiographic provinces except the Blue Ridge, where they may or may not also occur. These toads breed in ephemeral aquatic habitats such as temporary ponds, borrow pits, roadside ditches, and deep wheel ruts in dirt roads. When not breeding they occupy terrestrial habitats that vary from mixed pine-hardwood forests to upland pine stands. The common variable in these situations is usually the presence of sandy or loamy soils important for the subterranean habits of the frog.

Reproduction and Development

This species typically breeds April–October throughout most of its range in Georgia, although a Coastal Plain population was recorded breeding as early as March and as late as November. Individuals breed in explosive pulses associated with warm rains. Males call from well-concealed positions along the edges of breeding pools or from clumps of emergent vegetation within the pool itself. The call is reminiscent of the bleating of sheep, although it has a much higher pitch and a distinctly nasal quality (21_Eastern_narrow-mouthed_toad.mp3). Others have described the call as resembling an electric buzzer. The call typically lasts for 0.5–4 sec, but large choruses give the impression of a

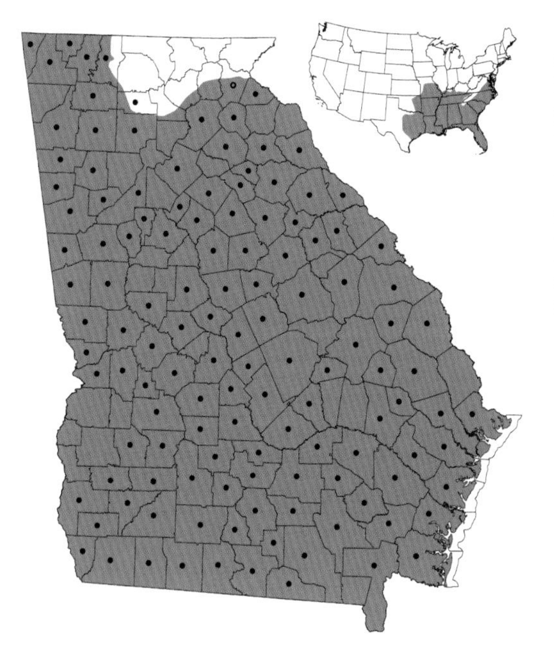

Eastern narrow-mouthed toad tadpole, Liberty County (Dirk J. Stevenson)

Adult eastern narrow-mouthed toad, Treutlen County (Walter W. Knapp)

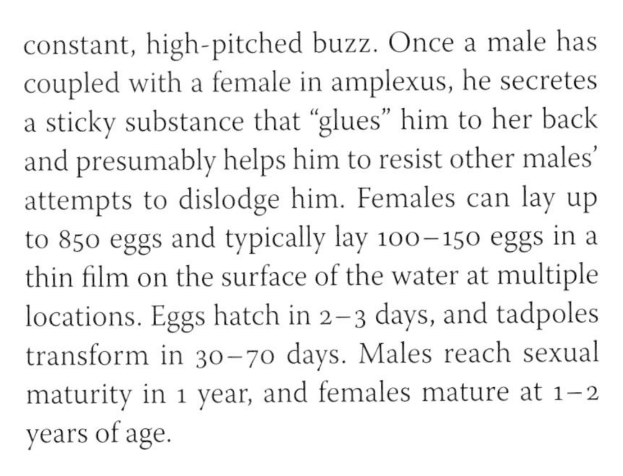

constant, high-pitched buzz. Once a male has coupled with a female in amplexus, he secretes a sticky substance that "glues" him to her back and presumably helps him to resist other males' attempts to dislodge him. Females can lay up to 850 eggs and typically lay 100–150 eggs in a thin film on the surface of the water at multiple locations. Eggs hatch in 2–3 days, and tadpoles transform in 30–70 days. Males reach sexual maturity in 1 year, and females mature at 1–2 years of age.

Habits

Eastern narrow-mouthed toads are fossorial for most of the year. When active on the surface they are almost always found in moist areas under logs or other debris; they are rarely encountered in the open except during the breeding period. These frogs are accomplished burrowers and are adept at disappearing under forest litter and soil very quickly. Ants and termites are their primary prey, together making up as much as 75 percent of the diet. Other prey items include small beetles and other arthropods. The tadpoles are suspension feeders, filtering organisms and organic particles out of the water column. Eastern narrow-mouthed toads secrete toxins from skin glands to deter predators. Copperheads (*Agkistrodon contortrix*) and common garter snakes (*Thamnophis sirtalis*) are among the few species reported to prey on them. The toxic secretions also play a role in deterring ant attacks as individuals forage near ant mounds. Like the adults, older stages of the larvae are toxic and unpalatable to many predators.

Conservation Status

Georgia populations of the eastern narrow-mouthed toad appear robust, and the species does not seem to be in decline.

W. Ben Cash

Spadefoots

FAMILY PELOBATIDAE

Georgia's only species (*Scaphiopus holbrookii*) is typical of the family, which gets its common name from the large black "spades" on the hind feet that are an adaptation to a burrowing lifestyle. Most pelobatids are 50–75 mm (2–3 in) in body length. All have vertically elliptical pupils. The tadpole's mouth has a frilled, unbroken border that runs along the edge of the lower lip. Spadefoots live in areas with friable soil into which they burrow, rear end first. They are explosive breeders with distinct breeding pulses that are triggered by heavy rainfall and are characterized by large numbers of individuals that gather and breed within a few days. The 11 species of this small family are distributed across North America and Europe.

Adult eastern spadefoot, Liberty County (Dirk J. Stevenson)

Eastern Spadefoot

Scaphiopus holbrookii

Description

The eastern spadefoot is a robust, relatively smooth skinned toad that reaches an adult body length of up to 71 mm (2.8 in). Females are larger than males and lack the black pads that are present on the thumbs and inner toes of mature males. The signature character of the genus *Scaphiopus* is the dark brown to black shovel-shaped protrusion (spade) on the inside of each hind foot. In fact, the genus name is derived from the Greek words for "shovel" and "foot." The base color varies from very dark brown or nearly black to gray or olive. Usually two yellow lines run from the eyes along the sides of the back, often suggesting the shape of a lyre. Compared with toads in the genus *Bufo,* the warts on spadefoots are less conspicuous, as are the enlarged parotoid glands behind the eyes. The pupil is vertical in spadefoots and horizontal in other native toads and frogs. Tadpoles reach a total length of nearly 50 mm (2.0 in), are dark brown to bronze, and have clear tail fins and no markings on the tail musculature. The belly skin is virtually transparent, making the gut clearly visible. The eyes are closer together than they are in any other native species. Tadpoles metamorphose at a body size of only 10–15 mm (0.4–0.6 in).

Adult eastern spadefoot, Liberty County (Matt O'Connor)

Eastern spadefoot tadpole, McIntosh County (Dirk J. Stevenson)

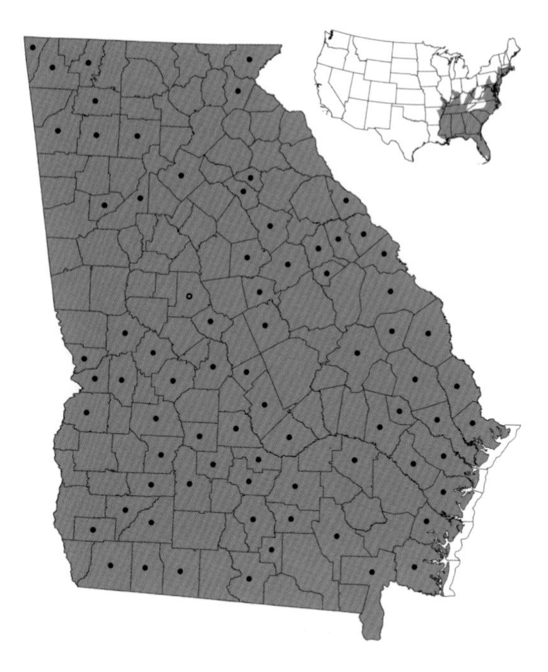

Taxonomy and Nomenclature

Also called the storm toad because adults are seen at times of flooding rains, the eastern spadefoot is Georgia's sole representative of the family Pelobatidae. Two subspecies have been described: the eastern spadefoot (*S. h. holbrookii*) and Hurter's spadefoot (*S. h. hurterii*). Some taxonomists consider these forms to represent distinct species, with only the eastern spadefoot being found in Georgia.

Distribution and Habitat

Eastern spadefoots occur in a variety of habitats throughout the state, although they appear to be more common in the Coastal Plain than in the physiographic provinces above the Fall Line. Adults are explosive breeders that use ephemeral wetlands such as pastureland pools, Carolina bays, and roadside ditches. The most important component of the terrestrial habitat may be the soil type; these toads prefer loose, well-drained, sandy soils in which they can burrow. They often inhabit relatively dry terrestrial habitats and can be found in pastures and farmland, pine plantations, forested lowlands, and coastal dunes.

Reproduction and Development

Unlike most amphibians, which tend to breed seasonally, eastern spadefoots may breed in any month of the year as long as the temperature is above 7° C (45° F). The primary cue for breeding is very heavy rain, often 50 mm (ca. 2 in) or more. Breeding choruses are most common at night but can occur in daylight. Males emit an explosive, nasal *errrrrh* at 3–10-sec intervals while floating on the surface of the water. Many liken the call to the sound of someone vomiting (22_Eastern_spadefoot.mp3). Females lay strings of 2,500 or more eggs, which they attach to twigs, vegetation, or leaves near the water's surface. Egg and larval development are temperature dependent and may be extremely rapid. Eggs laid in summer months may hatch within 24 hours; in winter, hatching may take 4–7 days. After hatching, tadpoles continue developing for 1–4 days before feeding. As tadpoles develop and grow, thousands of individuals may congregate in schools dozens of feet long, several feet wide, and up to a foot deep. The larval period is approximately 3 weeks in summer and 2 months in late winter and spring. Adults reach maturity in 2–3 years, although some males may reproduce at 1 year of age.

Habits

The eastern spadefoot may be the quintessential secretive amphibian. Individuals spend much of the year 50–300 mm (ca. 2–12 in) underground in burrows, often remaining there for months at a time. They generally emerge only on warm, humid, or rainy nights and tend to be more active in the spring and fall. The diet of adults includes beetles, ants, spiders, flies, caterpillars, and millipedes. Tadpoles may initially feed on plankton suspended in the water column, but after their teeth are fully developed they begin feeding on the film of material that adheres to submerged vegetation and debris; they are voracious feeders that eat anything they can catch. Their schools move through temporary ponds like locusts, feeding on suspended and surface materials as well as on smaller animals, amphibian eggs, and even other spadefoot tadpoles. Although home ranges have not been studied in Georgia, adult eastern spadefoots in Florida have a fairly small home range (10 m^2, or 108 ft^2) where they are likely to stay for years. After breeding, adults tend to migrate back to their burrows, which may be 400 m (1,312 ft) or more from the breeding habitat. If the eggs and tadpoles do not die due to cold or early pond drying, many tens or hundreds of thousands of newly metamorphosed toadlets can emerge from a single wetland. Observations of the mass movements of thousands of recently metamorphosed eastern spadefoots may be one source of the phrase "raining frogs and toads." Although their skin produces toxins that deter many predators, juveniles and adults may be eaten by raccoons, some snake species—particularly hognose snakes (genus *Heterodon*)—grackles, and other birds. Tadpoles are eaten by aquatic insects and salamanders. Average annual survivorship of adults is approximately 75 percent.

Conservation Status

The eastern spadefoot appears to be secure in Georgia, although its secretive nature makes assessment of its actual population status difficult.

David E. Scott

True Frogs

FAMILY RANIDAE

Georgia's nine ranid species include the state's largest frogs. Although this family includes the world's largest frog—Africa's goliath frog (*Conraua goliath*), which may be up to 300 mm (12 in) long—our ranids are relatively moderate in size, ranging from 50 to 180 mm (2.0–7.2 in). True frogs have relatively smooth skin and long legs. The eyes of a true frog tadpole are closer to the top of its head than those of treefrogs, and the mouth has a distinct notch on each side. The ranids are the typical frogs found around Georgia ponds and streams, although some, like the wood frog (*Rana sylvatica*), are highly terrestrial. Breeding habitats include both temporary and permanent wetlands. This family occurs statewide in Georgia and is represented on all of the world's continents except Antarctica. Approximately 630 species are found in the world. These are probably the frogs most familiar to humans. They have been incorporated into certain cultural activities such as frog-jumping contests and frog gigging, albeit only as reluctant participants. The large hind limbs are valued as culinary delicacies in many parts of the world, especially in countries and communities that delight in French cuisine.

Adult gopher frog, Liberty County (Dirk J. Stevenson)

Gopher Frog

Rana (Lithobates) capito

Description

These are moderate-sized frogs; adults are 64–112 mm (2.5–4.4 in) in body length, and females are larger than males. Compared with other members of the genus, gopher frogs have a large head and short hind limbs that give the body a squat appearance. The back color ranges from light brown or gray to dark brown. Dark (often black), irregular blotches are prominent over the back and sides, and on the legs. Warts of various sizes and shapes cover most of the back and often merge together along the sides of the body, giving it a wrinkled appearance. Two dorsolateral folds, often tinged with yellow or a brassy color, run down the back, each reaching from behind the eye to the hind limb. Most Georgia populations have a white or cream-colored belly with dark mottling. The rear part of the belly and inner surfaces of the hind limbs are typically washed in yellow. Breeding males have enlarged thumbs. Metamorphosing juveniles have little or no mottling on the belly. Tadpoles can reach 84 mm (3.3 in) in body length and are generally olive green. Dark spots, sometimes faded, are irregularly scattered over the upper parts of the body and tail. The snout is usually translucent and does not have a light line extending away from each corner of the jaw. The latter two features help differentiate gopher

Adult gopher frog, Baker County (Gabriel J. Miller)

Gopher frog tadpole, Taylor County (John B. Jensen)

frog tadpoles from the tadpoles of the southern leopard frog (*R. sphenocephala*). Newly metamorphosed juveniles measured in southeastern Georgia averaged 32 mm (1.3 in) in body length.

Taxonomy and Nomenclature

Gopher frogs were once assigned to *R. areolata*, a species name currently applied to the crawfish frog of central North America. Historically, two subspecies of gopher frogs were recognized in Georgia: the Carolina gopher frog (*R. c. capito*) and the Florida gopher frog (*R. c. aesopus*). Recent genetic studies suggest that these subspecies are invalid. Some taxonomists have suggested that most North American species in the genus *Rana*, including all of those in Georgia, be placed in the genus *Lithobates*.

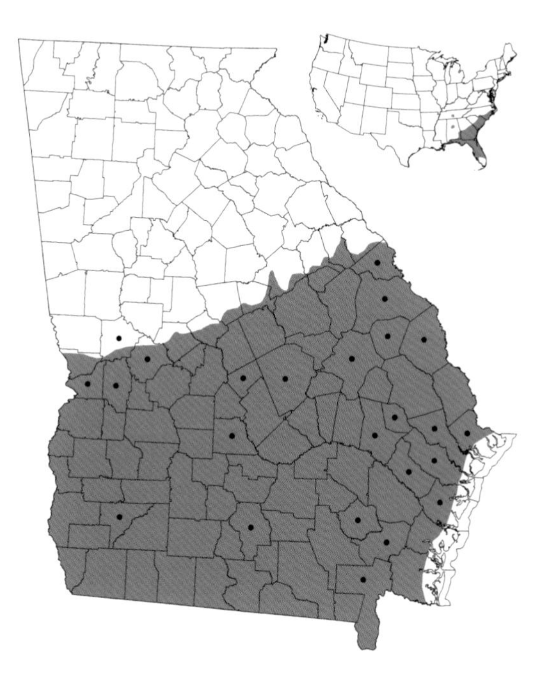

Distribution and Habitat

The gopher frog occurs throughout the Coastal Plain in Georgia, where it inhabits longleaf pine–wiregrass flatwoods and longleaf pine–turkey oak sandhills. Gopher frogs spend much of their nonbreeding time in gopher tortoise (*Gopherus polyphemus*) burrows, thus their common name. Gopher tortoise burrows make good retreats because they have relatively stable temperature and humidity, something particularly important in times of drought. Gopher frogs may also use burrows constructed by oldfield mice, other mammals, or crayfish; or they may use stump holes. Of central importance to these frogs is the presence of suitable breeding habitat in the form of temporary wetlands. Breeding ponds vary in water permanency and depth; typically, they dry annually or every few years and have a maximum depth of 1–2 m (3.3–6.6 ft). Most are either treeless or support a scattered canopy of pond cypress and swamp black gum; open-canopy areas with emergent herbaceous vegetation are an important component. Gopher frogs also breed in borrow pits that have these hydrologic and vegetative characteristics.

Reproduction and Development

Gopher frogs typically migrate to breeding ponds in the fall, winter, and early spring in association with heavy rains. Summer breeding, especially during the passage of tropical storm systems, has also been documented. Adults spend a variable amount of time at the breeding pond; males typically remain longer than females. Adult males call from shallow water along the shoreline, while floating on the surface of the water in deep areas, or while submerged. The significance of underwater calling has not been investigated. The call has a very low frequency and has been likened to a snore or low growl (23_Gopher_frog.mp3). Because of its low frequency the call can carry a considerable distance. Females attach large egg masses—up to the size of a grapefruit and containing 2,000 or more eggs—to emergent vegetation near the water's surface. Average length of time to hatching is approximately a week. Gopher frog populations in neighboring states have larval periods ranging from 87 days in South Carolina to as long as 215 days in Florida. Adults studied in Georgia began breeding in early January, and metamorphosing juveniles were captured exiting the pond from late May to late June. The larval period in that study was estimated to last 145–169 days.

Habits

Gopher frogs are among the most secretive of the true frogs, spending much of their lives underground. Adults migrate up to 2 km (1.2 mi) to and from breeding ponds. Outside the breeding season individuals remain very close to their retreats, often forming a distinctly worn pad from sitting outside their burrow at a particular spot. When disturbed or handled, gopher frogs place their front limbs over the top of their head and eyes in what may be a type of protective behavior or a warning posture. Adults feed on invertebrates and on other anurans, especially toads. Feeding occurs primarily at night. Adults and tadpoles are eaten by banded watersnakes (*Nerodia fasciata*), black racers (*Coluber constrictor*), Florida softshells (*Apalone ferox*), eastern mud turtles (*Kinosternon subrubrum*), and probably other vertebrate predators. Eggs are occasionally eaten by eastern newts (*Notophthalmus viridescens*), striped newts (*N. perstriatus*), and caddisfly larvae.

Conservation Status

Successful reproduction does not occur every year because breeding ponds may fail to fill at the appropriate time or may dry prior to metamorphosis. Even when reproduction is successful, the number of juveniles entering the population may be low. The loss of terrestrial habitats to development and agriculture, especially silviculture, coupled with the destruction of breeding wetlands make the gopher frog a species of great conservation concern in Georgia, where it is listed as Rare.

W. Ben Cash, John B. Jensen, and Dirk J. Stevenson

Adult male bullfrog, Liberty County (Dirk J. Stevenson)

Bullfrog

Rana catesbeiana (Lithobates catesbeianus)

Description

Bullfrogs are the largest frogs in the United States; adults are generally 90–154 mm (3.5–6.0 in) long. Males are smaller than females, have relatively wider tympanums, and usually have a yellowish tinge to their underside and a bright yellow throat. The greenish background color is often covered with a netlike pattern of gray, brown, or olive. The belly is generally white with gray or black mottling. Dark individuals that appear almost black from above and have a heavily mottled belly are occasionally present in some populations. Bullfrogs do not have dorsolateral ridges running along the sides of the back as the otherwise similar-looking green frog (*R. clamitans*) does. Bullfrogs are most likely to be confused with other large Coastal Plain frogs that lack dorsolateral ridges such as the pig frog (*R. grylio*) and river frog (*R. heckscheri*). They differ from pig frogs in having the end of the longest toe on the hind foot free of webbing. Pig frogs also have a light stripe running down the back of each thigh; bullfrog thighs are patterned but not striped. River frogs have rougher skin and a belly so dark that it gives the impression of a light pattern on a dark background; the impression usually given by the belly of the bullfrog is a dark pattern on a light background. River frogs also have obvious white spots on

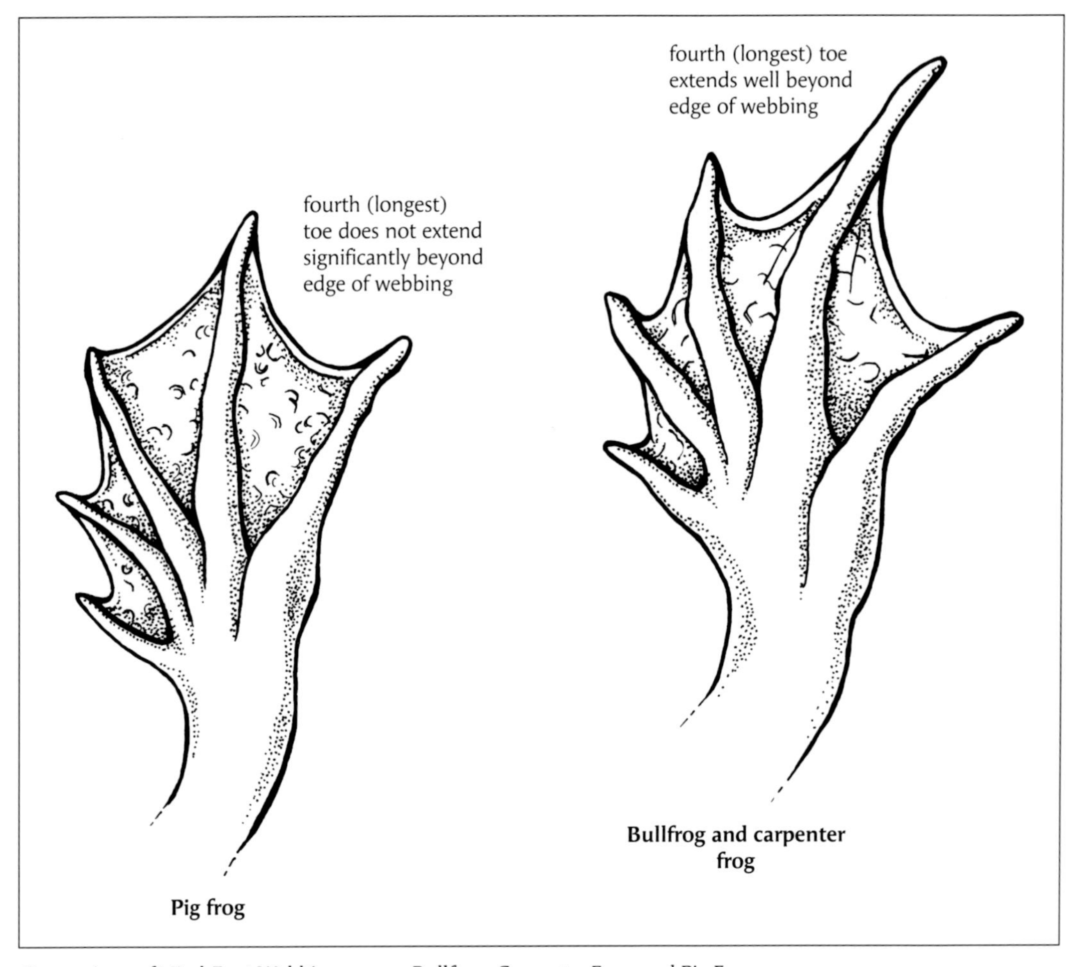

Comparison of Hind-Foot Webbing among Bullfrog, Carpenter Frog, and Pig Frog

their lips. Bullfrog tadpoles are mostly green or greenish brown, typically with numerous dark brown or black, even-sized flecks scattered along the body and tail. Tadpoles in many Coastal Plain populations lack the black flecks altogether. The tadpole's belly is usually yellowish. The tadpoles may reach 168 mm (6.6 in) in total length just before metamorphosis. Newly transformed individuals are 30–60 mm (1.2–2.3 in).

Taxonomy and Nomenclature

No subspecies are recognized. (See **Gopher Frog** account concerning the genus name *Lithobates.*)

Distribution and Habitat

Bullfrogs occur in every physiographic province in the state. They are semiaquatic and prefer ponds, lakes, reservoirs, and other permanent bodies of water to nonpermanent ones. Adults also frequent borrow pits, canals, and drainage ditches. They are common in residential areas, where they often inhabit ornamental garden ponds and swimming pools that are not regularly maintained. During the winter, bullfrogs generally remain close to their aquatic summer habitats. They spend the colder months beneath stones, logs, or piles of debris near or in the water or in tunnels beneath the banks of streams.

Bullfrog tadpole, Evans County (Dirk J. Stevenson)

Adult female bullfrog, Bibb County (John B. Jensen)

Reproduction and Development

In Georgia, breeding occurs from late spring until late summer. The advertisement call has been variously rendered as *jug-o'-rum, more rum,* and *bottle-o'-rum.* Often the call is a deep, throaty *br-wum* repeated at irregular intervals (24_Bullfrog.mp3). Other types of calls are sudden and of short duration and appear to be related to territorial behavior. Large males are fiercely territorial during the breeding season. Small satellite males sit quietly along the territorial boundaries hoping to intercept and amplex females that are attracted to the larger calling individuals; many are successful. Females lay large numbers of eggs; at one site near Savannah female bullfrogs laid between 10,000 and 20,000 eggs each. Eggs are deposited in a floating surface film that may measure 0.6 m (2 ft) across. The larval period is variable, and transformation may take up to 2 years. Sexual maturity is usually reached within 3 years of hatching.

Habits

Bullfrogs have a varied diet. In general, they feed on any smaller animal they can catch. Reported food items include crayfish, insects, fish, reptiles, mammals, and other amphibians, including other frogs. Bullfrogs in turn serve as food for many forms of wildlife. Leeches, aquatic salamanders, and fish are known predators of bullfrog eggs, although newts (*Notophthalmus viridescens*) and larval marbled salamanders (*Ambystoma opacum*) apparently find them distasteful. Many aquatic invertebrates, including the six-spotted fishing spider, eat the tadpoles. Snakes are probably major predators of adults; documented snake predators include northern (*Nerodia sipedon*), banded (*N. fasciata*), and red-bellied (*N. erythrogaster*) watersnakes; eastern ribbon snakes (*Thamnophis sauritus*); black racers (*Coluber constrictor*); and copperheads (*Agkistrodon contortrix*). Various bird species also eat bullfrogs.

Conservation Status

Bullfrogs are common throughout the state. Unlike many other species, they may have benefited from humans' activities in Georgia in the form of artificial lakes and ponds. There is certainly more permanent water above the Fall Line now than there was prior to European colonization. In many areas residents gig bullfrogs for food, and attempts to raise bullfrogs commercially for human consumption have been under way for decades.

Robert A. Moulis

Adult male green frog, Macon County (Dirk J. Stevenson)

Green Frog

Rana (Lithobates) clamitans

Description

Body length of this medium-sized frog averages 57–90 mm (2.2–3.5 in); males are smaller than females. Individuals in the northern part of the state may be larger than those in southern Georgia, which rarely exceed 77 mm (3 in). Adult males often have a yellowish throat, and their tympanum is larger in diameter than the eye. Green frogs are highly variable in coloration. Typical base colors range from brown or bronze to dull green. During cold weather, green frogs may be nearly black. Green frogs from northern Georgia often have a heavily mottled or spotted back and green face; those from southern Georgia usually have a uniformly colored back and less green on the face. A distinct dorsolateral fold runs from each eye down the edge of the back but does not extend the full length of the back. The dorsolateral folds distinguish green frogs from other true frogs. Bullfrogs (*R. catesbeiana*), pig frogs (*R. grylio*), carpenter frogs (*R. virgatipes*), and river frogs (*R. heckscheri*) do not have dorsolateral folds; pickerel frogs (*R. palustris*), southern leopard frogs (*R. sphenocephala*), gopher frogs (*R. capito*), and wood frogs (*R. sylvatica*) have dorsolateral folds that run the full length of the back. Newly metamorphosed green frogs, usually 25–35 mm (1–1.4 in) in body length, resemble adults but often have more dark blotches on the back. Tadpoles may be up to 90 mm (3.5 in) in length and are typically olive green or brown with small black spots and flecks.

Taxonomy and Nomenclature

Two subspecies are currently recognized: the bronze frog (*R. c. clamitans*) and the green frog (*R. c. melanota*); the former occurs in northern

Green frog tadpole, Bryan County (Dirk J. Stevenson)

Adult female green frog, Jasper County (John B. Jensen)

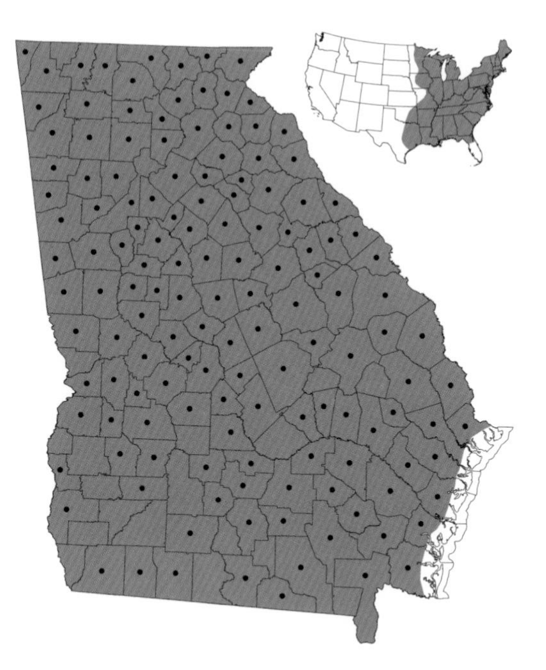

Georgia and the latter occurs below the Fall Line. Much of the Georgia Piedmont may be a zone of genetic exchange between the two subspecies. The characters defining the two are poorly distinguished, however, and the validity of the two forms has been questioned. (See **Gopher Frog** account concerning the genus name *Lithobates*.)

Distribution and Habitat

Green frogs are present throughout the state and are common in all physiographic regions. They occur in permanent as well as temporary ponds; along streams and rivers, especially in backwater areas; and in bogs and marshes. They are particularly abundant in permanently aquatic habitats such as farm ponds, beaver ponds, and the swampy ends of larger impoundments.

Reproduction and Development

The breeding season may be quite lengthy, beginning in March and extending to as late as July or August. Males establish territories and defend them against intruding males. The advertisement call has been likened to the plucking of a low-sounding banjo string, a *clung* or *c'tung* given as a single note or as two or three notes in rapid succession (25_Green_frog.mp3). This explosive call may carry as far as 0.8 km (0.5 mi). The eggs are laid in a thin film on the surface of the water and typically are attached to floating or shallow submerged vegetation. Clutch size usually ranges from 3,000 to 5,000 eggs. Females sometimes produce a second clutch. The eggs hatch in 3–7 days, depending on the water temperature. Food availability, temperature, and other environmental factors determine the length of the larval period, which ranges from 3 to 22 months. Adults frequently become more terrestrial after breeding and can return to familiar foraging areas. Georgia males reach maturity in 1 or 2 years; females become mature in 2 years. Adults may live for 5 or more years.

Habits

Green frogs frequently rest on floating vegetation or on the banks of ponds and streams and may surprise passersby by leaping into the water with a loud *squeenk.* Adults are mainly nocturnal. They are effective predators and actively forage for food as well as using ambush techniques. Prey consists primarily of invertebrates such as insects, snails, and worms, although small vertebrates may also be eaten. The author once observed a large green frog eating a young northern watersnake (*Nerodia sipedon*). Green frogs have a number of predators, including snakes, mammals, and wading birds. Like other true frogs, they leap to safety to avoid predators; when caught, either sex may emit a loud scream of distress. Tadpoles appear to be able to chemically sense predators like dragonfly larvae and may reduce their activities in habitats where such predators are abundant. Green frogs may migrate long distances (more than 500 m, or 1,640 ft) to overwintering sites, which tend to be flowing-water areas that do not freeze.

Conservation Status

Populations of the green frog appear robust throughout the state.

Gregory C. Greer

Adult female pig frog, Liberty County (Dirk J. Stevenson)

Pig Frog

Rana (Lithobates) grylio

Description

One of North America's larger frogs, the pig frog is second only to the bullfrog (*R. catesbeiana*) in size, with adult body lengths ranging from 83 to 162 mm (3.2–6.3 in). Adult males are smaller on average than females and often have a bright yellow throat. The male's conspicuous tympanum is larger in diameter than the eye. The back may be any shade from bright green to olive to dull brown, often with dark spotting toward the hind end and on the hind legs. The belly is white or yellowish white with considerable dark mottling toward the rear. A longitudinal white stripe or series of white spots suggesting a stripe extends the length of the back of the thighs. The head is more pointed and narrower than that of the bullfrog, and the eyes seem closer together and somewhat elevated. Juveniles are characterized by a pair of golden brown stripes that run along the border between the back and sides and suggest a dorsolateral ridge such as the one present in the green frog (*R. clamitans*) but lacking in adult pig frogs. The juvenile pattern resembles the markings on the carpenter frog (*R. virgatipes*). The pig frog can be distinguished from both bullfrogs and carpenter frogs by the extensive webbing of its hind feet. The webbing extends nearly to the tip of the fourth (longest) toe on the pig frog but ends above the terminal joint

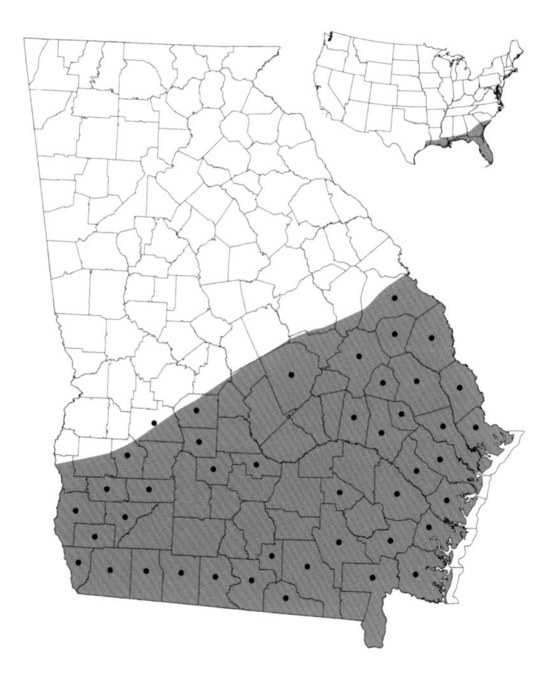

Pig frog tadpole, Long County (Dirk J. Stevenson)

Adult female pig frog, Chatham County (Dirk J. Stevenson)

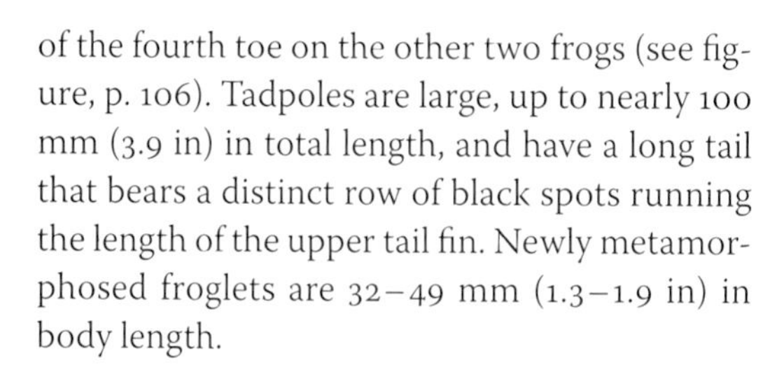

of the fourth toe on the other two frogs (see figure, p. 106). Tadpoles are large, up to nearly 100 mm (3.9 in) in total length, and have a long tail that bears a distinct row of black spots running the length of the upper tail fin. Newly metamorphosed froglets are 32–49 mm (1.3–1.9 in) in body length.

Taxonomy and Nomenclature

Although no rangewide survey of geographic variation has been conducted, individuals from western populations (Louisiana) tend to be darker and more extensively spotted (and their juvenile pattern more persistent) than those from eastern populations. (See **Gopher Frog** account concerning the genus name *Lithobates.*)

Distribution and Habitat

Pig frogs are found only in the Coastal Plain. These strongly aquatic frogs favor quiet, open bodies of water with emergent or floating vegetation. In Georgia they frequent Carolina bays, sloughs, swamps, sinkhole ponds, permanent roadside ditches, and similar aquatic habitats. Pig frogs are unquestionably the most common large frogs in the Okefenokee Swamp; they are particularly abundant in the swamp's wet prairies and cypress ponds.

Reproduction and Development

Pig frogs are characterized by a prolonged breeding season. A study conducted in southwestern Georgia near Albany found that males first established calling sites in late March, and chorusing activity continued into mid-September. Females contained mature eggs from April through July, the same time that male chorusing was peaking. Similar patterns of breeding activity have been observed for populations in the Okefenokee Swamp. Males issue a distinctive call that was first reported by the famous

18th-century naturalist William Bartram, who noted that "their voice is loud and hideous, greatly resembling the grunting of swine." Bartram was right: the call entails a burst of 1–13 short, piglike grunts (26_Pig_frog.mp3). In 1918, herpetologist Percy Viosca suggested that, given the call, "'pig' frog might prove an appropriate common name." Calling activity generally takes place at night but may continue throughout the day at the height of the breeding season. Eggs are deposited in a sheetlike film only one or two eggs thick whose shape is determined by the supportive material, usually vegetation, on which the film settles. Clutches contain 8,000–15,000 eggs, and a large female from the Everglades was reported to have some 34,000 eggs in her ovaries. Hatching occurs 2 or 3 days after deposition. In southwestern Georgia, tadpoles overwinter and undergo metamorphosis 9 or more months after hatching. No information is available on age at maturity for Georgia pig frogs; if they are similar to bullfrogs in this respect, they mature in approximately 3 years.

Habits

Often described as shy, the pig frog's highly aquatic nature and wary disposition make it difficult to approach. Males typically call some distance from shore, generally concealed in patches of maidencane or fragrant water lilies. The females surveyed in southwestern Georgia occupied shallow water along the shore of a large sinkhole pond year-round. Males used the same habitat during the nonbreeding season but moved out to deeper water some 40–50 m (131–164 ft) from the shoreline at the onset of chorusing. Males' survivorship is lower than that of females, perhaps a function of lower foraging rates and higher predation risks incurred while calling during the breeding season. Pig frogs are opportunistic feeders. Their diet consists largely of aquatic invertebrates, though they occasionally capture small fish, frogs, and salamanders. Several studies have identified crayfish as a major component of the diet, and dragonflies (both nymphs and adults) and aquatic beetles are also important prey items. Predators include frog-eating snakes such as banded (*Nerodia fasciata*) and red-bellied (*N. erythrogaster*) watersnakes, American alligators (*Alligator mississippiensis*), wading birds, and mammals such as raccoons. The younger tadpoles occasionally fall prey to aquatic insects.

Conservation Status

Pig frogs, like bullfrogs, are prized for their edible legs. Most frogs taken in Georgia are hunted recreationally, but many Florida populations are commercially harvested. This frog is relatively common in appropriate habitats throughout southern Georgia but no doubt benefits from extensive wetland protection in the Okefenokee National Wildlife Refuge.

Trip Lamb

Adult male river frog, Tattnall County (Andy Day)

River Frog

Rana (Lithobates) heckscheri

Description

Although similar in size to the bullfrog (*R. catesbeiana*) and pig frog (*R. grylio*)—some 83–155 mm (3.2–6.0 in) in total length—the river frog is distinguished from them by its uniformly dark color. The background color varies from greenish gray to chocolate brown with black spotting or mottling, and the back has a rough, textured appearance. The gray to dark gray belly is suffused with light spots that change to light reticulation toward the hind legs; a pale, crescent-shaped line often girdles the groin. The overall effect is essentially a "photo negative" of the belly patterns characterizing the bullfrog and pig frog. White spots show prominently on the frog's dark lips, especially along the margin of the lower lip. Males are smaller than females but have noticeably wider tympanums. The tadpoles, which may be up to 110 mm (4 in) long, are among the largest of any North American frog. They are gray to virtually black with extensive metallic flecking, and the tail fins, otherwise opaque to clear, have a distinct black border. The upper part of the tail musculature is also black. The conspicuous tail pattern makes these tadpoles among the few that can be readily identified in the field. Older tadpoles have brick red eyes, as do the transformed froglets.

Early stage river frog tadpole, Liberty County (Dirk J. Stevenson)

Late stage river frog tadpole, Liberty County (Dirk J. Stevenson)

Adult female river frog, Calhoun County (Aubrey M. Heupel)

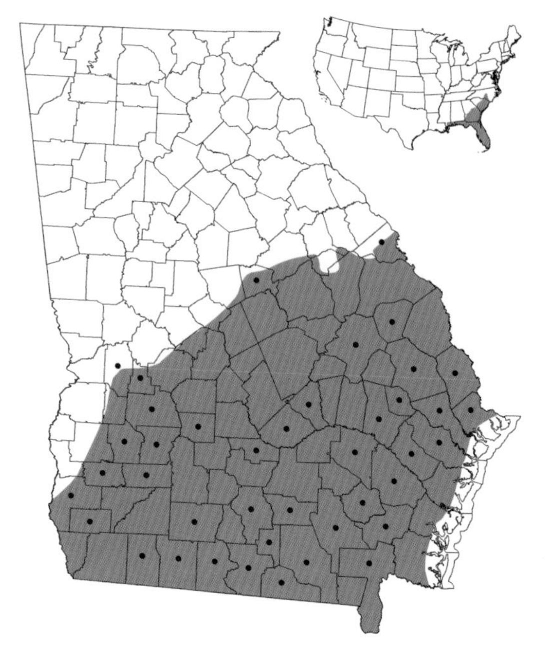

Distribution and Habitat

A strictly Coastal Plain species, the river frog's range in Georgia extends north to the Fall Line. The frog's common name is a bit of a misnomer. An earlier name, river swamp frog, is far more apt because this frog is truly a denizen of swampy backwaters along meandering streams and rivers. In addition to swamps, river frogs frequent oxbows as well as other lakes and sinkhole ponds. Banks and shorelines shaded by a mix of woody shrubs (e.g., fetterbush, titi, and tag alder) and trees (e.g., bald or pond cypress and swamp black gum) provide ideal habitat. Permanently flooded borrow pits are also used.

Taxonomy and Nomenclature

The last large frog to be recognized in the Southeast, the river frog was described in 1924 by Albert Hazen Wright. No subspecies are recognized. (See **Gopher Frog** account concerning the genus name *Lithobates.*)

Reproduction and Development

Males can be heard calling from April to August in Georgia. The advertisement call is a deep, roaring snore (27_River_frog.mp3) that is eerily suggestive of a bellowing alligator (*Alligator mississippiensis*). Males call from both shoreline and shallow-water sites. Eggs are deposited in a sheetlike film resembling the egg masses of bullfrogs and pig frogs; each mass may consist of several hundred eggs. Tadpoles overwinter and metamorphose the following spring or summer. Age at maturity is not known for Georgia river frogs; if they resemble bullfrogs in this regard, they are sexually mature at 3 years.

Habits

Despite its strong ties to still or slow-moving waters, the river frog is more typically observed sitting on shore or on fallen logs near shore than in the water proper. Of the 22 individuals captured during an amphibian survey of a large sinkhole pond in Dougherty County, only one, a calling male, was sighted in the water. Stomach content analysis of these individuals revealed a diet of various arthropods, 70 percent of which were terrestrial forms (primarily beetles). Less wary and more easily approached than other large frogs, river frogs offer little resistance on capture, often going completely limp in the hand. Complementing the tadpole's distinctive appearance is its unusual social behavior: individuals aggregate in large schools comprising hundreds of tadpoles. From a distance, such a school appears as a black sheet moving just below the water's surface and can actually be heard; a strange crackling sound emanates from the mass as dozens of tadpoles at a time surface to gulp air. Experimental evidence suggests that schooling may facilitate feeding activity, as foraging initiated by one individual elicits similar behavior in adjacent individuals. Schooling may also thwart predation. The scent of a common garter snake (*Thamnophis sirtalis*) reportedly caused a school of river frog tadpoles "to explode." The chaotic situation created by an abruptly disbanding school may startle potential predators. The tadpoles are also said to be distasteful to certain predators, providing another tenable explanation for schooling: predators may associate the dark mass of tadpoles with a former negative experience and thus avoid additional encounters. Like other members of this genus, river frogs are eaten by a number of predators, including snakes, birds, and mammals.

Conservation Status

In Georgia, river frogs appear to be relatively common in appropriate habitat.

Trip Lamb

Adult pickerel frog, Walker County (John B. Jensen)

Pickerel Frog

Rana (Lithobates) palustris

Description

This medium-sized frog may exceed 80 mm (3.1 in) in body length. Males are smaller than females and have swollen thumbs during the breeding season. Males also have a vocal sac on each side of the head between the ear and forelimb. A dorsolateral ridge runs backward from each eye along the margin between the back and side. The upper body is grayish with two rows of dark, somewhat square spots running down the back; the sides also have spots. The edges of the spots are often irregular, and the spots may coalesce to form rectangles or long bars. A single spot is often present above each eye and on the end of the snout. The inner surface of the hind legs is bright yellow or orange, and the upper surface is marked with bars. The belly is white or whitish. Southern leopard frogs (*R. sphenocephala*) are similar in appearance, but the spots on the back are round or oval instead of square and are scattered, not in two rows. Southern leopard frogs also lack bright yellow or orange on the legs. Pickerel frog tadpoles have a highly arched back fin that begins in front of the tail-body junction. A faint white line may be present along the lip. The body is purplish black or greenish with light flecking, and the tail fins are uniformly speckled; the belly is cream colored, and the eyes are yellow. Tadpoles may reach a

Pickerel frog tadpole, Watauga County, North Carolina (Robert Wayne Van Devender)

Adult pickerel frog, Hall County (Walter W. Knapp)

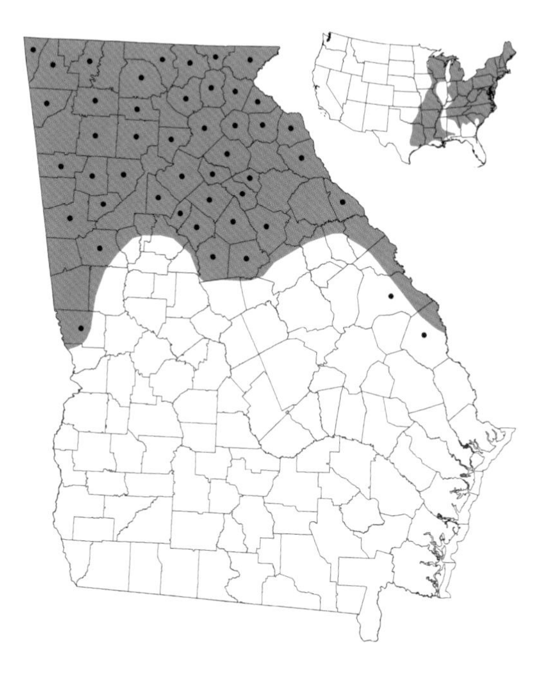

total length of 75 mm (2.9 in) prior to transformation. Newly transformed froglets are 19–27 mm (0.7–1.1 in) long.

Taxonomy and Nomenclature

No subspecies are currently recognized, although two subspecies, *R. p. palustris* and *R. p. mansuetii,* have been described. (See **Gopher Frog** account concerning the genus name *Lithobates.*)

Distribution and Habitat

The pickerel frog occurs in all physiographic provinces in the state but is most common in the Blue Ridge, Ridge and Valley, Cumberland Plateau, and upper Piedmont. Along the eastern border of its range, the species follows the Savannah River drainage south into the Coastal Plain, and in the west it follows the Chattahoochee drainage south to the Fall Line. Habitats include streams and pools with cool, clean water; sphagnum bogs; and marl (lime-rich) ponds in upland forests as well as grassy or weedy fields adjacent to these aquatic habitats. In cave-forming limestone (karst) regions, individuals are often found in and around springs and cave openings, often well back in the twilight zone.

Reproduction and Development

During the winter and spring breeding period the male produces a variable, low-pitched croak with a snorelike quality (28_Pickerel_frog.mp3), sometimes while underwater. Eggs are deposited in woodland pools or quiet pools in streams, with each female laying about 3,000 eggs in a globular mass up to 100 mm (3.9 in) in diameter. The eggs are brown or yellow rather than black. Each egg mass is attached to vegetation and is usually laid in shallow water. Breeding congregations may develop, with up to 15 pairs in an area less than 2 m^2 (2 yd^2). The larval period may last for 3 months, and sexual maturity is reached in 2 years.

Habits

Pickerel frogs feed on insects and other small invertebrates. Skin gland secretions give some protection from predators by making the frog distasteful, but a variety of animals eat pickerel frogs nevertheless, including snakes, birds, and mammals. The tadpoles are probably eaten by insects and other aquatic predators. Pickerel frogs associated with caves move into and out of these refuges according to seasonal reversal of temperature gradients between the cave interior and exterior.

Conservation Status

Populations appear to be fairly stable across most of the species' range in Georgia.

James C. Godwin

Adult southern leopard frog, Bryan County (Dirk J. Stevenson)

Southern Leopard Frog

Rana sphenocephala (Lithobates sphenocephalus)

Description

Adult southern leopard frogs are generally 50–130 mm (2.0–5.1 in) in body length. Males are smaller than females and have swollen thumbs during the breeding season. Males also have paired vocal sacs, one on each side of the head between the ear and forelimb. The base color of the back ranges from olive to brown. Most individuals have prominent dark (often black), irregular spots scattered over the back. The tympanum usually has a white or yellow spot in the center. The snout is distinctly more pointed than that of most other species in the genus, and most individuals have dark spots on the lower lips and light upper lips. Two dorsolateral folds, typically yellow, run from behind each eye to the insertion of each of the hind limbs. Dark, vertical bars are typically visible on the hind limbs when the frog is at rest, and the hind feet are extensively webbed. The belly is white or cream with no prominent mottling. This species is easily confused with the pickerel frog (*R. palustris*), but the latter species has squarish spots arranged in two distinct rows down the back. Tadpoles can reach 76 mm (3.0 in) in total length and are variably colored. Black blotches of irregular size and shape are scattered over the upper body and tail, but the tail fins may be clear. A light line extends backward from each corner of the jaw, and often another runs from the tip of the snout. Newly transformed froglets are 20–33 mm (0.8–1.3 in) in length.

Southern leopard frog tadpole, Taylor County (John B. Jensen)

Adult southern leopard frog, Liberty County (Andy Day)

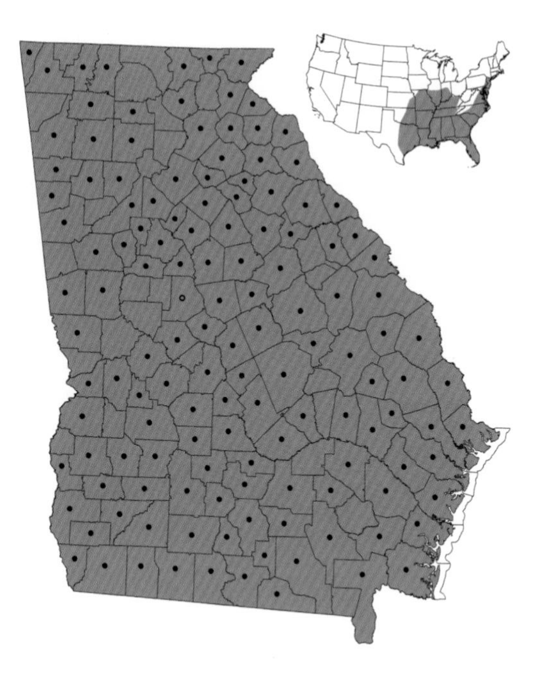

Taxonomy and Nomenclature

The southern leopard frog was once considered a subspecies of *R. pipiens* (*R. p. sphenocephala*), but we now know that leopard frogs form a species complex that ranges over much of North America. There has been some confusion regarding the appropriate scientific name for the southern leopard frog, and both *R. sphenocephala* and *R. utricularia* have been used in the literature to refer to this species. Two subspecies are recognized: the Florida leopard frog (*R. s. sphenocephala*) and the southern leopard frog (*R. s. utricularia*); the latter occurs throughout Georgia. (See **Gopher Frog** account concerning the genus name *Lithobates.*)

Distribution and Habitat

The southern leopard frog is one of the most common and widely distributed ranids in Georgia. Populations can be found both in hardwood cove forests in northern Georgia and in longleaf pine–wiregrass flatwoods in the southern part of the state. More indicative of the frog's ubiquitous nature are the habitats it uses for breeding. Southern leopard frogs breed in both temporary and permanent aquatic habitats, including roadside ditches, creek and river swamps, farm ponds, and isolated temporary wetlands. The basic requirement seems to be enough water for tadpoles to complete their development and little or no water flow. Open, herbaceous, submerged or emergent vegetation may also be important for egg mass attachment and to provide cover for tadpoles. These frogs are often found on land under debris and may use stump holes or the burrows of small mammals or crayfish as retreats.

Reproduction and Development

Southern leopard frogs may breed in any month of the year in Georgia; populations in northern

Georgia typically breed in the spring (March–June), and populations in southern Georgia breed during winter (December–February). Breeding is typically correlated with heavy rainfall. Limited data suggests that populations may breed multiple times each year if environmental conditions are favorable. Adult males call from shallow water along the shoreline, while floating on the surface of the water in deeper areas of the pond (often while grasping emergent vegetation), or while submerged. Individuals have even been observed calling from crayfish burrows. The call is most often described as a thumb running over a wet balloon or two balloons being rubbed together. Although that is true, this frog has a much more varied call than those descriptions suggest. Individual calls may be staccato or ratchetlike; some are smoother and reminiscent of rubber-soled shoes on a newly waxed floor. Sometimes the call has a chuckling quality (29_Southern_leopard_frog.mp3). Females usually attach large egg masses (up to the size of a baseball) to vegetation near the water's surface, but clutches laid in shallow water near the edge are often unattached. Individual egg masses may contain 1,000–1,500 eggs. Egg masses laid in colder water tend to clump together. Average length of time to hatching is approximately a week, but eggs subjected to warmer temperatures can hatch in as little as 3 days. The presence of predators may also accelerate hatching. The length of the larval period varies greatly but is usually at least 90 days. One study in a temporary wetland in the Georgia Coastal Plain estimated a larval period of 175–191 days. Metamorphosis in this study population corresponded to declining water level and the eventual drying of the pond. Little information is available on age at maturity in this species; related, similar-sized species of leopard frogs mature in 1–2 years.

Habits

Although this is one of the most abundant frogs in Georgia, there is not a wealth of information on its biology. Adult frogs migrate between breeding and terrestrial habitats, but the distance they travel is unknown. They feed on invertebrates, primarily arthropods, and occasionally on other frogs such as the little grass frog (*Pseudacris ocularis*). Documented predators of adults and tadpoles include eastern ribbon snakes (*Thamnophis sauritus*), banded watersnakes (*Nerodia fasciata*), cottonmouths (*Agkistrodon piscivorus*), black racers (*Coluber constrictor*), and numerous species of wading birds. The eggs are eaten by eastern newts (*Notophthalmus viridescens*), striped newts (*N. perstriatus*), caddisfly larvae, and other arthropods.

Conservation Status

Southern leopard frogs occur throughout most of the state in relatively large numbers.

W. Ben Cash

Adult wood frog, Banks County (John B. Jensen)

Wood Frog

Rana sylvatica (Lithobates sylvaticus)

Description

The wood frog is a small to medium-sized true frog. The average body length of adult males is 50–55 mm (2.0–2.2 in); adult females are larger, averaging 60–67 mm (2.3–2.6 in). The dark facial patch—often referred to as a robber's mask—that extends through each eye to the base of the forelimb is the key identifying feature. The mask is always visible, even on dark specimens. A white line borders the upper lip beneath the mask. Males are typically much darker than females, their background color varying from dark brown to chocolate. Females are usually tan to reddish pink or brown. Both sexes usually have prominent, widely spaced dark bars on the hind limbs and two pronounced dorsolateral ridges that extend along the edges of the back. Scattered dark markings may or may not be present on the back and sides, but a dark bar is usually present at the base of each forelimb. The tadpoles lack a distinct body pattern, although there may be small markings on the tail fin, which is rounded on top and tapers to a sharp point. Tadpoles hatch at a length of about 9 mm (0.4 in) and may reach 50–55 mm (2.0–2.2 in) prior to metamorphosis. Newly transformed individuals are approximately 16–18 mm (0.6–0.7 in) in body length.

Adult wood frog, Meriwether County (Sean Graham)

Wood frog tadpole, Banks County (John B. Jensen)

Taxonomy and Nomenclature

The two subspecies once accepted in the eastern United States, *R. s. sylvatica* and *R. s. cantabrigensis*, are no longer recognized. (See **Gopher Frog** account concerning the genus name *Lithobates*.)

Distribution and Habitat

Wood frogs are present primarily in the northeastern counties of the Blue Ridge physiographic province and in the adjacent areas of the upper Piedmont. Populations are also known from the lower Piedmont of western Georgia. Although the localities have yet to be documented with museum specimens, reputable biologists have reported seeing wood frogs or their eggs in the middle Piedmont (Fulton County) of western Georgia as well as in the Cumberland Plateau

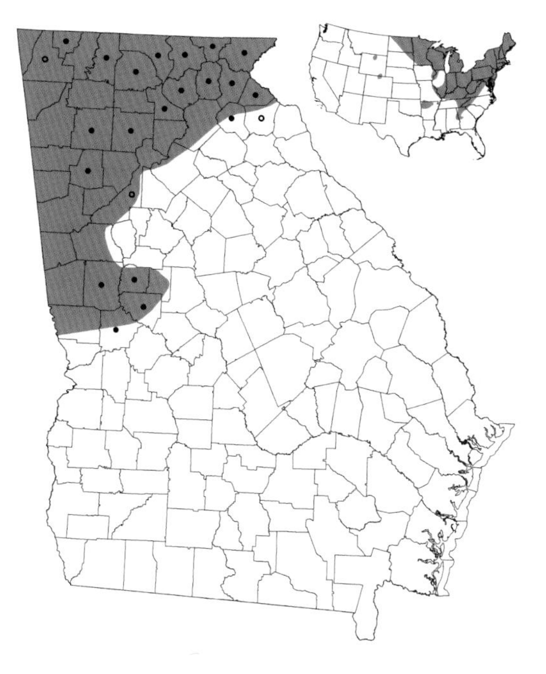

Province (Walker County). Considering the secretive nature of these frogs, the documented range probably underestimates their true distribution in the state. Georgia wood frogs are seen infrequently outside the breeding season, usually in moist, deciduous woods such as those along streams or on north-facing slopes. Individuals are occasionally seen long distances from water. Breeding ponds are usually temporary rain-filled woodland pools or permanent pools without fish predators. Wood frogs also breed in temporary pools in pastures adjacent to forested areas and in pools that form in the wheel ruts on dirt roads.

Reproduction and Development

In Georgia, breeding can begin with warm temperatures in January but typically occurs in early to mid-February. Wood frogs are explosive breeders; virtually all the population's reproduction for the season takes place during a short period. Large numbers of frogs congregate in the breeding ponds, and all the eggs are typically laid within a few days. Adults disappear after breeding and are seen infrequently thereafter. Breeding choruses are most common at night, but daytime choruses are not infrequent in isolated areas on warm days. The distinctive call of the male wood frog is usually described as sounding like a quacking duck (30_Wood_frog.mp3). Eggs are deposited as submerged globular masses, usually attached to vegetation. Wood frog eggs are easy to identify because the jelly envelopes are quite large (average diameter is more than 11 mm, or 0.4 in) and usually turn green after a few days due to the presence of microscopic green algae. A single egg mass is usually about the size of a grapefruit, and the egg masses are often deposited in communal aggregates of 10–80 or more clutches in a small area of the breeding pond. Average clutch size is about 550 eggs per mass. Tadpoles hatch within about 18–25 days, depending on water temperature. Metamorphosis is completed in about 115–130 days. Individuals usually reach sexual maturity in 1 or 2 years, although those at high elevations may take longer.

Habits

Wood frogs are seldom seen outside the breeding season, although adults are occasionally found on roads after rains. Little information exists on the natural history of this species in Georgia, but these frogs appear to be opportunistic terrestrial feeders that prey on insects, spiders, earthworms, and snails. The tadpoles are eaten by various predaceous insects and by larval marbled salamanders (*Ambystoma opacum*), which are often present in the same breeding ponds, especially those outside the Blue Ridge. Birds, snakes, and small mammals eat the adults.

Conservation Status

Georgia wood frogs are thought to be secure despite their limited documented distribution in the state. As mentioned earlier, the distribution of this species in Georgia is probably underestimated because wood frogs are so secretive.

Mark S. Davis

Adult carpenter frog, Baker County, Florida (Dick Bartlett)

Carpenter Frog

Rana (Lithobates) virgatipes

Description

The carpenter frog is one of the smallest members of the family Ranidae; adults are usually only 41–67 mm (1.6–2.6 in) long. Females are larger than males and have slightly smaller tympanums. The general body coloration is uniformly brownish with four yellowish or golden brown stripes running the length of the body—one on each side and two along the edges of the back. These frogs are similar to young pig frogs (*R. grylio*), which may be similarly striped. They can be distinguished by the webbing on their hind feet. The webbing of pig frogs extends to near the tip of the longest toe; in carpenter frogs the longest toe extends well beyond the webbing (see figure, p. 106). The hind legs have a series of alternating dark and light bars. The underside of the frog is yellowish white with dark brown or black spots. Carpenter frogs lack dorsolateral ridges. The tadpoles approach 100 mm (3.9 in) in total length. They are dark brownish olive above with a few large, widely scattered black spots. The tail is grayish and has a row of dark spots along the center of both the upper fin and the tail musculature. The upper part of the tail is irregularly edged in black down to the tip. Newly transformed froglets are 23–31 mm (0.9–1.2 in) in body length.

Taxonomy and Nomenclature

No subspecies are recognized. (See **Gopher Frog** account concerning the genus name *Lithobates.*)

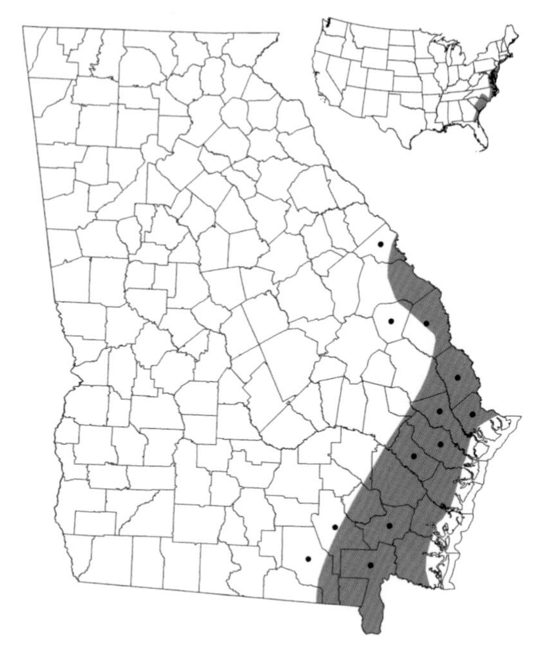

Carpenter frog tadpole, Berkeley County, South Carolina (Robert Wayne Van Devender)

200–600 eggs. The eggs are attached to aquatic vegetation at or submerged within 30 cm (11.7 in) of the surface of the water. Hatching usually occurs within 3 or 4 days. Larvae hatching in the spring or summer generally grow throughout the fall and winter. Metamorphosis usually takes place early the following spring, although some individuals may not transform until the summer. Spring-transforming individuals may reach sexual maturity by the end of the summer.

Distribution and Habitat

The carpenter frog is endemic to the eastern Coastal Plain, where it occurs in wetlands such as Carolina bays, cypress ponds, blackwater creek swamps, and open wet prairies, a habitat common in the Okefenokee Swamp. Areas of quiet water that are densely vegetated with floating or emergent plants are preferred, especially bog-type settings with an abundance of sphagnum moss.

Reproduction and Development

Males begin calling in early March and continue through the end of August. Small males have a higher growth rate and tend to call less frequently than larger ones. The advertisement call is often compared to the sound of a carpenter hammering (31_Carpenter_frog.mp3). Egg masses are oval to spherical, and each contains

Habits

Information concerning the natural history of this frog is scarce. The diet apparently consists of aquatic and flying insects, crayfish, spiders, and other arthropods. Little is known about the natural predators in Georgia, although watersnakes (genus *Nerodia*) and black racers (*Coluber constrictor*) have been reported as predators outside the state. Other snakes, turtles, various birds, and certain mammals may also prey on both tadpoles and adults. Several types of aquatic insects and fish may be important predators of larval stages and possibly of breeding adults.

Conservation Status

This is a poorly understood species in Georgia. Few records have been reported for the state, and large choruses are never encountered.

Robert A. Moulis

Salamanders

SALAMANDERS REMAIN AMONG the more unfamiliar animals to many Georgians. Nevertheless, this group is an important component of the state's fauna. In fact, more species of salamanders are present here than are species of any other group of amphibians or reptiles—including frogs, snakes, turtles, and lizards. Their importance is due to more than just the number of species, however. In some mountain forests, salamander densities exceed 18,000 per hectare (7,300 per acre). Such robust numbers are evidence that salamanders perform an extremely important ecological role in the forest floor–soil community. They participate in a complex food web, feeding on invertebrates that are involved in the breakdown of leaf litter and other plant material, and in turn providing a source of food for forest predators. They perform similarly important functions in Coastal Plain wetlands, where their densities can be equally impressive.

People's lack of familiarity with this group of amphibians is at least in part due to salamanders' silent and secretive nature. Salamanders are far less likely than frogs to be casually encountered, and they never make sounds by calling or jumping into the water. Nonbiologists who do encounter salamanders rarely refer to them as such, instead using colloquial names that obscure the true amphibian nature of these creatures. In the mountains of northern Georgia, for example, the stream-dwelling dusky salamanders (genus *Desmognathus*) are called spring lizards and the terrestrial woodland salamanders (genus *Plethodon*) are known as wood lizards. The brightly colored spring (*Gyrinophilus porphyriticus*) and red (*Pseudotriton ruber*) salamanders are sometimes referred to as red dogs. The gilled larvae, particularly those of the mole salamanders (genus *Ambystoma*), are called waterdogs. The large, short-limbed salamanders of the Coastal Plain (genera *Amphiuma* and *Siren*) may be called conger eels or ditch eels. Even the "scientifically accepted" common names of some species (e.g., hellbender, mudpuppy, newts, and sirens) fail to establish the connection that their bearers are actually salamanders.

Most species resemble lizards in overall body shape, and this similarity has undoubtedly given rise to much of the confusion involving common names. Unlike lizards, however, salamanders lack scales, claws, or shelled eggs. Several features set salamanders apart from other amphibians as well. One is the peculiar method by which they mate. In almost all of our native species, mating involves internal fertilization. Rather than copulating like most tetrapods, though, an aroused male salamander secretes a packet of sperm (called a spermatophore) onto the ground, and the female sits down on the packet and picks it up with her vent. Another unique feature of the life cycle of some species is paedomorphosis, in which larvae fail to metamorphose—or do so only partially—and reproductively mature adults retain larval characteristics such as external gills. Paedomorphosis may be obligate, meaning that it occurs regardless of external conditions, as is the case for the sirens (genera *Siren* and *Pseudobranchus*) and waterdogs (genus *Necturus*); or it may depend on environmental conditions, as is the case for

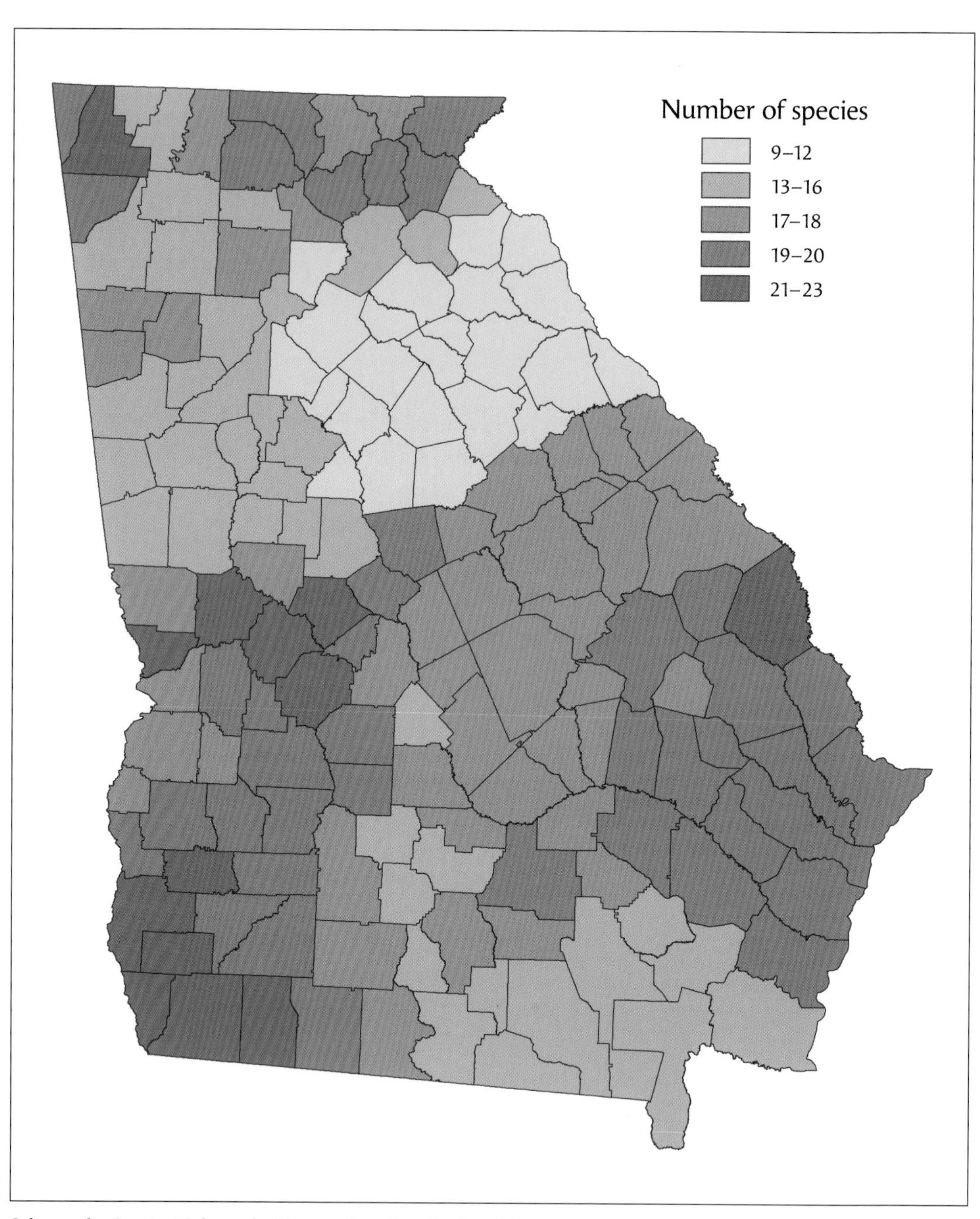

Salamander Species Richness by County, Based on Predicted Ranges

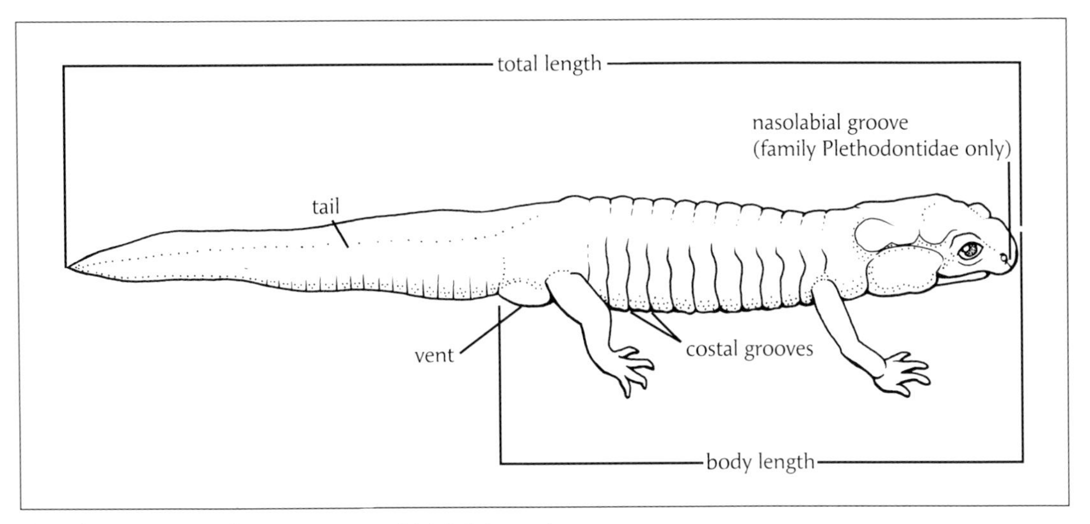

Identifying Features and Measurements of Adult Salamanders

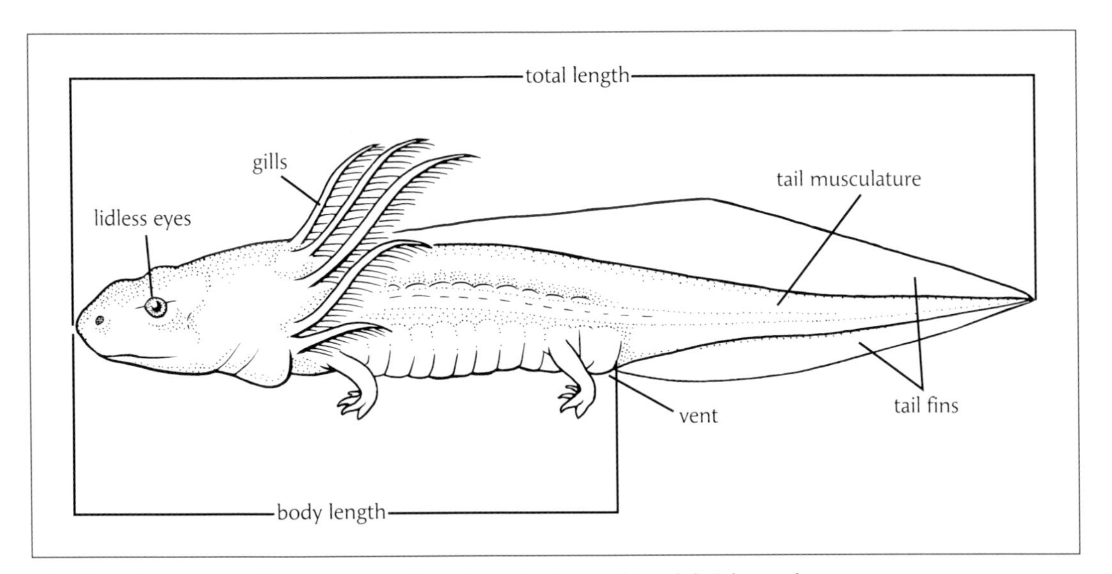

Identifying Features and Measurements of Larval and Paedomorphic Adult Salamanders

newts (genus *Notophthalmus*) and some mole salamanders. Amphiumas and the hellbender (*Cryptobranchus alleganiensis*) partially transform, losing their gills but not the spiracular opening associated with them, and never develop the eyelids that are present in fully transformed salamanders.

Salamander larvae exhibit adaptations for the type of environment they occupy. Those that live in ponds have large tail fins to help in swimming and extensive gills for maximizing oxygen uptake. Stream-dwelling larvae have small gills that are sufficient to exploit the higher oxygen content of streams and small tail fins because large fins are cumbersome in dealing with currents. Larvae of terrestrial salamanders such as the seepage (*Desmognathus aeneus*), green (*Aneides aeneus*), and woodland salamanders have poorly developed gills, never enter standing water, and metamorphose within a few days of hatching.

Mole Salamanders

FAMILY AMBYSTOMATIDAE

Five members of this family are present in Georgia. Mole salamanders are typically 7.5–15 cm (3–6 in) in total length, although the tiger salamander (*Ambystoma tigrinum*) may reach 30 cm (12 in). They have a thicker body, shorter tail, and larger head than other terrestrial salamanders. Some are beautifully marked with spots or blotches of yellow or silver. The larvae are short and stumpy with tall tail fins and bushy gills. This endemic North American family includes 30 or so species. The family is distributed statewide in Georgia, although more species occur in the Coastal Plain than elsewhere. Mole salamanders inhabit forests in situations ranging from hardwood mountain slopes and river floodplains to sandy south Georgia flatwoods. They usually occur in proximity to the temporary ponds they use for breeding. As the common name for the family implies, mole salamanders spend most of their lives underground. Although some species can be found under logs during the winter or spring, others are rarely observed aboveground except during the breeding season.

Adult flatwoods salamander, Liberty County (Dirk J. Stevenson)

Flatwoods Salamander

Ambystoma cingulatum

Description

This is a relatively small mole salamander; adults reach a total length of 13.5 cm (5.3 in) and a body length of 7.6 cm (3.0 in). Females can be up to 9 percent longer than males. Adults are black or dark gray (less commonly dark brown) with light gray or white flecks forming a netlike pattern on the back, sides, head, and tail. Individuals in some populations are less prominently patterned and take on a frosted appearance. The belly is dark gray or black with scattered white or light gray specks or spots. Breeding males do not exhibit the obviously swollen vent present in other species of mole salamanders. The head is fairly small, not noticeably wider than the neck or shoulder. The tail is quite stout and is taller than wide. Costal groove number ranges from 13 to 16, with an average of 15. The similar mole salamander (*A. talpoideum*), although dark colored and patterned with light flecking, is much stouter and lacks the reticulations or frosting characteristic of adult flatwoods salamanders. The aquatic larvae have a broad head, high tail fins, and bushy red gills. Hatchling larvae have a body length of 8–12 mm (0.3–0.5 in). The maximum total length of larvae is 96 mm (3.7 in), and the maximum body length is 47 mm (1.8 in). The larvae are generally brown and are more boldly patterned than any other mole salamander species in the southeastern United States. A prominent pale tan stripe extends along the side, and a dark brown or black stripe extends from the snout through the eye to the gills. Larval

Flatwoods salamander larva, Liberty County (Dirk J. Stevenson)

Gravid adult female flatwoods salamander, Okaloosa County, Florida (John B. Jensen)

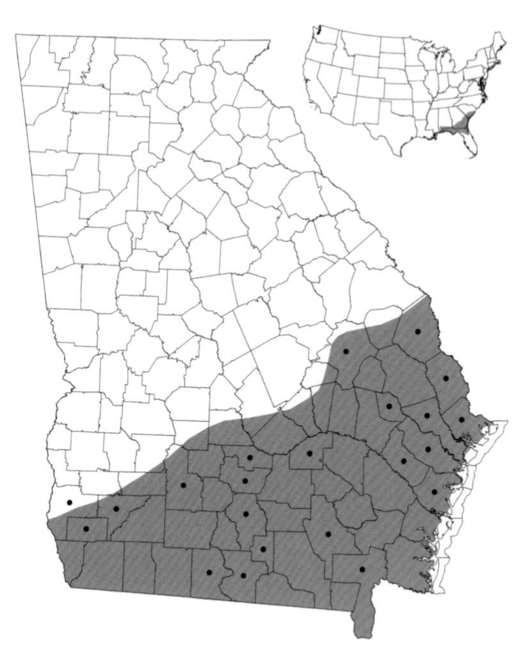

mole and tiger salamanders (*A. tigrinum*) may have ragged, light side stripes, but their darker coloration is mottled rather than uniform. Medium-sized and larger mole salamander larvae often have a dark midbelly stripe.

Taxonomy and Nomenclature

A recent proposal suggests that the populations on either side of the Flint River be considered two separate species because they are so genetically divergent.

Distribution and Habitat

This salamander is found in moist pine flatwoods and savannas of the Coastal Plain, a habitat that characteristically has a relatively open canopy of longleaf pine and a diverse ground cover dominated by wiregrass. Suitable flatwoods contain numerous shallow, isolated wetlands, which are required for breeding and larval development. Typically these wetlands have scattered forest canopies of pond cypress, swamp black gum, and slash pine and an abundance of emergent herbaceous vegetation. Breeding ponds generally lack large predatory fish.

Reproduction and Development

In response to rains associated with approaching cold fronts during the fall and early winter, adults migrate from their upland retreats to breeding ponds, which are often dry on their arrival. The flatwoods salamander is the only ambystomatid salamander other than the marbled salamander (*A. opacum*) known to practice terrestrial courtship and egg-laying. Females deposit up to 225 eggs singly or in small clumps in the leaf litter, under logs, at the base of grass clumps, or at the entrance to crayfish burrows, all within the pond basin or along its margin. Larvae begin development inside the egg and are induced to hatch when inundated by rising water. A developmental period of 11–18 weeks follows hatching, and larvae transform and leave the pond in March or April. Sexual maturity is reached in 1–2 years.

Habits

The majority of the adult's life is spent in terrestrial habitats, presumably underground, although very little is known about the upland habits of this species. There is some evidence that flatwoods salamanders use old crayfish tunnels. Terrestrial refuges may be up to 1.6 km (1.0 mi) from the adults' breeding wetlands. Adults are known to eat earthworms but probably consume other invertebrates as well. Larvae eat a wide variety of aquatic invertebrates, especially crustaceans. The common garter snake (*Thamnophis sirtalis*) is the only documented predator of flatwoods salamanders; however, other snakes, mammals, and wading birds likely prey on them as well. When handled, an adult flatwoods salamander will often tuck its head down and simultaneously raise and undulate its tail, exuding a milky substance from glands in the tail. This substance may be noxious and may ward off predators. The tail waving also likely draws a potential predator's attention away from the more vital parts of the salamander's body and toward the expendable tail.

Conservation Status

Populations of flatwoods salamanders have declined severely in conjunction with the rapid alteration and disappearance of the longleaf pine–wiregrass community. This alteration includes the suppression of fires necessary to prevent pine flatwoods habitats from being invaded by canopy-closing hardwoods. Periodic fires are also essential to maintain the open-canopied, grassy ponds where flatwoods salamanders breed. Many of the isolated wetlands necessary for breeding have been drained or filled. As a result, only a handful of widely isolated populations continue to exist in Georgia. Deservingly, the species is designated Threatened by both Georgia and federal law.

John B. Jensen and Dirk J. Stevenson

Adult spotted salamander, Appling County (Dirk J. Stevenson)

Spotted Salamander

Ambystoma maculatum

Description

This is a relatively large member of the genus *Ambystoma*. Adults are 15–25 cm (5.9–9.8 in) in total length and near 13 cm (5.1 in) in maximum body length. The tail is moderately flattened from side to side, and the head is round and dish shaped. Two distinct but somewhat irregular rows of rounded spots extend down the back from the head to the tip of the tail. Spots may vary in number (from 10 to more than 50), size, arrangement, and color. The spots are usually bright yellow but may be muted or absent in some individuals. The spots on the head of young salamanders are often bright orange, a color that may persist into adulthood for many individuals of mountain populations. The black, bluish black, or dark gray back grades into the slate gray of the sides and belly. There are typically 12 costal grooves on each side. Breeding males are noticeably swollen in the area around their vent and may have a more keeled tail than females; otherwise, differences between the two sexes are not obvious. Larvae are dull olive green without any conspicuous markings except for a clear but poorly defined line along each side of the body. They have large, fluffy gills and an extensive tail fin that extends forward on the back to the level of the front limbs. Hatchlings are approximately 14–15 mm (0.6 in) in total length, and larger larvae may reach 75 mm (2.9 in) prior to metamorphosis.

Spotted salamander larva, Bartow County (John B. Jensen)

Juvenile spotted salamander, Upson County (John B. Jensen)

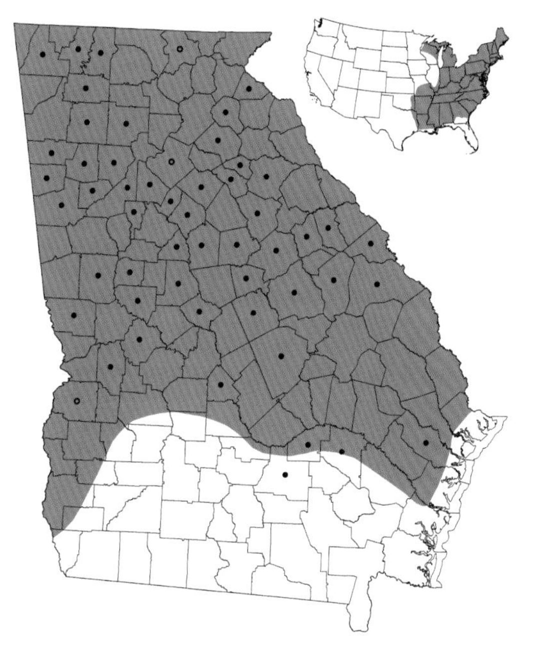

Taxonomy and Nomenclature

Two genetically distinct groups have been identified. Georgia falls between them, and which of the two groups is represented in the state is currently unknown. No subspecies have been recognized.

Distribution and Habitat

The spotted salamander is present throughout much of Georgia north of the lower Coastal Plain. Although populations are known from the Coastal Plain, this form is most abundant above the Fall Line. Spotted salamanders prefer bottomland hardwood forests in or adjoining floodplains, but they also may be common in upland hardwood forests if suitable breeding sites are available. Breeding ponds are typically temporary woodland pools, depressions in floodplains, rain-filled ditches, and wheel ruts in abandoned roads. Beaver ponds also are occasionally used.

Reproduction and Development

Adults migrate to breeding ponds during rainy, relatively warm nights. Some breeding may take place in January, but February is the peak month for breeding in Georgia. In northern Georgia this is one of the first amphibian species to emerge, its breeding often coinciding with that of the wood frog (*Rana sylvatica*). Courtship, which takes place at night in shallow water, involves mutual touching and nudging. The male deposits his spermatophore on the bottom of the pond, and his partner picks it up with her vent. Within 2–3 days of breeding, females deposit three or four egg masses containing as many as 250 eggs each. The masses are often attached to submerged twigs or aquatic plants by means of the jelly coat, which affords the eggs some protection from cold and pond drying. Egg masses within the interior of large aggregates stranded by receding water have been known to survive for 35 days. The jelly is initially clear or white and, unlike frog egg masses, is relatively firm to the touch. Within a few days of deposition the jelly is invaded by green algae, which provide oxygen via photosynthesis for the developing salamander embryos. Incubation may take as long as 2 months if temperatures are cool. Larvae metamorphose less than 6 months after hatching, and transformation occurs from late spring into fall. Individuals from Georgia populations require only 2 or 3 years to reach reproductive maturity. Some individuals have lived for more than 30 years.

Habits

Spotted salamanders are fossorial and are rarely seen aboveground except when breeding or when in migration to and from breeding sites, although some individuals may remain under logs near breeding ponds for several months after the breeding season. Nonbreeding individuals maintain small (usually less than 10 m^2, or 12 yd^2) home ranges. Although they may migrate several hundred meters to breeding ponds, individuals return to their initial home range after, often to the same burrow. Spotted salamanders feed on forest floor invertebrates such as earthworms, spiders, and insects. Individuals have been observed at night with their head protruding from their burrow, apparently waiting to ambush wandering invertebrate prey. Adults aggressively defend feeding sites from one another. The larvae often share pond habitats with larval marbled salamanders (*A. opacum*). The eggs of the latter species hatch several months earlier than spotted salamander eggs do, and breeding ponds often contain two distinct sizes of mole salamander larvae: small spotted salamanders and large marbled salamanders. Larval spotted salamanders feed on small aquatic invertebrates. Although the extensive jelly mass protects the eggs from some predators, they may still be eaten by eastern newts (*Notophthalmus viridescens*) and aquatic insects. Predators of larvae include newts, aquatic insects, and marbled salamander larvae. The skin of adults produces chemical secretions that are distasteful to many predators, and the bright yellow spots may represent aposematic coloration. Even so, predators such as small mammals and snakes do occasionally attack adults.

Conservation Status

Historical patterns of human settlement in northern Georgia centered on floodplains along rivers because these sites provided access to food and water as well as gentle topographic relief conducive to agriculture. Both Native Americans and Old World settlers modified these areas for their own use. Ditching streams and draining floodplain basins eliminated many, perhaps most, of the temporary isolated wetlands north of the Fall Line. It seems reasonable to assume that the loss of so much breeding habitat has reduced populations of spotted salamanders. This species is nevertheless common across the northern half of the state.

Thomas M. Floyd and Carlos D. Camp

Adult female marbled salamander, Jenkins County (Dirk J. Stevenson)

Marbled Salamander

Ambystoma opacum

Description

Marbled salamanders are stout-bodied salamanders; adults may be 7.5 cm (2.9 in) in body length and 12.5 cm (4.9 in) in total length. The size of reproductive adults is highly dependent on size at metamorphosis, which can occur at body lengths ranging from 2.5 to 5.8 cm (1.0–2.3 in). The head, body, and tail are black with white or gray markings. The pattern generally consists of two stripes, one running along each edge of the back, with uniformly spaced cross bands connecting the two stripes in a series running from the snout to the end of the tail. The markings on males are whiter than the silvery gray markings on females. The pattern is extremely variable; individuals whose cross bands connect across the back appear banded, and those whose bands are interrupted appear striped. The larvae have feathery gills and a row of small white dots along the lower edge of each side. Hatchling larvae are typically 10–14 mm (0.4–0.6 in) in total length.

Taxonomy and Nomenclature

No subspecies are recognized.

Distribution and Habitat

Marbled salamanders are common in a variety of wooded habitats throughout most of Georgia, although they are seldom encountered

Adult male marbled salamander, Appling County (Dirk J. Stevenson)

Marbled salamander larva, Liberty County (Dirk J. Stevenson)

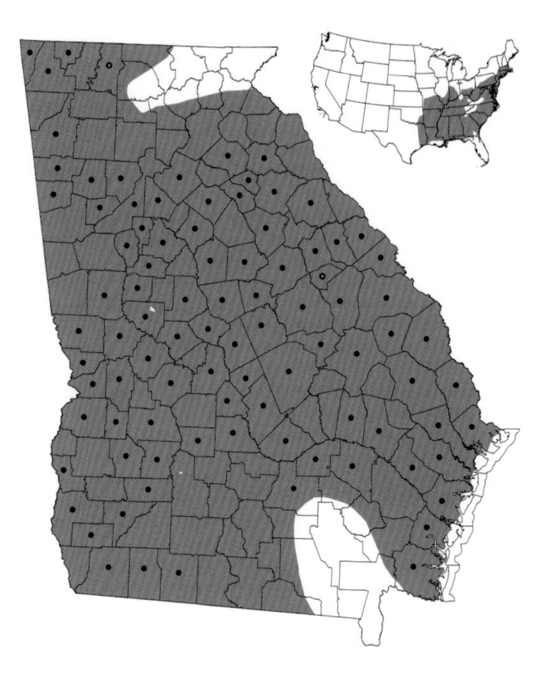

in mountainous areas. Adults disperse an average of 200 m (656 ft) from wetland breeding sites and consequently require intact terrestrial habitats. Mature deciduous forests such as floodplain bottomlands are preferred, although mixed hardwood-pine forests are also used. Within forests, individuals use leaf litter, logs, root holes, and small-mammal burrows for cover. Marbled salamanders breed in fish-free wetlands that hold water for only a portion of the year. Breeding sites include floodplain pools, sinkhole ponds, sagponds, and Carolina bays.

Reproduction and Development

The marbled salamander is one of the two species of mole salamanders that breed on land and is the only one known to exhibit parental care. Nesting sites are generally located in the dried beds of seasonal wetlands and the dry margins of reduced ponds. Adults migrate to breeding sites on rainy nights from September to November; males generally arrive first. Courtship and mating are terrestrial. Females construct nests and lay eggs under vegetation, logs, or leafy debris where the nest is likely to be flooded by subsequent rains. Some females brood their eggs until flooding occurs, which can be several months, but many abandon their nest within a few weeks. Larger females lay more eggs, with clutch size ranging from 30 to 250 eggs. Females may occasionally share nests. Female marbled salamanders lay fewer but larger eggs than do other mole salamanders of similar size. The eggs may remain viable for 3–4 months, but the embryos may die from freezing, dehydration, predation, or fungal infestation; hatching success depends on the timing of pond filling and the duration of female brooding. The larval period is 3–6 months, and larvae usually metamorphose from early April to early June. Larval growth and development depend largely on temperature, food resources, date of hatching, and larval density. The density of larvae in a wetland may sometimes be quite high (e.g., dozens of larvae per square meter). Catastrophic larval mortality may result from extreme cold during winter or early pond drying. On average, males must be 2–3 years old before breeding; females are able to breed at 3 or 4 years. The maximum life span estimated for individuals in the wild is approximately 11 years.

Habits

Marbled salamanders are generally seen in large numbers only during the few rainy nights of the fall breeding migrations. Breeding animals may represent a quarter or less of the true population size, however, because immature individuals and nonbreeding adults remain in adjacent forests. Newly metamorphosed individuals disperse from breeding wetlands on rainy nights. The period between metamorphosis and dispersal may be several weeks or more, depending on the timing of rainfall. Marbled salamanders are generally solitary when not breeding. As soils dry, they may retreat deeper into their underground burrows. Although the home range has not been determined for Georgia individuals, the average summer home range size for adults in an Illinois population was estimated to be 14.5 m^2 (17.4 yd^2). Terrestrial juveniles and adults eat millipedes, centipedes, spiders, insects, worms, and snails. Larvae feed primarily on zooplankton such as ostracods, cladocerans, and copepods but also eat amphibian eggs and larvae, including the larvae of other mole salamanders. Annual survivorship of nonbreeding adults appears to be quite high (more than 70 percent), although raccoons, opossums, skunks, and shrews are known to eat them. Glands in the skin produce noxious secretions that protect individuals from many predators. Salamanders, frogs, beetles, and possibly other arthropod species may eat marbled salamander eggs. Larvae are palatable to fish but usually do not inhabit ponds where fish occur. Larval marbled salamanders fall prey to many invertebrates such as the aquatic larvae of dragonflies.

Conservation Status

Marbled salamanders have no special legal status in Georgia and remain fairly common. Given their reliance on small seasonal wetlands and forested floodplains, however, their abundance has presumably declined as habitats have been eliminated or altered.

David E. Scott

Adult mole salamander, Liberty County (Dirk J. Stevenson)

Mole Salamander

Ambystoma talpoideum

Description

The mole salamander is a moderately small member of the genus *Ambystoma*. Many mole salamander larvae fail to metamorphose completely, so some adults are paedomorphic. Transformed terrestrial adults are 7.6–12.2 cm (3.0–4.8 in) in total length; approach 8 cm (3.1 in) in maximum body length; and have a short, stocky body with relatively large limbs and a disproportionately large head. The tail is short, less than half the total length. The back and sides are gray to brown with scattered light gray flecks that are sometimes inconspicuous. The belly is normally bluish gray with light flecks. Although the sexes typically do not differ significantly in appearance, the margins of the vent are swollen in sexually active males, and their tail is more keeled. There are 10 or 11 costal grooves on each side. Larvae have a cream-colored or dull yellow stripe along each edge of the belly and often have an additional stripe along each side of the body. These stripes may break into blotches near the tail. A dark stripe, often edged by two smaller cream-colored stripes, extends along the midline of the belly on older larvae, recently transformed juveniles, and gilled adults, and occasionally persists in terrestrial adults. Gilled adults generally resemble larvae but tend to be darker. Gilled adults are considerably smaller

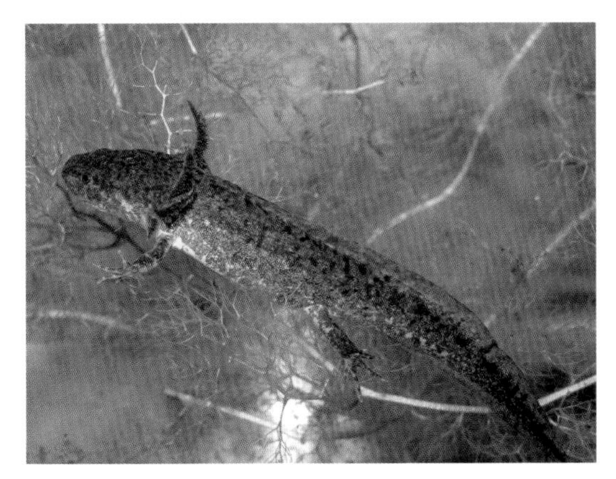

Paedomorphic adult mole salamander, Bryan County (Dirk J. Stevenson)

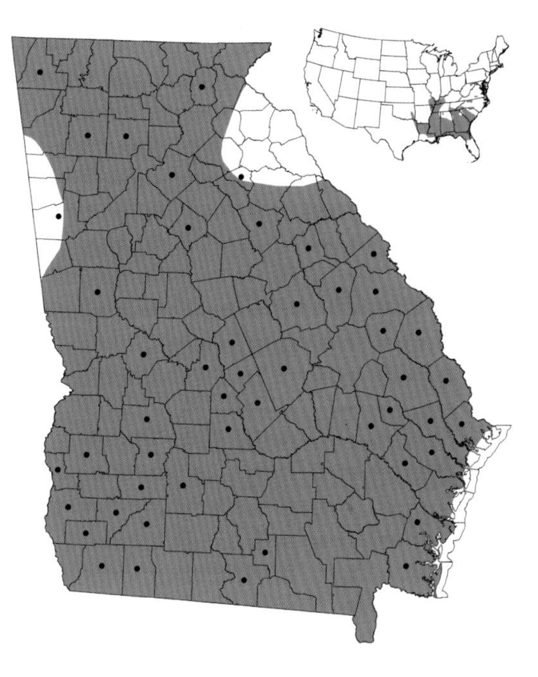

Mole salamander larva, Liberty County (Dirk J. Stevenson)

than their transformed counterparts, averaging about 4 cm (1.6 in) in body length. Recently metamorphosed individuals are brownish green, 5.6–7 cm (2.2–2.7 in) in total length, and begin developing adult patterning within 2–4 weeks of transformation.

Taxonomy and Nomenclature

The taxonomy of this species has been stable for the last 150 years. Differences in the mode of egg deposition and related life history parameters of Atlantic and Gulf Coastal Plain populations have been documented, but whether this pattern is related to genetic differentiation is unknown. No subspecies have been designated.

Distribution and Habitat

Until recently, this reclusive species was thought to be confined primarily to the Coastal Plain with isolated occurrences in other regions. Discovery of additional populations within the Piedmont, Blue Ridge, Ridge and Valley, and Cumberland Plateau provinces of Georgia has revealed that this salamander is much more widely distributed in the state. The mole salamander is the only member of its family known from Georgia's barrier islands. Coastal Plain populations are commonly associated with extensive floodplain forests or found near gum and cypress ponds. Suitable terrestrial habitats are unaffected by salt water and typically have loose sandy or loamy soils that facilitate burrowing. Populations outside the Coastal Plain usually inhabit upland hardwood or mixed pine-hardwood forests surrounding isolated wetlands. Mole salamanders breed in a wide variety of temporary and permanent habitats, including open grassy ponds, woodland ponds, Carolina bays, roadside ditches, borrow pits, and farm ponds, but rarely in wetlands containing large predatory fish.

Reproduction and Development

Although the timing of breeding varies geographically and with rainfall and temperature, adults migrate from upland habitats to breeding wetlands on rainy nights from fall to early spring. Gilled adults normally breed about 6 weeks earlier than terrestrial adults; however, courtship between the two forms occurs regularly. After breeding, terrestrial females in some Gulf Coastal Plain populations lay small clusters containing as many as 50 eggs, often attaching them to submerged twigs. Terrestrial females of at least some Atlantic Coastal Plain populations and gilled females in general scatter their eggs singly in ponds. Females require as many as 3 nights to lay their entire egg complement (200–700), which is generally higher in Atlantic Coastal Plain populations. Most terrestrial adults have migrated from breeding ponds into the surrounding uplands by early spring. Eggs hatch in late winter or early spring, and the larval period typically lasts 3–4 months. Newly metamorphosed individuals migrate away from the ponds as early as May. Some larvae become sexually mature gilled adults by September and breed the following winter, then metamorphose in the spring when 12–15 months old. A few larvae may overwinter as immature forms and either transform the next summer or mature into gilled adults. Individuals typically reach reproductive maturity at 2 years of age.

Habits

Larvae are more likely to become gilled adults in the absence of fish predators, when pond water levels are relatively constant, and when larval growth rates are low. Both gilled adults and larvae feed opportunistically on tadpoles, worms, insects, and other invertebrates. Migrating adults are significantly more active at night; those studied in one population dispersed an average of 178 m (584 ft) away from breeding ponds. Juveniles disperse shorter distances than adults. During the summer, terrestrial individuals live in underground burrows and tunnels about 5 cm (2 in) beneath the soil surface, either digging new burrows or enlarging existing cracks and crevices. Juveniles and adults have well-developed skin glands on the head and along the tail that produce noxious secretions. Although predators are poorly known, at least one snake species, the eastern ribbon snake (*Thamnophis sauritus*), is known to prey on mole salamanders. When they encounter a predator, mole salamanders assume a defensive posture by tucking their head down. Marbled salamanders (*A. opacum*) breed earlier than mole salamanders in ponds where both species breed, and the larger, better-developed marbled salamander larvae may feed on larval mole salamanders.

Conservation Status

Although the mole salamander is currently common in Georgia, extensive fragmentation of upland habitats and the isolation and alteration of temporary wetland habitats may threaten the long-term survival of local populations.

Thomas M. Floyd

Adult tiger salamander, Evans County (John B. Jensen)

Tiger Salamander

Ambystoma tigrinum

Description

The tiger salamander is the largest terrestrial salamander in Georgia. Adults seldom exceed 25.6 cm (10 in) in total length, although exceptionally large individuals may reach 35 cm (13.7 in). Body length is usually 7.5–15 cm (2.9–5.9 in). The body is stocky with stout limbs and a large, somewhat flattened head. Tiger salamanders typically have irregularly shaped yellow to beige spots on a black or dark brown background. The spots can be round to elongated and occur on the head, body, and tail, and sometimes extend onto the sides. The belly is typically yellowish with dark mottling. Adult males are the same size or slightly smaller than females but have a slightly longer tail with a higher keel. Breeding males have a noticeably swollen vent. Larvae have a large, round head with a blunt snout, and bushy external gills. Larvae are approximately 14 mm (0.6 in) in total length at hatching but may reach or exceed 10 cm (ca. 4 in) before transforming. Large larvae have a light, ragged lateral stripe bordered by dark, mottled bands above and below. Transforming individuals typically lack spots and are uniformly black or dark gray.

Taxonomy and Nomenclature

Seven subspecies of tiger salamanders are recognized—six in the continental United States—but the taxonomic status of this group is still very much in dispute. The eastern tiger salamander (*A. t. tigrinum*) is the most broadly distributed of the seven and is the only subspecies that occurs in Georgia.

Tiger salamander larva, Miller County. Note the wide head relative to the body. (John B. Jensen)

Tiger salamander larva, Long County (Dirk J. Stevenson)

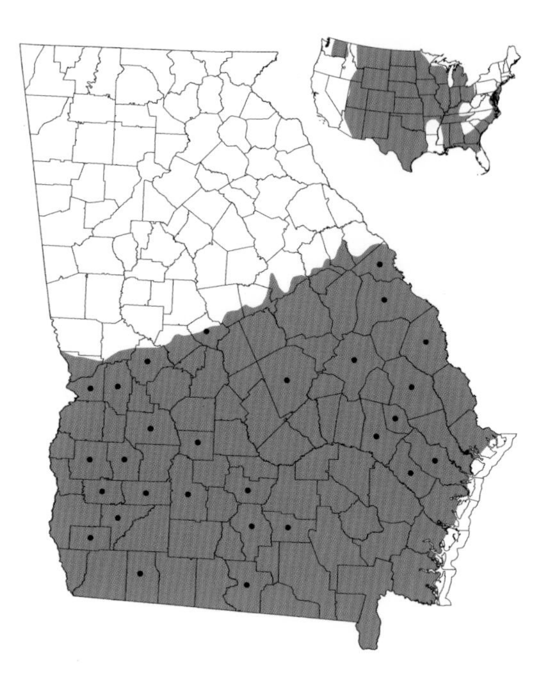

Distribution and Habitat

The tiger salamander occurs south of the Fall Line within most of the Coastal Plain, excluding the barrier islands and the Okefenokee Swamp. Based on known localities in Alabama, the species might also occur in Georgia's Ridge and Valley Province; a fossil record exists from that part of the state. These salamanders inhabit pine forests and open fields with loose, moderately well drained, sandy soils. They breed in open, grassy, usually temporary, ponds. Artificial ponds such as borrow pits are also used.

Reproduction and Development

Tiger salamanders migrate to breeding sites in fall and winter as ponds are filled by rainwater. These nocturnal migrations are cued by environmental variables such as temperature and rainfall. Males typically arrive at ponds in advance of females. Eggs are attached to twigs or rooted vegetation as discrete masses of 13–110 eggs. Incubation time is variable, but hatching can occur in 24–30 days. Larvae grow rapidly, typically transforming in 75–118 days. The entire clutch may be lost, however, if the pond dries early. Tiger salamanders can reach sexual maturity within a year.

Habits

Like other mole salamanders of the genus *Ambystoma*, tiger salamanders spend much of the year underground and are seldom observed except at breeding sites. They actively dig burrows for themselves but presumably also use burrows made by other vertebrates. Little is known about the terrestrial phase of this salamander's life; however, the burrow of one individual was discovered 12 cm (4.7 in) underground. Larvae eat a variety of prey, including assorted invertebrates, tadpoles, and other larval salamanders. In some populations the larvae are cannibalistic and have a specialized physical form that includes a broader, more flattened head and an extra row of teeth. Individuals of populations in Michigan and in some of the western states may develop into gilled adults, a phenomenon known as paedomorphosis, but this has not been documented in Georgia populations.

Conservation Status

Little is known about the status of tiger salamander populations in Georgia. Urbanization and the loss of small temporary ponds have certainly had an adverse effect on their populations, as has the introduction of predatory fish into breeding sites. Tiger salamanders are still common in some areas, though, and may be somewhat tolerant of small-scale disturbances.

Lora L. Smith

Amphiumas

FAMILY AMPHIUMIDAE

This family, represented in Georgia by two species, includes Georgia's longest salamander, the two-toed amphiuma (*Amphiuma means*), whose maximum size approaches 125 cm (4.1 ft). These strange-looking creatures have a greatly elongated body and four tiny limbs. Amphiumas have fewer toes than most salamanders (one, two, or three per foot, depending on the species). The larvae are similar to adults except for their small size and external gills. This family is endemic to the Coastal Plain of the southeastern United States. Its members burrow into mud or move about in weed-choked wetlands such as river swamps, roadside ditches, and cypress ponds. Amphiumas are voracious predators that will eat any small aquatic animal they can catch. These salamanders do not fully metamorphose but retain certain larval characteristics such as a lateral line system for detecting underwater movements. The model for the asteroid-living creature in George Lucas's film *The Empire Strikes Back* was an amphiuma.

Adult two-toed amphiuma, Liberty County. Note the pores of the lateral line system. (Dirk J. Stevenson)

Two-toed Amphiuma

Amphiuma means

Description

Two-toed amphiumas are among the largest salamanders in North America. Adults average 50–60 cm (19.5–23.4 in) in total length, and some individuals exceed 100 cm (39 in); body length averages about 40 cm (15.6 in). Amphiumas are generally slender, although larger animals may acquire substantial girth. The front and hind limbs are tiny. As the common name indicates, each limb usually has two toes, although variation in toe number is not uncommon. The body is a uniform dark gray with a lighter belly. Two-toed amphiumas have very small, lidless eyes; a shovel-shaped head; and a cylindrical tail that accounts for approximately 15–20 percent of the total length. Adults have gill openings on each side of the head but lack external gills. Visible pores along the sides of the body are evidence of the well-developed lateral line system. The lateral line system, gill openings, and lidless eyes are features of larval salamanders; two-toed and other amphiumas exhibit partial paedomorphosis. Two-toed amphiumas are similar in overall appearance to greater sirens (*Siren lacertina*) and are found in the same habitats, and the two species are commonly confused. Both in turn are confused with American eels and have been nicknamed ditch eels. The three can be easily distinguished,

Adult female two-toed amphiuma with eggs, Evans County (Karen Dyer)

Adult two-toed amphiuma, Liberty County (Dirk J. Stevenson)

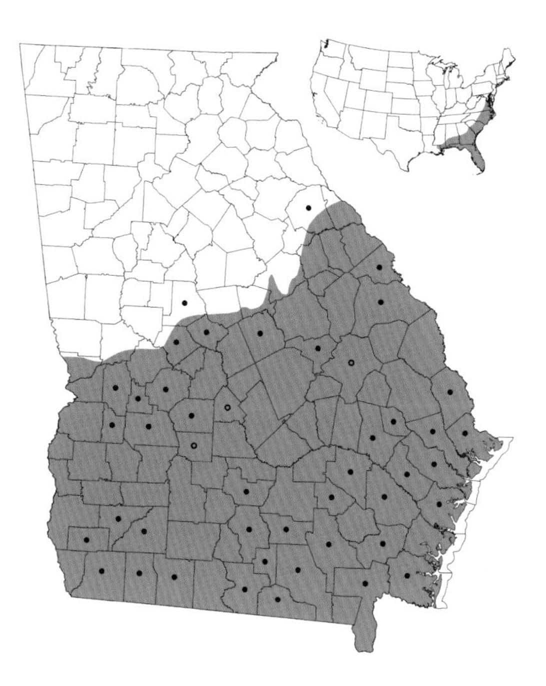

however. Sirens have external gills, substantial front limbs, and lack hind limbs. Eels are fish and have gills protected by a bony flap (opercle) and fins rather than legs. Hatchling amphiumas are approximately 50 mm (2 in) long and resemble adults except for the presence of wispy external gills.

Taxonomy and Nomenclature

No subspecies are recognized. In addition to ditch eel, amphiumas are sometimes called congo or conger eels.

Distribution and Habitat

Two-toed amphiumas are found throughout the Coastal Plain of Georgia and range north above the Fall Line into the Piedmont in a few areas. They inhabit aquatic and wetland habitats, including lakes, ditches, temporary ponds, slow-moving rivers, swamps, and canals. Amphiumas are most often associated with areas of thick muck and/or dead organic material.

Reproduction and Development

Time of reproduction can range from winter to summer and is probably influenced by environmental conditions. Two-toed amphiumas usually lay between 100 and 200 transparent eggs in a beadlike string that looks like a pearl necklace. The eggs are approximately 10 mm (0.4 in) in diameter. Eggs are typically deposited on land near the water's edge in a moist, hidden area such as under logs or vegetation. The mother remains coiled around the eggs until they hatch. Her presence keeps the eggs moist, and antibiotics produced by bacteria living on her skin may protect the eggs from fungi and harmful bacteria. Hatching appears to be stimulated by inundation; thus, incubation time is variable but may last several months. The external gills are absorbed within a few weeks of hatching. Age at maturity is unknown; the similar three-toed amphiuma (*A. tridactylum*) found in the Gulf Coast states matures in 4 years. Two-toed amphiumas have lived in captivity for more than 20 years.

Habits

Many of the two-toed amphiuma's anatomical features are adaptations for burrowing. The muscular body propels the animal through water, thick aquatic vegetation, and loose or mucky substrates. Amphiumas have well-developed lungs and may surface to gulp air when in poorly oxygenated water. The powerful jaws and sharp teeth are used to crush the shells of snails and crayfish and to subdue larger prey such as frogs and snakes. Amphiumas are voracious nocturnal feeders and are sometimes caught at night by fisherman using live bait. They can bite if handled carelessly. Wading birds, alligators, snakes, and large fish are all predators of two-toed amphiumas, and the mud snake (*Farancia abacura*) specializes on them. Amphiumas, like sirens, aestivate when their wetlands dry, burrowing in the muck and forming a cocoon from shed mucous and skin. They become dormant and can remain in the cocoon for months until water levels rise. During heavy rains they are capable of dispersing overland to nearby wetlands.

Conservation Status

Two-toed amphiuma populations appear to be stable throughout their range in Georgia, although no systematic population studies have been done.

Kristina Sorensen

Adult one-toed amphiuma, Liberty County, Florida (Suzanne L. Collins, CNAH)

One-toed Amphiuma

Amphiuma pholeter

Description

Adults of this slender, eel-like salamander have a total length ranging from 25 to 31 cm (9.8–12.1 in) and may be slightly wider than a pencil in diameter; body length reaches about 24 cm (9.4 in). The sexes do not appear to differ in body size. The tail length is about 22 percent of the total length, and the tail, unlike the uniformly cylindrical body, tapers to its tip. Although adults lack external gills, they retain other larval features such as lidless eyes, small gill openings, and pores of the lateral line sensory system. This amphiuma is unique among salamanders in having four tiny, single-toed legs, one pair just behind the small gill openings at the sides of the neck and the other pair just forward of the vent. Coloration is uniformly dark brown above and below, although the belly may be slightly lighter than the back. The one-toed amphiuma is easily confused with small two-toed amphiumas (*A. means*), but members of the latter species have one more toe per leg at all body sizes. Also, the head of the two-toed amphiuma is slightly flattened and that of the one-toed amphiuma is more cone shaped. Differences between the two sexes are subtle; the tissue surrounding the vent of males is slightly swollen. The larval stage is unknown, and neither eggs nor hatchlings have been described.

Taxonomy and Nomenclature

The one-toed amphiuma is one of only three species in the family Amphiumidae. This species was described in 1964 and has no subspecies.

Adult one-toed amphiuma, Grady County (Dirk J. Stevenson)

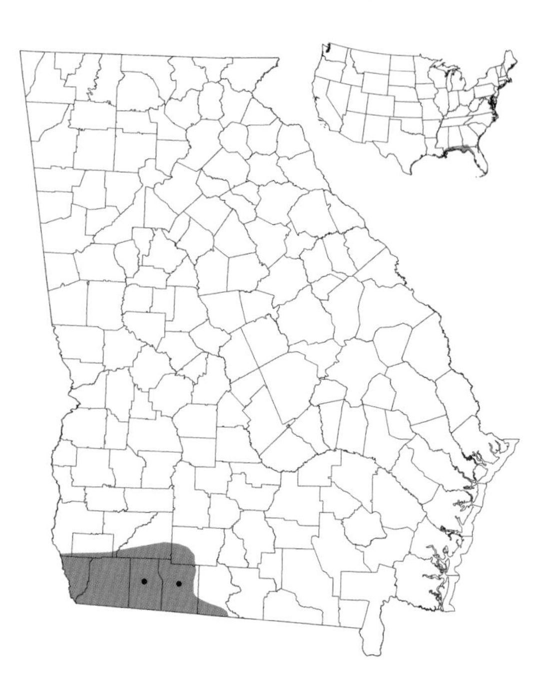

Distribution and Habitat

The one-toed amphiuma is known from two Coastal Plain localities in the Ochlockonee River drainage, but it may also occur in Mosquito Creek (Apalachicola drainage) and the Aucilla River drainage. The preferred habitat is decomposing organic litter, derived from swamp hardwoods and cypress, that has lost its fibrous nature and forms deep, soupy muck that is subject to periodic flooding and drying. Individuals may also be found in aquatic leaf packs in small streams having low to moderate gradient and under wet logs in seepage ravines. The one-toed amphiuma cannot swim through fibrous peat and is rarely found in shallow muck deposits less than about 15 cm (5.9 in) deep.

Reproduction and Development

Eggs apparently are laid during the summer in peat under the roots of vegetation growing adjacent to the muck beds in which these salamanders live. Because neither eggs nor larvae have been found, the duration of either of these stages of development is unknown. In captivity, very small young may reach adult size in 2 years; the time required from hatching to sexual maturity is unknown.

Habits

In one instance, individuals were excavated from the bottom of moist muck where they were seeking refuge from drought. Each was coiled in a small chamber made by its body. They were stimulated during excavation to move into small tunnels they appeared to have excavated. Individuals found in November–March are cold to the touch and very sluggish. They are less abundant in muck beds during winter, indicating that individuals may migrate to other sites. The guts of 100 one-toed amphiumas examined in the laboratory contained tiny clams, snails, aquatic earthworms, and a variety of arthropods. Notably lacking from the diet were dragonfly larvae, salamander larvae, tadpoles, and small fish that share the same mucky sites. Potential predators found in their muck beds include the common snapping turtle (*Chelydra serpentina*), mud turtles (*Kinosternon* spp.), and a variety of snakes such as the mud snake (*Farancia abacura*) and red-bellied watersnake (*Nerodia erythrogaster*). The one-toed amphiuma produces a noxious skin secretion that creates foam when an individual is placed in a container such as a plastic sandwich bag. This secretion is bitter tasting and slightly numbing, and is probably a deterrent to predation.

Conservation Status

Because of the scarcity of suitable habitats, this species' range is very limited in Georgia, where it is listed as Rare. A serious threat is the flooding of preferred habitat by the artificial damming of streams and rivers.

D. Bruce Means

Giant Salamanders

FAMILY CRYPTOBRANCHIDAE

This family of very large salamanders is represented in Georgia by one species, the hellbender (*Cryptobranchus alleganiensis*), which reaches a total length of 75 cm (29.3 in). The world's largest salamander, the giant Chinese salamander (*Andrias davidianus*), which grows to a length of 1.5 m (4.9 ft), is also a member of this family. The third of the three species that make up this family is found in Japan. Cryptobranchid salamanders are very flat and have conspicuously wrinkled skin along their sides. The larvae resemble the adults but are smaller and have external gills. Adults obtain most of their oxygen through their skin, which is folded to maximize its surface area. This respiratory system restricts these animals to turbulent, oxygen-rich streams.

Adult hellbender, Union County (Dirk J. Stevenson)

Hellbender

Cryptobranchus alleganiensis

Description

Hellbenders are among the largest amphibians in Georgia. Large individuals can reach a body length in excess of 50 cm (19.5 in) and a total length of nearly 75 cm (29.3 in); most adults are 30–60 cm (11.7–23.4 in) in total length. Females grow larger than males in some populations. Hellbenders are easily recognized by their flattened body; shovel-shaped head; large, paddle-like tail; and wrinkled folds of skin that extend along the sides of the body and the backs of the legs. These folds tend to obscure the costal grooves that are obvious in most other salamanders. Coloration varies a great deal. The base color ranges from light gray or black to reddish or even bright orange, and patterns can range from tiny spots to large, darker blotches. The belly is generally uniformly colored without any markings. Hellbenders are extremely slimy, making them very difficult to capture and handle. Males develop a swollen ring around the vent that is visible from mid-to-late summer through the fall breeding season. Unlike mudpuppies (*Necturus maculosus*), which may be found in the same habitats, adult hellbenders have small gill slits on the sides of the back of the head rather than gills. Hellbenders also have five digits on each of the hind feet whereas mudpuppies have only four. Newly hatched larvae are 25–30 mm (1.0–1.2 in) long, mostly black, and have large eyes; they lose their gills when they are about 13 cm (5.1 in) long.

Adult hellbender, Transylvania County, North Carolina (Jeff Humphries)

Hellbender larva, Sevier County, Tennessee (C. Kenneth Dodd)

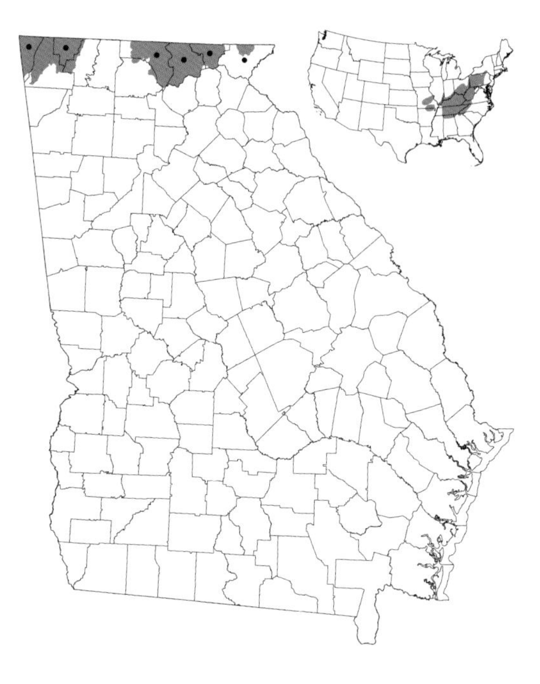

Taxonomy and Nomenclature

Hellbenders are the only members of the genus *Cryptobranchus*; their closest relatives are the giant salamanders (genus *Andrias*) of Japan and China. The eastern hellbender (*C. a. alleganiensis*) is the only subspecies that occurs in Georgia. As is true of many strange-looking and little-known animals, hellbenders have been given a variety of interesting and creative names; these include devil dog, mud devil, mollyhugger, grampus, and snot otter.

Distribution and Habitat

In Georgia, hellbenders live in streams within the Tennessee River drainage of the Blue Ridge, Ridge and Valley, and Cumberland Plateau provinces. These fully aquatic salamanders live in medium-sized to relatively large streams with cold, clear water and a rocky bottom, usually in water about 30–60 cm (11.7–23.4 in) deep.

Reproduction and Development

Nest guarding and egg deposition have been observed in Georgia during the first week of September; in other parts of the Southeast, nesting occurs from late August to early September. Each male finds and maintains a site under a rock for a nest, and one or more females deposit their clutches of several hundred eggs beneath the rock. Hellbenders differ from most other salamanders in having external fertilization. As the female deposits her eggs, the male sprays milt onto them while rocking his body up and down, creating a current in the water that apparently ensures fertilization for a maximum number of eggs. After the female has deposited her eggs, the male forces her out of the nest. A single nest may contain 1,000 or more eggs, each measuring about 18 mm (0.7 in) in diameter. The eggs in each clutch are connected to each other, forming a string. The male guards and aggressively defends the nest while the eggs develop. Time until hatching is unknown in Georgia, but incubation periods in New York and Pennsylvania range from 68 to 84 days. The larvae metamorphose after about 1.5–2 years and reach sexual maturity at about 5–8 years.

Habits

Hellbenders have lungs, but most respiration takes place through the highly folded skin. During the daytime, adults tend to remain under large rocks, often with their snout protruding, although populations in nearby North Carolina can be highly diurnal. At night, hellbenders can be seen stalking crayfish and other prey on stream bottoms. Larvae may be found beneath small rocks and cobbles, although little is known about their habitat requirements. During the fall breeding season, large numbers of adults congregate in nesting areas. Many adult males exhibit injuries and scars that may reflect their aggressively territorial nature during the breeding season. Hellbenders tend to have relatively small home ranges, usually staying within an area 28–500 m^2 (34–600 yd^2) over a period of several years, although long-term studies of movements have not been done. At night hellbenders may venture 10–20 m (33–66 ft) away from their cover rock in search of prey. Hellbenders are very long-lived; one captive individual lived for 29 years. Hellbenders feed almost entirely on crayfish, but they may also take small fish, snails, frogs, snakes, small mammals, and even smaller hellbenders. Predators may include river otters, large fish such as catfish, and turtles, although very large hellbenders are probably safe from all but the largest predators. Their main defense against predators is their cryptic coloration, which exactly matches the stream bottom, and the very slippery slime they exude when captured.

Conservation Status

Hellbender populations in the United States have drastically declined or disappeared in many areas. This species is listed as Threatened in Georgia, but very little is known about how many populations still exist in the state and how well they are doing. Habitat alteration, degradation by pollution and sedimentation, and overcollecting are the main reasons for this salamander's decline. Hellbenders are often caught on baited hooks and killed by anglers who mistakenly believe their bite is venomous; in fact, they rarely bite and are nonvenomous. Hellbenders are sensitive to degradation of the aquatic environment, and their presence is one of the best indicators of excellent water quality.

Jeff Humphries

Lungless Salamanders

FAMILY PLETHODONTIDAE

Most species in this family, including the 38 found in Georgia, are small; very few exceed 18 cm (7.2 in) in total length. Some, like the seepage salamander (*Desmognathus aeneus*), do not even reach 7.5 cm (3 in). Certain species patterned in bright red, yellow, or emerald green are among our most colorful forms of wildlife. The larvae typically have small tail fins and relatively short gills. This family accounts for 70 percent of Georgia's salamander species. It is also the largest salamander family in the world, with approximately 270 named species. Most are distributed across North, Central, and South America, although there are a few representatives in Europe.

Lungless salamanders exploit a wide variety of habitats, including streams, swamps, forests, and caves. They typically occur under cover along stream margins or on the forest floor. Some climb extensively on rocky outcrops, and some even occur in trees. Lungless salamanders exhibit parental care in that females brood their eggs until hatching. Nesting sites include cavities inside rotting logs, deep crevices in rocky outcrops, caves, and underground burrows. The larval period ranges from just a few days in the terrestrial species to 4 years in some of the streamside species. As the common name indicates, these animals do not have lungs and depend entirely on their skin for gas exchange. They also have unique grooves (naso-labial grooves) running from the nostrils to the lips or down the edges of snout projections called nasal cirri. Lungless salamanders investigate their environment by tapping on the surface with their chin or cirri, and tiny hairs in the naso-labial grooves then carry chemical information from the surface to the nostrils.

Adult green salamander, Walker County (John B. Jensen)

Green Salamander

Aneides aeneus

Description

The green salamander is a moderate-sized lungless salamander, with adults typically attaining a total length of 8–14 cm (3.1–5.5 in) and a body length of 4–7 cm (1.6–2.7 in). Females are slightly larger than males. Yellowish to green lichenlike patches extend along the back from the head to the tail, overlying a background color of black. The belly is light with little pigment. The body is noticeably flattened, and the toes are slightly expanded with flattened tips. These physical features represent adaptations for climbing and fitting into narrow spaces. There are 14 or 15 costal grooves. Juveniles hatch at a total length of approximately 19–20 mm (0.7–0.8 in) and, despite what many published descriptions say, are not "miniature replicas of adults." Hatchlings from the Blue Ridge and Cumberland Plateau provinces typically have distinctly yellow limbs, especially the upper parts, and the pattern on the head and back is usually duller and less extensive than that on adults.

Taxonomy and Nomenclature

Due to their spotty distribution, populations of the green salamander have long been suspected to represent more than one form. Recent unpublished genetic research seems to support this

Adult green salamander, Rabun County (Matt J. Elliott)

Green salamander hatchling, Rabun County (Andrew Grosse)

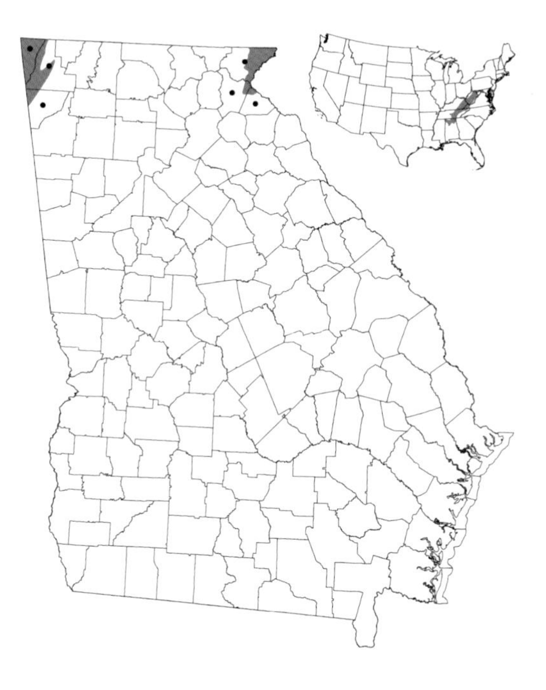

conclusion. The green salamanders found in the Cumberland Plateau province are divergent and may eventually be considered a separate species from those in the Blue Ridge. No subspecies are currently recognized.

Distribution and Habitat

Green salamanders are found throughout the Cumberland Plateau of northwestern Georgia. In the Blue Ridge and extreme upper Piedmont of northeastern Georgia they are apparently restricted to an area extending from the eastern half of Rabun County south to just below Tallulah Gorge in the Tugaloo River valley. Within this range populations tend to be somewhat scattered. Green salamanders are normally found on and near shaded rock outcrops or in trees within moist forests. Suitable outcrops contain abundant cracks and crevices.

Reproduction and Development

Females apparently breed every other year; males may breed every year. Breeding takes place in late spring in crevices of rock outcrops. Eggs are deposited in clutches of 10–30 within crevices in early or midsummer. Females remain with their eggs until hatching occurs in 10 weeks to 3 months. Terrestrial hatchlings remain with females for an additional 1–2 months and then disperse to nearby cracks and crevices. Green salamanders reach sexual maturity in 3 years.

Habits

In spring green salamanders emerge from hibernation within deep rock crevices and disperse either into other nearby rock crevices or perhaps into the surrounding forest, where they may be at least partly arboreal. During the warmer months they are most active at night and during rainy periods. Beginning in October, green salamanders return to hibernation outcrops and are generally more visible. From December to March they avoid cold temperatures by moving deep within the outcrops and thus are rarely seen. Green salamanders are frequently found in crevices that are higher and drier than other outcrop-dwelling species, which in Georgia include the seal salamander (*Desmognathus monticola*), Ocoee salamander (*D. ocoee*), southern gray-cheeked salamander (*P. metcalfi*), and species of the slimy salamander (*P. glutinosus*) complex. In northwestern Georgia green salamanders commonly share rocky outcrops with cave salamanders (*Eurycea lucifuga*), southern zigzag salamanders (*Plethodon ventralis*), and Pigeon Mountain salamanders (*P. petraeus*) but, unlike those three species, are rarely found in caves. Densities of green salamanders are relatively low on outcrops within the range of the Pigeon Mountain salamander; competition between the two has been suggested, but clear evidence is lacking. Males defend crevices against intruders of the same and other species. Predators include ringneck (*Diadophis punctatus*), common garter (*Thamnophis sirtalis*), and other small snakes. Green salamanders feed opportunistically on small insects, spiders, snails, and slugs.

Conservation Status

Several researchers have noted declines in green salamander populations in the Blue Ridge province since the 1970s. Reasons for such declines may include climatic factors, habitat loss, disease, and overcollecting. Populations of the green salamander on the Cumberland Plateau are considered stable. Animals from both regions are vulnerable to habitat modifications that alter outcrops or remove the forest canopy. The relative scarcity of the green salamander has resulted in a state listing of Rare, and it has been a candidate for protection under the federal Endangered Species Act.

Matthew J. Elliott

Seepage salamander, Gilmer County (John B. Jensen)

Seepage Salamander

Desmognathus aeneus

Description

The seepage salamander competes with dwarf (*Eurycea quadridigitata* complex) and Webster's (*Plethodon websteri*) salamanders for the title of Georgia's smallest salamander. Adults can reach a body length of 3.3 cm (1.3 in) but usually range from 2 to 3 cm (0.8–1.2 in); adult total length is 3.8–6.4 cm (1.5–2.5 in). The round tail is slightly shorter than the body. Males are 3–4 mm (0.1–0.2 in) longer than females. The top of the back and tail typically has a relatively straight-edged stripe extending along its length, sometimes with chevrons that produce a faint herringbone pattern. The stripe varies in color from brown to red to yellow. The back of the head is marked by a dark Y, and the top of each thigh commonly has a light oval spot. An obvious stripe runs from the eye to the rear angle of the jaw, a feature shared by other dusky salamanders (genus *Desmognathus*). There are 13–14 costal grooves. This species is most easily confused with the Ocoee salamander (*D. ocoee*) and the southern red-backed salamander (*Plethodon serratus*). The Ocoee salamander has a triangular tail and sometimes a stripe on the back that forms a bold zigzag, and the head lacks a Y-shaped mark. The southern red-backed salamander has neither a stripe behind the eye nor a Y-shaped mark; it also differs by having 20–22 costal grooves. Hatchling seepage salamanders are approximately 10–12 mm (0.4–0.5 in) in total length.

Adult female seepage salamander with eggs, Gilmer County (John B. Jensen)

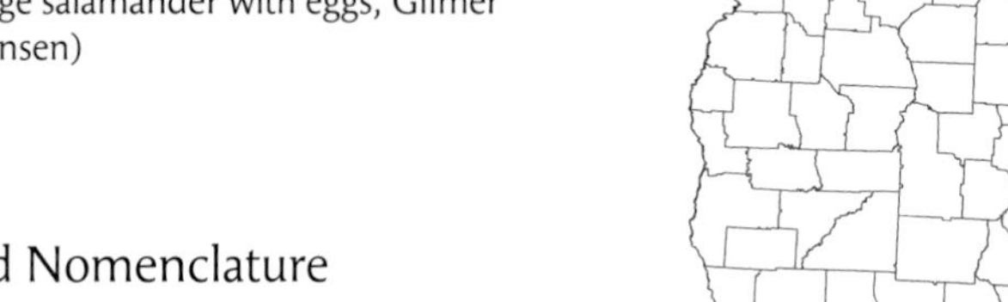

Taxonomy and Nomenclature

Two subspecies have been described but are no longer recognized. Genetic variation in seepage salamanders has not been studied in detail.

Distribution and Habitat

Populations of seepage salamanders are locally distributed across the Blue Ridge and in isolated locations in the Piedmont and Ridge and Valley provinces. Seepage salamanders are the most terrestrial of Georgia's dusky salamanders. They occur under wet leaf litter or moss near small streams and seepages or, given sufficient moisture, on the floor of hardwood forests, but are never found in water. Populations reach their greatest densities in cove forests.

Reproduction and Development

Mating apparently occurs in the fall. Females lay grapelike clusters of 6–17 eggs in late spring or summer. Favored nesting sites are under moss growing on rocks or downed logs near streams and seeps. Larvae, which hatch with small gills, are more similar to those of the terrestrial woodland salamanders (genus *Plethodon*) than to the aquatic larvae of other dusky salamanders. Transformation, including the loss of gills, takes place within the nest and occurs from a few days to a few weeks after hatching. Sexual maturity is reached in 2 years.

Habits

Members of this species are more secretive than most dusky salamanders and forage predominantly underneath the leaf litter. They feed on small invertebrates, including worms, mites, spiders, and small insects such as springtails. Natural predators have not been documented but probably include small snakes, mammals, birds, and large predaceous insects.

Conservation Status

Because populations of this species are often isolated, they are vulnerable to extinction. Seepage salamanders are sensitive to timber management practices that disrupt the moisture-preserving shade and eliminate food-rich leaf litter.

Carlos D. Camp

Juvenile Apalachicola dusky salamander, Early County (Aubrey M. Heupel)

Apalachicola Dusky Salamander

Desmognathus apalachicolae

Description

This small plethodontid salamander resembles the Ocoee salamander (*D. ocoee*) of northern Georgia but is slightly larger. Adult males are 3.5–5.2 cm (1.4–2.0 in) in body length, and adult females are 3.3–4.7 cm (1.3–1.8 in). The tail, when complete, is slightly longer than the body, round in cross section at the base, and tapers to a filament at the tip that may be round or slightly flattened from side to side. There are 14 costal grooves, and a light line runs from the back of each eye downward to the back of the jaw. Juveniles and adult females have five to seven pairs of round or diamond-shaped yellow, cream, tan, or reddish blotches on the back. These blotches coalesce down the midline to form a light stripe that is boldly bordered by dense black or brown pigment that may form scalloped or zigzag lines. Old adult males often lose the light stripe and appear uniformly dark brown or black. Color pattern is highly variable among individuals and can also vary in response to temperature and background. Individuals found in cold black seepage litter in winter appear black, for example, and those on white sand or clay appear very light. The larvae, which have five to seven pairs of discrete blotches, are small, less than 14 mm (0.6 in) in body length.

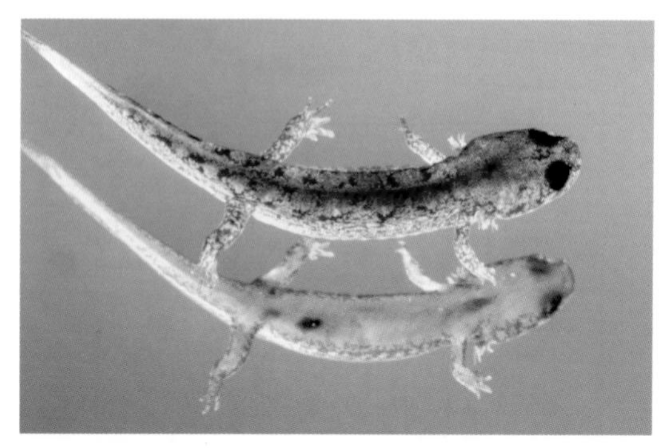

Apalachicola dusky salamander larva, Liberty County, Florida (D. Bruce Means)

Adult Apalachicola dusky salamander, Early County (John MacGregor)

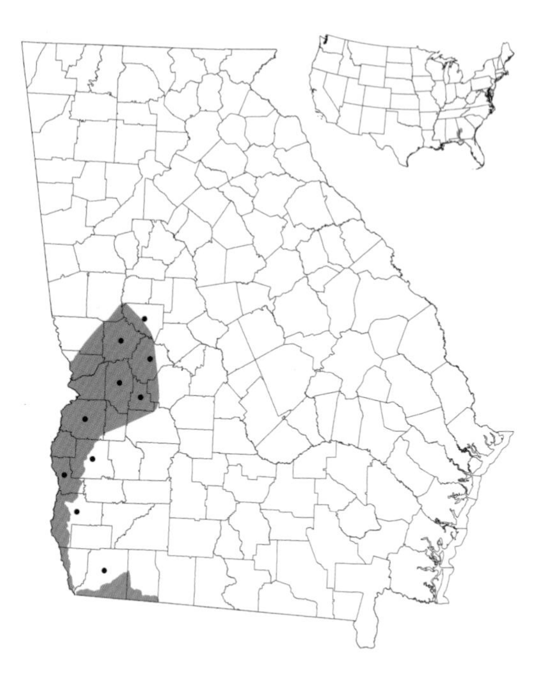

Taxonomy and Nomenclature

Until 1989 the Apalachicola dusky salamander was called by various names in the scientific literature; at one time or another it was considered to be the same species as the northern dusky (*D. fuscus*), spotted dusky (*D. conanti*), and Ouachita dusky (*D. brimleyorum*) salamanders. No subspecies are presently recognized.

Distribution and Habitat

The range in Georgia is mostly confined to the Coastal Plain, principally to ravines in headwater tributaries of the Chattahoochee River below the Fall Line. Populations are also known from ravines in the Ochlockonee River basin and along the Pelham Escarpment, a northeast–southwest line of topographic uplift south of Albany that drains into the lower Flint River. Other populations are found just above and below the Fall Line in small streams and seeps of the upper Flint River drainage. The principal habitats are first-order streams in areas of high topographic relief such as the heads of gully-eroded ravine valleys or other types of moist ravines. Metamorphosed individuals are rarely found more than 0.5 m (1.6 ft) from water. Larvae occupy very shallow, flowing seepage water, usually less than 1 cm (0.4 in) deep.

Reproduction and Development

Eggs have been found from the third week in May until the second week in November, and larvae from the second week in July until the second week in March. Egg clutches are attended by the female, who lays them in a small chamber under debris in a seep. The larval period is 9–10 months. Age at sexual maturity may be 2 years for males and 3 years for females. No studies of clutch size and growth have been published.

Habits

Adults can be found in the daytime under debris, especially logs, rocks, and leaf packs in moist ravines. Individuals are usually situated at the interface of the ground, water, and air, lying under debris with the body immersed in water and the head protruding into the air. When suddenly exposed, they usually dive into water to escape. At night, individuals forage in leaf litter or over the moist soil at the edges of seeps and on the banks of small streams. Adult males are almost never found under debris with other members of the same species, except for an occasional adult female. Behavioral experiments indicate that adult males actively defend their daytime retreats from other males. Apalachicola dusky salamanders are least commonly found during daytime in midsummer months when air temperatures are high, especially during droughts. It is presumed that they retire to deeper recesses in the ground during these times, although individuals are active on warm summer nights, especially during and after rains. Individuals found in the winter are sluggish but become active immediately upon warming. In seepage water, which is relatively warm in midwinter, individuals are as active as at other times of the year. Two other plethodontid salamanders—the red salamander (*Pseudotriton ruber*) and the southern two-lined salamander (*Eurycea cirrigera*)—always occur with the Apalachicola dusky salamander. Although no studies of their foraging ecology have been published, Apalachicola dusky salamanders probably prey on small arthropods and other invertebrates. Potential predators include snakes—particularly watersnakes (genus *Nerodia*) and eastern ribbon snakes (*Thamnophis sauritus*)—turtles, and various mammals.

Conservation Status

Apalachicola dusky salamander populations seem to be healthy throughout their range, probably due to the species' proclivity for deep, shaded, wet ravines, which up to now have been considered unsuitable for human development.

D. Bruce Means

Adult southern dusky salamander, Bryan County (Dirk J. Stevenson)

Southern Dusky Salamander

Desmognathus auriculatus

Description

As its common name implies, this is a dark salamander. Adults reach up to about 12 cm (4.7 in) in total length and 6 cm (2.3 in) in body length. The hind legs are noticeably stouter than the forelegs, and many individuals have a patch of light brown to reddish pigment reaching from the base of the eye backward to the corner of the mouth. Each side of the body has 14 costal grooves. The tail in cross section is progressively bladelike toward the tip and does not taper into a slender filament as that of the Apalachicola dusky salamander (*D. apalachicolae*) does. The back is dark brown to black, and the sides are often flecked with tiny white specks. The coal black belly is usually peppered with small white or silvery flecks. Two rows of small, round light spots are usually evident along the upper and lower sides, respectively, between the front and hind legs, both rows continuing onto the tail. These spots and a lighter stripe down the tail crest are often reddish. The best external characters for distinguishing this species from the closely related spotted dusky salamander (*D. conanti*), with which it is most likely to be confused, are the tail, which is more bladelike in cross section in the southern dusky, and the black belly with white flecks. The larva is tiny, usually 9–13 mm (0.4–0.5 in) long, and has bushy black external gills.

Underside of adult southern dusky salamander, Bryan County (D. Bruce Means)

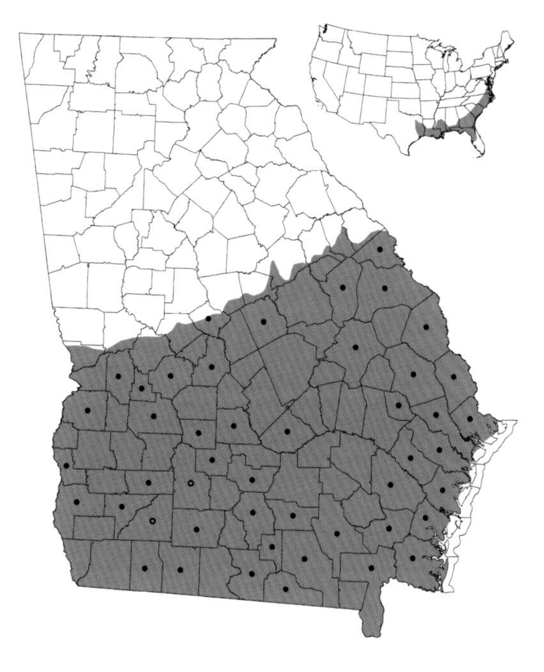

Southern dusky salamander larva, Liberty County, Florida (D. Bruce Means)

Taxonomy and Nomenclature

This species was originally named from a series of specimens collected near Riceboro in Liberty County, Georgia. It was once considered a subspecies of the dusky salamander (*D. fuscus*) and is called *D. f. auriculatus* in the older scientific literature. No substantial racial variation has been noted in Georgia, and no subspecies are recognized. The wide geographic range of this species makes significant genetic divergence similar to that seen in other widely distributed members of the family Plethodontidae a possibility.

Distribution and Habitat

In Georgia the southern dusky salamander is confined to the Coastal Plain and is most common in mucky, swampy, blackwater streams of the lower Coastal Plain. The preferred habitat is under logs, leaf litter, and debris at the edge of water in swampy, mucky, saturated floodplain sloughs; swales associated with rivers and larger streams; and sometimes along swampy lake and pond margins. During low-water stages and drought, the southern dusky salamander follows the lowering water table down into peat. Larvae are found in shallow water along edges of swamps and sloughs and along slow-moving streams.

Reproduction and Development

Females lay one cluster of about 6–20 eggs under logs and debris in moist soil at or near the edge of water, and remain with them until they hatch. Eggs with developing embryos have been found from early September to late October. Very tiny hatchlings have been found in October, and older larvae have been found in all months except July, August, and September, suggesting a larval period of approximately 7 months. Metamorphosis can take place as early as January during drought, but larvae normally remain aquatic until late March or early April, at which time they metamorphose into tiny miniatures of the adults. Age at maturity is probably at least 3 years. Little else is known about the reproduction and development of this species.

Habits

In swampy habitats subject to extremes of temperature, southern dusky salamanders are less commonly found in winter than in spring and fall and are cold and sluggish when handled; however, very active individuals were found in air temperatures of −2–0° C (28–32° F) in Richmond County. When exposed, individuals wriggle vigorously to escape into water or down into peat via crayfish burrows. The southern dusky salamander is almost always associated with the mud salamander (*Pseudotriton montanus*), dwarf salamander (*Eurycea quadridigitata*), three-lined salamander (*E. guttolineata*), many-lined salamander (*Stereochilus marginatus*), and one-toed amphiuma (*Amphiuma pholeter*) in geographic areas where these species occur. Food items reported in the diet include larval and adult insects, arachnids, and annelid worms. Only the banded watersnake (*Nerodia fasciata*) and the redfin pickerel (*Esox americanus*) have been reported as predators; other potential predators are two-toed amphiumas (*A. means*), other watersnakes (genus *Nerodia*), garter and ribbon snakes (genus *Thamnophis*), aquatic turtles, and feral pigs. Oophagy, the practice of eating one's own eggs, and cannibalism have both been reported for this species. Until recently, this species was always the most abundant salamander found in its habitat, but because it lives at the edge of water full of soggy organic debris, biologists often overlooked it.

Conservation Status

This species has experienced a widespread decline throughout most of its range since about the mid-1970s. The cause of the decline is currently unknown.

D. Bruce Means

Adult spotted dusky salamander, Catoosa County (John B. Jensen)

Spotted Dusky Salamander

Desmognathus conanti

Description

The spotted dusky salamander is a moderate-sized dusky salamander; adults are 3.5–6.5 cm (1.4–2.5 in) in body length and reach a maximum total length of 13 cm (5.1 in). Little size information is available on specimens from Georgia, but in Alabama, adult males average 10 percent larger than females in body length. The extreme amount of variability found in *Desmognathus* salamanders is perhaps most evident when attempting to identify this species. The back may be tan to dark brown; it may have no pattern, or there may be six to seven pairs of round, oval, or rhomboid spots extending down its length. These spots may fuse to form a continuous stripe with a jagged or smooth edge. The spots (or stripe) vary in color from pale yellow to dark reddish brown and are typically edged in dark brown. Very old individuals may have only vestiges of the pattern in the form of dark brown lines or curves. The belly's background color varies from cream to yellowish tan overlain with irregular dark brown mottling. The color of the belly merges on the sides with that of the back in a gradual or irregular manner and does not form the clear, two-toned effect seen in most seal salamanders (*D. monticola*). Each side has 14 costal grooves. A nearly white to reddish brown stripe extends from the eye to the rear angle of the jaw. The tail is flattened from side to side and has a moderate keel; it is normally taller than

Adult spotted dusky salamander, Floyd County (John B. Jensen)

Spotted dusky salamander larva, Walker County (Robert Wayne Van Devender)

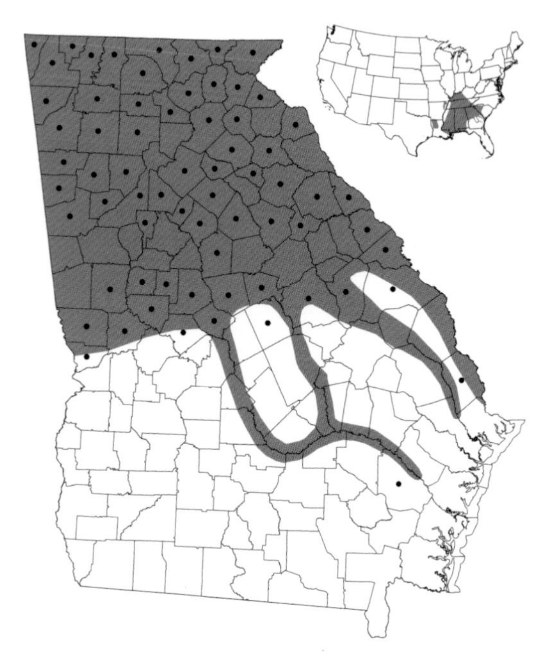

wide at its base. The tail of the Ocoee salamander (*D. ocoee*) is usually flatter (wider than tall at the base) and lacks a distinct keel, although that species exhibits considerable variation in this trait. The pattern on the larvae consists of six or seven pairs of alternating spots. Metamorphosis occurs at a body length of 13–19 mm (0.5–0.7 in).

Taxonomy and Nomenclature

The spotted dusky salamander was originally described as *D. f. conanti*, it being, along with the northern dusky salamander (*D. f. fuscus*), one of the subspecies of the widely distributed dusky salamander (*D. fuscus*). Genetic analysis has shown it to be distinct from its northern relative, and because the ranges of the two contact with only limited hybridization, both are now recognized as species. What is now considered *D. conanti* consists of several distinct genetic forms that may eventually be split into different species. Populations from South Carolina and northeastern Alabama (and presumably the intervening populations in Georgia) are genetically different from populations from northwestern Alabama and points farther north. Some authors have speculated that the northern dusky salamander may enter Georgia in the extreme northeastern corner, but there is no evidence for that.

Distribution and Habitat

The range of the spotted dusky salamander includes all of the physiographic provinces north of the Fall Line as well as the uppermost part of the Coastal Plain. Even so, these salamanders are generally restricted to regions of moderate topography. They are most abundant across the Piedmont, Ridge and Valley, and Cumberland Plateau provinces, and occur only rarely in mountainous regions. This species and the Ocoee salamander appear to replace each other ecologically, so that in places where one species occurs, the other is usually absent. Spotted dusky salamanders are found along or near the water's edge in a number of habitat types including swamps, seeps, springs, small streams, and temporary floodplain pools. They are intermediate in their moisture requirements among the dusky salamanders, being more terrestrial than the larger seal, black-bellied (*D. quadramaculatus*), dwarf black-bellied (*D. folkertsi*), and shovel-nosed (*D. marmoratus*) salamanders and more aquatic than the smaller seepage (*D. aeneus*) and Ocoee salamanders. They do not climb onto the surfaces of wet cliffs as often as members of the latter species do.

Reproduction and Development

Most of the information available on reproduction is from Alabama and Tennessee populations. Mating is thought to occur during the late summer and fall. Eggs are laid from July into early October. Females brood clutches of 13–37 eggs. Nests are usually in depressions in soil covered by leaves, wood, or a rock. Hatching occurs in 5–7 weeks, and the larval period varies from 7 months to a year. Maturity apparently occurs at about 2 or 3 years of age.

Habits

Like other dusky salamanders, spotted dusky salamanders hide during the day and emerge at night to forage. The diet encompasses a variety of invertebrate groups, including earthworms and arthropods such as spiders, mites, centipedes, millipedes, and insects. Densities of this species are often high, and dozens of individuals may occur within just a few meters (yards) along the margin of an optimal stream. Spotted dusky salamanders are palatable to a wide variety of predators, which undoubtedly include various amphibian-eating snakes such as the ringneck snake (*Diadophis punctatus*), common garter snake (*Thamnophis sirtalis*), eastern ribbon snake (*T. sauritus*), and watersnakes (genus *Nerodia*), as well as birds and small mammals such as shrews. Larvae and very small juveniles are vulnerable to predaceous insects. Like other salamanders of this genus, individuals squirm and jump to escape predators, often fleeing into the water. The tail may break as part of the struggle, and when all else fails, the salamander will turn to bite its attacker.

Conservation Status

Populations of spotted dusky salamanders are robust across much of the species' range, particularly in small streams in the Piedmont, Ridge and Valley, and Cumberland Plateau provinces. The stream-disturbing impacts of urbanization have been detrimental to populations near cities such as Atlanta, however.

Thomas M. Floyd and Carlos D. Camp

Adult dwarf black-bellied salamander, Union County (John B. Jensen)

Dwarf Black-bellied Salamander

Desmognathus folkertsi

Description

Adults of this moderately large dusky salamander species average 6.8–7 cm (2.7 in) in body length, with a range of 5.6–8.2 cm (2.2–3.2 in). Total length typically averages 11.5–12 cm (4.5–4.7 in) and may reach 15 cm (5.9 in). Males average 8–10 percent larger than females. The tail is strongly flattened from side to side, producing a sharp-bladed keel down the back. As in most other dusky salamanders, there are 14 costal grooves and a light line running from the eye to the back of the jaw. This species shares several features of its color pattern with the black-bellied salamander (*D. quadramaculatus*). These include two rows of evenly spaced, small, light-colored spots running down each side; adults with a completely black belly; juveniles with a white belly at metamorphosis that becomes mottled with gray and white prior to turning black; and at least occasionally a back pattern of alternating blotches, although this occurs more often in the dwarf species. The differences between these two forms are subtle, and they are sometimes nearly impossible to differentiate. The major difference is body size, with dwarf black-bellied salamanders being roughly 30 percent smaller at every comparable stage of development. For example, dwarf black-bellied salamanders are nearly mature and have a black belly when their body length is about 5 cm (2.0 in); black-bellied salamanders—at least

Dwarf black-bellied salamander larva, Union County (John B. Jensen)

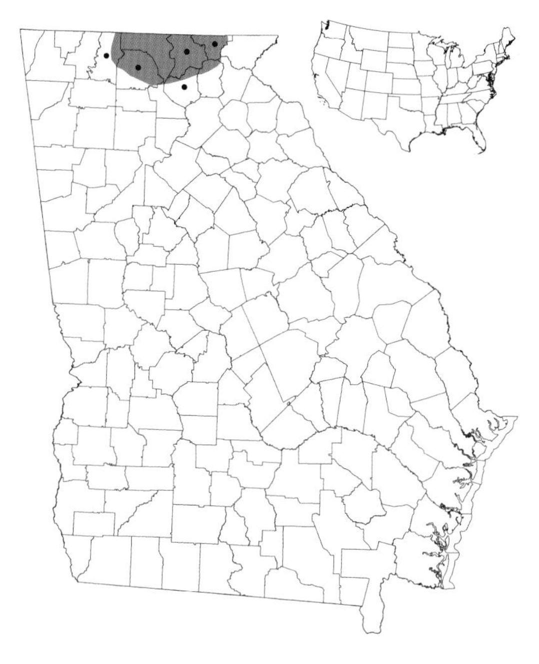

those that occur in the same streams with dwarf black-bellied salamanders—are still larvae or white-bellied juveniles at that size. Immature black-bellied salamanders have a reddish stripe down the tail fin and a body color that is often olive green; dwarf black-bellied salamanders have neither. The larvae are similar to those of the black-bellied salamander, having two rows of alternating spots extending the length of the back and two rows of tiny spots corresponding to the lateral line pores on each side. Older larvae can be distinguished from those of the black-bellied salamander by the absence of a reddish stripe running down the top of the tail. Newly metamorphosed juveniles average 35–36 mm (1.4 in) in body length.

Taxonomy and Nomenclature

The dwarf black-bellied salamander is a recently described species. Despite its similarity to the black-bellied salamander, the two occur together without interbreeding. They are closely related to each other, however, and to the shovel-nosed salamander (*D. marmoratus*), although the genetic relationships among the three are not yet fully understood.

Distribution and Habitat

The geographic distribution is incompletely known. The species is currently known from the Conasauga, Hiwassee, Nottely, Toccoa, Chattahoochee, and Savannah river systems, both north and south of the Blue Ridge divide. Dwarf black-bellied salamanders are abundant in small to moderate-sized high-gradient streams with an abundance of strewn rock. They are most at home in the same kinds of habitats that the larger black-bellied salamander favors.

Reproduction and Development

Females lay an average of 35–40 eggs in the summer. Although actual nests have not been documented, they are likely placed under rocks in running water much as those of black-bellied and shovel-nosed salamanders are. The larval period is 2 years, and another 2–3 years is likely necessary to reach sexual maturity.

Habits

Individuals hide under rocks or other types of cover during the day and emerge at night to forage. The diet has not been determined, but it undoubtedly includes an array of small invertebrates such as insects and worms. Although dwarf black-bellied salamanders live with and among much larger black-bellied salamanders, an extensive study found no evidence that the larger species preys on the smaller. Natural predators probably include snakes, birds, and small mammals.

Conservation Status

Although its known geographic range is small, the dwarf black-bellied salamander is very abundant where it occurs. Its small size makes it ideal fish bait, and "spring lizard" hunters collect these and other dusky salamanders for that purpose. The overall effect of such collecting is unknown, although harmful techniques such as using bleach to drive salamanders out of their holes could have severe consequences for local populations.

Carlos D. Camp

Adult shovel-nosed salamander, White County (John B. Jensen)

Shovel-nosed Salamander

Desmognathus marmoratus

Description

Adults of this moderately large dusky salamander average approximately 6.5 cm (2.5 in) in body length; the maximum length known is 8 cm (3.1 in). Total length is 8–15 cm (3.1–5.9 in). Mature males are on average about 10 percent larger than adult females. The tail is strongly flattened from side to side, with the upper margin forming a sharp-edged keel. The brown to black back is typically marked with two rows of irregularly edged yellowish, reddish, or gray blotches. The belly is dark gray or smoky black, never the ebony black seen in black-bellied (*D. quadramaculatus*) and dwarf black-bellied (*D. folkertsi*) salamanders. The sides usually have two rows of small, very regular spots that correspond to pores of the lateral line organ. Most individuals have at least the suggestion of a spot or stripe between the eye and the back of the jaw, although it is not as intense as in other dusky salamanders. There are 14 costal grooves, and the toe tips are usually black. Unlike other dusky salamanders, shovel-nosed salamanders have closed internal nostrils. This is a difficult character to see, however, unless the specimen is dead and was preserved with its mouth open. The black larvae have tall, paddlelike tail fins and white gills. Larvae are about 12–13 mm (0.5 in) in body length at hatching, and juveniles metamorphose at about 40 mm (1.6 in).

Taxonomy and Nomenclature

This species until recently formed its own genus, *Leurognathus*, primarily because of its unique internal nostrils. Genetic evidence has shown that it is more closely related to the black-bellied salamander than either species is to the other dusky salamanders, and it was moved into the genus *Desmognathus*. The shovel-nosed, black-bellied, and dwarf black-bellied salamanders apparently form a group of closely related species. Shovel-nosed salamanders from Georgia and South Carolina are genetically distinct from populations in other areas and have been proposed as a separate species (*D. aureatus*). A more complete resolution of taxonomic issues involving shovel-nosed salamanders awaits the results of ongoing research, which is also investigating

Shovel-nosed salamander larva, White County (Thomas M. Floyd)

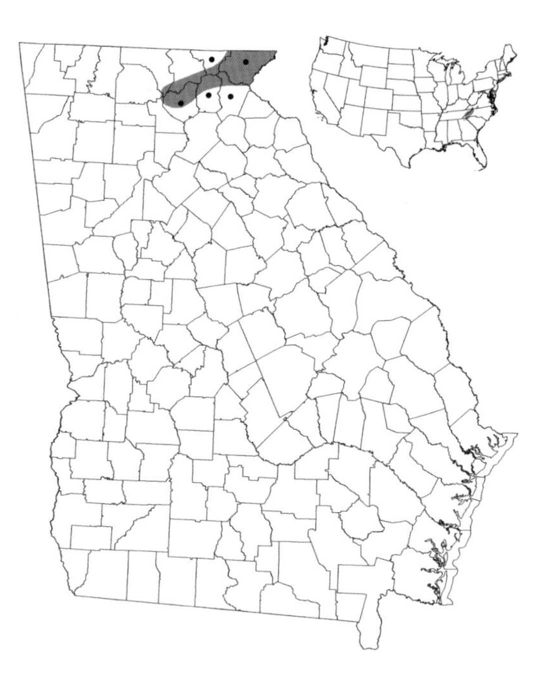

the relationships among shovel-nosed salamanders, black-bellied salamanders, and dwarf black-bellied salamanders. The description of new cryptic species from this group is possible.

Distribution and Habitat

This salamander is restricted to the eastern portion of the Blue Ridge province. It is Georgia's most aquatic lungless salamander that metamorphoses, rivaled in this respect only by the many-lined salamander (*Stereochilus marginatus*) of the Coastal Plain. Shovel-nosed salamanders occur in mountain streams ranging in size from small, first-order streams to large trout streams. Although they are occasionally found in quiet pools or in shallow water along stream edges, they are most abundant in high-current riffles, where they take cover under rocks on the bottom. Larvae, juveniles, and adults occupy similar habitats.

Reproduction and Development

The mating season has not been documented. Females lay 20–65 eggs in late summer, attaching them to the undersurface of stones in flowing water at a depth of 8–36 cm (3.1–14 in). Females remain with their eggs until hatching, which occurs in 10–12 weeks. The larval period is 2–3 years. Although the exact age at sexual maturity has not been determined, individuals remain sexually immature for at least a year following metamorphosis.

Habits

Individuals take cover under rocks on the bottom of streams during the day and emerge at night to forage on aquatic insects. In this regard their foraging strategy is more similar to that of sculpins—small bottom-dwelling fish—than to other salamanders. In turn, they are preyed on by fish, including trout. Small individuals may be eaten by large aquatic insects.

Conservation Status

Many populations have been eliminated by dams because these salamanders require fast-flowing water. Siltation caused by development-induced erosion has undoubtedly harmed many populations by plugging spaces under and among bottom rocks with sand. This species is also sensitive to pollution, and populations in neighboring states have disappeared as a result of chemical runoff. Introduced species of trout feed on shovel-nosed salamanders, but they do not appear to significantly affect populations.

Carlos D. Camp

Adult seal salamander, Union County (Dirk J. Stevenson)

Seal Salamander

Desmognathus monticola

Description

Seal salamanders are fairly large dusky salamanders; adults average 6.2–6.5 cm (2.4–2.5 in) in body length (range = 4.5–8 cm, 1.8–3.1 in) and are 7.5–15 cm (2.9–5.9 in) in total length. Males are approximately 10 percent larger than females. The front part of the tail is round, but the back two-thirds is strongly flattened from side to side, creating a sharp-edged keel along the top. The back is usually gray or brown, often with dark, wormy markings on a lighter background. Some individuals, particularly older ones, are uniformly dark. The belly is very pale, often white, and is never mottled like that of the spotted dusky salamander (*D. conanti*). The dark color of the back usually abruptly meets the light color of the belly on the sides to produce a two-toned effect. The row of irregularly shaped white spots that runs along the lower portion of each side between the legs is clearly a different pattern from the two rows of distinctly round spots seen on the black-bellied (*D. quadramaculatus*) and dwarf black-bellied (*D. folkertsi*) salamanders. A stripe runs from the eye to the back of the jaw. Each side has 14 costal grooves, and the toe tips are typically black. Larvae have white gills and vary in total length from an average of 20 mm (0.8 in) at hatching to 25 mm (1 in) at metamorphosis. The larvae have four or five pairs of spots extending along the back between the fore and hind limbs (as opposed to five to

Adult seal salamander, Gilmer County (John B. Jensen)

Seal salamander larva, Banks County (Carlos D. Camp)

Adult seal salamander, White County (Dirk J. Stevenson)

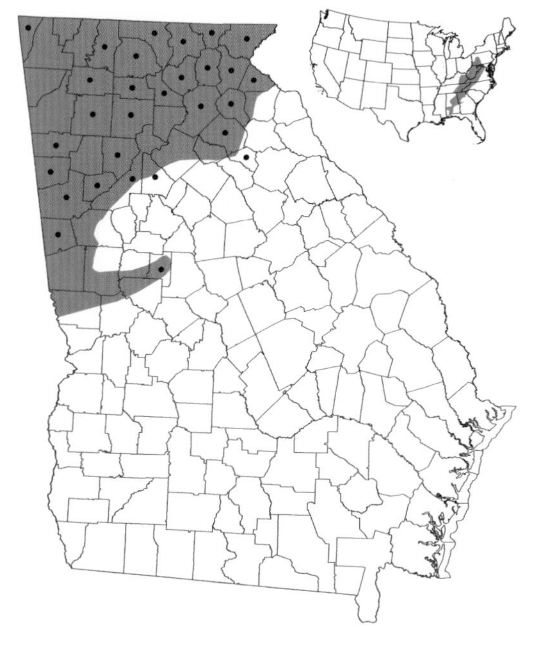

seven on Ocoee salamanders, *D. ocoee*); the pattern persists in juveniles but is lost with age. Juveniles also often have a reddish stripe extending along the top of the tail.

Taxonomy and Nomenclature

Research published to date has found relatively little genetic variation among seal salamander populations throughout the species' range.

Distribution and Habitat

Seal salamanders are an upland form and are most abundant in the Blue Ridge and adjacent areas of the upper Piedmont provinces. They are especially abundant in the western half of the Blue Ridge, often outnumbering all other streamside salamanders in their primary habitats, which include seeps and small streams in coves and ravines. This species can also be found along the edges of larger streams. Seal salamanders are more aquatic than spotted dusky or Ocoee salamanders. Their refuges tend to be under rocks or logs near the water's edge. Although they occasionally wander out into the forest under rainy conditions, they do not remain there as Ocoee salamanders do. Juveniles frequently climb on the walls of wet cliffs.

Reproduction and Development

In North Carolina, mating takes place during late summer and fall. Georgia females nest during August and deposit an average of 17 eggs, attaching them to the roof of nesting cavities under rocks, sticks, or moss or in narrow crevices of wet cliffs. The eggs hatch approximately 45 days later. As is the case with other lungless salamanders, the female attends the eggs until hatching. There are no data on the larval period of Georgia populations, but in other states it lasts 9–11 months. Individuals become sexually mature following a juvenile period of 4–6 years.

Habits

Seal salamanders are so named because they hold their head up in a manner reminiscent of a seal when resting. Individuals typically seek refuge during the day but emerge at night to forage in and alongside small streams. Larger salamanders often remain in their refuge and ambush unwary prey. The diet includes a broad array of invertebrates, primarily insects. Resident salamanders aggressively defend their refuge against intruders of their own and other salamander species. Very few natural predators have been documented but undoubtedly they include snakes, small mammals, and birds. Young seal salamanders probably also fall prey to spring salamanders (*Gyrinophilus porphyriticus*). Defense mechanisms include squirming and jumping to escape and biting as a last resort. The tail may break in a struggle, and if the number of individuals with missing or regenerating tails is an accurate indication, predators must often be satisfied with eating just that.

Conservation Status

Populations of seal salamanders are robust across Georgia's mountains and adjacent portions of the Piedmont, and the species does not appear to be threatened by the current rate of habitat alteration in the region. This species is commonly collected and sold for bait under the name spring lizards, and although populations are not likely to be significantly damaged by this practice, habitat-disturbing collection techniques may be locally harmful to this and other stream-dwelling organisms.

Carlos D. Camp

Adult Ocoee salamander, Gilmer County (John B. Jensen)

Ocoee Salamander

Desmognathus ocoee

Description

Among the smaller dusky salamanders, adult Ocoee salamanders average 4 cm (1.6 in) in body length (range = 3–5.5 cm, 1.2–2.2 in); total length is 7–11 cm (2.7–4.3 in). Males are generally 10–15 percent larger than females. The body is somewhat flattened from top to bottom, and the tail, which is triangular in cross section, usually lacks a distinct keel along the upper surface. This is a variable character, however, and a keel is approached in individuals of some populations, making it more difficult to distinguish them from spotted dusky salamanders (*D. conanti*). The base of the tail is usually wider than tall and is flattest in older adults. The tip of the tail may be extended and threadlike. There are 14 costal grooves. The color of the back is extremely variable and may include various hues of gray, brown, red, orange, and yellow. Some individuals, especially older males, are all black or dark gray. In certain high-elevation populations, a large proportion of the adults are black. There is usually an obvious pattern on the back consisting of alternating spots; these often merge into a zigzag stripe. As in other dusky salamanders, there is an obvious stripe running from the eye to the rear angle of the jaw. The red cheek patches or red legs found on some individuals may mimic the aposematic coloration of the Jordan's salamander (*Plethodon jordani*) complex. The belly is typically mottled with

Adult Ocoee salamander, Towns County (John B. Jensen)

Ocoee salamander larva, Banks County (Carlos D. Camp)

Adult Ocoee salamander, Rabun County (John B. Jensen)

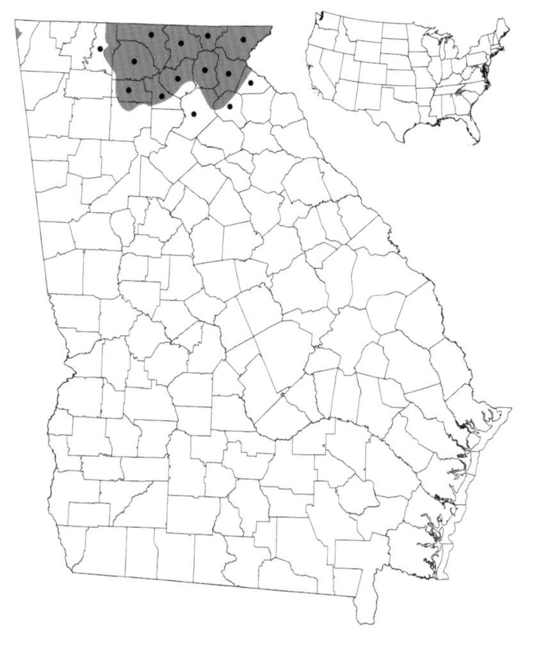

various mixtures of brown, gray, red, or yellow. Larvae are small, having total lengths of 12–18 mm (0.5–0.7 in) and 17–27 mm (0.7–1.1 in), respectively, at hatching and metamorphosis. Each larva has five to seven pairs of alternating spots extending along its back between the two sets of limbs; as in adults, the spots sometimes touch. The gills are white.

Taxonomy and Nomenclature

This species is a member of the mountain dusky salamander (*D. ochrophaeus*) complex, which has been divided into several species on the basis of genetic evidence. The species now recognized as the Ocoee salamander consists of a group of genetically divergent populations and may ultimately be split into several species.

Distribution and Habitat

Ocoee salamanders are abundant in the Blue Ridge province and in adjacent areas of the upper Piedmont. They likely occur in the Cumberland Plateau as well because there are nearby populations in the Cumberland areas of both Alabama and Tennessee. This species occurs at elevations from 225 m (738 ft) in the Piedmont to 1,400 m (4,592 ft) or higher on the tallest mountains such as Rabun Bald and Brasstown Bald. Ocoee salamanders are associated with wet substrates and occur along small streams, in seeps, on

Adult Ocoee salamander, Union County (John B. Jensen)

permanently wet cliffs, and well away from streams in the very moist forests found at high elevations and in mountain coves.

Reproduction and Development

North Carolina populations mate in late summer or fall. Females deposit an average of 15–16 eggs on the roof of nesting cavities within or beneath moss clumps, under rocks in streambeds, or beneath or within rotting logs near small streams or seeps. They may also nest in narrow crevices in wet cliffs. Eggs are laid from late July to September in Georgia populations, and the female attends the eggs until hatching. The incubation period in the North Carolina mountains may be as long as 70 days, and the larval period lasts 9 or 10 months. In the upper Piedmont of Georgia, incubation lasts 32–39 days and the larvae metamorphose in 7 months. Sexual maturity is usually reached 2–3 years after metamorphosis.

Habits

Ocoee salamanders are most active in the warmer months, although they may be found along streams even during the winter. Home ranges are small. Individuals in a population of marked animals in North Carolina moved an average of only 44–45 cm (17.2–17.6 in) between observations, which were taken approximately every 5 days over the course of 2 months. Individuals take cover in rocky crevices, under rocks and fallen logs, and in moss beds during the day, and emerge at night to forage. The diet consists of a wide variety of invertebrate prey, especially small insects. Ocoee salamanders actively defend their territories against other similar-sized salamanders. This species appears ecologically to replace the more-lowland spotted dusky salamander (*D. conanti*), although the reasons for this are unclear. Ocoee salamanders have a wide variety of predators, including ringneck snakes (*Diadophis punctatus*), northern watersnakes (*Nerodia sipedon*), common garter snakes (*Thamnophis sirtalis*), spring salamanders (*Gyrinophilus porphyriticus*), birds, and small mammals. Threatened individuals flip and jump in their efforts to flee, and turn and bite their attacker when grasped. Close combat with a predator may result in the loss of the tail, which then wiggles to attract the predator's attention away from the head and body.

Conservation Status

In spite of being among the most abundant amphibians in Georgia's mountains, Ocoee salamanders are sensitive to habitat changes that result in the loss of either substrate moisture or the food-rich layer of leaves on the forest floor. Timber removal is thus detrimental to populations of this species. The lack of wide-ranging movements by individuals combined with the rugged, isolating topography of the Blue Ridge has led to wide genetic divergence among populations. This suggests that genetically unique populations may have very tiny geographic ranges.

Carlos D. Camp

Adult black-bellied salamander, Towns County (John B. Jensen)

Black-bellied Salamander

Desmognathus quadramaculatus

Description

When its weight and length are considered together, the black-bellied salamander is the largest lungless salamander in the world. Although adult size varies from one population to the next, the largest individuals of this species occur in Georgia. Adult body length typically averages approximately 7.5–8 cm (2.9–3.1 in) but may reach 12 cm (4.7 in). Total length normally ranges from 9 to 20 cm (3.5–7.8 in) with a maximum of 22 cm (8.6 in). Males generally grow about 10–15 percent larger than females. The tail is strongly flattened from side to side and has a sharp-edged keel extending along the top. Two rows of small, light-colored, round, regularly spaced spots extend the length of each side, corresponding with pores of the lateral line. There are 14 costal grooves. The back of adults is typically dark brown, olive, or nearly black with a few irregular black markings. Occasional individuals have two rows of alternating, irregularly edged brown blotches down the back. Young individuals are often olive or reddish brown and have a dark red stripe extending along the top of the tail. The belly is white at metamorphosis, lightly to heavily mottled in young juveniles, and ebony black in older juveniles and adults. The similar shovel-nosed salamander (*D. marmoratus*) never has a completely black belly. Most adult black-bellied salamanders can be distinguished from dwarf black-bellied salamanders (*D. folkertsi*) by their greater body length (above 8 cm, or 3.1 in). Young black-bellied salamanders can often be positively identified by their reddish tail stripes, but older juveniles that have lost

Black-bellied salamander larva, Union County (John B. Jensen)

Juvenile black-bellied salamander, Union County (John B. Jensen)

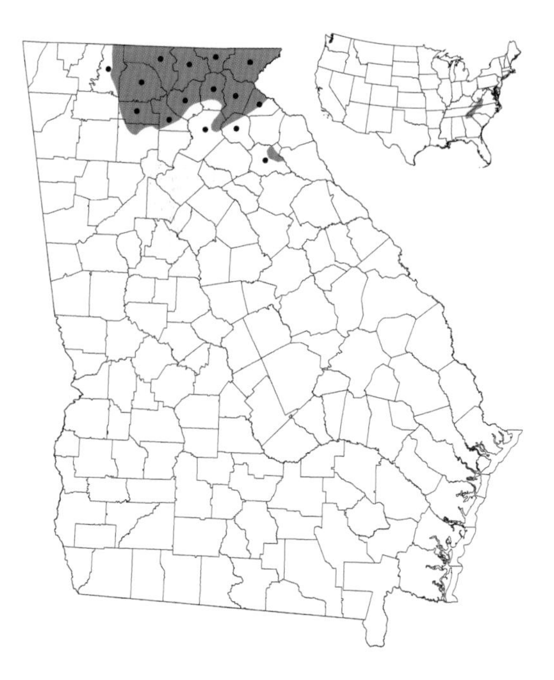

the tail stripe can be exceedingly difficult to distinguish from dwarf black-bellied salamanders. Larval black-bellied salamanders have white gills and range in color from blackish to gray or brown with two rows of alternating blotches extending the length of the back. The reddish stripe along the upper surface of the tail can be used to distinguish older larvae from larvae of the dwarf black-bellied salamander in areas where both occur. The lateral line pores are obvious on the larvae, which range in body length from about 15 mm (0.6 in) at hatching to as much as 50 mm (2.0 in) prior to metamorphosis.

Taxonomy and Nomenclature

As a species, the black-bellied salamander was long thought to represent a fairly stable form. Large dusky salamanders with all-black bellies in Union County, however, were determined to comprise two species, and the smaller one was then described as the dwarf black-bellied salamander. Furthermore, preliminary genetic analysis has indicated that some black-bellied populations are more closely related to shovel-nosed salamanders than to other black-bellied salamanders. These two facts suggest that the evolutionary relationships among black-bellied, dwarf black-bellied, and shovel-nosed salamanders have yet to be fully resolved, and the discovery of more cryptic species within this group is certainly possible.

Distribution and Habitat

These are salamanders of turbulent mountain streams in the Blue Ridge and upper Piedmont provinces. Records of specimens from the middle and lower Piedmont have been presumed to represent animals introduced through the widespread use of this species as fish bait. While most of these Piedmont records are old and the current status of any potential populations at those localities remains uncertain, a Madison County population of unknown origin persists. Although their favored habitat is small,

Underside of adult black-bellied salamander, Towns County (John B. Jensen)

permanent streams with an abundance of loose rock, black-bellied salamanders can also occur along the margins of large streams and, rarely, in small seeps and headwater streams. Individuals are occasionally encountered in large crevices of wet cliffs, particularly those kept wet by the spray of waterfalls.

Reproduction and Development

Mating probably occurs in the late summer or fall. Clutches averaging 50–80 eggs are laid during the spring or summer, attached to the bottom surface of flat stones in flowing water. Females remain with the eggs until hatching, which takes place in 1–2 months. The length of the larval period is 3–4 years. Sexual maturity is reached 3–5 years after metamorphosis. This is a long-lived species; individuals have been documented to reach at least 15 years of age.

Habits

Black-bellied salamanders can be active any time of the year if ambient temperatures are relatively mild. Individuals take cover during the day under rocks in or along the edges of streams, in burrows in the banks of streambeds, or in crevices along wet cliffs. Young salamanders often leave their retreats at night to forage along the stream. Adults tend to stick only their heads out of their refuges, preferring to ambush passing prey. Larvae feed on aquatic insects; metamorphosed salamanders eat a wide variety of aquatic and terrestrial arthropods, including mayflies, spiders, and bees. Although black-bellied salamanders have often been portrayed as predators of other salamanders, diet studies indicate that they eat other salamanders only rarely. These salamanders aggressively defend their refuge against intruders of the same species and respond to predators such as shrews, watersnakes (genus *Nerodia*), and fish by biting, twisting, and fleeing. The tail can break off in a struggle with a predator or an aggressive competitor.

Conservation Status

This is typically the most abundant salamander in small, fishless, turbulent streams. Individuals are commonly collected for use as fish bait, but this activity has not been harmful on a large scale; however, the occasional use of liquid bleach to drive salamanders from their refuges undoubtedly harms local populations.

Carlos D. Camp

Adult male southern two-lined salamander, Banks County (John B. Jensen)

Two-lined Salamanders

Eurycea bislineata complex

Description

These small, slender salamanders usually have an adult body length of approximately 3.7 cm (1.4 in) and a total length of 6.5–12 cm (2.5–4.7 in). Most have a yellowish or brown band flecked with dark spots on the back; this band is bordered by dark brown or black stripes that extend onto the tail. The underside is yellow. Depending on the population, sexually active males may have a pair of fleshy nasal cirri extending downward from the upper lip. The two recognized members of this complex that occur in Georgia are visibly different, although the tail makes up 50 percent or more of the total body length in both the southern two-lined salamander (*E. cirrigera*) and the Blue Ridge two-lined (*E. wilderae*) salamander. The southern two-lined salamander tends to be larger than the Blue Ridge two-lined salamander, but there is considerable overlap in size. The form that some recognize as the brownback salamander (*E. aquatica*) is shorter and stockier, and its tail length does not exceed 50 percent of the body length. Although color and pattern vary markedly among populations, the typical southern two-lined salamander is yellow-brown and has an unbroken stripe extending the length of the tail. Each side has 14 costal grooves and, except for many in the Cumberland Plateau and Ridge and Valley provinces, the sides are mottled with a row of light circles. Individuals from swamps

Blue Ridge two-lined salamander larva, Habersham County (Carlos D. Camp)

Adult male Blue Ridge two-lined salamander, Gilmer County. Note the cirri extending down from the upper lip. (John B. Jensen)

Southern two-lined salamander larva, Evans County (Dirk J. Stevenson)

Adult southern two-lined salamander, Early County (John B. Jensen)

of blackwater streams in the Coastal Plain may be uniformly dark brown with the light side circles appearing as small pinholes; those from the Cumberland and Ridge and Valley provinces tend to be yellow and lack the light circles. The Blue Ridge two-lined salamander is yellow to orange, has a broken tail stripe that does not extend the length of the tail, lacks the row of light circles on each side, and has 13–16 costal grooves. The sides of the brownback salamander are dusky black and have 13–14 costal grooves. Young of all the members of this complex hatch as gilled larvae 7–9 mm (0.3–0.4 in) in body length. The drab larvae have a blunt, squared-off snout; bushy reddish gills; and pairs of light spots on the back. Most larvae metamorphose before they reach a body length of 40 mm (1.6 in).

Taxonomy and Nomenclature

Two recognized members of this complex occur in Georgia: the southern two-lined salamander and the Blue Ridge two-lined salamander. Genetic data from the contact zone between these two species shows little gene exchange, and there is evidence of character displacement. A third species that may occur in Georgia, the brownback salamander, has been described, but many researchers consider it a local variant of the southern two-lined salamander. Recent genetic research supports the recognition of the brownback salamander as a separate species but restricts its range to Alabama. This complex is genetically diverse, and the future description of new species is likely.

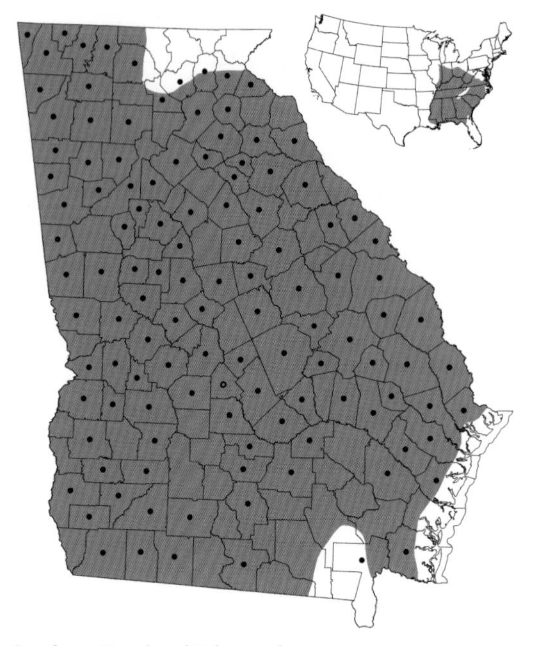

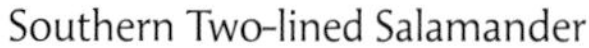

Southern Two-lined Salamander

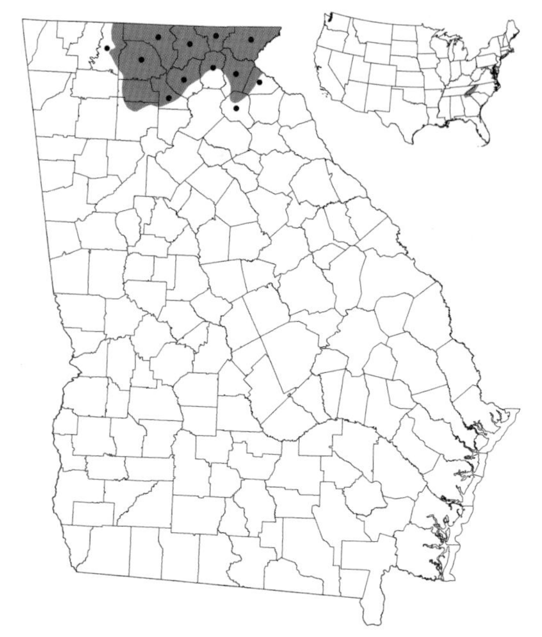

Blue Ridge Two-lined Salamander

Distribution and Habitat

Individuals of this group occur in all physiographic provinces in Georgia. As the common name suggests, the Blue Ridge two-lined salamander is largely restricted to the Blue Ridge province. The southern two-lined salamander occurs throughout the rest of the state. The two species occur together in a narrow zone in the upper Piedmont of northeastern Georgia. Brownback salamanders may occur in geologic springs of the Ridge and Valley province. Two-lined salamanders are associated with small streams, springs, seeps, swamps, and the hardwood forests surrounding such sites. Adults typically migrate to streams to reproduce in late winter and spend the rest of the year in underground burrows far from running water, although a few individuals may be found along stream margins throughout the year. Individuals in populations associated with geologic springs may be found in or near the water throughout the year. Two-lined salamanders depend on rocks and logs for cover but use sticks and vegetation when their preferred cover is scarce. Larvae are found most often in shallow areas with abundant rocky cover such as riffles.

Reproduction and Development

Courtship is not well documented, but at least some adults are thought to mate on land in the months prior to their breeding migration. Preliminary observations of some Georgia populations suggest that breeding males found out of water have long nasal cirri; those found underwater lack cirri and have a swollen head. The significance of this difference remains unknown. Migrants arrive at breeding streams, depending on the population, from late December to March. Eggs are typically attached singly to the underside of submerged rocks—or to sticks and vegetation if insufficient rocks are available—in shallow running water. Clutch size averages 50 but varies widely. Individuals generally nest singly, but multiple brooding females have been observed under the same rock. The female attends the nest until the eggs hatch, and the nest will not survive if abandoned. Eggs can be found from January to April, depending on the population, and hatching occurs after about a month. Larvae are entirely aquatic and metamorphose in approximately 1.5 years, although this may vary between 1 and 3 years. Sexual maturity is attained at 2–4 years of age.

Habits

Adults occupy underground burrows during much of the nonbreeding season, although individuals can occasionally be found under logs on the forest floor. During the mating season they are aggressive toward one another. Two-lined salamanders are most active at dusk and feed on a wide range of invertebrates, primarily terrestrial ones. They feed year-round, although intake is reduced during winter. Larvae forage on stream bottoms for small invertebrates. Adults and larvae are prey for a variety of snakes, fish, salamanders, birds, and mammals.

Conservation Status

Members of the two-lined salamander complex are common, and their populations appear to be stable.

Stacy N. Smith

Adult three-lined salamander, Wayne County (Dirk J. Stevenson)

Three-lined Salamander

Eurycea guttolineata

Description

Adults of this comparatively large brook salamander (genus *Eurycea*) are approximately 4.5–7.5 cm (1.8–2.9 in) in body length and 10–18 cm (3.9–7.0 in) in total length. Adult females are slightly larger than adult males. The back and tail are yellow or tan. Three black or dark brown stripes run the length of the body, one on each side and one down the middle of the back. The belly is heavily mottled with black (or dark gray) and white markings. The sides of the tail are darkly mottled. Each side of the body has 14 costal grooves. The larvae are streamlined with tail fins that do not reach forward past the back limbs. The pattern on the back of the larvae is similar to that on adults except that small larvae lack the dark middle stripe. The larvae can be distinguished from those of the two-lined salamander (*E. bislineata*) complex by the absence of paired light-colored spots running down the back. Three-lined salamander larvae range in body length from 11–12 mm (0.4–0.5 in) to approximately 30 mm (1.2 in).

Taxonomy and Nomenclature

For many years this species was considered a subspecies of the long-tailed salamander (*E. longicauda*). Although the ranges of the two forms approach each other in northern Alabama and northwestern Georgia, investigations of both genetic and physical characteristics of populations in this region failed to detect evidence of interbreeding.

Three-lined salamander larva, Tattnall County (Dirk J. Stevenson)

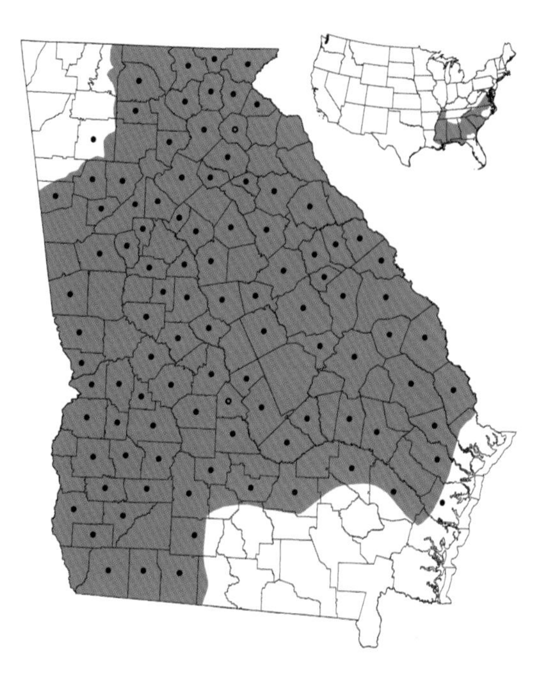

Distribution and Habitat

Three-lined salamanders occur in all of the physiographic provinces of Georgia except the Cumberland Plateau and Ridge and Valley, where they are replaced by long-tailed salamanders. They rarely occur at elevations above 900 m (2,952 ft). Individuals are occasionally found under cover along rocky streams in hardwood forests or in adjacent terrestrial sites such as crevices in rock outcrops. The greatest densities occur in bottomland forests along the saturated edges of seeps, springs, swampy streams, and ponds; in floodplains; or in other low-lying areas. The larvae are most commonly associated with weedy or leafy zones in swampy streams, ponds, or bogs.

Reproduction and Development

Courtship and mating have not been described. Eggs are laid in swampy, sometimes temporary, wetlands during the fall or early winter. Egg clutches have only rarely been observed. Larvae metamorphose within 3 months to a year after hatching, typically during the warmer months in Georgia populations.

Habits

Like other lungless salamanders, three-lined salamanders hide under logs, rocks, or other cover objects during the day and on relatively dry nights. They emerge from cover when the humidity is high to hunt and feed on various invertebrates, mostly insects. Researchers who have worked with territorial behavior in salamanders have dubbed this species "the gentleman salamander" because of the lack of territorial aggressiveness exhibited among individuals. When attacked by a predator, a three-lined salamander may curl to hide its head while waving its tail as a potential distraction. The extent of predation on these salamanders is not really known. Natural predators of transformed individuals may include snakes, small mammals, and birds; aquatic insects likely eat the larvae.

Conservation Status

Although abundant in many areas, timber harvesting in bottomland hardwoods has undoubtedly reduced the numbers of this species.

W. Ben Cash, Carlos D. Camp, and Whit Gibbons

Adult long-tailed salamander, Walker County (John B. Jensen)

Long-tailed Salamander

Eurycea longicauda

Description

This slender salamander reaches a maximum total length of about 18.5 cm (7.2 in), with the tail making up 60–65 percent of that. Body length of adults averages approximately 4.5–5 cm (1.8–2.0 in). The body is yellow, yellowish orange, or yellowish brown with scattered black or dark brown spots. Along each side the spots may coalesce and form the suggestion of a broad, irregular stripe. A series of chevronlike markings produces a herringbone pattern along the sides of the tail, and this character can be used to distinguish long-tailed salamanders from cave salamanders (*E. lucifuga*). The belly is immaculate pale yellow or cream. There are 13 or 14 costal grooves. Sexually active males have thin cirri extending downward from the snout. Larvae are approximately 10–12 mm (0.4–0.5 in) in body length and 17–19 mm (0.7 in) in total length. Their streamlined body and short gills are traits characteristic of salamander larvae that develop in streams. A tail fin is present and does not extend forward beyond the hind limbs. Recently hatched individuals are cream colored and have an immaculate belly. Older larvae develop dark mottling on the sides of the body. Recently metamorphosed individuals have an olive gray back with black sides and take on the coloration and pattern of adults within a few months.

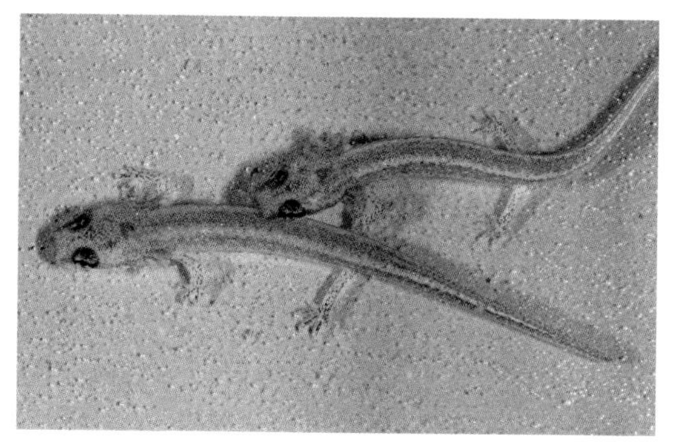

Long-tailed salamander larvae, Wise County, Virginia (John MacGregor)

Adult long-tailed salamander, Walker County (John B. Jensen)

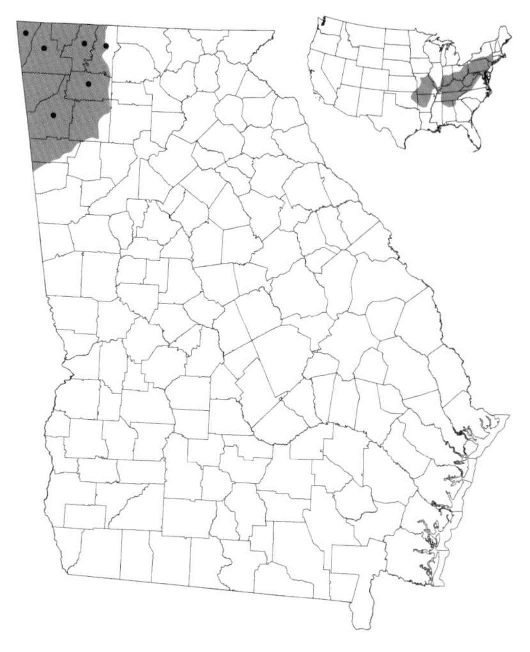

Taxonomy and Nomenclature

In the past, *E. longicauda* was considered to encompass three subspecies, two of which were recognized in Georgia: the long-tailed salamander (*E. l. longicauda*) and the three-lined salamander (*E. l. guttolineata*). *Eurycea l. guttolineata* is now considered a distinct species, *E. guttolineata*. *Eurycea l. longicauda* is the only currently recognized subspecies of the long-tailed salamander found in Georgia and shares the common name of the species.

Distribution and Habitat

As with the cave salamander, the range of the long-tailed salamander largely conforms to the distribution of karst topography in the eastern United States. In Georgia, it is generally confined to the extreme northwestern counties on the Cumberland Plateau and in the Ridge and Valley province. While long-tailed salamanders use caves, they are more likely to be found in seeps and springs and along damp margins of streams or ponds. They may also be found in forests some distance from water, especially during periods of rainy weather. Larvae have been reported from waters in caves, sinkholes, and surface streams.

Reproduction and Development

Courtship and mating take place from October through February, and females breed annually. Following mating, and continuing into early spring, eggs are deposited singly in subsurface streams, seeps, or stream-fed ponds, and may be attached to submerged substrates or to objects hanging over water. Each female lays up to 100 eggs in small groups. Because the females lay eggs in several areas, nest guarding is considered unlikely. Hatching occurs 4–12 weeks after egg deposition. The larval period is typically less than a year, although a small percentage of individuals may overwinter and transform the following summer. June and July are the months in which most larvae transform to the terrestrial stage. Sexual maturity is reached about 2 years after transformation.

Habits

Adults and juveniles make seasonal movements among streams, ponds, caves, mines, and forests. During summertime they may remain near streams, while in winter they move to underground retreats. Adults take refuge under rocks and logs during the day and emerge to forage on rainy or humid nights. Nocturnal activity peaks in the first few hours after dark. The diet includes a wide variety of invertebrates such as worms, spiders, mites, ticks, millipedes, centipedes, and insects.

Conservation Status

Although long-tailed salamanders occur only in a relatively small area of the state, they are quite common there. Their association with caves (a habitat that is very sensitive to disturbances) makes those populations potentially vulnerable.

James C. Godwin

Adult cave salamander, Walker County (John B. Jensen)

Cave Salamander

Eurycea lucifuga

Description

This slender salamander attains a total length of about 18 cm (7.0 in) and averages approximately 5 cm (2.0 in) in body length. The background color is orange, reddish orange, or yellowish orange, and black spots are scattered over the entire body, legs, and tail except for the belly, which is an unmarked yellow. Physical adaptations for living in crevices and climbing on rock faces include a somewhat flattened body; long legs; and a long, prehensile tail that makes up about 60–65 percent of the total length. The eyes bulge, and there are 14 or 15 costal grooves. Transformed cave salamanders may be confused with long-tailed salamanders (*E. longicauda*), but the latter species has a herringbone pattern along the sides of its long tail. Young cave salamander larvae are uniformly darkly pigmented with three lines of light spots on each side. As they age, they become yellowish with the typical black spotting. Transformation occurs when larvae are around 31–37 mm (1.2–1.4 in) in body length and 58–70 mm (2.3–2.7 in) in total length.

Taxonomy and Nomenclature

No subspecies are recognized.

Distribution and Habitat

The range of the cave salamander correlates with the distribution of karst topography in the eastern United States. In Georgia, it appears to be confined to the most northwestern counties

Cave salamander larva, Dade County (John B. Jensen)

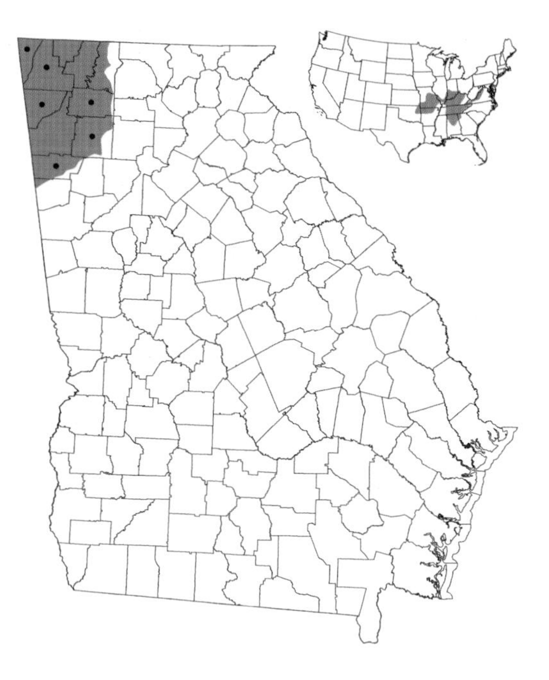

in the Cumberland Plateau and Ridge and Valley provinces. There is an old (ca. 1900) record of this species from Habersham County in the northeastern Piedmont; no specimens have been observed there since, and this record is regarded as dubious. Cave salamanders are most frequently associated with caves, sinkholes, and associated rock walls and ledges, but they may also be found in hardwood forests surrounding these karst habitats. The use of different habitats is strongly associated with weather and season. Individuals move away from caves during periods of high moisture and moderate temperatures. In Georgia, densities in caves are highest during late spring and summer. Adults in caves are generally found near the entrance within the twilight zone, but individuals may be found hundreds of meters from the cave entrance. Larvae occur in shallow, clear pools in caves and in surface streams.

Reproduction and Development

No information is available regarding egg laying in Georgia, but evidence from other states indicates that egg deposition takes place from September to February, peaking between October and January. Mating precedes egg-laying, presumably in summer and early fall. Eggs are laid singly in rimstone pools or quiet stream pools within or outside caves and are attached by a stalk to the substrate. Females may lay 49–120 eggs, each approximately 2–3 mm (0.1 in) in diameter. The larval period may be as short as 6 months or as long as 18, with larvae transforming in late summer or overwintering and transforming the following summer. Larvae have also been noted to transform in fall, winter, and spring. Sexual maturity is reached within 2 years.

Habits

Temperature and humidity shifts near cave mouths affect movements of adults and juveniles both within and into and out of caves. Dry conditions drive salamanders deeper into caves. Given sufficient moisture and moderate temperatures, Georgia cave salamanders often occur in crevices of rocky outcrops or under cover on the forest floor. They are particularly abundant in such sites during late winter and spring. On Pigeon Mountain in Walker County, individuals marked within a cave during summer were found on nearby outcrops the following February. Nocturnal activity is the norm, particularly for individuals outside caves, and they are most active on rainy nights. During the day, these salamanders take refuge in rocky crevices or under bark and debris. Cave salamanders consume a wide variety of primarily invertebrate prey, including flies, crickets, beetles, springtails, moths, mites, and other arthropods, as well as the occasional salamander. Individuals that feed in the twilight zones of caves have a greater prey base than individuals from deeper within caves. Noxious secretions from glands in the tail provide some degree of protection from predators. When attacked, cave salamanders often coil the body, elevate and wave the tail, and tuck the head beneath the body for protection. Researchers who found a negative correlation between densities of cave salamanders and long-tailed salamanders (*E. longicauda*) in areas where they occur together suggested competition between the two, but the two species occur together in certain caves in northwestern Georgia without showing such a negative correlation, indicating that the presumption of competition is premature.

Conservation Status

The status of the cave salamander in Georgia and elsewhere in its range is uncertain. Because of its close association with the fragile cave environment, however, this species may be vulnerable to disturbance.

James C. Godwin

Adult dwarf salamander, Long County (Dirk J. Stevenson)

Dwarf Salamanders

Eurycea quadridigitata complex

Description

The two species of dwarf salamanders that occur in Georgia are similar-looking, slender salamanders that resemble the southern two-lined salamander (*E. cirrigera*) but are smaller and slimmer. Adults are 2.2–3.6 cm (0.9–1.4 in) in body length and 5.5–9 cm (2.2–3.5 in) in total length. Chamberlain's dwarf salamander (*E. chamberlaini*) averages slightly smaller (10 percent) than the dwarf salamander (*E. quadridigitata*) in body length. Individuals have only four toes on each hind foot, a condition shared with only one other member of the family Plethodontidae, the four-toed salamander (*Hemidactylium scutatum*). The latter species is easily identified by its chunkier proportions and alabaster white belly with black spots. The back of the dwarf salamander has a broad, yellowish brown to chestnut stripe extending onto the top of the tail. This broad stripe is bordered by darker brown or black pigment that forms two parallel stripes that run along the upper parts of the sides and then widen to cover the sides of the tail. The back of Chamberlain's dwarf salamander is golden bronze and often has faint thin, dark chevrons like inverted Vs running across it. The parallel stripes along the upper sides are not as bold as in the dwarf salamander. The belly and surface underneath the tail are silvery white in the dwarf salamander and bright yellow-orange in Chamberlain's dwarf salamander, but a wash

Underside of adult dwarf salamander, Long County (Dirk J. Stevenson)

Dwarf salamander larva, Liberty County (Dirk J. Stevenson)

Underside of adult Chamberlain's dwarf salamander, Meriwether County (Sean Graham)

Adult Chamberlain's dwarf salamander, Muscogee County (John MacGregor)

or smudge of darker pigment may occur over the belly of either species. Chamberlain's dwarf salamander has slightly longer limbs and fewer costal grooves (15–16 versus 17–18) than the dwarf salamander. Larvae of both species have bushy red gills and a bladelike tail. Larvae hatch at around 7 mm (0.3 in) in body length and generally exceed 20 mm (0.8 in) prior to metamorphosis. Small larvae are dark gray with a row of bold white dashes along the lower sides between the fore and hind limbs; scattered white dashes on the upper sides; and small, round white specks on the sides of the tail. The pattern of older, larger larvae is similar to that of adults.

Taxonomy and Nomenclature

Studies of protein variation indicate that at least two species of this complex of possibly four species are found in Georgia. The recently described Chamberlain's dwarf salamander was once thought to represent a light-colored form of the dwarf salamander. No subspecies are presently recognized for either species.

Distribution and Habitat

The dwarf salamander is present throughout the Coastal Plain of Georgia but is most common in the swampier, low-lying southern half. Adults are most often found beneath logs, pine litter, and other debris around the margins of ponds; in swampy floodplains of rivers and streams; and in sphagnum moss and vegetation associated with acidic wetlands such as the Okefenokee Swamp. Little is known about the habitat and distribution of Chamberlain's dwarf salamander in Georgia. The species is presently known

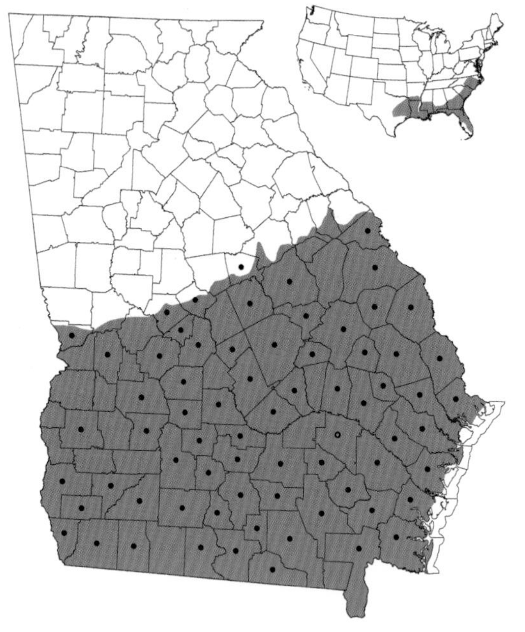

Dwarf Salamander

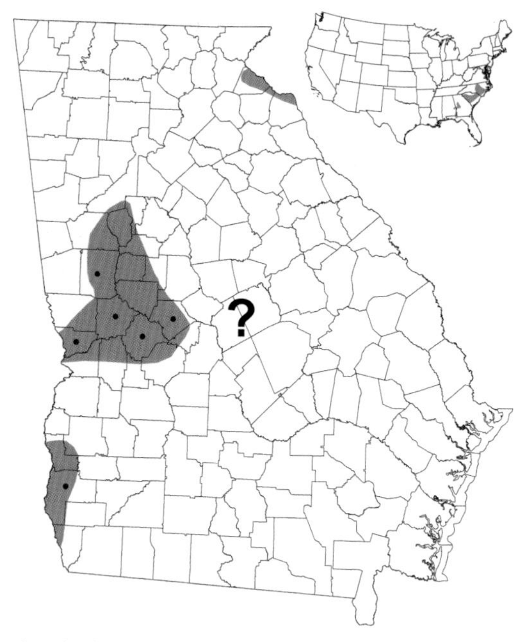

Chamberlain's Dwarf Salamander

from ravines along the lower Chattahoochee River. Individuals also have been found crossing roads during late winter near the Fall Line in the western part of the state. Records from adjacent Alabama and South Carolina indicate that this salamander is probably found farther inland in Georgia than the dwarf salamander, principally in areas of considerable topographic relief along the Fall Line in the upper Coastal Plain and lower Piedmont. Chamberlain's dwarf salamander prefers larger ravine streams with seeps along the valley sidewalls, and at least some sites are vegetated by hardwood beech-magnolia forest with sweetbay magnolia and black gum occurring in seeps.

Reproduction and Development

The breeding biology of the two species appears to be quite different. Dwarf salamanders migrate to and from breeding sites from July to October, and Chamberlain's dwarf salamanders make that migration from mid-November to late February. The smaller Chamberlain's dwarf salamander lays more eggs on average than the dwarf salamander (45 versus 40). Hatchling dwarf salamanders have been found in late December and early January. Larvae develop quickly and may metamorphose at a small size if breeding ponds dry up; if water remains in the ponds, the larval period may last 5–6.5 months. Individuals reach sexual maturity in 1 year, but females probably do not produce their first clutch until their second year. Dwarf salamander females may lay eggs in small groups, and evidence suggests that they may not brood them; females of Chamberlain's dwarf salamander brood their clutches in February. Little else is known about the growth and reproductive biology of Chamberlain's dwarf salamander.

Habits

Many aspects of the biology of this complex of salamanders that have been reported in the scientific literature may be a mixture of information involving two or more species. The dwarf salamander is commonly found with the southern dusky salamander (*Desmognathus auriculatus*) and mud salamander (*Pseudotriton *montanus*) in swampy streams and may breed in the same ponds with the flatwoods salamander (*Ambystoma cingulatum*), striped newt (*Notophthalmus perstriatus*), and ornate chorus frog (*Pseudacris ornata*). Other salamander species found with Chamberlain's dwarf salamander are the southern two-lined salamander (*E. cirrigera*), southeastern slimy salamander (*Plethodon grobmani*), and Apalachicola dusky salamander (*D. apalachicolae*). Few predators have been documented, among them pig frogs (*Rana grylio*), but such small salamanders undoubtedly have many enemies, including snakes, small mammals, and large predaceous arthropods. These salamanders feed on small invertebrates such as ants, springtails, mites, and spiders.

Conservation Status

The dwarf salamander is one of the most commonly found salamanders in southern Georgia, but little is known about the distribution, life history, or status of Chamberlain's dwarf salamander.

D. Bruce Means

Adult Tennessee cave salamander, Walker County (Matthew Niemiller)

Tennessee Cave Salamander

Gyrinophilus palleucus

Description

This large lungless salamander has a maximum body length of approximately 9.5 cm (3.7 in) and a total length of 15.5 cm (6.1 in). Tennessee cave salamanders are stygobitic and paedomorphic; that is, they inhabit only subterranean water and retain gills and other larval features throughout life. The overall coloration of the body is salmon to pale pink. Because so few specimens have been found in Georgia, the range of patterns in the state is not fully known. Individuals from Tennessee often have dark spots, but those from Alabama typically do not. The belly is pale without any markings. The eyes are small, and the head is long with a squared-off, spatulate snout. The gills are typically reduced, stubby, and pale, although under certain conditions they may be bright red and bushy. A well-developed lateral line system with noticeable unpigmented sensory pores is present, and the tail is flattened from side to side and has well-developed fins. Mature males may be as small as 6.6 cm (2.6 in) in body length, but no obvious physical differences between the sexes have been reported. Metamorphosis in the wild is rare but has been stimulated in the laboratory. Transformed individuals absorb their gills and tail fin, and the head becomes extremely flattened; the eyes are well pigmented but become enlarged and bulging. The eyes are smaller than they are in the related spring salamander (*G. porphyriticus*). The two species can be distinguished by eye diameter, which is less than 25 percent of the length of the snout in the Tennessee cave salamander and usually 30 percent or more of the snout length in the spring salamander. Newly hatched individuals are approximately 10 mm (0.4 in) in body length.

Adult Tennessee cave salamander, Walker County (Brad Glorioso)

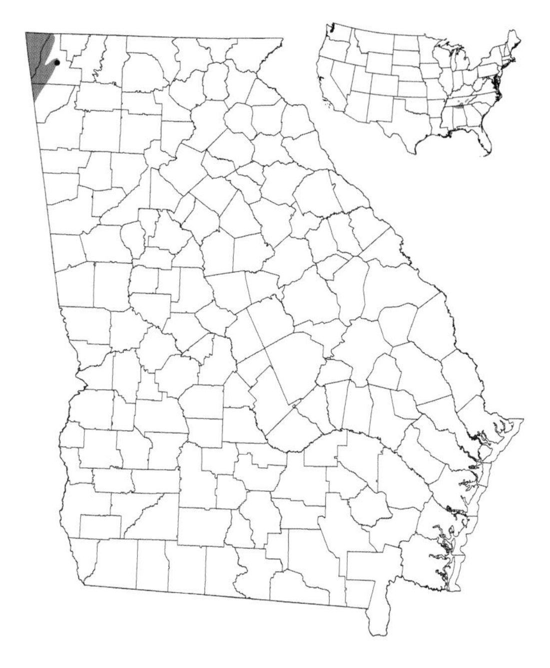

Taxonomy and Nomenclature

The Sinking Cove cave salamander (*Gyrinophilus p. palleucus*) is geographically the nearest of four subspecies to the Georgia populations, but the Georgia populations have yet to be assigned to any particular subspecies.

Distribution and Habitat

The Tennessee cave salamander is tied to the karst topography of the Cumberland Plateau of northwestern Georgia. The species has been reported from only two caves in Georgia, one on Lookout Mountain and one on Pigeon Mountain. Caves and sinkholes with an underground aquifer or stream are the typical habitats. Specimens have been reported in springs outside caves in other states, but rarely. Survival success outside the cave environment is poorly known. Within caves individuals occupy stream and rimstone pools. Substrates in these habitats are typically gravelly and rocky but may be sand or soft mud.

Reproduction and Development

Very little is known about the reproduction or development of the Tennessee cave salamander. Fragmentary data indicates that the species breeds in the fall or early winter when water levels tend to be low, thus concentrating individuals in pools. Females may not breed every year. Animals that are adapted to cave life typically have low metabolism, slow growth, a protracted juvenile stage, low reproductive potential, and a relatively long life span. The Tennessee cave salamander may also exhibit these life history characteristics. Individuals that have been induced to metamorphose are short lived.

Habits

Because there are no living plants to produce energy by photosynthesis, cave ecosystems depend on an outside source of energy, which is frequently in the form of loose organic debris (detritus) such as leaves and woody vegetation, bat guano, or the occasional animal carcass. Animals that live in this environment have to compensate for this restriction. Consequently, populations of cave animals tend to be at lower densities than aboveground populations; this appears to be the situation with the Tennessee cave salamander. Movements of Tennessee cave salamanders are cyclical; in the summer and fall they occupy cave pools and are consequently isolated from other individuals. In the winter and spring, when waters in the streams rise, the salamanders move to deeper areas. On one occasion, an individual in an Alabama cave was seen awkwardly crawling over a gravel streambed from one pool to another. Tennessee cave salamanders are not selective predators; they eat anything they can capture and swallow. The list of prey reported includes aquatic insects, crustaceans, and even other Tennessee cave salamanders. Predators include surface-dwelling fish that move upstream into caves as well as crayfish and frogs. When alarmed, these salamanders seek refuge beneath rocks or submerged ledges. Ecological associates include crayfish, other freshwater crustaceans, and the southern cavefish.

Conservation Status

As an aquatic and obligate cave-dwelling organism, the Tennessee cave salamander is particularly vulnerable to impacts that threaten the cave environment such as subterranean water pollution from agricultural runoff or septic tanks. One of the populations in Georgia is under the ownership of a private cave conservancy, which recognizes and values the biological integrity of the cave and limits access. Outside Georgia the species has shown declines. In recent surveys in Alabama this species was seen in about one-half of the historical localities that have been surveyed. The U.S. Fish and Wildlife Service has been petitioned to list the Tennessee cave salamander as a federally Threatened species. The state of Georgia lists it as Threatened.

James C. Godwin

Adult spring salamander, Union County (John B. Jensen)

Spring Salamander

Gyrinophilus porphyriticus

Description

Spring salamanders are among the larger members of the lungless salamander family. Adults typically average 7.3–8.1 cm (2.9–3.2 in) in body length and are 11–21 cm (4.3–8.2 in) in total length. Those in northwestern Georgia are larger and more slender than those in the Blue Ridge and Piedmont provinces. The tail is approximately the length of the body and strongly flattened from side to side, producing a distinct keel along its upper edge. Body color varies from reddish orange or yellowish brown to purplish brown; the latter color is more prevalent among large individuals in the northwestern corner of the state. There are small black spots or chevrons along the back. A distinct canthus rostralis runs from the eye to the nostril. This ridge is often boldly outlined with yellow and black stripes, although it is rarely pigmented in individuals from northwestern Georgia. Costal grooves range in number from 17 to 19. Larvae are long and cylindrical, reaching up to 60–65 mm (2.3–2.5 in) in body length before metamorphosis. Larval color varies from light gray to pinkish, with none of the black spotting typical of most other larval salamanders. The tail fins are relatively low and reach forward only to the hind limbs. Larvae in northwestern Georgia have a broad head and flattened, squared-off snout. In this they somewhat resemble Tennessee cave salamanders (*G. palleucus*), but the eyes of the spring salamander are much larger.

Taxonomy and Nomenclature

Four subspecies are currently recognized, and two of them occur in Georgia. The northern

Spring salamander larva, Walker County (Matthew Niemiller)

Adult spring salamander, Dade County (John B. Jensen)

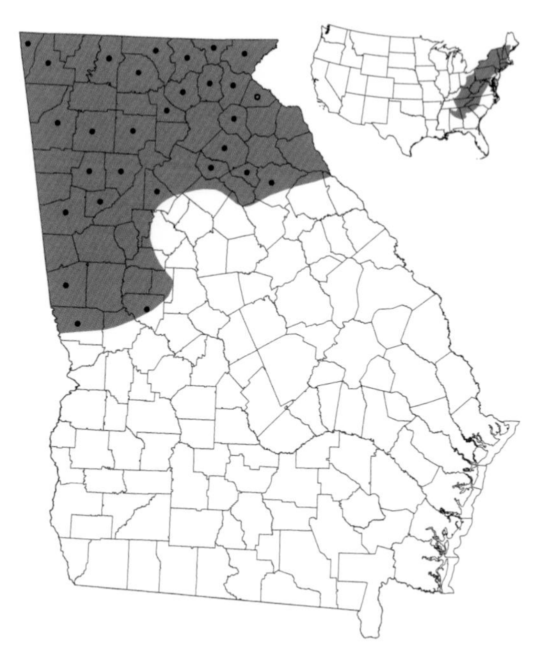

spring salamander (*G. p. porphyriticus*) occurs in extreme northwestern Georgia, and the Carolina spring salamander (*G. p. dunni*) is found in the Blue Ridge and Piedmont provinces. Populations of spring salamanders occupying various regions of the mountains of North and South Carolina differ in size and shape as well as in their reproductive life history. Salamanders from these different populations are sexually incompatible with each other, suggesting that what is now considered a single species may actually be a complex of several cryptic species.

Distribution and Habitat

Although this species occurs throughout the state above the Fall Line, it is most commonly encountered in mountainous areas. Spring salamanders inhabit springs, seeps, and small streams that flow through moist hardwood forests. In rainy weather they can be found under logs on the forest floor near such sites. The greatest densities occur in small streams in high-elevation cove forests. These salamanders also occur in caves in the northwestern part of the state.

Reproduction and Development

Information from adjacent states indicates that mating probably occurs during the winter and/or spring. Females attach their 15–100 eggs to the underside of rocks on the bottom of small streams, sometimes in deep recesses below the surface. Eggs are laid during summer and usually hatch by the onset of fall. Larval development may take as long as 4–5 years. In some populations sexual maturity is reached shortly after metamorphosis; others have a year-long juvenile period.

Habits

Spring salamanders feed on a variety of prey, including various arthropods and other invertebrates. Unlike other lungless salamanders, however, they are significant predators of other salamanders. The most commonly taken species are two-lined (*Eurycea bislineata* complex) and Ocoee (*Desmognathus ocoee*) salamanders. Cannibalism has also been observed in Georgia. Larvae are more apt to feed on arthropods, but they too occasionally eat other salamanders. Known predators of spring salamanders include northern watersnakes (*Nerodia sipedon*) and common garter snakes (*Thamnophis sirtalis*). Individuals tuck the head under the body and wave the raised tail when attacked. The skin of adults secretes noxious chemicals that repel small mammals such as shrews. The noxious quality of these secretions and the bright coloration of many individuals suggest that they belong to a group of sympatric, aposematically colored noxious species that includes the red salamander (*Pseudotriton ruber*) and the terrestrial juvenile form (eft) of the eastern newt (*Notophthalmus viridescens*). Systems like this in which several noxious aposematic species look similar are known as Müllerian mimicry complexes.

Conservation Status

This species appears to be nowhere abundant. Whether that is because individuals are highly secretive or because densities are actually low is unknown. Known locally as red dogs, spring salamanders are prized as fish bait. The fish bait trade is largely unregulated in Georgia, but bait collecting probably has no lasting effect on spring salamander populations. The greatest threat to this species is probably habitat destruction as the hardwood forests in which they live are harvested and the resulting erosion increases sediment loads in small streams. The sediment fills the spaces among and beneath stones that spring-dwelling salamanders need for cover and nest sites.

Carlos D. Camp and W. Ben Cash

Adult Georgia blind salamander, Jackson County, Florida (Dick Bartlett)

Georgia Blind Salamander

Haideotriton (Eurycea) wallacei

Description

This small, stygobitic salamander is well adapted for life in subterranean waters. Relatively few adults have been collected; the largest was 7.8 cm (3 in) in total length and 4.2 cm (1.6 in) in body length. The body is somewhat translucent and ranges from pinkish to silvery white. Small traces of dark pigment are typically expressed as tiny specks or very faint blotches. The legs are long and slender, the head is slightly flattened, and the eyes are reduced to small black spots buried under the skin. The head with its squared-off snout is proportionately longer than that of other salamanders. Adults retain larval characters including bushy red gills, skin pores associated with a lateral line sensory system, and a flattened tail fin. There are 11–13 costal grooves on each side. Although the largest specimens collected were females, differences between adult males and females have not been reported. No other similar salamander occurs in Georgia. Most collected specimens have been sexually immature individuals measuring 13–50 mm (0.5–1.9 in) in total length. Juveniles are similar to the adults except that the dark pigment is generally more pronounced in the smallest individuals.

Taxonomy and Nomenclature

Georgia blind salamanders are uniformly similar throughout their range in Florida and Georgia. No subspecies are recognized. Some taxonomists have suggested placing this species in the genus *Eurycea*.

Adult Georgia blind salamander, Jackson County, Florida (Barry Mansell)

Juvenile Georgia blind salamander, Jackson County, Florida (Robert Wayne Van Devender)

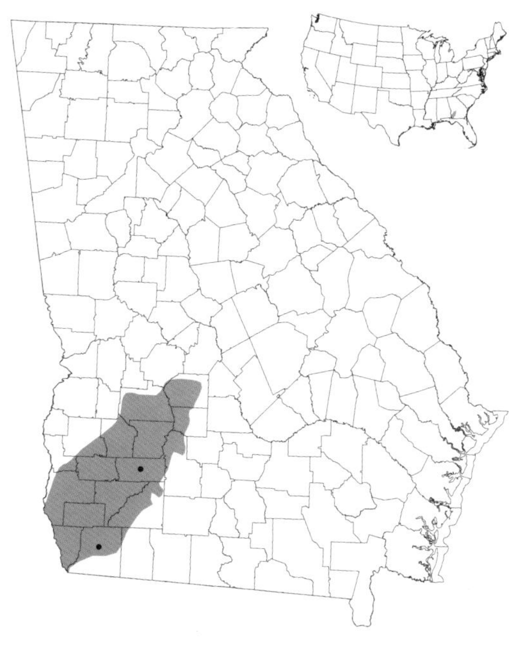

Distribution and Habitat

This completely aquatic salamander is restricted to subterranean waters of the Coastal Plain. The first specimen available to science was pumped from a 61-m-deep (200 ft) well in Albany in Dougherty County in 1939. The persistence of this species in underground waters of the Dougherty Plain was confirmed as recently as 1999 in a Decatur County cave and in two springs along the Flint River in Dougherty County. Knowledge of populations of the Georgia blind salamander is biased toward places where humans have easy access to subterranean waters in air-filled caves and sinkholes. Populations should be expected in tunnels producing spring discharge from the Floridan Aquifer into creeks and rivers of the Dougherty Plain, such as the Flint River, Spring Creek, Fishpond Drain, Ichawaynochaway Creek, and Lake Seminole. Limestone forms the walls and ceilings of the underground habitat, and fine red clay and silt lie deep on the floors. The water is usually about 18–21° C (64–70° F) and crystal clear but becomes turbid during heavy rains.

Reproduction and Development

This species is permanently paedomorphic. Very little is known about its life cycle. Eggs and hatchlings have not been discovered, although females carrying enlarged eggs with diameters up to 2 mm (0.1 in) have been found in May and November. No studies have been conducted on growth rates or on age or size at maturity.

Habits

Individuals observed in pools and underground streams move about slowly, rest on bottom sediments, or climb on underwater sidewalls and ledges. They bolt when the water is physically disturbed and commonly swim rapidly in an upward spiral, then become motionless and resettle to the bed of the pool or subterranean passageway. Georgia blind salamanders are commonly observed in caves where bats defecate over or near the water and are less common in subterranean tunnels away from cave or sinkhole entrances. These observations suggest their dependence on outside energy sources such as washed-in leaves and sticks or bat guano. This salamander species appears to feed predominantly on aquatic invertebrates. Tiny, cave-adapted crustaceans formed the bulk of the food items found in a survey of 40 stomachs; other prey items included a mite and some beetles. The Dougherty Plain blind crayfish is almost always found with the Georgia blind salamander and may represent a significant predator. Several species of fish, including the American eel, brown bullhead, and Florida chub, are also commonly found in these same subterranean waters and may prey on this salamander as well.

Conservation Status

Threats to Georgia populations of the Georgia blind salamander are pollution of the underground aquifer from agricultural runoff and drawdown of the aquifer by water use demands from center-pivot irrigation and municipalities. This species is listed as Threatened by the state of Georgia.

D. Bruce Means

Adult four-toed salamander, Walker County (John B. Jensen)

Four-toed Salamander

Hemidactylium scutatum

Description

This small salamander attains a total length of 5–10 cm (2.0–3.9 in) and a body length of 2.5–5 cm (1.0–2.0 in). Females average about 15 percent larger than males. Adults are yellowish to reddish brown along the back and tail, gray along the sides, and have an unmistakable bright white belly with black spots. The tail may be more brightly colored than the body and is constricted at the base. These salamanders are also distinctive in having four toes on each hind foot; most other salamanders have five. Four-toed salamanders have 13 or 14 costal grooves. The aquatic larvae are about 12–13 mm (0.5 in) long, yellowish brown, and have a compressed, wedge-shaped trunk and tail. Young terrestrial juveniles resemble the adults.

Taxonomy and Nomenclature

This is the only species in the genus *Hemidactylium*, and there are no recognized subspecies. Analyses of genetic and other variation across the range of the four-toed salamander, which extends across most of the eastern United States and into southern Canada, have not yet been conducted.

Distribution and Habitat

Once believed to have a very patchy distribution in Georgia, four-toed salamanders are now known to occur throughout most of the state except for much of the lower Coastal Plain. During the breeding season, adults can be found in and around boggy, forested wetlands.

Four-toed salamander larva, Menifee County, Kentucky (John MacGregor)

Adult four-toed salamander, Walker County (John B. Jensen)

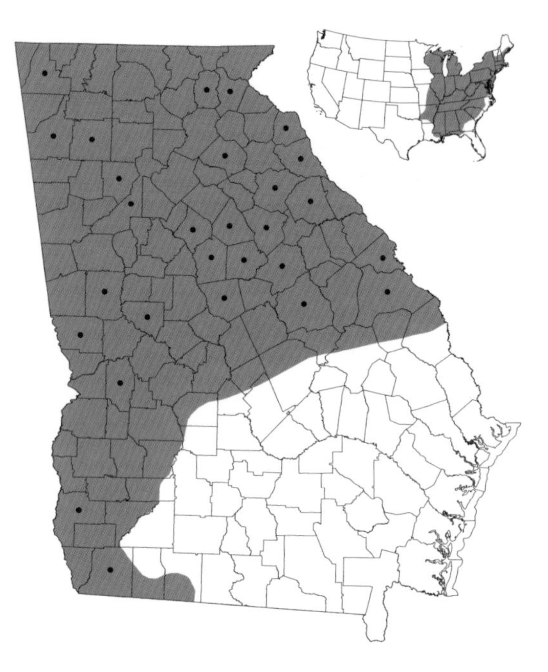

They are normally associated with shallow pools of standing water, and often the ground nearby is covered with sphagnum or other mosses. Nesting habitats include sites adjacent to or in the vegetated edges of bogs, beaver ponds, seasonal pools, older man-made lakes, and small sluggish streams. After the breeding season, four-toed salamanders lead a fossorial existence in adjacent upland deciduous or mixed forests.

Reproduction and Development

Mating probably takes place in the fall. Females migrate to breeding ponds in winter and lay eggs between January and early March. Nests contain between 20 and 80 eggs and are placed in rotting logs or moss clumps adjacent to or over water. Females often share nesting sites, creating a communal or joint nest. Those that lay eggs in solitary nests usually remain to brood them; in joint nests, some females remain and others leave their eggs behind. The incubation period is 38–62 days. Hatchlings wriggle through moss and organic debris and enter the water. The larval period lasts 3–6 weeks. Sexual maturity is reached in the second year.

Habits

For much of the year, four-toed salamanders are very secretive. Adults move away from breeding ponds in spring and spend most of their time underground until the following winter. Between January and April they may be readily found in the proper habitat under logs and other forest floor debris. Occasional individuals are found on rainy nights at other times of the year. Because their aboveground activities are mostly confined to cold-weather months, four-toed salamanders are often described as sluggish. When warmed up, however, they become more active. Recently transformed juveniles probably migrate away from breeding ponds between mid-April and June. Foods consist of small invertebrates such as insects, spiders, and mites. Information on competition with other species and predation is scarce, but predators probably include small mammals, snakes, and birds. Four-toed salamanders often use their tail as part of a defensive strategy. Individuals may wave it as a distraction, secrete noxious chemicals from it, or even voluntarily break it off. Most other salamanders only break their tail if it is grasped. The eggs are unpalatable to at least some insect predators.

Conservation Status

Four-toed salamanders are uncommon and rather spotty in their occurrence but are not as rare in Georgia as was previously believed.

Matthew J. Elliott

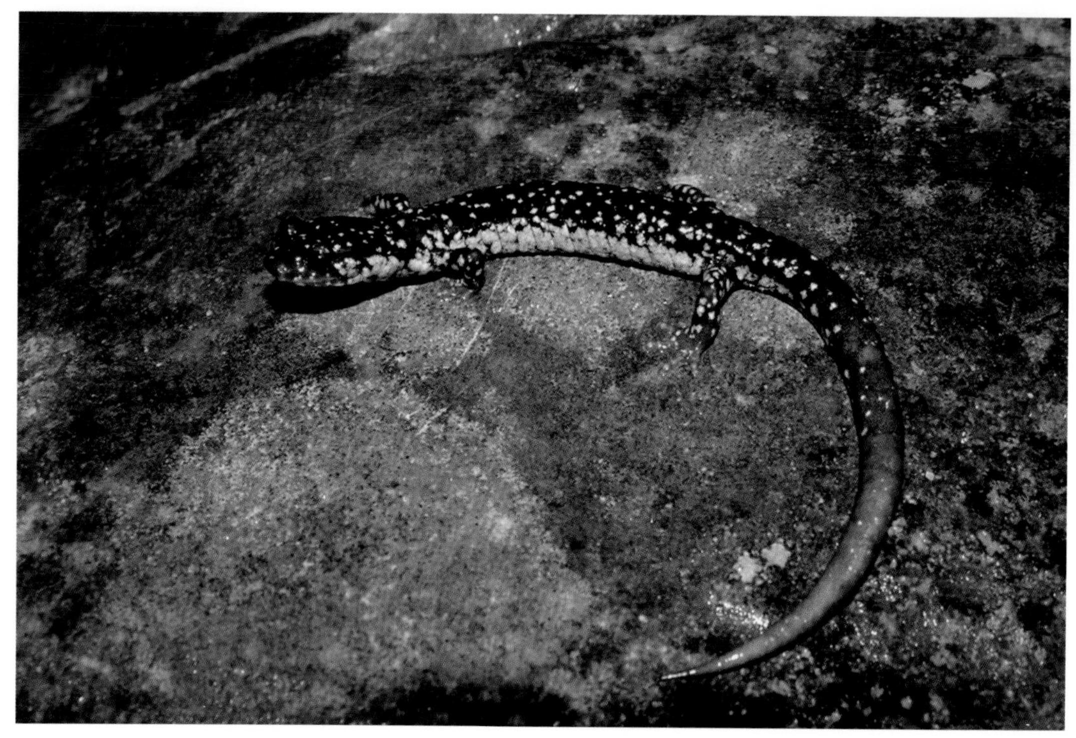

Adult northern slimy salamander, Walker County (Matthew Niemiller)

Slimy Salamanders

Plethodon glutinosus complex

Description

The slimy salamanders include some of the largest members of the genus *Plethodon*, reaching body lengths in excess of 9 cm (3.5 in) and total lengths of nearly 20 cm (7.8 in). Adult size varies from population to population and depends on the age at which individuals mature, which in turn depends on environmental temperatures. Those in upland areas become larger than those living in the Coastal Plain, and the largest ones occur in the mountains in the extreme northeastern corner of the state. The tail is round and longer than the body. The back is black and usually has small, scattered, white or brassy spots. The sides have 16 costal grooves and may be variously marked with white, gray, or yellow spots that sometimes coalesce into lichenlike bands. Those living in mountainous regions, especially at higher elevations, may have small red spots on their legs. Mature males have prominent, saucer-shaped mental glands on the chin and average slightly smaller in size than females. Young hatch as gilled larvae but quickly metamorphose into miniature versions of the adults. Hatchlings are approximately 15–18 mm (0.6–0.7 in) in body length.

Taxonomy and Nomenclature

Analysis of protein variation has shown that slimy salamanders form a series of genetically distinct neighboring groups that look similar but have been named as distinct species. Eight of them occur in Georgia:

Adult Chattahoochee slimy salamander, Union County (Thomas M. Floyd)

Adult Savannah slimy salamander, McDuffie County (Suzanne L. Collins, CNAH)

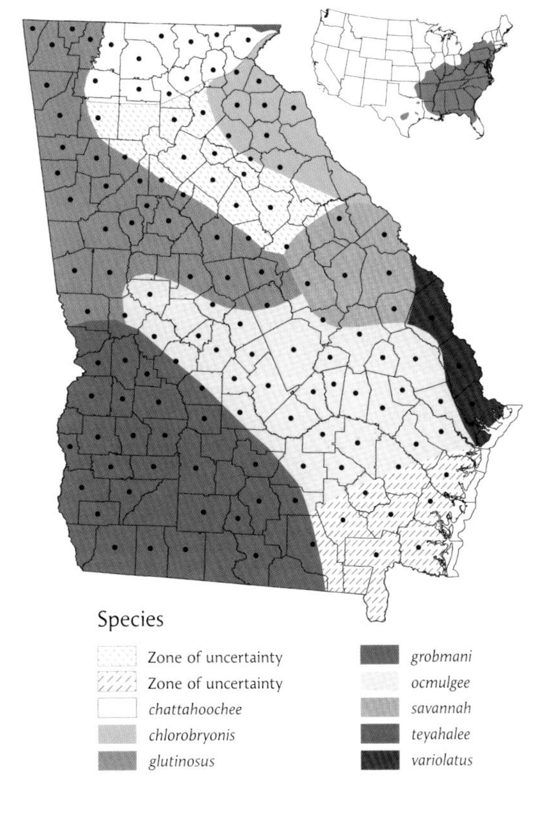

Chattahoochee slimy salamander
Plethodon chattahoochee
Atlantic Coast slimy salamander
Plethodon chlorobryonis
northern slimy salamander
Plethodon glutinosus
southeastern slimy salamander
Plethodon grobmani
Ocmulgee slimy salamander
Plethodon ocmulgee
Savannah slimy salamander
Plethodon savannah
southern Appalachian salamander
Plethodon teyahalee
South Carolina slimy salamander
Plethodon variolatus

Several of these forms readily hybridize with each other, and some of the mountain forms (i.e., *P. chattahoochee* and *P. teyahalee*) hybridize with members of the Jordan's salamander (*P. jordani*) complex. This may explain the occasional red spots on the legs of individuals at high elevations.

Distribution and Habitat

Slimy salamanders are present in hardwood forests throughout the state. Although they do occur in rather dry woods of oak mixed with pine, they are most abundant in moist forests associated with coves, north-facing slopes, and on the higher terraces of floodplains. They cannot swim and avoid areas that frequently flood. Although they spend a large part of their time underground, they are often active on the surface and depend on woody debris such as fallen logs for protective cover. Very young juveniles are rarely seen and apparently spend most of their first year underground.

Adult Atlantic Coast slimy salamander, Banks County (John B. Jensen)

Adult Ocmulgee slimy salamander, Bibb County (John B. Jensen)

Adult southern Appalachian salamander, Rabun County (John B. Jensen)

Adult South Carolina slimy salamander, Chatham County (C. Kenneth Dodd)

Reproduction and Development

No information is available on the timing of courtship in Georgia populations, but some Florida populations apparently mate during the summer, and Alabama slimy salamanders mate in the spring. Females lay an average of 15–16 heavily yolked eggs in the summer and/or fall, attaching them to the roof of a protective terrestrial cavity, usually underground. Females remain with the eggs until hatching, which occurs in 2–4 months, depending on temperature and egg size. Metamorphosis takes place within the nest and occurs 2 or 3 days after hatching. Very few egg clutches have been seen in Georgia. A newly laid clutch observed in late October in a Walker County cave consisted of 16 eggs, each 6 mm (0.2 in) in diameter. This clutch began hatching after an incubation period of 128 days. Individuals in northwestern Georgia reach sexual maturity 3 years after hatching; Coastal Plain populations in other states mature in 2 years, and this may be the case with slimy salamanders in Georgia's Coastal Plain as well.

Habits

Individuals may be active on the forest floor at any time during the year given sufficient moisture and mild temperatures. On mild, rainy nights they crawl about the forest floor or on the lower surfaces of trees or rocky outcrops foraging for insects and other invertebrates. They escape inclement conditions by moving into underground burrows, and it is likely that a significant portion of the population is

Adult southeastern slimy salamander, Baker County (John B. Jensen)

underground at any given time. In northwestern Georgia, many slimy salamanders escape the heat of summer by moving into caves. Individuals have small home ranges, sometimes no larger than 9 or 10 m (30–33 ft) in diameter. Slimy salamanders are aggressive toward each other and defend their territory against salamanders of the same and other species. The main defense against predators is a gluey, noxious substance released from glands in the tail, from which the name *glutinosus* derives. Even so armed, slimy salamanders are occasionally eaten by a variety of predators including small mammals, snakes, and birds.

Conservation Status

Generally speaking, populations throughout the state appear to be stable, and slimy salamanders are often among the most common amphibians in hardwood forests.

Carlos D. Camp

Adult red-legged salamander, Towns County (Brian D. Todd)

Jordan's Salamanders

Plethodon jordani complex

Description

Among the larger members of the genus *Plethodon*, adult Jordan's salamanders reach a body length of 7.8 cm (3.0 in) and a total length of up to 17.2 cm (6.7 in). Females are slightly larger at sexual maturity (5.3–6.1 cm, 2.1–2.4 in) than males (4.8–5.3 cm, 1.9–2.1 in). Sexually mature males develop prominent saucer-shaped mental glands underneath the chin. Georgia specimens have either large, irregular red blotches on the upper surface of their legs (red-legged salamander, *P. shermani*) or lack red markings entirely (southern gray-cheeked salamander, *P. metcalfi*). Some individuals of the latter form have pronounced white spotting along the sides of the body and head; others are uniformly black. Red-legged salamanders have some white lateral markings and a light belly. Southern gray-cheeked salamanders in Georgia have a lightly pigmented chin and dark gray belly that shades to lighter bluish black or gray above. Individuals of both species have 15 or 16—usually 16—costal grooves. Small juvenile Georgia specimens are similar in appearance to adults. The limbs of very young red-legged salamanders are marked with white rather than red. Hatchlings have not been described from Georgia but are probably similar to those of slimy salamanders (*P. glutinosus* complex), which have a body size of about 15 mm (0.6 in) and short gills that are quickly absorbed.

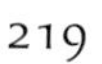

Adult southern gray-cheeked salamander, Rabun County (John B. Jensen)

Taxonomy and Nomenclature

This species complex and the closely allied slimy salamander complex together form a closely related group of species that often hybridize. Recent genetic analysis has identified seven distinct species of Jordan's salamanders. Two are known from Georgia—the southern gray-cheeked salamander and the red-legged salamander—although there is some question whether genetically pure red-legged salamanders occur in the state. The red-legged salamander hybridizes extensively with the Chattahoochee slimy salamander (*P. chattahoochee*) along the Georgia–North Carolina border on the southern periphery of North Carolina's Standing Indian Mountain, the southernmost location of purportedly pure red-legged salamanders. Extensive hybridization has been documented between North Carolina populations of the red-legged salamander and the southern Appalachian salamander (*P. teyahalee*) as well. In Georgia, the southern gray-cheeked salamander occurs together with the southern Appalachian salamander without interbreeding, but it does hybridize with the Atlantic Coast slimy salamander (*P. chlorobryonis*) just south of Rabun Bald.

Distribution and Habitat

Jordan's salamanders occur in relatively cool, moist hardwood forests in mountainous terrain in extreme northern Rabun and adjacent Towns counties. They are typically associated with high elevations but are found at their lowest elevations in Georgia and South Carolina,

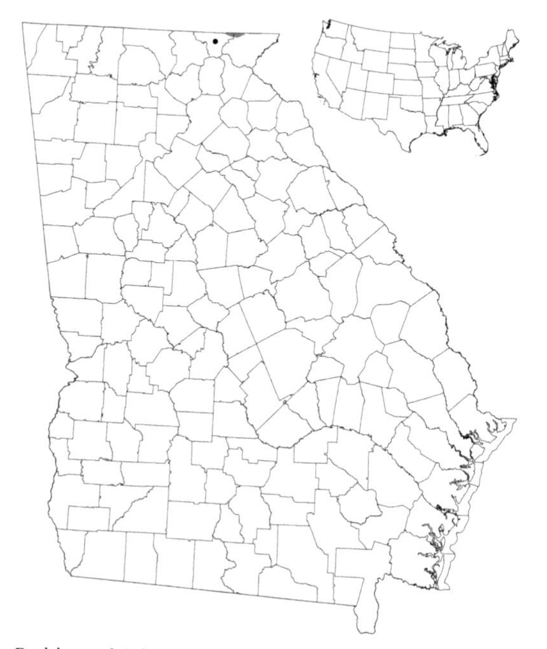

Red-legged Salamander

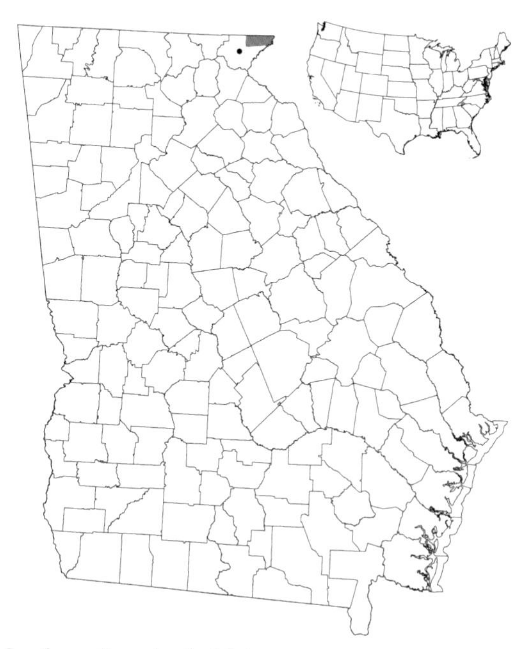

Southern Gray-cheeked Salamander

where deep gorges and high annual precipitation provide suitable conditions. Although these salamanders spend much of their lives in underground burrows, they depend on woody debris such as fallen logs and on accumulated leaf litter on the forest floor for cover and foraging.

Reproduction and Development

Courtship of members of this complex has been observed from mid-July through early October. Females are thought to lay eggs underground late in the spring or in early summer; many females appear to lay eggs in alternate years. The average number of eggs per clutch has not been reported. Typical of the genus *Plethodon*, embryos undergo virtually all of their development within the egg, hatching as gilled larvae and metamorphosing in the nest into miniature versions of adults within a few days. Sexual maturity occurs at approximately 3 years of age.

Habits

Jordan's salamanders remain in retreats underneath rocks or logs on the forest floor or in underground burrows during the day. They generally restrict their surface activity to moist forest floor microhabitats, emerging at night during rainy weather or when the relative humidity is sufficiently high to allow them to forage on invertebrates within the leaf litter. These salamanders also regularly forage above the forest floor in humid weather by climbing plants, and have been collected on trunks and low branches of shrubs up to 1 m (3.3 ft) from the ground. Individuals ambush prey from their burrow entrances when conditions are too dry to forage outside. Very young juveniles are rarely seen and apparently remain underground for most of the first year of life. Prey items include ants, millipedes, centipedes, spiders, springtails, mites, snails, and a variety of insects. Salamanders of this complex apparently have smaller home ranges in areas where they are found with members of the slimy salamander complex. Like many other salamanders of this genus, they defend their territories. The tail produces slimy secretions that are noxious and deter predators. Some researchers have suggested that individuals with red markings are exhibiting aposematic coloration. Nonetheless, these salamanders are occasionally eaten by a variety of predators, including birds, snakes, small mammals, and rarely other salamanders.

Conservation Status

Despite their restricted range and their rather peripheral distribution within Georgia, our populations are apparently secure. Georgia populations of both species occur largely on federal land (Chattahoochee National Forest) and consequently enjoy some degree of protection.

Thomas M. Floyd

Adult Pigeon Mountain salamander, Walker County (John B. Jensen)

Pigeon Mountain Salamander

Plethodon petraeus

Description

Adults of this large woodland salamander (genus *Plethodon*) reach a body length of 5.6–8.5 cm (2.2–3.3 in) and a total length of 11.5–18 cm (4.5–7.0 in). Mature males are approximately 1 cm (0.4 in) smaller in body length than females and have prominent round or oval mental glands on the chin. The entire upper surface is black with small white and brassy spots overlain by an irregularly edged brown "stripe" that typically extends from the top of the head along the length of the back and onto the tail. This reddish brown to olive brown "stripe" may be very narrow, covering only the center of the back, or as wide as the body and spreading down onto the sides. Many specimens resemble a slimy salamander (*P. glutinosus* complex) drizzled with liquid caramel or thick brown paint. The back of young juveniles less than 50 mm (2.0 in) long is marked with 3–12 alternating or opposite, brown, slightly brassy spots, which gradually coalesce with age into the brown pattern of adults. This species is unique among woodland salamanders in having expanded toe tips that aid in climbing and is similar to the green salamander (*Aneides aeneus*) in this regard. The body of Pigeon Mountain salamanders is slightly flatter and thinner than that of slimy salamanders, and the legs are longer and more slender. As with other woodland salamanders, the tail is round and is longer than the body. Most individuals have 16 costal grooves, although a few have only 15. Hatchlings have not been observed.

Juvenile Pigeon Mountain salamander, Walker County (John B. Jensen)

Adult Pigeon Mountain salamander, Walker County (John B. Jensen)

Adult male Pigeon Mountain salamander, Walker County. Note the round mental gland on the chin. (John B. Jensen)

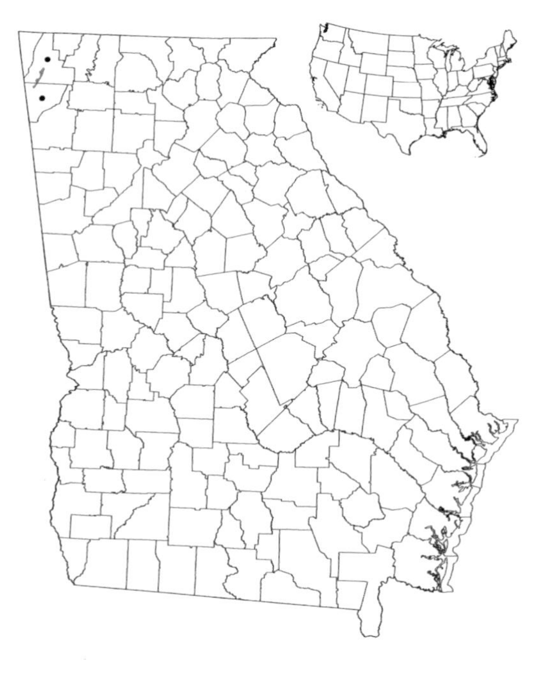

Taxonomy and Nomenclature

This species was initially discovered by Carol Ruckdeschel in 1972 while she was an employee of the Georgia Department of Natural Resources. It was described in 1988 following its rediscovery. The Pigeon Mountain salamander is a member of the slimy salamander "species group," although it is not particularly closely related to any other member.

Distribution and Habitat

Pigeon Mountain salamanders are currently known only from the eastern side of Pigeon Mountain in the Cumberland Plateau province. Populations are generally located in deep incisions in the side of the mountain known locally as gulfs. The covelike gulfs provide relatively moist conditions and are filled with extensive forests dominated by oaks, hickories, and other hardwoods. Scattered throughout the area are caves and outcroppings of sandstone and limestone. These outcroppings consist of rocks of varying sizes, often stacked atop one another, and exposed cliffs. Small, linear crevices are common in both types of outcroppings, and it is within these crevices that individuals often take refuge. During relatively warm, rainy weather these salamanders can occasionally be found under logs on the forest floor. During summer, unless the season is abnormally rainy, most individuals retire to the cool, humid atmosphere of caves. In fact, the densest aggregations appear centered along extensive rocky outcrops that connect to caves either by large openings or by small vents. Crevices associated with the sandstone caprock at the top of the mountain are also densely populated. Pigeon Mountain salamander populations are fragmented into patches coincident with the patchy distribution of optimum habitat, a spatial arrangement that ecologists refer to as a metapopulation.

Reproduction and Development

Mating apparently occurs in spring and summer, and egg-laying in summer or fall. Clutch size averages 18–19 eggs. Actual nests have yet to be discovered, but they are probably underground. Young are presumed to be similar to other members of the genus in having no aquatic larval period and in losing their gills 2–3 days after hatching. Males apparently reach sexual maturity in 3 years and females in 4 years.

Habits

These salamanders may be active anytime during the year given moderate temperatures and sufficiently moist conditions. Individuals actively forage on rainy nights, climbing across boulders and moving out into the leafy zone of the forest floor, where they feed on invertebrates such as insects, spiders, and worms. On dry nights individuals remain in their crevices with their head protruding. They readily take food when offered by human observers, and thus probably also occasionally ambush unwary invertebrates that wander too close. Individuals defend their territory against invaders of their own and other species. Northern slimy salamanders are abundant on the forest floor that fills the areas between the habitat patches occupied by Pigeon Mountain salamanders. Northern slimy salamanders are aggressively superior in laboratory trials, and their ubiquitous presence on the forest floor may interfere with the dispersal of Pigeon Mountain salamanders among habitat patches. Predators have not been documented but probably include the same types of animals that feed on other woodland salamanders: snakes, birds, and small mammals. Like slimy salamanders, individuals secrete a noxious, sticky secretion from the tail as a means of defense.

Conservation Status

Although this species is abundant in patches of optimum habitat, it is protected by the state of Georgia as a Rare species because of its tiny geographic range. The majority of the known distribution occurs within the Crockford–Pigeon Mountain Wildlife Management Area.

Carlos D. Camp and John B. Jensen

Adult southern red-backed salamander, Walker County (John B. Jensen)

Southern Red-backed Salamander

Plethodon serratus

Description

This is a very small, slender salamander. Adults range in body length from 3.5 to 5 cm (1.4–2.0 in); mature females are slightly larger than males. Males can be recognized by the small nasal cirri that project downward from the tip of their snout. Some individuals are a uniform dark gray or brown (the lead phase); others have a stripe that extends along the top of the back and continues at least partway down the tail. The stripe is usually dark brick red, although the stripes of individuals in the Ridge and Valley and Cumberland Plateau provinces may be gray, yellowish tan, or coppery brown. Unlike those of the southern zigzag (*P. ventralis*) or Webster's (*P. websteri*) salamanders, the stripe is regularly edged with small serrations that tend to alternate with the 20–22 costal grooves. The belly is speckled with black and white, producing a salt-and-pepper effect. The round tail is slightly shorter than the body. Hatchlings, which average 12.5 mm (0.5 in) in body length, spend a couple of days as gilled larvae and then rapidly develop into tiny versions of the adults.

Taxonomy and Nomenclature

At one time, Georgia populations of red-backed woodland salamanders were thought to represent a subspecies of the widely distributed red-backed salamander (*P. cinereus*), but genetic

Adult southern red-backed salamanders, Walker County (John B. Jensen)

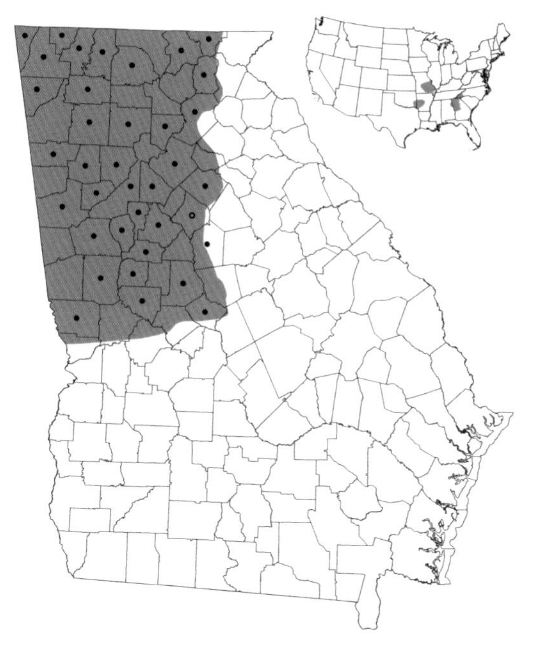

evidence indicates that populations inhabiting Georgia and other areas of the southern and central United States represent a distinct species.

Distribution and Habitat

Southern red-backed salamanders are found primarily in the western half of the Piedmont, although they are also known from the Cumberland Plateau, the Ridge and Valley Province, and the western portion of the Blue Ridge Province. They occur in hardwood forests that range from dry, oak-dominated ridgetops to moist slopes covered by oaks, yellow poplar, American beech, and other hardwoods. They are common in the Pine Mountain range of the western Piedmont in very dry oak forests that have scattered longleaf pines. Surface cover in the form of downed trees and limbs is important to provide refuge during dry conditions. During periods of extended rainfall, individuals may be found at the edges of open areas such as pastures if suitable cover is available. During extended dry periods during the winter, individuals may be found in wet leaf packs in ravines and gullies. They spend the warmer months in underground retreats.

Reproduction and Development

Mating takes place in Georgia populations during the winter. Females lay an average of five eggs in an underground cavity in the summer. Each clutch is suspended as a mass from the roof of the nesting cavity by a single stalk, and the female remains with it until hatching. The eggs hatch in 2 months. Although the hatchlings are technically larvae, they never enter water and metamorphose within 2–3 days. Individuals probably reach sexual maturity 2 years after hatching.

Habits

Southern red-backed salamanders are strongly seasonal in their activity patterns. During the late fall, winter, and early spring they are active on the forest floor, where they forage on a variety of small arthropods such as ants, termites, springtails, spiders, and mites. They move into underground retreats during April and resurface again in late October or November. Immature individuals surface before, and submerge later in the spring than, adults. Oddly, although they may be common in northwestern Georgia where caves are numerous, this species is rarely encountered in caves. Individuals aggressively defend their territory during the surface-active period but lose their aggressive tendencies during the summer. Although the geographic range broadly overlaps that of the similar-sized Webster's salamander, the two are rarely found together, possibly due to competitive exclusion, although southern red-backed salamanders do occur with southern zigzag salamanders in northwestern Georgia. Neither predators nor antipredator responses have been documented for this species, but likely predators include small snakes, birds, and small mammals such as shrews.

Conservation Status

This is an abundant salamander where it occurs. Like other terrestrial salamanders, southern red-backed salamander populations depend on the moisture provided by intact forests. A century ago, most of the tillable land of the Georgia Piedmont had been cleared for cotton monoculture. When the boll weevil epidemic brought down King Cotton in the 1920s, large tracts were abandoned and left fallow. Much of this land returned to a forested state, and a large part of it is now hardwood. As a result of this increase in available habitat, Georgia populations are probably more robust and widespread than they were 100 years ago, although some areas of suitable habitat may not yet be recolonized.

Carlos D. Camp

Adult southern zigzag salamander, Walker County (John B. Jensen)

Southern Zigzag Salamander

Plethodon ventralis

Description

This small woodland salamander (genus *Plethodon*) normally measures approximately 3.5–5 cm (1.4–2.0 in) in body length; large individuals may reach a total length in excess of 10 cm (3.9 in). The southern zigzag salamander is similar to Webster's salamander (*P. websteri*) in having 18 costal grooves and in being more stoutly built than the southern red-backed salamander (*P. serratus*). Like both of these other species, individual southern zigzag salamanders may or may not have stripes extending along the center of the back and the upper surface of the round tail. If present, the stripe usually has irregularly lobed edges, and these are often more pronounced on the front half of the body. Stripe color ranges from yellowish brown to brick red. Unlike Webster's salamander, whose stripe color is usually most intense at the base of the tail, the stripe of this species is often of similar intensity throughout its length. The background color of both striped and unstriped individuals is brown or gray and is marked with tiny, scattered, silvery and brassy flecks that may give the salamander a frosted look. The belly is mottled with varying amounts of black, white, and red. The proportion of individuals with stripes varies among populations. For example, stripeless individuals predominate on portions of Pigeon Mountain

Adult southern zigzag salamander, Walker County (John B. Jensen)

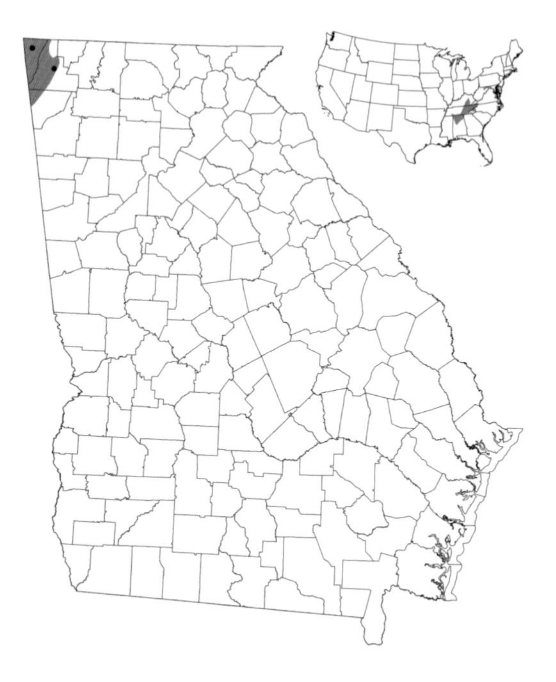

in Walker County while striped individuals are common on nearby Lookout Mountain. Males have prominent round mental glands underneath the chin and short nasal cirri projecting downward from the snout like a small mustache. Although hatchlings have not been described, those of other members of this species complex are 10–15 mm (0.4–0.6 in) in body length.

Taxonomy and Nomenclature

This species was recently split from the northern zigzag salamander (*P. dorsalis*) on the basis of genetic differences. Because the two readily hybridize across a narrow contact zone in Kentucky, however, some researchers do not recognize the southern zigzag salamander as a distinct species.

Distribution and Habitat

This species is restricted to the Cumberland Plateau, where it is abundant in hardwood forests. Individuals are most often discovered beneath woody debris such as fallen logs and limbs, but they may take cover in crevices of rocky outcrops if sufficient moisture is present there. They spend the warmer months underground and are commonly found in caves during the summer. On relatively mild, rainy, or foggy nights in the winter they forage in the leaf litter of the forest floor, sometimes climbing on the lower stems of trees and shrubs or on the vertical surfaces of boulders and rock outcrops.

Reproduction and Development

The reproductive life history of this species has not been studied in Georgia or in adjacent states. Eggs of the closely related northern zigzag salamander were found during the summer in a Kentucky cave. Southern zigzag salamanders presumably have nesting habits like those of other salamanders of this genus, whose members lay eggs in terrestrial sites. Gilled larvae, which never enter water, metamorphose into miniatures of the adults 2–3 days after hatching.

Habits

The surface activity pattern of this species is similar to that of southern red-backed and Webster's salamanders. Individuals are active on the forest floor during late fall, winter, and early spring, and move to underground retreats, including caves, for the warmer months. The diet has not been reported, but it probably consists of small invertebrates such as insects, mites, spiders, and worms. Southern zigzag salamanders occur together with southern red-backed salamanders on Pigeon Mountain, but the nature of potential interactions between members of the two species is unknown. Interestingly, southern zigzag salamanders are commonly seen in Pigeon Mountain caves where southern red-backed salamanders are rarely seen. Predators probably include small snakes such as ringneck snakes (*Diadophis punctatus*), small mammals, and woodland birds.

Conservation Status

Southern zigzag salamanders may be abundant in intact woodlands. Like other terrestrial salamanders, however, they are sensitive to forest removal.

Carlos D. Camp

Adult Webster's salamander, Upson County (John B. Jensen)

Webster's Salamander

Plethodon websteri

Description

This species rivals the seepage salamander (*Desmognathus aeneus*) and members of the dwarf salamander (*Eurycea quadridigitata*) complex for the title of Georgia's smallest salamander. Adults are normally 3.2–4.5 cm (1.3–1.8 in) in body length and reach a total length of only 6–8.5 cm (2.3–3.3 in). The tail is round and slightly shorter than the body. There are 18 costal grooves, and because the grooves are directly related to the number of vertebrae in the backbone, this species has a stockier build than the southern red-backed salamander (*P. serratus*), which has 20–22 costal grooves. Like Georgia's other small woodland salamanders (genus *Plethodon*), individuals may or may not have a stripe extending along the top of the back and tail. Unlike southern red-backed and southern zigzag (*P. ventralis*) salamanders, however, the unstriped form of this species is rare relative to the striped form. The stripe usually has irregularly wavy edges, particularly on the front half of the body, and never has the regular serrations characteristic of the southern red-backed salamander. The stripe ranges in color from yellowish brown to orange-red and is usually brightest over the hind legs and on the front part of the tail. The back and sides are usually brown with scattered silvery and brassy

Adult Webster's salamander, Upson County (John B. Jensen)

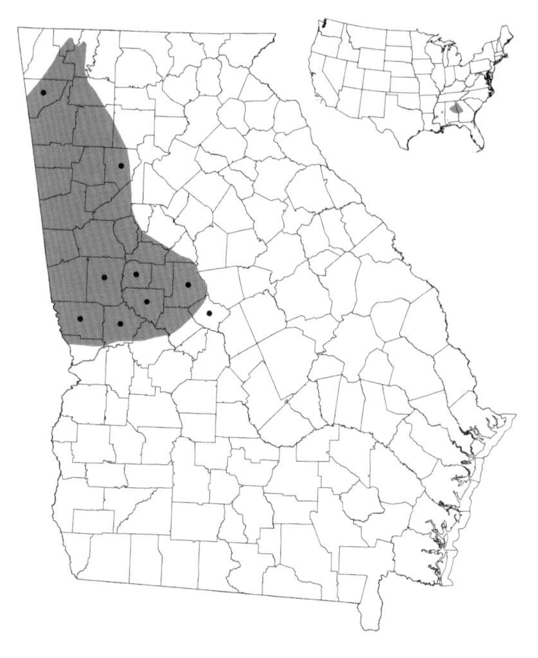

flecks. The belly is mottled with orange, white, and black. Males and females are approximately the same size, but males can be distinguished by their distinct round mental glands and short, lobelike nasal cirri. Newly hatched young have not been observed, but juveniles are 12–13 mm (0.5 in) in body length when they emerge onto the forest floor, presumably a month or more after hatching.

Taxonomy and Nomenclature

This species was first scientifically recognized in 1979. Prior to that, it and the southern zigzag salamander were thought to be the same species as the zigzag salamander (*P. dorsalis*).

Distribution and Habitat

Webster's salamanders are distributed across the western Piedmont and southern Ridge and Valley provinces. Although this species' range broadly overlaps that of the southern red-backed salamander, its distribution is more localized. It usually occurs in relatively dry, rocky woodlands dominated by hardwoods, although there may be a significant number of pines as well. Like other woodland salamanders, Webster's salamanders spend a lot of time underground. During the surface activity season they depend on fallen logs, limbs, and loose rock for protective cover. Individuals are most abundant in such sites in the winter during relatively warm periods when the ground and overlying leaf litter are wet from rainfall.

Reproduction and Development

Most of what is known about the reproductive life history is based on populations living in the western Piedmont of South Carolina. Mating occurs during the winter, and females lay an average of five or six eggs. Egg clutches have not been observed, but eggs are presumably laid during the summer in underground retreats. Both males and females reach sexual maturity 2 years after hatching. Although hatchlings have not been observed, they presumably have gills for a couple of days and then metamorphose into small versions of the adults without ever entering water.

Habits

The pattern of surface activity is strongly seasonal. These salamanders can be found on the forest floor from October into May; they spend the warmer months in underground retreats.

The diet is fairly generalized, consisting of various small invertebrates, particularly small arthropods such as mites, ants, and springtails. Individuals are aggressive toward each other during the surface-active period but become tolerant during the summer. Webster's salamanders are rarely found in the same woodlands as the sympatric southern red-backed salamander, suggesting the possibility of competitive exclusion. Predators have not been documented but probably include small snakes, shrews, and birds.

Conservation Status

Although Webster's salamanders are typically abundant, populations are localized. The rocky, often rugged terrain where they usually occur has not historically been suitable for row-crop agriculture and was thus spared the widespread deforestation associated with Georgia's former cotton-based economy.

Carlos D. Camp

Gravid adult female mud salamander, Macon County (John B. Jensen)

Mud Salamander

Pseudotriton montanus

Description

This slender to moderately stout salamander has an adult body size of approximately 5–12 cm (2.0–4.7 in) and a total length of 7.4–20.6 cm (2.9–8.0 in). Mud salamanders in the Piedmont province are larger than those in the Coastal Plain. The background color is orange-brown or reddish brown, and most individuals have round, well-separated black spots on the back and sides. Individuals from above the Fall Line have larger spots, and those from extreme southern Georgia may lack spots altogether and have irregular dark mottling or streaks instead. The ground color darkens with age, causing the spots to become obscure. The salmon to pink belly is noticeably lighter than the back. Mud salamanders in northern Georgia typically have dark markings on the belly; those from the Coastal Plain have an unmarked belly. The snout is short and blunt, and the slightly keeled tail is short relative to that of most lungless salamanders. The eyes are black or dark brown. Each side has 16 or 17 costal grooves. There are no apparent differences between the sexes. Adult mud salamanders are easily confused with red salamanders (*P. ruber*) but differ in having (1) fewer and more nearly circular spots that do not tend to touch or merge; (2) greater contrast between the back and belly; (3) dark eyes as opposed to the yellow or brassy eyes of red salamanders; and (4) in the Coastal Plain, at least, a

Mud salamander larva, Long County (Dirk J. Stevenson)

Adult mud salamander, Evans County (Dirk J. Stevenson)

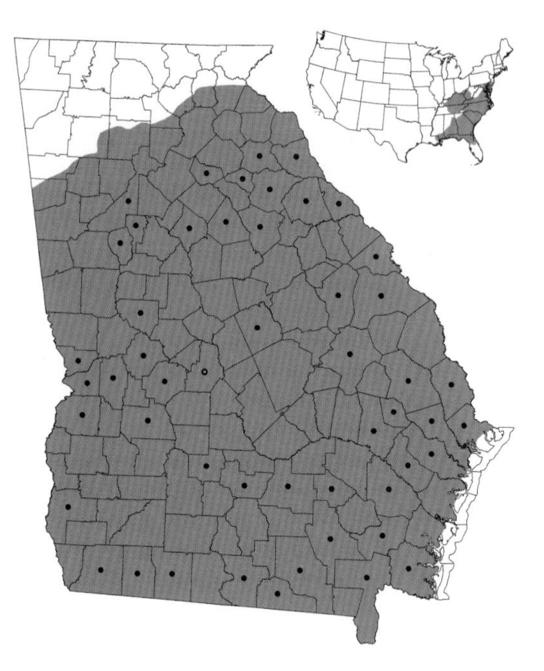

clear belly with no pattern, whereas the belly of red salamanders from this region is flecked with black. In northern Georgia, the mud salamander may also be confused with the orange to reddish spring salamander (*Gyrinophilus porphyriticus*). Mud salamander adults are smaller, have a round as opposed to angular snout, and lack the distinct canthus rostralis characteristic of spring salamanders. Larvae are light brown with small, faint spots on the back; flecks or streaks that may form reticulations along the sides; and an unpigmented belly. Hatchlings are 8–13 mm (0.3–0.5 in) in body length, and transformation occurs at 35–44 mm (1.4–1.7 in). Red salamander larvae are similar but are weakly mottled, streaked, or marked with vermiculations and typically lack the distinct spots normally found on mud salamanders. In the Coastal Plain, larvae of the many-lined salamander (*Stereochilus marginatus*) may superficially resemble mud salamander larvae in having a pattern of fine, dark longitudinal lines along the sides of the body, but many-lined salamander larvae are smaller, are often dull yellow in overall ground color, and have more costal grooves (18–19).

Taxonomy and Nomenclature

Three subspecies of mud salamanders are known from Georgia. The eastern mud salamander (*P. m. montanus*) occurs in the Piedmont. The Gulf Coast mud salamander (*P. m. flavissimus*) and the rusty mud salamander (*P. m. floridanus*) inhabit the Coastal Plain, where the rusty mud salamander is confined to the lower tier of Georgia counties bordering the Florida state line.

Distribution and Habitat

The mud salamander is widespread in the Coastal Plain and Piedmont provinces. Populations are associated with mucky seepage areas, springs, sphagnum bogs, and small streams running through hardwood forests. This lowland species is typically found within stream floodplains at elevations below 700 m (ca. 2,300 ft). Populations in the Coastal Plain are often found in blackwater creek swamps. Bay swamps and seepage areas situated at the base of extensive sandhills also harbor this species. These latter habitats usually support a canopy of mixed hardwoods with a lush carpet of sphagnum rimming the edges of seeps and rivulets. The water table is very near the ground surface in such areas, and black, saturated soils are the norm. Larval mud salamanders inhabit mucky seeps or mats of leafy organic detritus in small streams.

Reproduction and Development

Mating occurs during late summer and fall. Females' egg complement of 77–192 eggs is large for a lungless salamander. The white eggs are laid in the winter in shallow, trickling water and are attached by gelatinous stalks to dead leaves, sphagnum, or tiny roots. The female may attend the eggs. In the Piedmont and Coastal Plain, most larvae transform at 15–17 months of age, but some require an additional year to complete larval development. Males reach sexual maturity soon after metamorphosis at 2–3 years of age. Females lay eggs for the first time when 4–5 years old.

Habits

Mud salamanders are fossorial and semiaquatic. Adults are occasionally found under moss or under logs embedded in muck, especially during the winter when the water table is near the surface of the ground. In the Coastal Plain this species is often associated with the many-lined salamander and the southern dusky salamander (*Desmognathus auriculatus*). Worms and other small invertebrates are the primary prey of adults. Larvae feed on small crustaceans and other aquatic invertebrates. Documented predators include common garter snakes (*Thamnophis sirtalis*) and watersnakes (genus *Nerodia*); various birds and small mammals probably eat them as well.

Conservation Status

This salamander is relatively common, albeit somewhat locally distributed, in the Coastal Plain. There are fewer documented populations in the Piedmont of Georgia.

Dirk J. Stevenson

Adult red salamander, Catoosa County (John B. Jensen)

Red Salamander

Pseudotriton ruber

Description

Red salamanders are moderately large lungless salamanders. Adults are 10.2–18 cm (4.0–7.0 in) in total length and near 10 cm (3.9 in) in maximum body length. Body size varies geographically, and the largest individuals in Georgia are in the northern half of the state. The basic color pattern is red, orange, dull purple, or brown adorned with numerous irregularly shaped black spots on the back and sides. Color fades with age, and older individuals are often a uniform purplish brown. Individuals in the Blue Ridge have heavy black flecking on the chin. Coastal Plain specimens are generally purplish brown with tiny white flecks, especially on the snout. The sex of individuals usually cannot be determined using external characters, although females tend to be slightly larger than males. There are 16–17 costal grooves on each side. The red salamander can be distinguished from the similar mud salamander (*P. montanus*) by its yellow or brassy eyes and the irregular nature of its spots, which often touch or coalesce. The spots on the back of mud salamanders are regular in shape and usually do not touch, and the eyes are brown or black. Red salamanders lack the canthus rostralis found on spring salamanders (*Gyrinophilus porphyriticus*), and the latter species is never bright red. Red salamander larvae are uniformly light brown or lightly mottled above and dull white below. They have distinctly red gills and may reach 50 mm (2.0 in) in body length prior to metamorphosis. Hatchlings are approximately 11–14 mm (0.4–0.6 in) in body length.

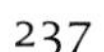

Red salamander larva, Walker County (Matthew Niemiller)

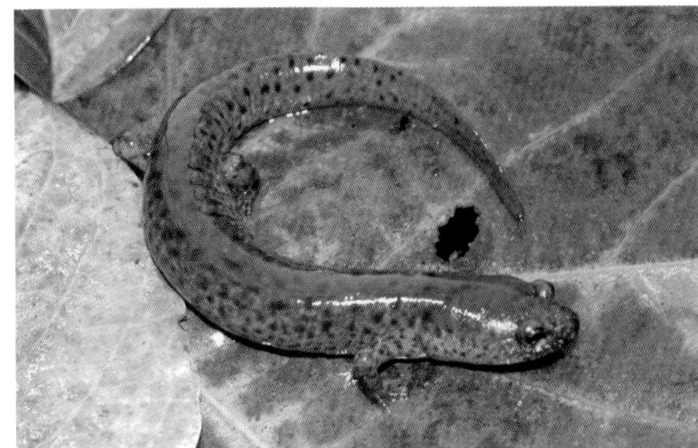

Old adult red salamander, Bibb County (John D. Willson)

Adult red salamander, Union County (John B. Jensen)

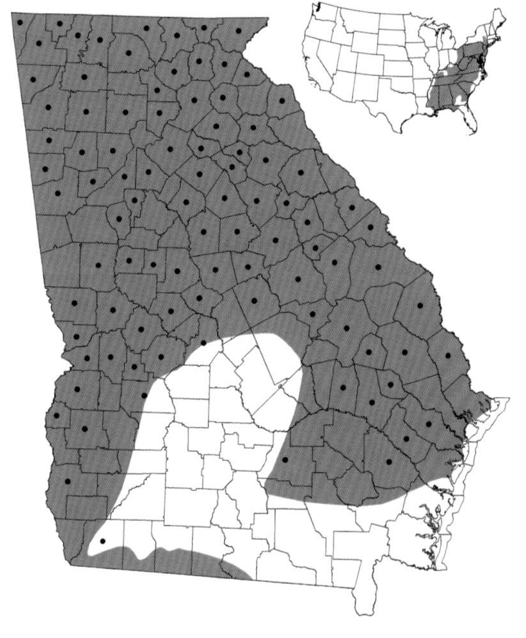

Taxonomy and Nomenclature

Three subspecies are generally recognized from Georgia. All were described prior to 1930, however, and no detailed analysis of geographic or genetic variation has ever been conducted, causing some researchers to question their validity. The black-chinned red salamander (*P. r. schencki*) and the southern red salamander (*P. r. vioscai*) occur in Georgia's Blue Ridge and Coastal Plain provinces, respectively. The northern red salamander (*P. r. ruber*) occupies the remainder of Georgia north of the Fall Line and is reported to intergrade with the southern red salamander in the Piedmont.

Distribution and Habitat

The red salamander is present throughout Georgia above the Fall Line as well as in much of the upper and middle Coastal Plain. Transformed individuals are found in both terrestrial and aquatic habitats in or near small headwater streams, slow-moving permanent springs and seeps, swamps, and spring-fed bogs. Individuals may be locally abundant in thick leaf litter in streams or swamps and within burrows along stream sides; they may also be found beneath logs or debris near aquatic sites and in adjoining forests or meadows, especially during wet weather. In northwestern Georgia, red salamanders can be common in streams that are or originate within caves.

Reproduction and Development

Courtship occurs during summer or fall. Females lay an average of 70 eggs in early fall, attaching each to the underside of a submerged rock or debris, and remain with them until they hatch in early December. Communal nests attended by several females were found under stream rocks within a Walker County cave in October; other likely nesting sites include springs, seeps, and stream banks. After hatching, larvae aggregate in slow-moving sections of springs, in seeps, and in stream pools, and are often abundant in leaf packs. The larval period varies geographically but generally lasts between 2 and 3 years. Populations in the mountains have longer larval periods than those in the Coastal Plain, which may metamorphose at 18 months. Individuals reach sexual maturity at 4–6 years of age, and males mature more rapidly than females.

Habits

Larval red salamanders feed on worms, insects, and other aquatic invertebrates. Transformed individuals are most active at night, feeding opportunistically on small animals, chiefly invertebrates. There are a few records of red salamanders feeding on smaller salamanders. During periods of rainy weather, individuals may move considerable distances—140 m (459 ft) or more—from the nearest water. Natural predators are poorly documented but presumably include woodland birds, shrews, and other mammals as well as amphibian-eating snakes such as watersnakes (genus *Nerodia*). Red salamanders produce toxic skin secretions, and the bright red or orange color of many individuals has been suggested to be aposematic. The red coloration frequently found in red and spring salamanders and in the juvenile eft stage of eastern newts (*Notophthalmus viridescens*)—all of which produce noxious secretions—has been hypothesized to represent Müllerian mimicry.

Conservation Status

The red salamander is common in Georgia and is in no immediate need of special conservation efforts.

Thomas M. Floyd

Adult many-lined salamander, Long County (David E. Scott)

Many-lined Salamander

Stereochilus marginatus

Description

This is a small, slender, relatively nondescript salamander. Adults are 3.3–4.7 cm (1.3–1.8 in) in body length and 6.4–11.4 cm (2.5–4.5 in) in total length. The small head is narrow, pointed, and flattened. The tail is short relative to that of most other plethodontid salamanders. The basic color pattern is brown or dull yellow with narrow, alternating light and dark longitudinal lines along the lower sides of the body that break up on the tail into a netlike pattern. The yellowish belly has scattered dark flecks. Like other salamanders with aquatic tendencies, the adults retain lateral line pores normally characteristic of larvae. Costal grooves number 18. There are no obvious differences between the sexes. Hatchling larvae are about 8 mm (0.3 in) in body length, and are brown above and yellow below; there are small whitish yellow spots on the head and body. Large larvae have bushy gills, possess the lateral streaks characteristic of adults, and may be mottled on the back.

Taxonomy and Nomenclature

No subspecies are recognized, and no studies of geographic variation have been conducted.

Distribution and Habitat

The many-lined salamander is found in the lower Atlantic Coastal Plain of Georgia. This species is unusually aquatic for a plethodontid salamander; the only other Georgia species that is completely aquatic as a metamorphosed adult is the shovel-nosed salamander (*Desmognathus marmoratus*) of the Blue Ridge Province. The

Many-lined salamander larva, Long County (Dirk J. Stevenson)

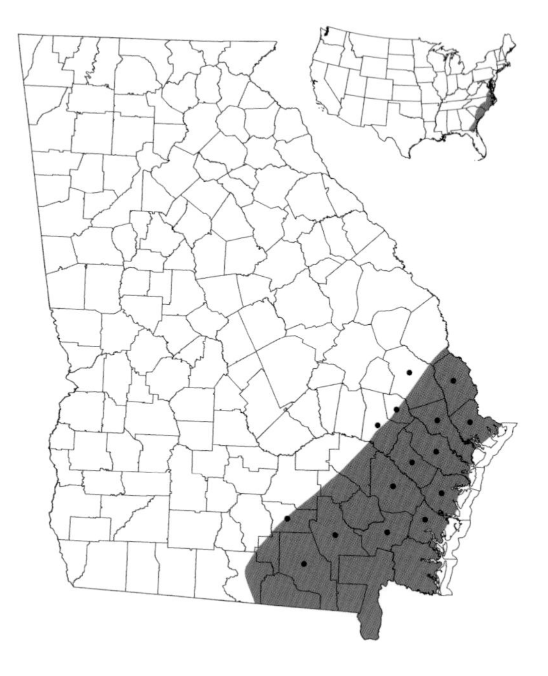

many-lined salamander inhabits forested swamps fringing slow-moving blackwater streams, shallow ditches choked with aquatic vegetation, and mucky seepage areas. Wetlands supporting populations of these salamanders are highly acidic and often have profuse growths of sphagnum. Individuals find shelter and forage in accumulations of leafy debris or in mats of sphagnum. When swamps go dry, these salamanders move under rotten logs or sphagnum mats in damp situations.

Reproduction and Development

Females typically lay approximately 20–40 eggs (range = 6–92) during the winter. The eggs are attached singly or in small adherent clumps to small roots, moss, aquatic plants, or within or under rotten logs. Nest sites are in shallow water or just above the water. Females sometimes brood their eggs, which hatch in early spring. The aquatic larvae hide in debris and vegetation and transform during the spring and summer months after a larval period of 1–2 years. Males reach sexual maturity soon after transforming; females breed for the first time at 3–4 years of age.

Habits

This small, secretive, aquatic salamander is likely to be encountered only by salamander aficionados. Larvae and adults feed on small invertebrates, including arthropods and worms. Very little is known about their longevity, movements, and many other aspects of their ecology. Mud salamanders (*Pseudotriton montanus*) and southern dusky salamanders (*Desmognathus auriculatus*) are often found at the same sites as many-lined salamanders. Predators have not been documented but probably include aquatic insects, snakes, and fish.

Conservation Status

This salamander appears to be fairly common in Georgia, although its range in the state is somewhat limited.

Dirk J. Stevenson

Waterdogs

FAMILY PROTEIDAE

Waterdogs are fairly large salamanders—the mudpuppy (*Necturus maculosus*) approaches 50 cm (20 in)—that retain gills and other larval features throughout their lives. They have four toes per hind foot rather than the five toes normal for most other salamanders. Three of the family's six species are found in Georgia. Two of the other three live in North America, and the third lives in Europe. Although they are not found statewide in Georgia, waterdogs do occur in river systems in the Coastal Plain, Piedmont, and mountains. They are permanently aquatic and are associated primarily with streams.

Adult Alabama waterdog, Sumter County (John B. Jensen)

Alabama Waterdog

Necturus cf. *beyeri*

Description

This large salamander may reach 21.6 cm (8.4 in) in total length and around 15 cm (5.9 in) in body length. Like other members of the genus *Necturus*, the Alabama waterdog does not metamorphose and retains into adulthood many characteristics normally associated with larval stages such as bushy external gills and a distinctly finned tail. Only four toes are present on each hind foot. The snout is somewhat squared-off when viewed from above. Several series of small pores on the sides of the body and on the head are part of the lateral line system. Adults may be uniformly dark above, although they are more typically lighter brown with numerous dark spots. The midline of the belly is white. Although not prominent, there are typically 16 costal grooves on each side of the body. The larvae are reddish brown with several lighter, pigment-free blotches on the tail fins. Large waterdogs are easily identified; small individuals, however, may be confused with the larvae of red (*Pseudotriton ruber*) and mud (*P. montanus*) salamanders. Waterdogs can be distinguished by counting the toes on the hind feet; the two *Pseudotriton* species have five.

Taxonomy and Nomenclature

The Alabama waterdog was long known by the scientific name *N. alabamensis*, but that name

Juvenile Alabama waterdog, Sumter County (Dirk J. Stevenson)

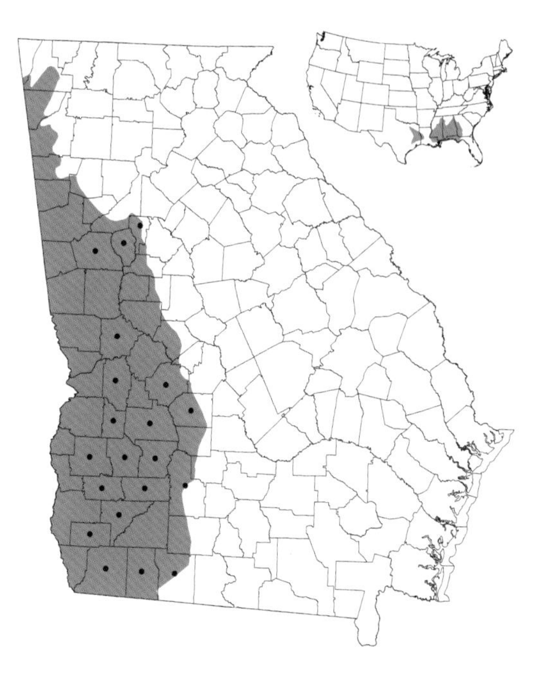

Underside of adult Alabama waterdog, Sumter County (John B. Jensen)

now refers to another species that is restricted to the Cumberland Plateau in northern Alabama. The form found in Georgia has not been given a scientific name, and until its taxonomic status can be clarified it is grouped with the closely related Gulf Coast waterdog, *N. beyeri*. Loding's waterdog is a common name proposed by some authorities, in part to avoid confusion with *N. alabamensis*.

Distribution and Habitat

Although most locality records lie below the Fall Line, this stream-dwelling salamander has been collected in the Flint River basin as far north in the Piedmont as Clayton, Coweta, and Fayette counties. In Georgia, it is known only from the Ochlockonee, Flint, and Chattahoochee river drainages. A record from the upper Coosa River in northeastern Alabama suggests that this species might also be found in the Coosa drainage in northwestern Georgia.

Reproduction and Development

Little data specific to this species is available, but the reproductive behavior is thought to be similar to that of other species of *Necturus*. Adults collected in November–February are typically in breeding condition, and courtship and mating are assumed to occur during this period. As in most other salamanders, mating results in the transfer of a sperm packet (spermatophore) that provides internal fertilization without actual copulation. Nesting takes place April–June, with each clutch consisting of 40–70 eggs that are attached to the underside of logs, rocks, or other debris. By the following December, young waterdogs are often common in accumulations of leafy debris along the stream margins. The distribution of size classes indicates that this species requires several years to reach sexual maturity.

Habits

This permanently aquatic salamander is encountered most frequently during the winter months, when it may be caught on hook and line or collected from submerged accumulations of leafy debris. Early in the fall, when leaf packs consist of recently fallen leaves, they support few invertebrates and this species is rarely found in them. As the season progresses, though, invertebrates become increasingly abundant among the leaves, and waterdogs of all sizes likewise become increasingly numerous. Alabama waterdogs are rarely encountered during the warmer months, even in seemingly suitable leaf packs. Related species are active year-round but venture from protective cover only when water temperatures are sufficiently cold (below 12° C, or 54° F) to inhibit feeding by predaceous fish, and Alabama waterdogs may do the same. Alabama waterdogs feed on crayfish, other invertebrates, and small fish. They may be eaten by a variety of predators, although their secretive nature probably provides considerable protection. An adult northern watersnake (*N. sipedon*) captured in an underwater leaf pack along the edge of Kinchafoonee Creek in Sumter County had three young Alabama waterdogs in its stomach. In a series of 14 northern watersnakes examined as part of a study of the diet of this species in Kinchafoonee Creek, 3 individuals contained the remains of Alabama waterdogs. These observations suggest that at least one species of predator forages for waterdogs within leaf packs, but the extent of such predation is not known.

Conservation Status

This species is often abundant in sand-bottomed Coastal Plain streams.

Paul E. Moler

Adult mudpuppy, Loudon County, Tennessee (Dick Bartlett)

Mudpuppy

Necturus maculosus

Description

The mudpuppy is the largest of the North American aquatic salamanders collectively known as mudpuppies and waterdogs. Adults in the northern part of the range can reach a total length of 43.6 cm (17.0 in), but those in the southern part (presumably including Georgia) rarely exceed 33 cm (12.9 in). The smallest adults are about 20 cm (7.8 in) long. The body length is typically 13.5–22 cm (5.3–8.6 in), and the tail length is 30–35 percent of the total length. Adults are rusty red to brownish gray, usually with some bluish black on the back that may form faint stripes that are traces of the juvenile pattern. A dark facial stripe runs through each eye, and the snout is bluntly squared. The belly is speckled gray with a few large, dark spots. Colors fade with age, and some older animals may be almost black. The adults are paedomorphic. Their conspicuous external gills, located just behind the head, are somewhat floppy and suggestive of canine ears, and the colloquial names mudpuppy and waterdog are thought to have sprung from this resemblance. The gills are large, maroon, and feathery in individuals from poorly oxygenated water; those from more oxygenated water have smaller, less conspicuous gills. The well-developed, slender legs have four toes on each foot. All other salamanders occurring in streams of northern Georgia have five toes on each of the hind feet. The body of the mudpuppy is cylindrical and not as flattened as that of some members of the genus, but the tail is strongly flattened from side to side. There are 15–16 relatively indistinct costal grooves on

Juvenile mudpuppy, Cannon County, Tennessee (Matthew Niemiller)

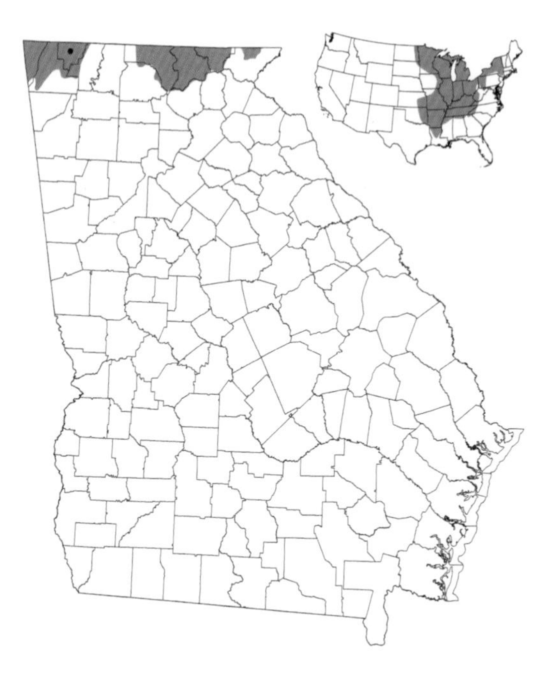

each side. Several series of small pores on the sides and head are part of the lateral line organ. Young mudpuppies have broad, dark stripes down the back that are bordered by yellow stripes. New hatchlings are 22–23 mm (0.9 in) in total length.

Taxonomy and Nomenclature

The genus *Necturus* has been a challenge to herpetologists and taxonomists, and at least one Coastal Plain waterdog remains to be named, but there is little question as to the validity of this species. Two subspecies are currently recognized; the form in Georgia is the common mudpuppy (*N. m. maculosus*). The common name for the mudpuppy can be confusing because in addition to "spring lizard," rural people have long used the terms "mudpuppy" and "puppy dog" to refer to salamanders in general, and pet shops sometimes sell larval tiger salamanders (*Ambystoma tigrinum*) as mudpuppies or waterdogs.

Distribution and Habitat

Georgia has not been adequately surveyed for this species, and there is only one documented record—from Little Chickamauga Creek. Based on its habits in adjacent states, the mudpuppy is expected to occur in the larger streams of the Tennessee River drainage in the northernmost tier of counties. Mudpuppies are totally aquatic and may exploit a variety of habitats ranging from free-flowing streams to sluggish backwaters to artificial impoundments. In northern portions of the range, mudpuppies occur in weedy ponds, but Georgia populations are probably limited to larger streams and associated reservoirs. Mudpuppies require shelter such as logs, rocks, and leaf packs and may be found in deeper waters than those in which other waterdogs can live. The deepest documented occurrence is 27.4 m (90 ft) in Lake Michigan.

Reproduction and Development

Courtship occurs in the fall, and fertilization is internal. In the spring, females firmly attach each egg to the underside of a rock in small clutches of 30–100. The female remains with the nest until hatching, which occurs in 5–9 weeks. Maturity is reached about 4–6 years after hatching. This is a long-lived species, and individuals may be reproductively active for more than 25 years.

Habits

Mudpuppies are most active from late fall through spring, and individuals are infrequently encountered during the summer. Although they can swim well, mudpuppies tend to live on or near the bottom, burrowing under rocks and logs and hiding in crevices. Submerged leaf packs may be important foraging and cover habitat for juveniles. Mudpuppies are generally nocturnal, although they may be active during the day in murky water. Foraging activity tends to increase following rain, and mudpuppies are known to roam beyond the normal stream banks during high water to forage in flooded areas. These salamanders are opportunistic feeders and eat whatever they can catch, including crayfish, aquatic insects, worms, and small fish, relying heavily on their sense of smell to find prey. Predators include river otters, watersnakes (genus *Nerodia*), aquatic turtles, and large fish.

Conservation Status

The status of populations in Georgia is currently unknown. Although they can be locally common in other parts of their range, mudpuppies have disappeared from many locations in eastern North America. Water pollution and siltation have adversely affected the habitats of virtually all native aquatic fauna throughout the Southeast. Mudpuppies are sometimes caught by fishermen and then killed because they are either believed to be detrimental to fishing or assumed to be dangerous.

Mark A. Bailey

Adult dwarf waterdog, Crawford County (John B. Jensen)

Dwarf Waterdog

Necturus punctatus

Description

The smallest of the North American waterdogs, adult dwarf waterdogs reach a body length of 7.5–12 cm (2.9–4.7 in) and a total length of 11.5–19.1 cm (4.5–7.5 in). This permanently aquatic salamander has three pairs of conspicuous reddish gills throughout life. The head and snout are noticeably flattened from top to bottom, and there are four toes on all four feet. This moderately slender waterdog typically lacks a distinct pattern or markings. Some individuals have numerous pale or golden flecks over the back and sides; otherwise the back is uniformly slate gray to brown, black, or purplish black. Dwarf waterdogs are gray and unspotted below, with the central part of the belly being a clear bluish white. There are 14–16 costal grooves. Breeding males have a swollen vent and two enlarged cloacal papillae in the vent. Females have proportionately longer tails than males. Young larvae, which have been found as small as 28 mm (1.1 in) in total length, are brown above but are otherwise similar to adults.

Taxonomy and Nomenclature

No subspecies are recognized. A spotted form of this salamander known from several river drainages in the Carolinas is genetically similar to the typical unspotted form.

Distribution and Habitat

The dwarf waterdog inhabits small to medium-sized (i.e., first- to third-order) perennial streams and rivers of the Coastal Plain, ranging from the Savannah River watershed south to the Altamaha River drainage. In North and South Carolina, the range extends into the lower Piedmont, and Georgia populations may be similarly distributed. This species inhabits

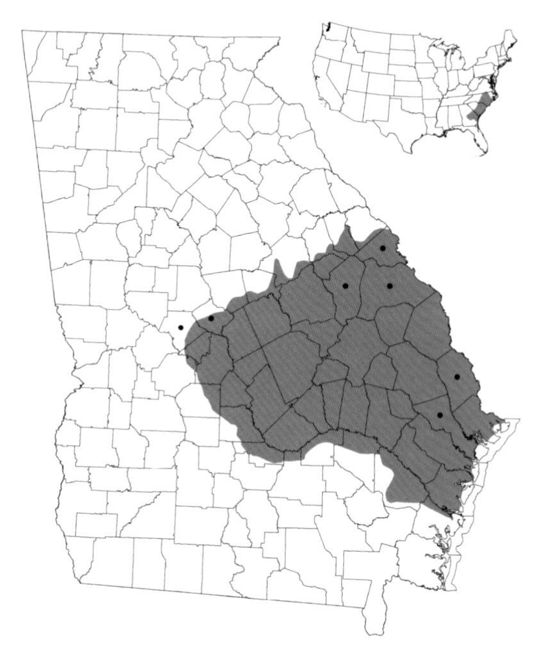

Adult dwarf waterdog, Richmond County (Ken H. Boyd)

slow-moving, meandering streams that have a substrate of sand or a mix of silt, sand, and leaf litter. They will use sandy sections but occur most frequently in parts of the stream that are deep and mud bottomed. Individuals shelter in places where debris accumulates including leaf packs, logjams, and beneath undercut banks.

Reproduction and Development

The reproductive biology of this species is poorly known. Dwarf waterdogs breed in the winter, and females lay 15–55 (average = 20–40) eggs in the spring. No nests have been discovered in the wild. Clutch size is positively correlated with female body size. If they are similar to other waterdog and mudpuppy species, the females attach their eggs to the bottom of submerged logs or objects embedded in the stream bottom and remain with them until they hatch in 2–3 months. Sexual maturity is reached in 5 years. Because other salamanders that mature at that age live at least 10–15 years, it is likely that this species does as well.

Habits

During the winter, dwarf waterdogs are prone to shelter in areas where stream debris accumulates, particularly leaf packs. Other mudpuppy species reduce their activity to coincide with increased activity by predatory fish; this may also be true for the dwarf waterdog. Adults and juveniles of this permanently aquatic species feed on a wide variety of aquatic invertebrates that inhabit the leafy debris in which they live. Prey items include crustaceans, earthworms, snails, aquatic insects, and occasionally other salamanders.

Conservation Status

This is one of Georgia's least-known salamanders. There are very few museum specimens and few known localities for this species in the state. The primary threats are pesticide and herbicide runoff, increased siltation, and other anthropogenic factors that negatively affect water quality and stream integrity.

Dirk J. Stevenson

Newts

FAMILY SALAMANDRIDAE

Georgia's two species are small salamanders, usually not reaching 13 cm (5.2 in) in total length. Newts differ from other salamanders in having rough skin and lacking the vertical costal grooves on the sides. The juvenile forms, called efts, can be bright red or orange. Like those of other salamanders, the larvae have external gills. This family occurs throughout Georgia.

The more than 50 species that make up the Salamandridae are distributed in North America, Europe, northern Africa, and Asia. Adults are aquatic pond dwellers, but the efts are terrestrial and live in a variety of habitats near the ponds where adults are found. Newts are known for their toxic skin secretions and aposematic coloration. Some populations may be paedomorphic.

Adult striped newt, Bryan County. Note the near absence of a tail fin and slightly granular skin, indicating that this individual only recently entered the wetland. (Dirk J. Stevenson)

Striped Newt

Notophthalmus perstriatus

Description

The striped newt is a relatively small salamander; adults range from 3 to 4.2 cm (1.2–1.6 in) in body length and from 5.1 to 10.4 cm (2.0–4.1 in) in total length. Metamorphosed adults are olive green or brown with a red stripe extending along each side of the body. The stripes are usually continuous on the trunk but may be broken on the head and tail. A series of red spots often runs along each side of the body below the stripe like a row of portholes. The belly is yellow with scattered black spots. The costal grooves are indistinct. Transformed adults are aquatic when their pond habitats have water, and during this time have relatively smooth skin and pronounced tail fins. When ponds dry, adults lose the tail fins and their skin becomes rough and grainy. Breeding males differ from females in having a swollen vent and large hind legs with black pads called excrescences on the thighs and toe tips. Gilled adults may have a pattern similar to that of transformed individuals; however, the red pigment, arranged as either stripes or portholes, is lacking on gilled adults in some Georgia populations. The only salamander apt to be confused with the striped newt is the eastern newt (*N. viridescens*). Where the two species occur together, the eastern newt is larger, has red pigment organized as spots rather than stripes, and has more black spots on the belly. Striped newt larvae are smooth skinned, have bushy gills, and have a distinct dark stripe extending from the snout through each eye to the gills; the upper and lower portions of the tail fin are flecked with black. Larvae of the eastern newt lack heavy black flecking on the tail fin. Hatchling larvae of the striped newt are about 8 mm (0.3 in) in total length. Like the eastern newt, the striped newt may have a terrestrial juvenile eft stage. Striped

Adult striped newts in amplexus (male above, gravid female below), Taylor County (John B. Jensen)

Paedomorphic adult striped newt, Taylor County (John B. Jensen)

Striped newt larva, Bryan County (Dirk J. Stevenson)

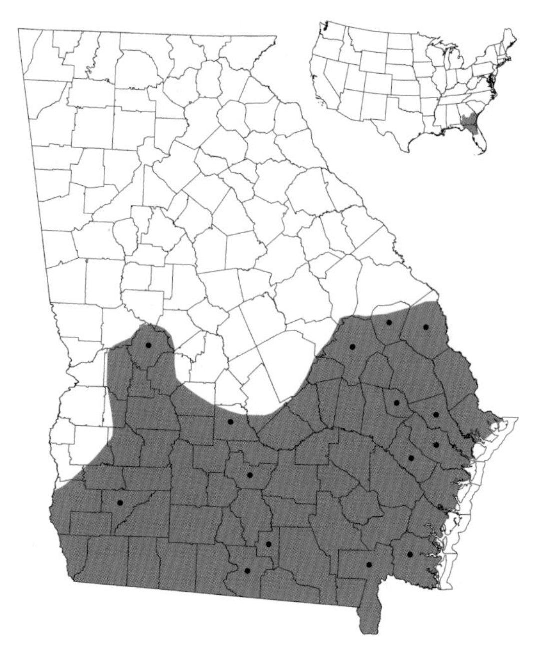

newt efts are generally 40–50 mm (1.6–2.0 in) in total length, have rough skin, are orange-red in overall coloration, and have red stripes like those of the adults.

Taxonomy and Nomenclature

Biochemical studies of the genus *Notophthalmus* have revealed the striped newt's closest relative to be the black-spotted newt (*N. meridionalis*) of southern Texas and Mexico. Striped newts occur in distinct eastern and western groups, but no subspecies are recognized.

Distribution and Habitat

The striped newt ranges over much of the Coastal Plain, but its distribution is spotty and very local. In Georgia, striped newt populations are presently known from only seven widely separated locations. This species is restricted to longleaf pine–wiregrass sandhills and flatwoods, where it breeds in isolated wet depressions. Many of the known Georgia populations are associated with dry sand-ridge communities. Although some breeding ponds are semipermanent, they are more typically small, ephemeral wetlands that are grassy and treeless or have a relatively open canopy of pond cypress and swamp black gum. This newt occasionally breeds in small borrow pits, but not nearly as often as does the eastern newt. Striped newts use both upland and wetland habitats that require periodic burning to maintain ecosystem integrity and function.

Reproduction and Development

This species has a complex life history that includes an aquatic larva, a terrestrial juvenile eft, and an adult that may result from one of two alternative reproductive strategies. Terrestrial efts may mature into adults in upland habitats and then migrate to aquatic breeding sites; or aquatic larvae may skip the eft stage and develop directly into gilled, paedomorphic adults. Migration of adults to breeding ponds from surrounding sandhills and flatwoods corresponds with rainfall, particularly winter rains. This species has a lengthy breeding season that extends from midwinter through spring and summer. Following aquatic courtship, females lay their eggs one at a time, attaching them to aquatic plants; the total number they lay has not been determined. After breeding, adults may migrate back into the surrounding uplands. Hatchling larvae first appear in early spring, and the larvae usually transform into efts later the same year, anytime from April through December, depending on pond water levels. If water remains in breeding ponds, the larvae may continue growing and may reach sexual maturity as gilled animals. After their first breeding event, paedomorphic individuals undergo metamorphosis and move into the uplands.

Habits

Efts commonly disperse up to 500 m (more than 1,600 ft) from their natal pond and may be active on the surface during wet weather. Otherwise, the ecology of the secretive efts is poorly known. Striped newts, like other amphibians that breed only in ephemeral wetlands, experience major fluctuations in reproductive success from year to year. In some years, thousands of larvae transform and join the population; in other years, few or no efts are produced. Breeding ponds monitored from 1992 to 2004 in southeastern Georgia remained dry for 7 of the 13 years. Adult striped newts return to the same site to breed year after year. In Georgia, this species often shares its breeding sites with the eastern newt and various species of mole salamander (genus *Ambystoma*). The nature, extent, and consequences of interactions that possibly occur among these species are not fully understood. Larvae of the striped newt feed on small aquatic invertebrates, including tiny forms suspended in the water. Terrestrial efts eat small terrestrial invertebrates. Adults are known to feed on invertebrates, including fairy shrimp, and amphibian eggs. Natural predators are unknown, but like other newts, this species produces noxious skin secretions that may deter predation.

Conservation Status

The state of Georgia lists the striped newt as Threatened. The distribution of this species in the state is very localized due to habitat fragmentation. This longleaf pine–wiregrass habitat specialist has declined due to habitat lost or degraded by fire suppression, development, ditching and draining of wetlands, commercial forestry practices, and various agricultural activities.

Dirk J. Stevenson and W. Ben Cash

Adult male eastern newt, Rabun County (Dirk J. Stevenson)

Eastern Newt

Notophthalmus viridescens

Description

Adult eastern newts are 2.9–5.1 cm (1.1–2.0 in) in body length and 5.8–14 cm (2.3–5.5 in) in total length. Transformed adults are olive green on the back, usually with scattered red spots that vary in number (up to 45) and location; the largest spots form a distinct row on each side of the body. Eastern newts in northern Georgia are relatively large and have varying numbers of red spots with bold black margins. Coastal Plain newts are smaller, and spots, when present, are small and usually lack black borders. The belly is yellow, ranging from bright lemon yellow to straw colored, and has many small black spots. Unlike other salamanders, newts have rough, granular skin and lack obvious costal grooves. Aquatic adults have slightly granular skin and tail fins, which are large in breeding males. Breeding males also differ from females in having a swollen vent and large hind legs with black excrescences on the inner surfaces of the thighs and on the toe tips. When ponds dry, adults absorb their fins and their skin becomes rough and grainy. Except for having prominent bushy gills, adults that fail to metamorphose are similar to transformed adults. Terrestrial juveniles, or efts, have rough, granular skin that ranges in color from brown in the Coastal Plain to reddish orange in northern Georgia, with red spots similar to those of adults. Larvae, which normally are 7–9 mm (0.3–0.4 in) in total length at hatching, are relatively smooth skinned and have bushy gills and a distinct dark stripe from the snout through each eye to the gills. Transformation from larva to eft occurs at a body length of 19–21 mm (0.7–0.8 in). The eastern newt can be distinguished from the striped newt (*N. perstriatus*) by its slightly

Adult female eastern newt, Rabun County (Dirk J. Stevenson)

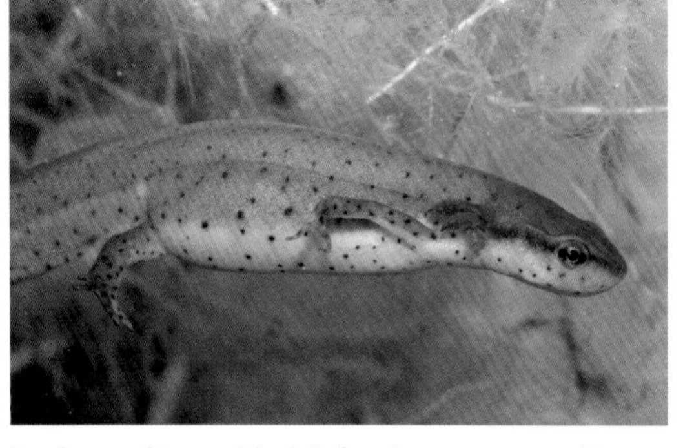

Paedomorphic gravid adult female eastern newt, Bryan County (Dirk J. Stevenson)

Eastern newt larva, Bryan County (Dirk J. Stevenson)

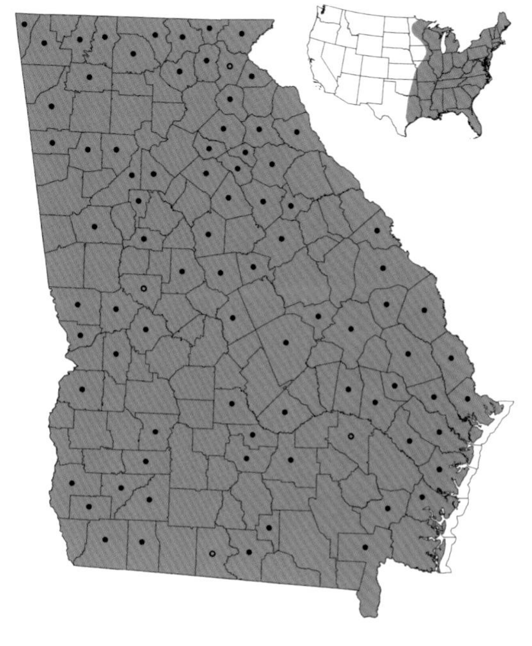

larger size, the presence of small red spots on the back, and more numerous black spots on the belly. The striped newt has a red stripe extending along each side of the back. Larvae of the eastern newt lack the heavy black flecking on the tail fin characteristic of the striped newt.

Taxonomy and Nomenclature

Four subspecies are recognized, two of which occur in Georgia. The red-spotted newt (*N. v. viridescens*) occurs north of the Fall Line, and the central newt (*N. v. louisianensis*) inhabits the Coastal Plain. Presumably the two intergrade near the Fall Line. Variation in Georgia newts is chaotic, however, and does not fit the presumed pattern of a northern versus southern subspecies separated by a narrow zone of intergradation. Studies of enzymes from throughout the range show that patterns of genetic variation do not match the currently recognized subspecies.

Distribution and Habitat

The eastern newt is found throughout the state and is known from virtually all types of standing freshwater habitats, including both permanent and ephemeral ponds. Prime habitats are shallow wetlands (less than 1 m, or 3.3 ft, deep) with an abundance of aquatic vegetation. Mountain lakes, beaver ponds, grassy marshes, floodplain swamps, Carolina bays, cypress ponds, and man-made habitats such as farm ponds and borrow pits often support sizable populations.

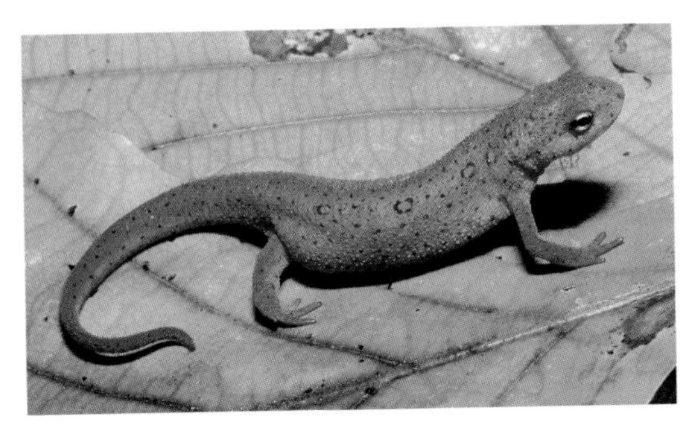

Eastern newt eft, Walker County (Suzanne L. Collins, CNAH)

Reproduction and Development

The eastern newt has a complex life history that generally includes an aquatic larva, a terrestrial juvenile eft, and an aquatic adult. Breeding takes place from fall to early summer, primarily during winter and spring. Newts use their sense of smell and environmental cues when migrating to breeding sites, which may be more than 0.8 km (0.5 mi) from their upland territories. Mature adults exhibit fidelity to their breeding ponds, returning to the same ponds year after year. The size of the breeding population may fluctuate from year to year, with males typically outnumbering females. During mating, the male may amplex the female, grasping her behind the head with his enlarged hind limbs. The male fans his tail to waft vent secretions through the water toward the female. Amplexus may last for many hours and culminates with the male dismounting and depositing one to several spermatophores on the bottom of the pond. The female picks up the spermatophore with her vent and then lays up to several hundred eggs over the ensuing weeks. She lays a few eggs each day, singly on aquatic plants or submerged leaf litter. Hatching occurs 3–5 weeks later followed by a typical larval period of 2–5 months. The number of other newts in the pond may influence both the length of the larval period and the size at metamorphosis. In the Coastal Plain, rapid larval growth combined with stable flooded conditions within the pond may result in eastern newts attaining sexual maturity while retaining larval characteristics such as gills. When such individuals ultimately transform, usually after they have bred once, they are already adult sized. The eft stage lasts longer in northern Georgia (up to 7 years) than in the Coastal Plain (up to 2 years or not at all).

Habits

Larvae consume numerous types of aquatic invertebrates, including crustaceans, snails, and insect larvae. Both efts and adults feed on a variety of invertebrates, including arthropods and worms. Adults also eat leeches, clams, and both the eggs and larvae of amphibians. In fact, dense newt populations may have a profound influence on the amphibian communities of some ephemeral ponds by interfering with the successful reproduction of some species. Eastern newts have powerful toxins concentrated in the skin of the back. Their bright orange color advertises their toxicity. Efts, particularly those in northern Georgia, frequently wander the forest floor in broad daylight after rain showers, apparently oblivious to would-be predators such as birds. Adults are less toxic than efts, and adult newts occasionally fall prey to predators, including turtles and large frogs.

Conservation Status

This species is abundant throughout most of the state. The propensity of landowners to create small farm ponds, often with an abundance of aquatic vegetation, may have greatly increased the breeding habitat, particularly above the Fall Line, relative to the amount available at the time of European colonization.

Dirk J. Stevenson and Carlos D. Camp

Sirens

FAMILY SIRENIDAE

This family has three representatives in Georgia, including the state's second longest salamander, the greater siren (*Siren lacertina*), which can reach 95 cm (3.1 ft) in total length. The sirens are bizarre in completely lacking hind limbs. They are also paedomorphic and have bushy gills throughout life. The family currently consists of only four named species, all endemic to the Coastal Plain of the southeastern United States. They are permanently aquatic and live in isolated wetlands, swamps, backwaters, and sluggish streams.

Adult northern dwarf siren, Liberty County (Dirk J. Stevenson)

Northern Dwarf Siren

Pseudobranchus striatus

Description

This slender, eel-like salamander with two front legs and no back legs is smaller than other Georgia sirens. Adults are 10–20 cm (3.9–7.8 in) in total length and 6–12 cm (2.3–4.7 in) in body length. Adults retain larval characteristics such as bushy gills, a single gill slit on each side of the head, and lidless eyes throughout life. The northern dwarf siren is unique among Georgia's salamanders in having three toes on each foot. A broad, dark stripe extends along the middle of the back. Bordering this stripe and separating it from another dark stripe along each side is a distinct, sharply defined yellow stripe; a narrower yellow stripe extends along the lower margin of each dark side stripe. The belly is gray to black, with variable amounts of light mottling. In populations south of the Altamaha River, three faint, light stripes run within the dark central stripe, although these are often masked by greenish gray pigment. Hatchlings look very similar to adults but usually have more intense black and an overall reddish cast. The stripe pattern and absence of back legs distinguish this species from all other Georgia salamanders except larval greater sirens (*Siren lacertina*), which are also striped but have four toes on each foot.

Taxonomy and Nomenclature

Until recently dwarf sirens were considered to comprise only one species with five subspecies, but two species are now recognized. Only the northern dwarf siren occurs in Georgia. That species includes at least two valid subspecies, both of which also occur in Georgia. The broad-striped dwarf siren (*P. s. striatus*) occupies the Atlantic coastal lowlands and the basin associated with the Okefenokee Swamp. The slender dwarf siren (*P. s. spheniscus*) is found in southwestern Georgia. An upland region called the Tifton and Vidalia Uplands forms a broad northeast–southwest zone across the middle of Georgia's Coastal Plain with little available habitat for northern dwarf sirens. Genetic studies indicate that little or no genetic exchange occurs between populations of the two subspecies, which occur east and west of this region, respectively.

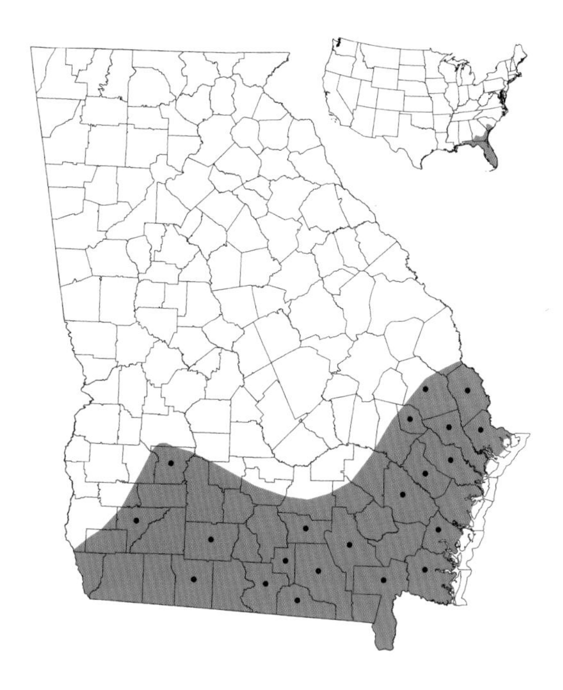

Distribution and Habitat

The northern dwarf siren is restricted to the region of the Coastal Plain that lies east and southeast of the Tifton and Vidalia Uplands, including Atlantic coastal lowlands and the Okefenokee Basin, and the drainages of the Ochlockonee and Flint rivers in southwestern Georgia. Northern dwarf sirens inhabit a variety of aquatic habits, including floating mats of vegetation, margins of permanent bodies of water, and seasonally flooded ponds and marshes. This species often occurs in cypress ponds associated with low pine flatwoods.

Reproduction and Development

The reproductive biology of sirens in general is poorly known. Because sirens lack the internal structures characteristic of salamanders that have internal fertilization, fertilization is generally assumed to be external, but this has not been documented. Northern dwarf sirens typically scatter eggs in aquatic vegetation singly or in small groups. The range of clutch sizes is unknown, but there is a record of a female northern dwarf siren laying 11 eggs. The young grow rapidly and reach sexual maturity within their first year.

Habits

Little is known about the ecology of this species. The northern dwarf siren is completely aquatic, and individuals spend most of their time in thick vegetation or burrowing through mucky sediments. If the pond or marsh where it lives dries up, a dwarf siren simply burrows into the mud, secretes mucous to form a cocoon that is highly resistant to water loss, and remains dormant until the basin refills. These sirens feed on small worms and a variety of other small invertebrates. Predators likely include a variety of carnivorous animals such as fish, aquatic turtles, snakes, and wading birds.

Conservation Status

Despite its secretive nature, this species is often abundant in suitable habitat. Many populations have been adversely affected by the extensive ditching and draining of wetlands in flatwoods habitats throughout its range.

Paul E. Moler

Adult lesser siren, Liberty County (Dirk J. Stevenson)

Lesser Siren

Siren intermedia

Description

This long, slender, eel-like salamander with two front legs and no back legs looks very much like a diminutive version of its relative, the greater siren (*S. lacertina*). Adults are approximately 15–38 cm (5.9–14.8 in) in total length and 10–25 cm (3.9–9.8 in) in body length. This species does not metamorphose, and adults retain larval characteristics throughout life. These include bushy gills and a lateral line system that is visible on the skin as several series of light-colored pores. Like the front feet of most other salamanders, each foot has four toes. The body is dark brown to black with small, faint darker spots on the head and at least the forward parts of the body. Although lesser sirens may have some light speckling along the sides, they lack the golden dashes often seen on greater sirens. Large males may have a swollen, muscular head, but the sexes are otherwise similar in appearance. In Georgia, lesser sirens have 31–34 costal grooves along each side of the body between the leg and vent; greater sirens have 36–39. Small juvenile lesser sirens are dark and have a red or yellow triangle on the snout. New hatchlings are approximately 11–12 mm (0.5 in) in total length.

Taxonomy and Nomenclature

Three subspecies are currently recognized, but only one, the eastern lesser siren (*S. i. intermedia*),

Juvenile lesser siren, Liberty County (Dirk J. Stevenson)

Adult lesser siren, Liberty County (Dirk J. Stevenson)

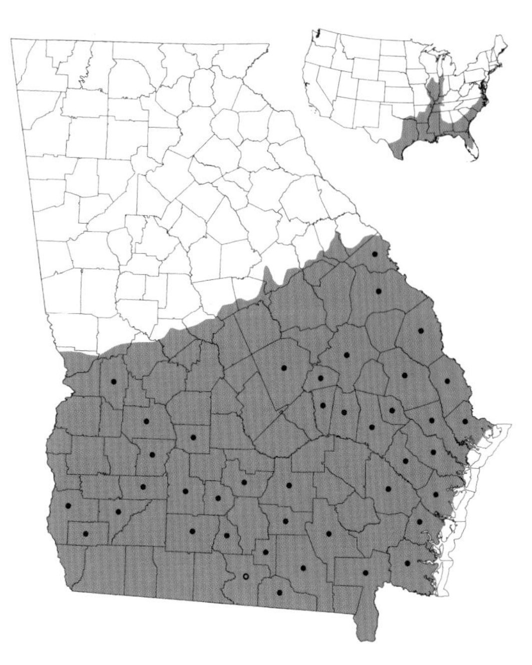

occurs in Georgia. A variety of genetic data including differences in chromosomes indicates that Georgia's lesser siren populations likely comprise at least two species.

Distribution and Habitat

Lesser sirens are found in swamps, cypress ponds, marshes, and blackwater streams throughout the Georgia Coastal Plain.

Reproduction and Development

The reproductive biology of sirenid salamanders is poorly understood. Furthermore, much of the literature on the biology of the lesser siren is based on the western lesser siren (*S. i. nettingi*) and may not be applicable to Georgia populations. Although fertilization is generally assumed to be external, this has not been documented. Nesting typically occurs from February through April, but reproduction is likely delayed in ephemeral wetlands that are dry during this period. Each female lays 130–380 eggs in clusters in vegetation or depressions in shallow water and attends them during incubation. Such females often bear bite marks, but whether these were inflicted by aggressive males during courtship or by other sirens attempting to eat the eggs is unknown. Growth is rapid, and lesser sirens reach sexual maturity during their second year.

Habits

Lesser sirens are active primarily at night. They spend the day sequestered in heavy vegetation or buried in muck. Those living in blackwater streams often frequent submerged beds of decaying leaves during the winter months. As with other sirens, this species is able to survive prolonged dry periods by aestivating—lying buried in muck within a moisture-retaining cocoon formed from shed mucous and skin. An individual can remain in this dormant state awaiting the return of water for many months. The diet includes a wide variety of small crustaceans, insects, snails, and worms. Specimens collected during the late winter or early spring often have siren eggs in their stomach. The gut of lesser sirens generally contains significant amounts of filamentous algae and other plant material, but it is not known whether these materials are consumed intentionally or incidentally. Predators probably include aquatic snakes, aquatic turtles, fish, wading birds, and two-toed amphiumas (*Amphiuma means*).

Conservation Status

Drainage of ephemeral wetlands in the coastal flatwoods and other parts of the Coastal Plain has reduced the availability of habitat for this species, but it is often common where habitat remains.

Paul E. Moler

Adult greater siren, Evans County (Dirk J. Stevenson)

Greater Siren

Siren lacertina

Description

Greater sirens are large, aquatic salamanders that have greatly reduced front limbs and no hind limbs. They are among the largest salamanders in North America; adult total length averages 62–77 cm (24–30 in), and some individuals exceed 92 cm (36 in). Body length averages 36–41 cm (14–16 in) and is a more accurate measure of size because tail injuries and/or loss are common in wild-caught individuals. Males and females look identical and cannot be distinguished externally. Greater sirens' substantial girth and mass give them a robust appearance. They use their tiny front limbs to crawl along the bottom, but the flattened, muscular tail is their main means of propulsion. Greater sirens spend their entire lives in water and have large external gills. General coloration is highly variable, ranging from olive green to grayish brown and becoming lighter toward the belly. Older, larger individuals are not as bright in coloration, and smaller individuals tend to have a more mottled appearance. Distinctive gold flecks often found on juveniles and adults can help distinguish this species from the lesser siren (*S. intermedia*), although juvenile greater sirens and adult lesser sirens can be difficult to tell apart. Lesser sirens are generally more slender, have a more pointed tail, are more uniformly dark, and sometimes have dark spots on the head and sides. Although it can be difficult with a live specimen, counting the number of costal grooves can positively identify each of these species; greater sirens have 36–39, and lesser sirens—in Georgia—have 31–34. In addition, recently hatched greater

Adult greater siren, Evans County (Dirk J. Stevenson)

Juvenile greater siren, Putnam County, Florida (Dick Bartlett)

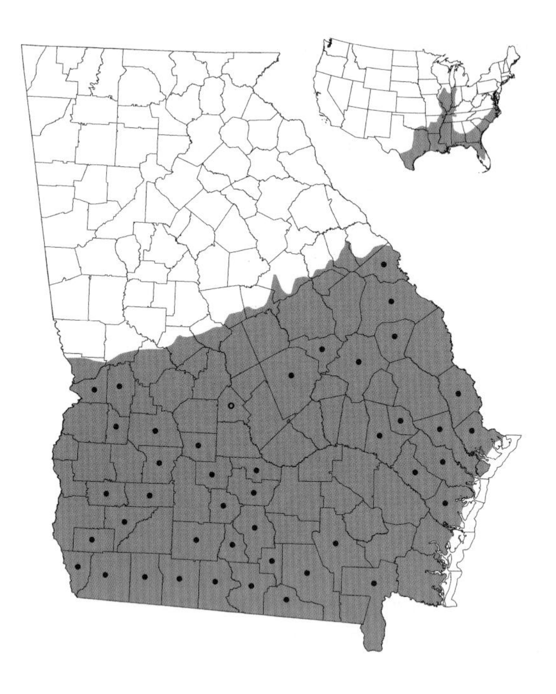

sirens are striped, while recently hatched lesser sirens are not striped and have a red or yellow triangle on the snout. Newly hatched young of the greater siren are approximately 15 mm (0.6 in) in body length.

Taxonomy and Nomenclature

The eel-like appearance of this salamander has led to colloquial names such as ditch eel and congo eel. No subspecies have been described.

Distribution and Habitat

Greater sirens are found throughout the Coastal Plain of Georgia in a variety of wetland habitats, including ditches, marshes, lakes, temporary ponds, slow-moving rivers, and canals. They are most often associated with areas of thick vegetation and/or muck. Juveniles are frequently found in dense mats of floating vegetation.

Reproduction and Development

Very little is known about the reproductive biology of the greater siren. Breeding has not been observed, and the mode of fertilization is not known. Females are capable of laying a large number of eggs; typically these are deposited individually or in small groups. The eggs, which resemble a cluster of small grapes, are sticky and can adhere to one another or to surrounding vegetation. Timing of reproduction is almost certainly a function of environmental conditions and may not be strictly seasonal. Information on age at maturity is not available.

Habits

Sirens are primarily nocturnal. They exhibit a strong negative response to light, and thus spend most of the day near the bottom, in burrows, or buried in dense vegetation. They become active at night and can sometimes be seen foraging near the surface of the water after dark. They have a well-developed lateral line system. Although they rely on gills for some of their gas exchange, they also have lungs and may surface to gulp air;

they can respire through their skin as well, making them well adapted to life in ephemeral wetlands of the southeastern United States. Their diet is varied and includes crayfish, snails, other invertebrates, small fish, and tadpoles. Their gut generally contains significant amounts of plant matter too, but whether plant consumption is intentional or incidental is unknown. Large fish, wading birds, snakes, and American alligators are the main predators of greater sirens. When captured, sirens often emit a faint yelping sound, which may have inspired their common name. Aggression within the species has been noted, and individuals commonly bear bite marks. It is uncertain whether biting is a result of male versus male aggression or male-female courtship. Greater sirens are well adapted for tolerating drought and are able to survive prolonged dry periods by aestivating—burrowing in muck and forming a moisture-retaining cocoon from shed mucous and skin. They can remain in this dormant state for more than a year until rains refill the wetland. They also are capable of dispersing short distances over land during periods of heavy rains to colonize nearby wetlands.

Conservation Status

This species appears to be common and even locally abundant in some areas. Little is known about its population status throughout the state. Habitat alteration through eutrophication and introduction of exotic weeds such as water hyacinths appears to have no adverse impact on greater sirens and may even create more favorable habitat by increasing the local density of wetland plants.

Kristina Sorensen and Paul E. Moler

Reptiles

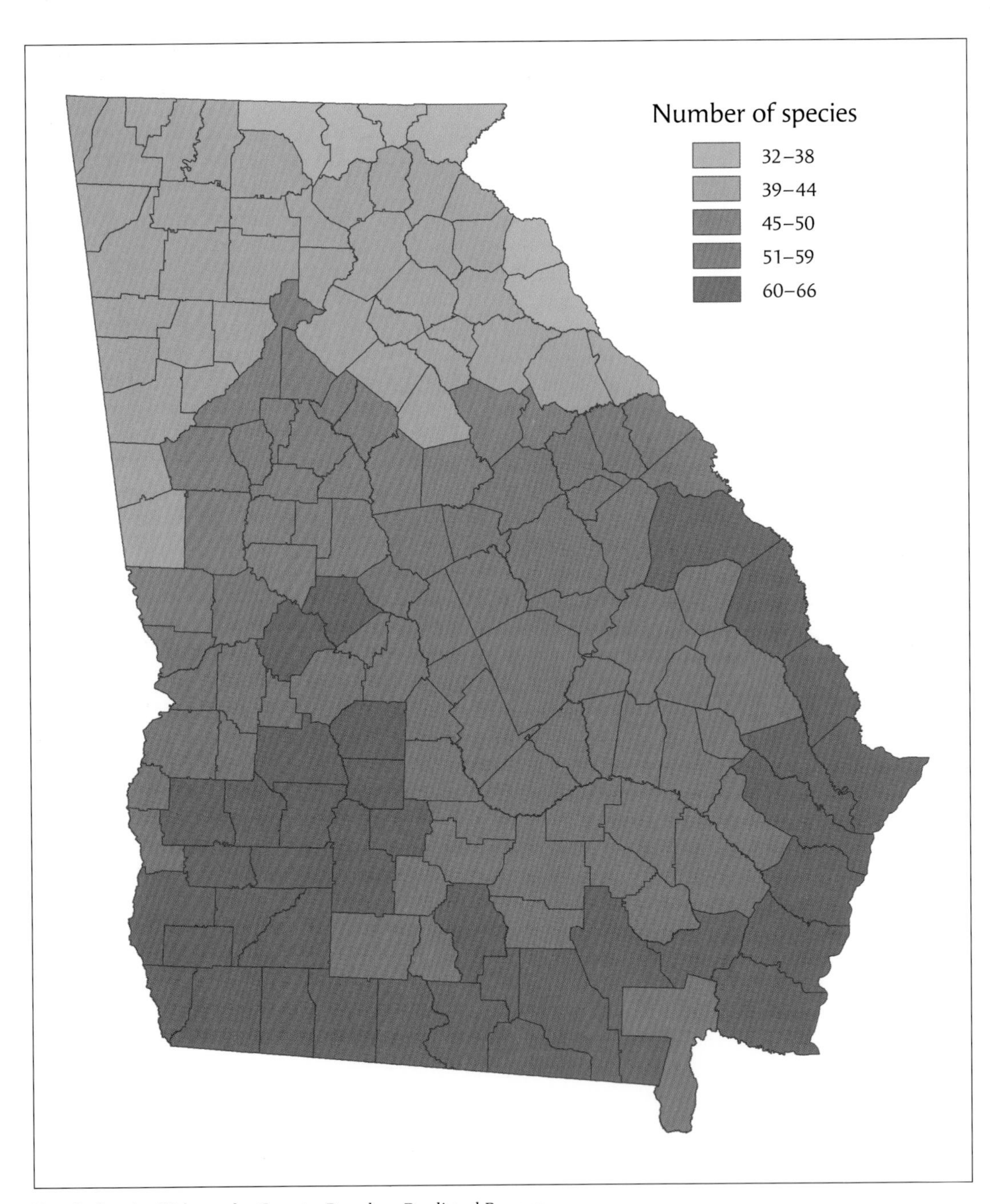

Reptile Species Richness by County, Based on Predicted Ranges

A REPTILE IS a backboned animal that has dry, scaly skin; an ectothermic ("warm-blooded") physiology; and hatches from an amniotic egg—a membrane-rich, usually shelled egg that is designed to survive out of water. For most of the past 100 years this rather straightforward definition satisfied both lay individuals and professional taxonomists. Living reptiles were distributed among several orders within the class Reptilia, all but one (the Rhynchocephalia, housing only the tuatara of New Zealand) having representatives in Georgia (see Chart 1). Recent advances in genetics-based research have given us a better understanding of relationships among animal groups, and the result has been a significant shift in what modern taxonomists consider a reptile to be. Groups that are no longer called reptiles include the earliest vertebrates with amniotic eggs (stem amniotes, once called stem reptiles) and the ancestors of modern mammals. On the other hand, birds, historically recognized as a distinct class (Aves), are now included in the Reptilia; their closest living relatives are alligators and crocodiles (see figure below). The taxonomic organization of subgroups within the Reptilia is undergoing a similarly radical reorganization. For example, the traditional placement of lizards, snakes, and amphisbaenians into three distinct suborders within the order Squamata is rapidly becoming passé. The lizards, in particular, do not form a natural evolutionary group, and therefore cannot be housed within a single taxonomic unit (e.g., suborder Lacertilia).

Although we do not dispute these conclusions, we have chosen to define and group reptiles along traditional lines. We recognize that some of the groups thus delineated no longer have taxonomic meaning. On the other hand, they do continue to have value in communication. Everyone, taxonomists included, knows what we mean when we say "lizard." Junking the term in favor of "iguanian," "scincimorph," and "anguimorph," while more taxonomically correct, is not particularly effective in communicating with a broader audience. In fact, the

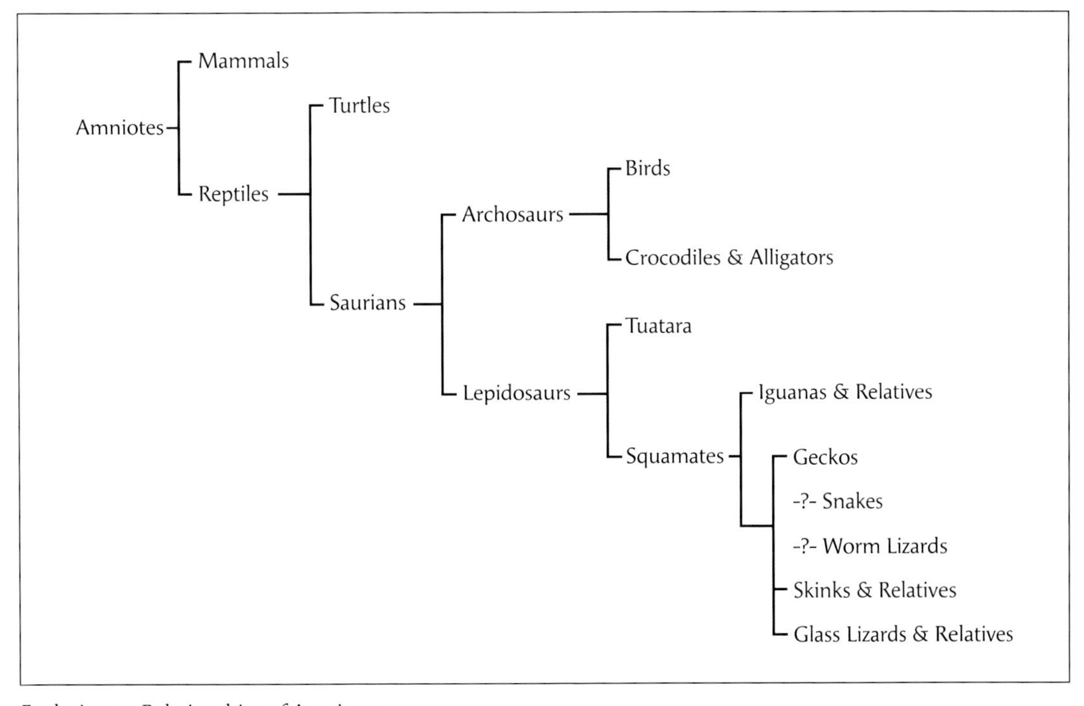

Evolutionary Relationships of Amniotes

very idea of publishing a book on reptiles today that does not include birds is a concession to the convenience of communicating within an already familiar framework. As a comfortable and widely appreciated point of reference, then, we consider "reptiles" to be backboned animals with scaly skin, an ectothermic physiology, and development from an amniotic egg.

It was, in fact, the development of the amniotic egg that enabled nature's most successful experiment in the evolution of land-dwelling vertebrates. This specialized egg in combination with a scaly, waterproof hide enabled reptiles to exploit environments largely closed to amphibians such as dry forests, grasslands, deserts, and salt water. Their radiation into diverse habitats coincided with reptiles' rapid development into a staggering array of fascinating creatures whose like has not been seen since. The diversity of reptilian forms reached its zenith during the Mesozoic, an era that paleontologists have appropriately dubbed the Age of Reptiles—a time brought to vivid life in the book and movie *Jurassic Park.* This was the time of dinosaurs, flying pterosaurs, and large marine reptiles such as plesiosaurs and ichthyosaurs. Most reptilian groups were snuffed out by the atmospheric aftermath of an asteroid collision approximately 65 million years ago. The survivors are represented in Georgia by turtles, snakes, lizards, and a single species each of crocodilian and amphisbaenian.

Crocodilians

OTHER THAN BIRDS, alligators are Georgia's closest living links to dinosaurs. The crocodile-like ancestors of alligators were, in fact, contemporaries of dinosaurs, first appearing more than 200 million years ago. Unlike their star-crossed cousins, however, these large reptiles have survived and currently include up to 25 species distributed across the world's warmer climates.

Most crocodilians are very large, and Georgia's own American alligator (*Alligator mississippiensis*) has been reported to reach a length of 5.8 m (19 ft). The crocodilian body shape is somewhat lizardlike, and the neck, body, and tail are armored on top with body plates covered with tough skin. Crocodilian hides have long been valued for their high quality as leather, the pursuit of which took several species to the brink of extinction.

All crocodilians are semiaquatic, and the tail is flattened from side to side to enhance swimming abilities. Most live in freshwater habitats, although a few can withstand long periods in salt water. They are powerful predators, often representing the top of the food chain in the ecosystems in which they live. Although several crocodile species are known to eat humans, attacks by American alligators on people are uncommon. Traditionally considered reptiles because of their ectothermic ("cold-blooded") physiology and scaly hide, crocodilians also have certain features that link them to birds, their closest living relatives. These features include a four-chambered heart and a strong tendency for parental care of eggs and offspring.

Alligators

FAMILY ALLIGATORIDAE

Georgia's American alligator (*Alligator mississippiensis*) has a body form and size typical of crocodilians. Alligators are distinguished from crocodiles by the relatively broad snout and certain peculiarities of the teeth and skull. This is the only crocodilian family that is restricted to temperate regions. Its two species occur in freshwater habitats in the southern United States and eastern China, respectively.

Adult American alligator, Charlton County (Jim Couch)

American Alligator

Alligator mississippiensis

Description

The American alligator is the largest nonmarine reptile in Georgia. Although Georgia alligators reach adulthood at around 1.8 m (5.9 ft), large males can reach 4–4.5 m (13.1–14.8 ft) in total length and can weigh in excess of 270 kg (597 lbs). Females are much smaller and rarely exceed 3 m (9.8 ft) in total length. Adult alligators are almost solid black or olive drab on their back and sides. The snout is broad, and the upper jaw overlaps the lower jaw so that no teeth in the lower jaw are visible when the mouth is closed. The eyes and nostrils are elevated compared with the rest of the skull, and alligators are capable of submerging their entire body with only these two areas exposed. Alligators have five toes on each front foot and four on each hind foot, with webbing between the toes. This webbing, coupled with the vertically flat tail, makes them excellent swimmers. On top of the large tail there is a series of keels on each side that converge into a single series halfway toward the tip. Beneath the thick skin on the back alligators have bony plates known as osteoderms, which serve as body armor. Hatchlings are approximately 20–25 cm (7.8–9.8 in) in total length and have yellow bands on a predominantly black background. These yellow bands fade with age.

Taxonomy and Nomenclature

The American alligator is the only member of the family Alligatoridae inhabiting North America. No subspecies are recognized.

American alligator hatchling, Long County (Dirk J. Stevenson)

Juvenile American alligator, Baker County (Gabriel J. Miller)

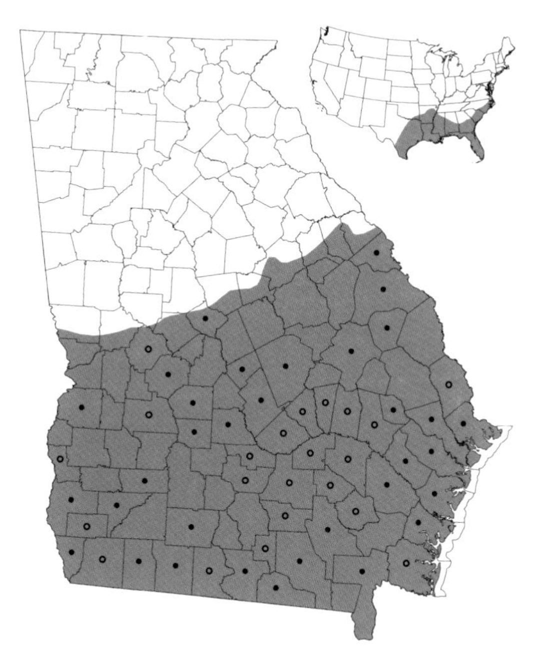

Distribution and Habitat

Alligators have been documented in more than 50 of Georgia's 159 counties, primarily south of the Fall Line. Individuals are occasionally encountered above the Fall Line, but no breeding population has ever been documented in the Piedmont. Habitats occupied include brackish and freshwater marshes, ponds, swamps, inland lakes, rivers, creeks, and drainage canals.

Reproduction and Development

Both sexes communicate in courtship rituals using sight, hearing, touch, and the sense of smell. Head slapping and bellowing (33_American_alligator_adult_male.mp3) are common courtship behaviors. The courtship and breeding season in Georgia begins in April, with peak egg-laying occurring in late June or early July. After breeding, the female seeks a secluded location for her nest. Most nests are located near water, and often they are close to an alligator den beneath the bank. Nests are dome-shaped mounds constructed above the ground to keep the eggs from being flooded while they are incubating. Georgia nests average nearly 2 m (6.6 ft) wide and approximately 0.5 m (1.6 ft) high. After completing the mound, the female makes a depression in the top, lays her eggs, and then covers them with more material. Alligator eggs are white and measure about 76 mm (3.0 in) in length and 46 mm (1.8 in) in diameter. Clutch size in Georgia is 13–55, with an average of 35–40. The incubation period is around 65 days. Female nest-guarding behavior ranges from aggressive defense of the nest to occasional visits to no attendance at all. Hatching rates in populations studied in Georgia ranged from 41.9 percent to 70 percent. In high-quality habitat, young alligators may grow up to 30 cm (11.7 in) per year for the first few years.

This growth rate slows considerably when the alligator reaches adulthood; the age at sexual maturity varies among populations.

Habits

Alligators may be seen throughout the year but enter a period of dormancy when water temperatures drop below 13° C (55° F). They stop feeding when the water temperature drops below 23° C (73° F). Adult males may occupy territories as large as 26 km^2 (10 mi^2) during the breeding season; females have smaller home ranges. Drought conditions show how important alligators are to other species that share their ecosystem. As ponds dry up, alligators wallow out deep areas called gator holes that provide habitat for individuals of many species that otherwise might die. Young alligators spend their first couple of years in close proximity to their mother, often emitting a series of grunted *umphs* to alert her of their whereabouts or of danger (32_American_alligator_hatchling.mp3). The hatchlings form groups called pods and may remain together for up to 3 years. During their first few years, young alligators fall prey to various predators, including mammals, wading birds, fish, snakes, and larger alligators. During that time they feed on snails, crayfish, frogs, insects, and other invertebrates. As alligators grow, their prey items begin to include fish, crabs, wading birds, waterfowl, turtles, snakes, mammals, and smaller alligators.

Conservation Status

Because of their valuable hide, alligators were hunted and trapped—legally and then illegally—across the southeastern United States until their existence was in peril. Georgia and many other states experienced their lowest alligator population numbers in the 1960s. The U.S. Fish and Wildlife Service listed the alligator as Endangered in 1967. This protective status and increased protection by the states resulted in a gradual increase in alligator populations in Georgia and elsewhere. The population comeback resulted in the reclassification of the American alligator to Threatened by Similarity of Appearance in 1987. Alligators have become so numerous that most state agencies now have some type of program to remove nuisance alligators from the population. Georgia and several other states have classified the alligator as a game animal and have specific hunting seasons. Several states also have large-scale ranching or farming operations where captive alligators are grown to specific sizes and harvested or returned to the wild to enhance native populations. Today's modern harvest programs have strict controls and quotas to prevent overharvesting and adverse effects on alligator populations and to ensure their continued survival and existence throughout Georgia and the Southeast.

D. Gregory Waters

Worm Lizards

WORM LIZARDS ARE not well known, and many herpetologists have never seen one, not to mention the average Georgian. They share certain features with lizards and snakes, including paired hemipenes for mating and Jacobson's organ, a specialized chemosensory organ located in the roof of the mouth. Other characteristics, however, are unique, and many are adaptations for a burrowing lifestyle. For example, the surface of the body and tail forms a series of narrow, circumferential rings, giving the superficial appearance of an earthworm. The eyes are greatly reduced, and most worm lizards do not have limbs, although a few species have two front legs. The lower jaw fits inside the upper, and the teeth of the two jaws alternate, so that the bottom teeth fit between the upper teeth.

Florida Worm Lizard

FAMILY RHINEURIDAE

Georgia's only species is a strange, limbless little reptile with skin that forms distinct rings encircling the body and the short, blunt tail. There are no external ear openings, and most individuals lack external eyes. The family is restricted to the southeastern United States, where it occurs in Georgia and Florida. The Florida worm lizard is the sole species in this family, and it is also the only species of worm lizard known from the United States.

Adult Florida worm lizard, Hernando County, Florida (Kevin Enge)

Florida Worm Lizard

Rhineura floridana

Description

This bizarre legless animal may reach a total length of 38 cm (14.8 in), but the average adult is approximately 25 cm (9.8 in) long. Worm lizards look more like earthworms than reptiles. They are pink or grayish, somewhat opalescent, and lack any pattern. The body scales are arranged in rings that look very similar to the segments of earthworms. The lower jaw is countersunk under the snout, an adaptation for burrowing. The rounded tail is short and flattened, and numerous small tubercles cover its upper surface. Very young individuals and some adults have tiny, poorly developed eyes, but most adults lack eyes. Unlike Georgia's "true" lizards, worm lizards have no ears. Hatchlings are 80–100 mm (3.1–3.9 in) long and look similar to adults, but their tail lacks the calloused upper surface.

Taxonomy and Nomenclature

Although commonly called a lizard, the worm lizard is not a "true" lizard. It belongs to a group of reptiles whose relationships to lizards and snakes are poorly understood. This is the only worm lizard in the United States. No subspecies are recognized, although recent genetic data suggests that multiple taxonomic forms may eventually be named.

Distribution and Habitat

Only two Florida worm lizards have ever been found in Georgia. One was excavated from a pecan orchard near Lakeland (Lanier County), and the exact locality of the other was not reported. Habitats reported for the species in Florida include hardwood hammocks, sandhills, dry pine flatwoods, and fallow fields.

Adult Florida worm lizard, Lanier County (Jerry Payne)

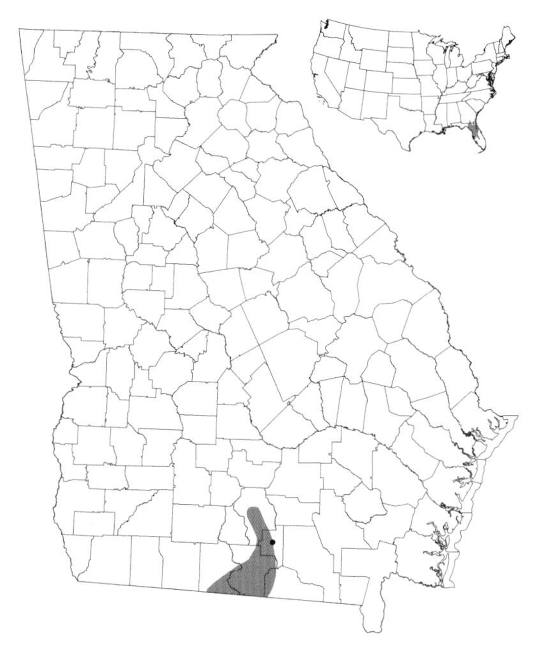

Reproduction and Development

Very little is known about the reproductive biology of this secretive reptile. Females produce one to three white, elongated, and very thin-shelled eggs. The eggs are laid in the summer and hatch during late summer or fall. They have been found in sandy soil at depths of 20–50 cm (7.8–19.5 in).

Habits

The Florida worm lizard is an obligate burrower, which is certainly why it is very rarely seen. Individuals are usually found by people digging, gardening, or plowing in sandy soil. Occasionally, heavy rains that fully saturate the soil bring Florida worm lizards to the surface, where they often drown in puddles. Unlike most other native reptiles, they cannot swim. Earthworms, spiders, and termites are known prey. Other than the loggerhead shrike, predators are unknown. Florida worm lizards do not break their tail as a defense against predation as "true" lizards do.

Conservation Status

With only two records for the state, the most recent in 1974, it is hard to consider this species anything other than extremely rare. However, the worm lizard's fossorial nature makes it very difficult to survey and properly assess its true status and distribution in Georgia.

John B. Jensen

Lizards

ALTHOUGH APPROXIMATELY 3,000 species of reptiles have traditionally been called "true" lizards, Georgia has only 13 native species and 2 introduced species. Even so, lizards are familiar members of Georgia's fauna because several species are commonly active in the same times and places as humans.

Most lizards should be easy to recognize by their typical elongate lizard shape with a head, four legs, and a tail. Because salamanders share those characters, however, people sometimes have difficulty distinguishing them. "True" lizards are reptiles with dry, scaly skin and, at least in those with feet, toes bearing claws. Our native lizards also have eyes that are recessed into their sockets so that they are flush with the skull and do not "bug out" as they do in most salamanders.

Some lizards, Georgia's glass lizards (genus *Ophisaurus*) among them, do not have limbs and superficially resemble snakes. Unlike snakes, though, Georgia's native "true" lizards have closable eyelids, external ears, and small belly scales. Glass lizards are mostly tail. The tail is easily broken, and the broken pieces appear to be alive because they continue to wriggle. This characteristic of glass lizards gave rise to the legend of the joint snake, whose body parts supposedly grow back together after having been chopped into pieces by a hoe.

Another Georgia lizard myth is that skinks, called scorpions in some parts of the state, are venomous. In fact, none of Georgia's lizards is dangerous to humans in any way, and all of them should be welcomed in human-occupied landscapes because of their insect-eating ways.

Reproductively, lizards are similar to snakes. Mating is accomplished by copulation using the male's dual sex organs, the hemipenes. Most lizards lay eggs, although females of some species in the western United States and on other continents retain the eggs inside the body and give birth to fully developed young.

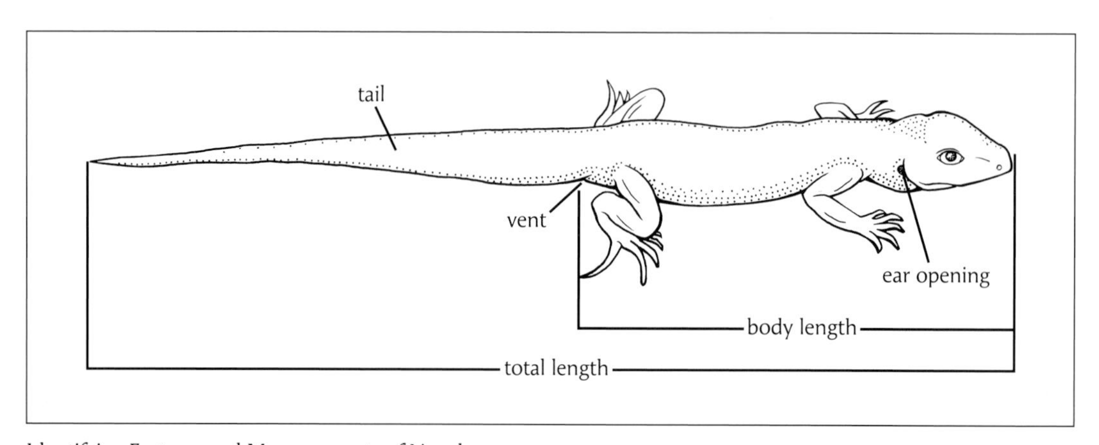

Identifying Features and Measurements of Lizards

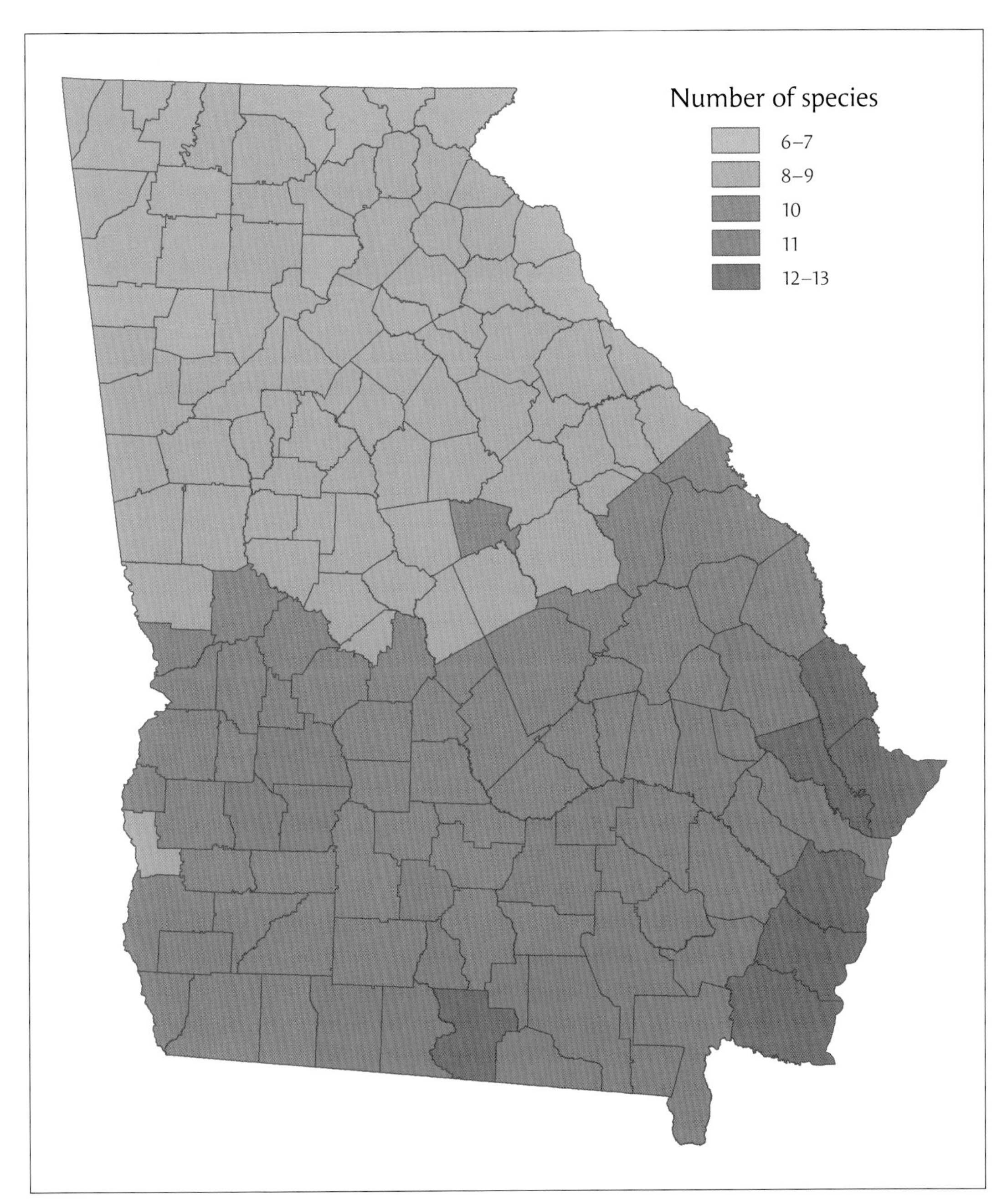

Lizard Species Richness by County, Based on Predicted Ranges

Glass and Alligator Lizards

FAMILY ANGUIDAE

The glass lizards are limbless and superficially resemble snakes, but all have external ears, movable eyelids, and small scales on the belly. Their western relatives, the alligator lizards, have legs. All members of the family have a distinct fold of skin extending along each side of the body. Some species may reach 100 cm (40 in) or more in total length, although the majority of this length consists of an extremely fragile and breakable tail, hence the common name.

Four species of glass lizards occur in Georgia; their combined ranges cover the state. Alligator lizards are found in the western United States, and other members of the family are found in Central and South America as well as Eurasia. Habitats exploited by glass lizards include open, grassy areas, forests, and coastal sand dunes.

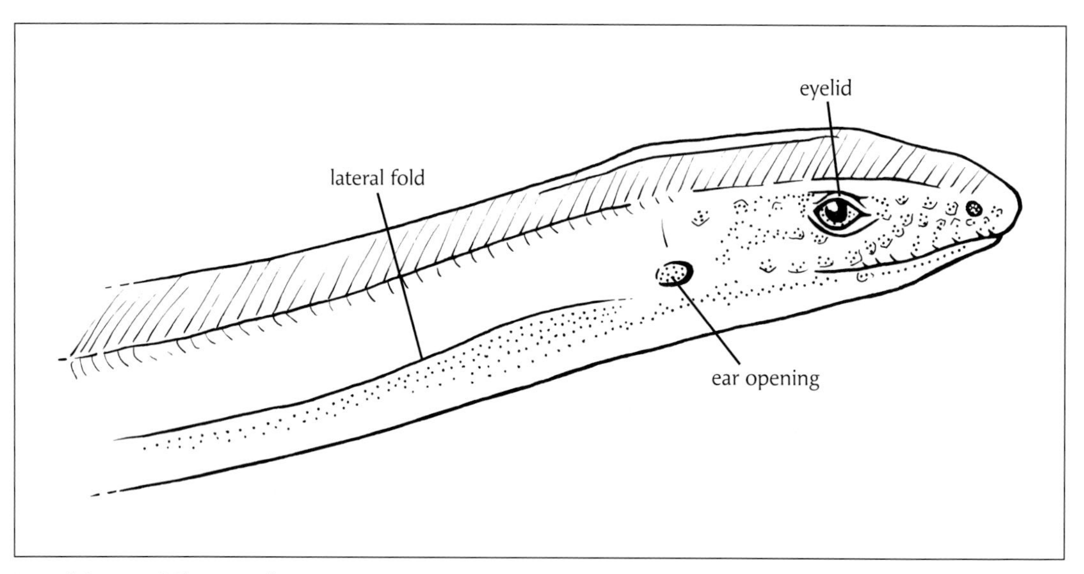

Lateral Aspect of Glass Lizards

Adult slender glass lizard, Long County (Dirk J. Stevenson)

Slender Glass Lizard

Ophisaurus attenuatus

Description

Slender glass lizards may be up to 118 cm (46 in) in total length, and about two-thirds of that is tail in specimens with an intact original tail. Males are larger than females. Slender glass lizards lack legs and are much less flexible than snakes. They can be distinguished from snakes by their movable eyelids, small belly scales, and external ear openings. Most slender glass lizards are brown to golden brown or tan and have dark, narrow lengthwise stripes below the lateral fold that extends along the sides of the body. There are also narrow stripes under the tail. The body and tail stripes are very dark in young specimens but may become lighter in older ones. Younger animals also have a dark stripe or series of dashes down the center of the back. Old males may develop horizontal blotches on the back and have a salt-and-pepper coloration. Scales along the lateral fold number greater than 97, which helps distinguish the slender glass lizard from the much smaller and rarer mimic glass lizard (*O. mimicus*), which has fewer than 97 scales. The other two glass lizards native to Georgia do not have stripes below the lateral fold. Hatchlings are 150–220 mm (5.9–8.6 in) in total length.

Taxonomy and Nomenclature

Two subspecies are currently recognized, with one, the eastern slender glass lizard (*O. attenuatus attenuatus*), occurring in Georgia. Slender and other glass lizards are sometimes called legless lizards, glass snakes, joint snakes, or horn snakes.

Adult slender glass lizard, Long County (Dirk J. Stevenson)

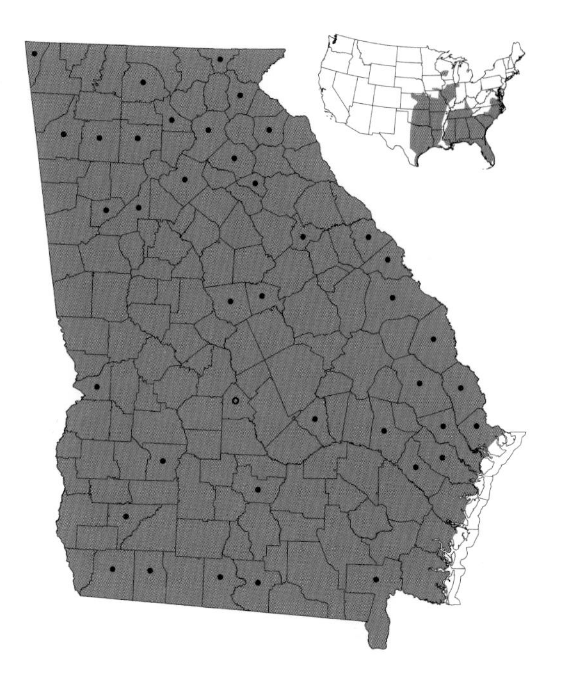

Distribution and Habitat

Slender glass lizards occur statewide in Georgia, or nearly so. They are found primarily in well-drained upland habitats such as those dominated by scrubby oaks and pines, and individuals prefer open woodlands or grasslands to more heavily forested areas. They tolerate dry, open areas and are occasionally found in vacant lots or clear-cut areas.

Reproduction and Development

Little is known about the mating behavior of glass lizards, but slender glass lizards are thought to mate in the early spring. Between late spring and midsummer, females lay 4–19 eggs in a shallow depression under a log or clump of vegetation. The female stays with her eggs, which hatch in about 2 months, and while she does not actively guard them, she will regather them if they are scattered. Sexual maturity is reached 3 or 4 years after hatching.

Habits

All glass lizards spend most of their lives under cover or underground, but slender glass lizards are not thought to burrow as readily as eastern glass lizards (*O. ventralis*) are known to do. They are most active in the early morning and late afternoon, and like many species of reptiles are often seen moving about on sunny mornings after a previous evening's rain. Slender glass lizards are carnivorous and eat a variety of invertebrates as well as other lizards, small snakes, young mice, and eggs of ground-nesting birds. Predators include kingsnakes (genus *Lampropeltis*) and undoubtedly other lizard-eating snakes, birds, and mammals. If grasped by a predator, a slender glass lizard can quickly break off its tail along fracture planes in the vertebrae, even if the tail is not touched. This is presumably a predator avoidance mechanism allowing the vital part of the lizard to escape while the predator is occupied with the wiggling, expendable tail. The regrown tail never quite matches the original in length or color; it has a stumpy appearance with a pointy end, which is the source of the colloquial name horn snake. Slender glass lizards seldom bite when captured.

Conservation Status

Slender glass lizards do not occur in high densities but are still fairly common throughout parts of Georgia that have not been heavily altered by urbanization.

Judith L. Greene and Tony Mills

Adult island glass lizard, Glynn County (John B. Jensen)

Island Glass Lizard

Ophisaurus compressus

Description

This legless, snakelike lizard reaches a maximum total length of about 61 cm (23.8 in); roughly two-thirds of the length is tail. A distinct lateral fold extends along each side from the neck to the vent. Unlike other glass lizards, this species lacks fracture planes in the tail vertebrae, and specimens with entire tails are more common than they are in other species. Like all glass lizards, the island glass lizard has plates known as osteoderms embedded in the body scales that make the body less flexible than that of a snake. These lizards also differ from snakes in having movable eyelids, external ear openings, and small scales on the belly. This species and the mimic glass lizard (*O. mimicus*) have fewer than 97 scale rows along the lateral fold while the eastern (*O. ventralis*) and slender (*O. attenuatus*) glass lizards have more than 97. The island glass lizard can be distinguished from all other Georgia glass lizards by the single dark stripe, bordered below by yellow or cream, extending along each side toward the rear of the body and along most of the tail. Other species have multiple thin, dark lines along the tail. The light-colored belly surface ranges from pinkish to yellowish and has no markings. The neck area takes on a brownish coloration in older specimens. Large males may develop conspicuous bars or speckling on the back and sides. Hatchlings average 40 mm (1.6 in) in body length and 136 mm (5.3 in) in total length.

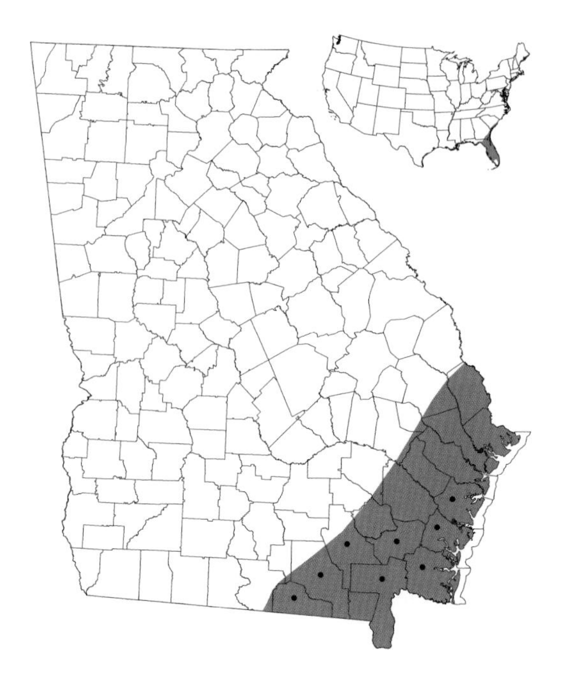

Taxonomy and Nomenclature

No subspecies are recognized. Island glass lizards are also called glass snakes, joint snakes, and legless lizards.

Distribution and Habitat

In Georgia, this species is found primarily in coastal habitats and throughout the Okefenokee basin. It is reported to be especially common on certain coastal islands, where individuals may even be found among tidal wrack. Inland populations are often associated with pine flatwoods.

Reproduction and Development

Breeding occurs in the spring. During mating, the male bites a receptive female just behind her head and holds on while he arches his body to the side in order to bring their vents together. Copulation may last as long as 4 hours. Egg deposition occurs about 6 weeks later. Reported clutch sizes range from 4 to 18 eggs. Incubation lasts about 6 weeks, and females remain with their eggs during this period. Information on development and maturity in island glass lizards in Georgia is unavailable.

Habits

Like other glass lizards, the island glass lizard is active during the day, but little else is known about the behavior and general ecology of this secretive and poorly studied lizard. It hunts insects, spiders, and other small invertebrates. Predators presumably include birds, mammals, and lizard-eating snakes such as black racers (*Coluber constrictor*), common kingsnakes (*Lampropeltis getula*), and eastern coral snakes (*Micrurus fulvius*).

Conservation Status

Many of the coastal islands where this species occurs are protected. The status of this species in Georgia is unknown.

Paul E. Moler

Adult mimic glass lizard, Okaloosa County, Florida (John B. Jensen)

Mimic Glass Lizard

Ophisaurus mimicus

Description

This long, slender, limbless lizard reaches a maximum total length of 65.7 cm (25.6 in). About two-thirds of that length is tail, at least in individuals with an intact tail. Males grow slightly larger than females. Like all four Georgia glass lizard species, the mimic glass lizard superficially looks like a snake. The scales of mimic and other glass lizards are reinforced with tiny osteoderms that make them considerably more rigid than snakes. Mimic glass lizards are distinguished from snakes by their small belly scales, movable eyelids, external ear openings, and lateral folds. The back is tan or brown and may or may not have light and dark speckling. A dark stripe is usually present down the middle of the back and most of the tail, but it may be faint on some individuals. Narrow stripes on the side of the tail and above the fold on each side of the body are typical. A few stripes are present below the fold in adults, but these are quite faint. Vertical bars may be present behind the ears. The belly is pale. Hatchlings have not been described, and few descriptions of juveniles are available. Juvenile specimens less than 90 mm (3.5 in) in body length have only a single stripe above the lateral fold; rows of dots, which presumably develop into the multiple stripes of the adults, are present between the stripe and the fold. Island glass lizards (*O. compressus*) are similar in size but have a very prominent dark stripe on each side of the body above the fold. The eastern (*O. ventralis*) and slender (*O. attenuatus*) glass lizards grow to be much larger, reaching body lengths in excess of 30 cm (11.7 in). Counting scale rows along the lateral fold is the most reliable way of distinguishing mimic glass lizards from smaller individuals of the latter two species; mimics have fewer than 97 scale rows, and eastern and slender glass lizards have more than 97.

Taxonomy and Nomenclature

This species was described in 1987 on the basis of differences detected primarily in museum specimens that were originally identified as other species. Glass lizards are known colloquially as legless lizards, glass snakes, or joint snakes.

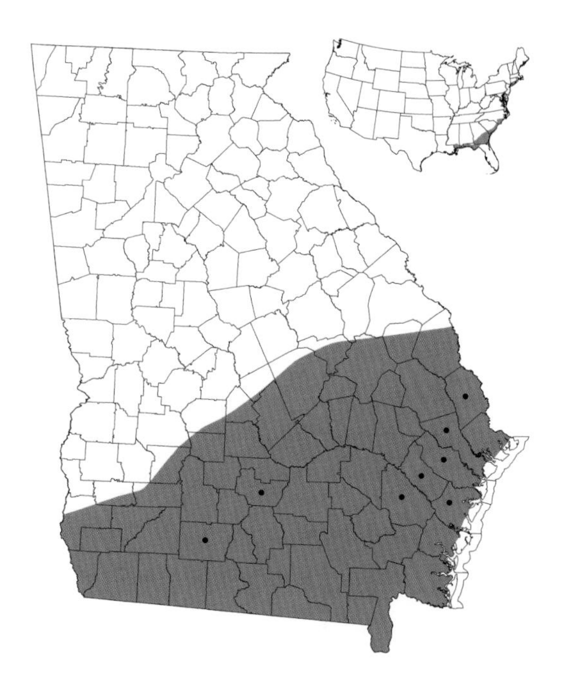

Distribution and Habitat

Mimic glass lizards are known from only a few sites in the lower Coastal Plain of Georgia. They are apparently strongly associated with the longleaf pine–wiregrass community of this region, but very little habitat information exists. Specific habitat types identified in other parts of the range include mesic pine flatwoods, savannas, and pitcher plant bogs. An intact ground cover dominated by grasses is characteristic of most, if not all, sites where mimic glass lizards occur.

Reproduction and Development

A female from Mississippi contained 11 enlarged eggs. Other than that, little is known about the reproductive biology. Other glass lizard species typically mate in the spring and nest from early summer to midsummer, with eggs hatching in late summer. Females usually attend the eggs, which they deposit in depressions in moist soil under logs or similar shelters or in rotten logs. Whether these traits also pertain to mimic glass lizards, however, remains to be determined.

Habits

This is a secretive and poorly known lizard, especially in Georgia. The only documented predator is the black racer (*Coluber constrictor*), but many other carnivorous animals undoubtedly eat mimic glass lizards. Fracture planes in their tail vertebrae allow the tail to break off easily, often in more than one piece, when grasped by a predator. While the dislodged tail wiggles and distracts the predator's attention, the glass lizard escapes with all its vital parts. The diet probably consists of a variety of invertebrates, as has been documented for other glass lizard species. Limited information suggests that this species is active primarily during the day and for a brief period just before sunrise and just after sunset.

Conservation Status

The last confirmed mimic glass lizard collected in Georgia was found in 1978, prior to its recognition as a distinct species. Very few other records exist. It is quite possible that its similarity to other glass lizards may have led to the misidentification of observed or captured specimens. Regardless, this is a very uncommon species in the state; it may be significantly imperiled or may even have been extirpated. Drastic habitat alteration in the lower Coastal Plain, including fire suppression and loss of the longleaf pine–wiregrass community, has left the mimic glass lizard and many other endemic reptiles and amphibians without much suitable remaining habitat. This lizard is state listed as Rare.

John B. Jensen

Juvenile eastern glass lizard, Long County (Dirk J. Stevenson)

Eastern Glass Lizard

Ophisaurus ventralis

Description

Eastern glass lizards can be up to 108 cm (42.1 in) in total length, and as is true of the other glass lizards, the tail accounts for two-thirds of the length. Males are larger than females. Eastern glass lizards are usually brown or tan above and pale yellowish underneath. Most large adults are greenish, and some have a speckled appearance. These legless lizards can be distinguished from snakes by their movable eyelids, external ear openings, and small belly scales. They differ in appearance from slender glass lizards by having no narrow, dark lengthwise stripes below the lateral fold that extends along the side of the body and no dark stripe down the middle of the back. They generally do have one or more dark stripes above the fold. These stripes are especially bold in young individuals, which are usually less than 180 mm (7 in) in total length at hatching. The presence of more than 97 scale rows along the lateral fold is helpful in separating this species from the mimic (*O. mimicus*) and island (*O. compressus*) glass lizards, both of which have fewer than 97 scale rows.

Taxonomy and Nomenclature

No subspecies are recognized. Like other glass lizards, eastern glass lizards are colloquially known as legless lizards, glass snakes, horn snakes, and joint snakes.

Adult eastern glass lizard, Chatham County (John B. Jensen)

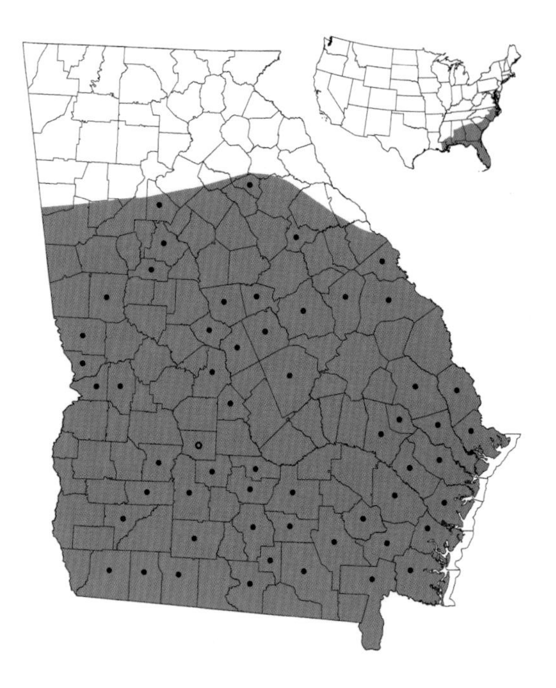

Distribution and Habitat

Eastern glass lizards are not as widely distributed in the state as slender glass lizards but are found in suitable habitat from the Piedmont down to the coast. They occur in pine flatwoods, cut-over areas, and other dry habitats, but they tolerate, and perhaps even prefer, much wetter habitats. They are often found near the edges of wetlands and creeks and in grassy areas near marshes. Eastern glass lizards are especially common in open, grassy areas on barrier islands, where they are said to frequent the burrows of ghost crabs.

Reproduction and Development

This species of glass lizard is thought to mate in the early spring. Between late spring and midsummer the female lays a clutch of 5–15 eggs in a shallow depression in the soil under a log or clump of grass or other vegetation. The female attends the clutch, which hatches in about 2 months. Like her relative the slender glass lizard, she will regather the eggs if they are scattered. Age at maturity is not known for this species.

Habits

Eastern glass lizards spend most of their lives under cover or underground. They are active burrowers but forage aboveground in the early morning and again from late afternoon to dusk; they are often seen moving about on sunny mornings after a previous night's rain and are occasionally seen crossing roads at night as well. Eastern glass lizards are carnivorous and eat a variety of insects and other invertebrates such as snails as well as lizards, including the six-lined racerunner (*Cnemidophorus sexlineatus*); small snakes; young mice; and the eggs of ground-nesting birds. Predators include black racers (*Coluber constrictor*), coachwhips (*Masticophis flagellum*), mole kingsnakes (*Lampropeltis calligaster*), common kingsnakes (*L. getula*), eastern coral snakes (*Micrurus fulvius*), and probably a variety of raptors and carnivorous mammals. The tail of eastern and most other glass lizards may break along fracture planes when grasped. This species may try to bite when handled but is calmer than the slender glass lizard (*O. attenuatus*).

Conservation Status

Eastern glass lizards are local in occurrence, but populations are thought to be common and stable in Georgia.

Judith L. Greene and Tony Mills

Geckos

FAMILY GEKKONIDAE

Most geckos, including Georgia's Mediterranean gecko (*Hemidactylus turcicus*), are less than 10 cm (4 in) in body length when fully grown, although some may reach 37 cm (14.8 in). The body is typically covered with small, granular scales and sometimes with warty protuberances as well. The toes are often expanded with tiny, intricate folds that adhere to vertical surfaces during climbing. Geckos are native to much of Eurasia, Africa, Australia, and the tropical Americas. Georgia's species was likely inadvertently introduced into the United States from the Mediterranean via imported cargo. Some gecko species are parthenogenetic. Unlike Georgia's native lizards, geckos are generally nocturnal. They feed on small invertebrates. Many are perpetual climbers, living on trees, cliffs, or the walls of buildings.

Adult Mediterranean gecko, Clayton County (John B. Jensen)

Mediterranean Gecko

Hemidactylus turcicus

Description

Adult Mediterranean geckos are small lizards averaging 10–13 cm (3.9–5.1 in) total length. They are the only lizards in Georgia with vertical pupils; no eyelids; rows of numerous small, warty tubercles on the back and tail; and small, sticky toe pads. The skin is gray, white, or pinkish with darker blotches. The tail is round in cross section, has many tubercles, and has alternating light and dark bands. The scales are tiny and granular, and the skin is relatively thin, fragile, and nearly translucent, especially the white belly skin. Hatchlings are 20–30 mm (0.8–1.2 in) in body length.

Taxonomy and Nomenclature

The subspecies present in Georgia and elsewhere in the southeastern United States is the Turkish house gecko (*H. t. turcicus*), also commonly called the Turkish gecko.

Distribution and Habitat

This Old World invader is native to nearly all the countries in the Mediterranean region, including southern Europe and northern Africa, and has been introduced into many tropical and temperate areas, including Mexico, Belize, Panama, Cuba, and Puerto Rico. It was

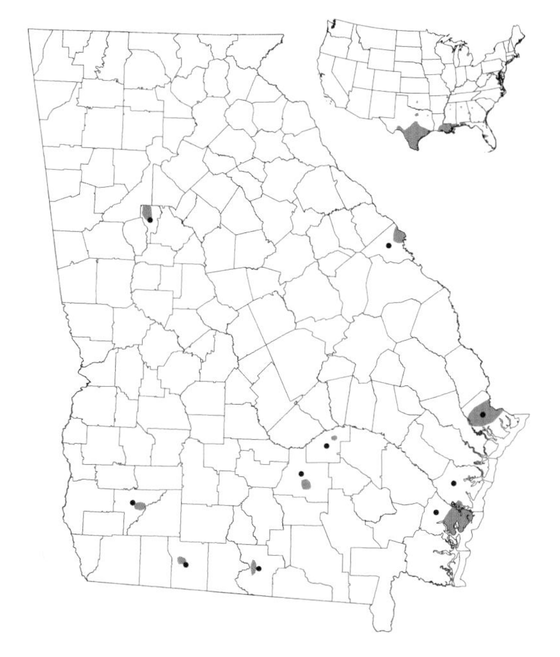

introduced into a number of U.S. port cities along the Gulf of Mexico in the early 1900s and is now widespread in the southern half of the country. In Georgia, this species has been found in Augusta, Brunswick, Darien, Douglas, Hazelhurst, Newton, Saint Simons Island, Savannah, Thomasville, Valdosta, and the Atlanta area, and it will possibly be found in most urban areas of Georgia in the near future. Although these nocturnal lizards rarely venture far from lighted buildings, they are sometimes found under the bark of nearby trees. If present, they can be numerous on walls around light fixtures after dark. They are most abundant at sites with many cracks and crevices, especially those that allow access to warm areas under light fixtures.

Reproduction and Development

Mediterranean geckos breed from April through September. Females lay up to three clutches of one or two white eggs, placing them in cracks and crevices. Before they are laid the eggs are visible through the female's translucent skin. Hatchlings grow rapidly and reach adulthood by late winter or early spring of the following year.

Habits

Males are territorial and defend a small area around outdoor lights where flying insects congregate at night. They compete vigorously for food resources, and the dominant individuals are generally found nearest the light source. No information on predators is currently available. This species has a faint, mouselike squeaking vocalization.

Conservation Status

The Mediterranean gecko is rapidly increasing in numbers and expanding its geographic range, but its impacts on native species and systems are largely unknown and likely restricted to urban and other areas of human habitation.

Todd S. Campbell

Fence and Horned (Spiny) Lizards

FAMILY PHRYNOSOMATIDAE

Most of these lizards—including Georgia's only native species, the fence lizard (*Sceloporus undulatus*)—are fairly small, although a few may approach 20 cm (8 in) in total length. The body is relatively stocky and covered with keeled scales that give the skin a rough appearance. Horned lizards are flat and wide bodied, with horny protuberances scattered on the skin and the back of the head that give the animal the appearance of a tiny dinosaur. Fence lizards are native to forested habitats throughout Georgia, and there are a few records of Texas horned lizards (*Phrynosoma cornutum*) here as well, apparently escaped pets. Other phrynostomatid lizards are found in western and central North America, down through Central America, and on some Caribbean islands.

Adult male eastern fence lizard, Baker County (Gabriel J. Miller)

Eastern Fence Lizard

Sceloporus undulatus

Description

Moderate in size and relatively drab in color, adult eastern fence lizards are fairly stout animals that range from 10 to 18.5 cm (3.9–7.2 in) in total length, with females generally being somewhat larger than males. As a member of the spiny lizard family, the eastern fence lizard has strongly keeled scales that give it a rough or "spiny" appearance. Indeed, these rough scales easily distinguish this species from all other native Georgia lizards. Eastern fence lizards are most commonly grayish brown but may range from nearly black to brown. The grayish body color renders them almost invisible against hardwood tree bark or rocks. Most females have dark horizontal patterning on the back; males are more uniformly colored. Adult males have bright blue patches on their chin and underside that intensify during the spring breeding season. Adults from northern Georgia tend to be lighter in color and to have smaller scales than their southern counterparts. Young eastern fence lizards look identical with adults, but younger males lack the bright blue patterning. Hatchlings measure 40–65 mm (1.6–2.5 in) in total length.

Taxonomy and Nomenclature

Of the ten subspecies of eastern fence lizards found in the United States, two occur in Georgia:

Underside of adult eastern male fence lizard, Liberty County (Dirk J. Stevenson)

Adult female eastern fence lizard, Coffee County (Dirk J. Stevenson)

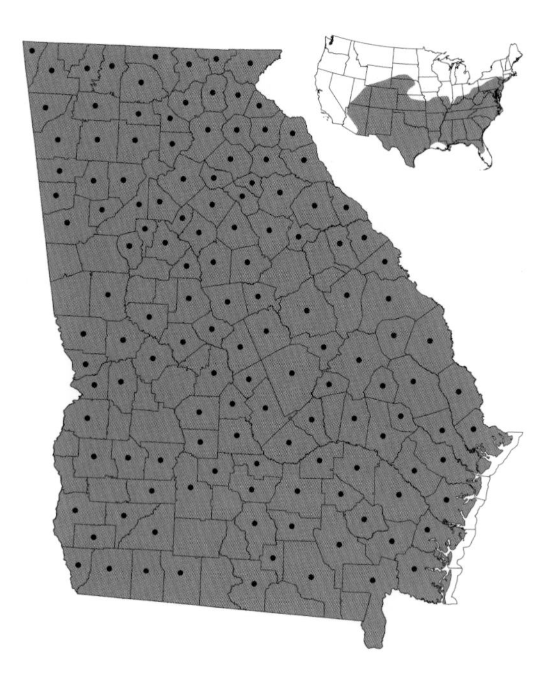

the southern fence lizard (*S. u. undulatus*) in the southern Coastal Plain and the northern fence lizard (*S. u. hyacinthinus*) in the mountains. These subspecies are similar in appearance, and intermediate individuals can be found throughout much of central Georgia. In any case, recent genetic studies have called into question the validity of the subspecies. Eastern fence lizards are known as swifts in some parts of their range.

Distribution and Habitat

Eastern fence lizards are present throughout Georgia and are common in many habitats from the mountains to the coast. They are not particularly common on barrier islands. Seldom found far from trees, eastern fence lizards are most common in dry, open forests and along forest edges, particularly sunny areas with abundant logs and stumps where they can hide. This species is less common in thick, closed-canopy forests and low, swampy habitats.

Reproduction and Development

Male eastern fence lizards are highly territorial and actively defend territories with bouts of head bobbing and push-ups. If such displays fail to repel a rival, males occasionally resort to outright combat. Courtship and mating apparently occur in spring; and females lay up to 16 (average = 8 or 9) eggs in the late spring or early summer, burying them under several inches of loose soil or in rotten wood. The eggs hatch about 70 days later. Sexual maturity is reached in less than 2 years.

Habits

Like all native Georgia lizards, eastern fence lizards are active only by day and may be observed basking in the sun even in cool spring and fall weather. Although sometimes found on the ground, they are among our most arboreal lizards and usually dash for the nearest tree when approached. When pursued, they stay on the opposite side of the tree from a predator, just as a squirrel does. Eastern fence lizards eat a wide variety of insects, spiders, and other invertebrates, generally capturing them with sit-and-wait predatory tactics. Reported predators include snakes, such as black racers (*Coluber constrictor*), rat snakes (*Elaphe obsoleta*), copperheads (*Agkistrodon contortrix*), mole kingsnakes (*Lampropeltis calligaster*), and milk snakes (*Lampropeltis triangulum*), as well as domestic dogs and cats, but this species is undoubtedly also eaten by various birds and mammals. For defense, eastern fence lizards rely on their cryptic coloration, arboreal habits, and an easily broken tail to distract a predator.

Conservation Status

Eastern fence lizards are common throughout Georgia, particularly in open forests and along field edges. This species is relatively tolerant of habitat alteration by humans, and populations usually persist as long as some forest patches remain.

John D. Willson

Anoles

FAMILY POLYCHROTIDAE

Anoles tend to be relatively small lizards; Georgia's species are typically no longer than 20 cm (7.8 in). They are covered with granular scales, and most, like Georgia's native green anole (*Anolis carolinensis*), can change their skin color between green and brown. Georgia's other species, the brown anole (*Anolis sagrei*), is a nonnative invader from the Caribbean. Anoles are abundant in much of Central and South America as well as on Caribbean islands.

Anoles are unique among lizards in having ovaries that can be in continuous production under suitable environmental conditions. Eggs are laid one at a time rather than during a restricted reproductive season.

Adult green anole, Walker County (John B. Jensen)

Green Anole

Anolis carolinensis

Description

The green anole is a small to medium-sized, slender-bodied, green or brown lizard. Adults have a total length of 12.5–20.3 cm (4.9–7.9 in); the tail makes up about 60–67 percent of that length. The maximum body length is 7.5 cm (2.9 in). The green anole has a long, pointed head with a ridge between each eye and nostril as well as two less obvious ridges on top of the head. Adhesive pads on the toes allow these agile lizards to climb on smooth surfaces. Males are about 15 percent larger than females and have a dewlap that is pink or reddish and three times larger than the female's, which is white. Juveniles are 52–67 mm (2.0–2.6 in) in total length at the time of hatching. Hatchlings and small juveniles have a relatively wider head, shorter tail, and less obvious ridges on the head than older individuals. The introduced brown anole (*Anolis sagrei*) differs in that it is never green, it often has bands or spots, and the female has a light stripe down the center of the back.

Taxonomy and Nomenclature

Two subspecies are recognized, but only the northern green anole (*A. c. carolinensis*) occurs in Georgia. Sometimes the green anole is mistakenly referred to as a chameleon because of its ability to change color rapidly. True chameleons, which are not present in the Americas, have extraordinary color-changing abilities that far exceed the brown and green of the green anole.

Adult green anole, Baker County (Anna Liner)

Adult male green anole, Monroe County (Todd Schneider)

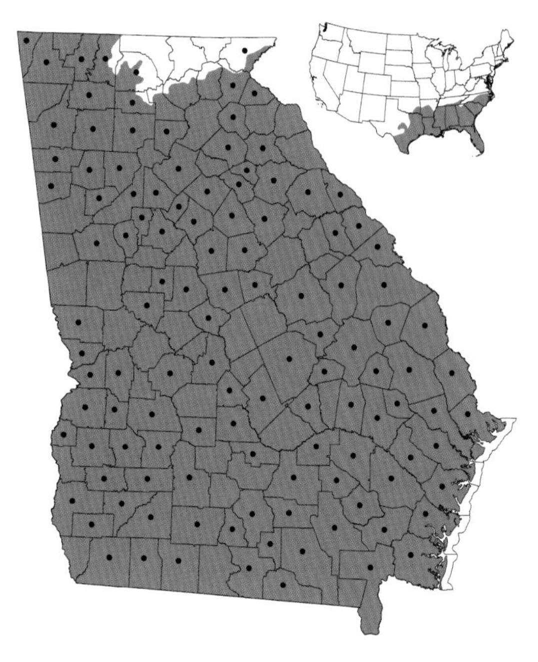

Distribution and Habitat

Green anoles are present throughout Georgia with the exception of much of the Blue Ridge province. Although they are occasionally found on the ground, they are more often arboreal. They prefer moist forests with brushy clearings and forest edges with an established shrub layer or vine component. They are also quite common in disturbed areas such as roadsides, forest edges, and old building sites that have plenty of sunlight and shrubs.

Reproduction and Development

The breeding season begins in early spring and lasts through late summer. During this time, territorial males patrol the area and perform visual displays—extending the dewlap and bobbing the head—to defend their territory and the females within it. Only one egg develops at a time, and the ovaries alternate in egg production. A female can produce 1 egg every 2 weeks during the prolonged breeding season. Eggs are soft shelled and are usually laid in a shallow depression in moist soil, leaf litter, rotten wood, or a tree hole. Measurements of 20 eggs from nests in North Carolina yielded average dimensions of 12.5 mm by 9.3 mm (0.49 by 0.36 in). Eggs hatch after 5–7 weeks, from late May to early October, and hatchlings do not receive any parental care. The juveniles mature in about 8 months.

Habits

Green anoles are diurnal and are active every month of the year in Georgia, with peak activity occurring in spring and fall. In winter, they are active when the weather is sunny and unseasonably warm. Their diet consists primarily of insects, spiders, and other arthropods. Major predators include broadhead skinks (*Eumeces laticeps*), snakes, and predatory birds. An anole's tail can break off, wiggle, and distract a predator while the lizard escapes. A new tail grows in place of the one that was lost. Males are strongly territorial and use their dewlap for both territorial and courtship displays. When a male approaches another male's territory, the defending male compresses his body, extends his dewlap, bobs his head, and attempts to chase the rival away. If the intruder does not leave, the two may fight aggressively. Despite popular belief, the green anole cannot change color to match its background. Its color varies depending on the anole's body temperature, stress, activity, and behavior. When it is active or in bright light, a green anole is usually light emerald green, but in damp, cool, or dark conditions, when it is generally less active, it changes to dull olive, brown, or gray.

Conservation Status

Populations of this lizard are quite common, and individuals are abundant throughout their range in Georgia. Their ability to live in disturbed areas is one reason for their abundance. Where the nonnative brown anole has invaded habitats occupied by the green anole, the former has influenced habitat use by the latter. The long-term effects of this introduction on populations of green anoles are unknown.

Gabrielle J. Graeter

Adult male brown anole, Taylor County, Florida (John B. Jensen)

Brown Anole

Anolis (Norops) sagrei

Description

This relatively small lizard has a maximum body length of less than 7 cm (2.7 in). In central and southern Florida, adult males are 6.4–6.8 cm (2.5–2.7 in) and females are 4.6–4.8 cm (1.8–1.9 in) in body length; both sexes are slightly larger in northern Florida and Georgia. Individuals vary dramatically in color and pattern. Brown anoles are similar in overall appearance to our native green anoles (*A. carolinensis*), but certain characters help distinguish the two. Brown anoles are stockier than green anoles and have a shorter, wider head; longer legs and toes; and smaller toe pads. Individual brown anoles can change color—for example, males can range from gray or nearly white to orange, brown, or black—but they are never bright green. Brown anoles may also have a highly variable pattern of chevrons or block-shaped spots on the back as well as a series of light markings. Males have a bright orange or brick red dewlap, which they extend during territorial displays. Territorial males engaged in combat are usually black with small light spots. Unlike that of green anoles, the dewlap of brown anoles is often edged in white or yellow. Males may also have a fleshy crest that can extend along the back. Females exhibit a color range similar to that of males but also have a distinct light line bordered by a wavy pattern along the midline of the back; some have a bright red head. Hatchlings are 16 mm (0.6 in) in body length and differ little from the adults in general coloration and appearance.

Gravid adult female brown anole, Lowndes County (John B. Jensen)

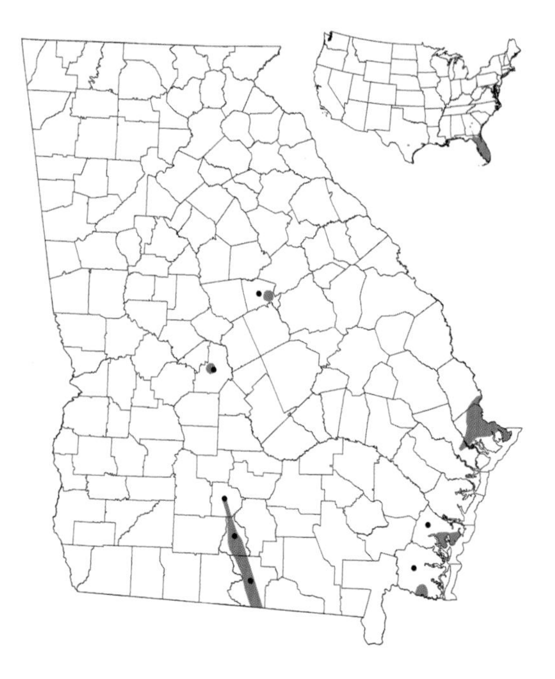

Taxonomy and Nomenclature

Although two subspecies, the Bahamian anole (*A. s. ordinatus*) and the Cuban anole (*A. s. sagrei*), were introduced into Florida, the Bahamian anole has been genetically overtaken by the Cuban form. Some taxonomists suggest that this species should be placed in the genus *Norops*.

Distribution and Habitat

A native of Cuba, the Bahamas, and their satellite islands, the brown anole was established in southern Florida by the early 1940s. By 2002 brown anoles were present in every county in peninsular Florida. This species is continuing to expand its range in the southeastern United States. Brown anoles first became established in Georgia in the late 1980s at hotels, plant nurseries, residences, and businesses along Interstates 75 and 95 and nearby coastal towns. This "fencepost lizard" prefers open, disturbed habitats such as forest edges, vacant lots, agricultural areas, residential areas, and subdivisions, and can be found in even the most sparsely vegetated urbanized or industrial areas. However, it also occurs in treeless or otherwise open natural habitats such as coastal scrub and enters forested areas along roads, bike paths, boardwalks, and even narrow hiking trails where the tree canopy has been thinned.

Reproduction and Development

Brown anoles reproduce from about April to September in central Florida, but their reproductive season may be more restricted in Georgia because of our longer winter. Males defend territories encompassing a few females by displaying their dewlap, performing push-up displays, and even erecting the crest on the back. About every 6 days females deposit a single pea-sized egg in organic soil, leaf litter, or under the roots of a cabbage palm. Hatchlings emerge in 25–30 days. In Florida, brown anoles quickly grow to adulthood in their first summer of life, reproduce throughout their second summer, and die the following fall or winter, essentially making them "annuals" with about an 18-month life span in the wild. Individuals are known to live longer in captivity, however.

Habits

Brown anoles are efficient migrants and are most often introduced to new locations by people unaware of their presence. Individuals catch rides on shipments of lumber and potted ornamental plants and even jump up into the undercarriage and bumpers of vehicles. Brown anole populations expand most rapidly along transportation corridors, and at one point in the 1990s they were present at only the northbound rest stops along I-75, indicating they were being transported unintentionally on vehicles. They are, however, also intentionally introduced by irresponsible pet owners. Once established, brown anole populations spread very quickly due to their high reproductive output. These arboreal lizards commonly perch head-down at about waist height or lower on large-diameter perches such as tree trunks, fence posts, and brush piles, where they sit and wait for prey animals to walk by. The prey is quickly pounced on, subdued, and carried back to the perch. The diet consists mostly of ground-dwelling arthropods, including cockroaches, beetles, and earwigs, but brown anoles prefer ants to all other foods. They will consume any moving organism they can handle, though, including hatchling green anoles and even hatchling brown anoles. Because they reach such high densities (up to one per square meter [1.2 yd^2] on disturbed sites), brown anoles on a small plot of land can literally have tens of thousands of arthropods in their stomachs at a given time and could disrupt natural food webs. Likely predators include lizard-eating snakes, mammals, and birds.

Conservation Status

Brown anoles are rapidly increasing in number and expanding their geographic range. They compete with native green anoles for food, but because they also eat green anole hatchlings they are capable of reducing green anole populations within a few years of their arrival.

Todd S. Campbell

Skinks

FAMILY SCINCIDAE

Although most of the six skink species native to Georgia are much smaller, some species found elsewhere in the world reach 35 cm (13.8 in) in total length. Skinks typically have small scales that give the skin a glossy appearance. The young of many species—including the blue-tailed lizards familiar to many Georgians—have a brightly colored tail. Skinks are present statewide in Georgia. The 1,200 or so species of this family are scattered across most of the warm regions of the globe, including the Americas, Eurasia, Africa, and Australia. Skinks occur in a variety of environments; Georgia's forms exploit habitats ranging from forests to wetlands. Some skinks give birth to fully developed young, although most species, including all of Georgia's, lay eggs. The females of many species exhibit parental care by coiling around the egg clutch until hatching. As would be expected from such a large family, skinks occupy a variety of ecological niches. Georgia counts ground-dwelling forms, burrowers, and climbers among its skink fauna.

Adult coal skink, Habersham County (Alex Pyron)

Coal Skink

Eumeces (Plestiodon) anthracinus

Description

This moderate-sized skink is usually 5–6.5 cm (2–2.5 in) in body length and 12.7–17.9 cm (5–7 in) in total length. Mature females are slightly larger than males. The smooth scales give the skin a glossy appearance. This species is unique among Georgia's skinks in having only a single postmental scale in the midline of the lower jaw immediately behind the chin (mental) scale. Other skinks of this genus have two such scales. There are normally six or seven labial scales on each side of both the upper and lower lips. Four thin, light lines extend from the head to the hind limbs or onto the tail. Two of these lines begin along the upper lips or at the ear openings and extend along the lower sides; the other two begin just above the eyes and extend along the edges of the back and sides. The back color varies from olive gray to brown and may have thin, dark lines extending along it between the light lines. The dark brown or black spaces between the light lines on the sides give the coal skink its common name. The belly is gray or bluish gray. Coal skinks from northern Georgia may also have a light line along the center of the back. Those from the Coastal Plain have lip scales with a light center and dark edges; specimens from northern Georgia normally have uniformly colored lip scales. There are no whitish or light gray lines on the head, although juveniles may have orange lines; except for old individuals, other skinks of this genus have whitish lines on their heads. The lips and lower parts of the face of adult males turn bright orange during the breeding season, and the orange

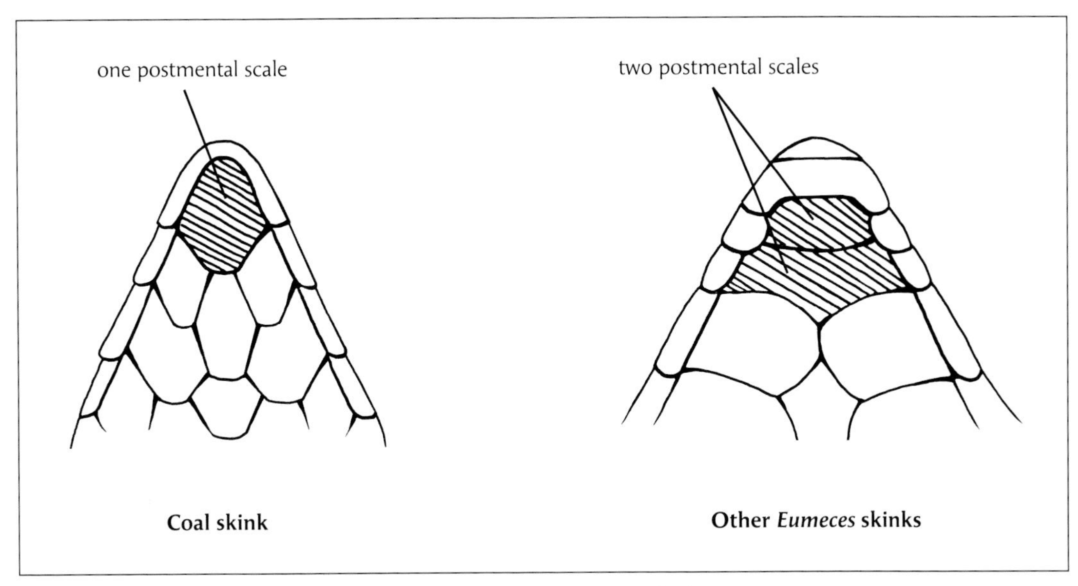

Comparison of Scales on the Chin of Coal Skinks and Other *Eumeces* Skinks

may extend back to the front limbs. Those areas of the head and shoulders are often salmon colored in adult females. Unlike other skinks in the genus *Eumeces*, the head of adult males is never enlarged and remains small and conical even during the breeding season. Juveniles are darker than adults, and those from the Coastal Plain may be virtually black without any markings. The tail of juveniles from the Coastal Plain is blue, and the tail of juveniles from northern Georgia is often iridescent and may vary from indigo to violet. Hatchlings average 20–25 mm (0.8–1.0 in) in body length.

Taxonomy and Nomenclature

Two subspecies are recognized. The southern coal skink (*E. a. pluvialis*) occurs in the Coastal Plain. Populations in northern Georgia show characteristics of both that subspecies and the northern coal skink (*E. a. anthracinus*) and are thought to be intermediate between the two. The coal skink is not closely related to Georgia's other *Eumeces* species. Scale and color patterns indicate that it is more closely allied to the four-lined skink (*E. tetragrammus*) and prairie skink (*E. septentrionalis*) of the western United States. Some taxonomists believe that most North American species in the genus *Eumeces*, including all of those in Georgia, should be placed in the genus *Plestiodon*.

Distribution and Habitat

The distribution is very spotty and probably underestimated given the secretive nature of this skink. Populations are known from the Blue Ridge and adjacent upper Piedmont in the northeastern corner of the state and from the upper Coastal Plain in western Georgia. This lizard requires moist conditions. In northern Georgia it occurs in hardwood forests and may venture out onto the edges of lawns or road cuts to bask on warm days during the winter. Coastal Plain populations are usually associated with wetlands or oak-pine forests near water.

Reproduction and Development

The mating season begins during the winter, usually by late January, and extends until early April. Males treat their mates rather rudely during courtship, biting them on the back of the neck and shaking them violently. Females

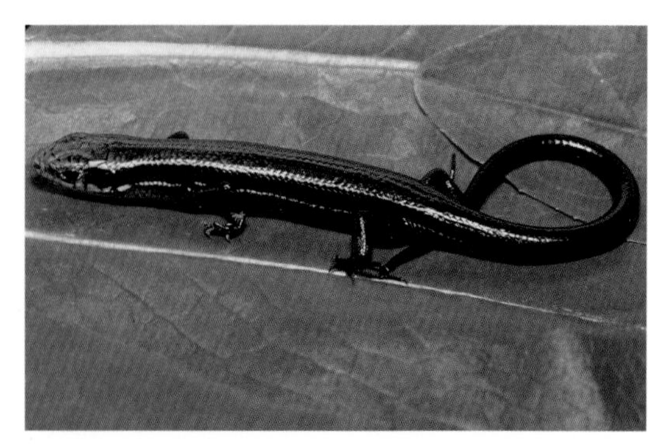

Juvenile coal skink, Habersham County (Alex Pyron)

typically lay four to eight eggs in the spring under cover or underground and remain with them until they hatch in midsummer. Sexual maturity is reached in 2 years.

Habits

Coal skinks are active throughout the year. An ecological study of a population in Habersham County involved monitoring a drift fence with pitfall traps over the course of 16 months. Coal skinks were caught in the traps during every month except October and December. Mature individuals are most active in February and March, which coincides with the peak of the breeding season. During the winter, coal skinks come out to bask when air temperatures exceed approximately 13° C (55° F). They are ground-dwelling lizards, and unlike Georgia's other members of this genus typically do not climb. They forage for small arthropods in the leaf-litter zone of the forest floor and disappear underground when alarmed. Individuals may also enter water to escape predators. At least in northern Georgia, this species appears to be much more fossorial than most published accounts of its behavior indicate. Documented predators include snakes, large lizards, and mammals. The tail breaks off easily when this lizard is attacked.

Conservation Status

This is a very secretive and seldom-encountered lizard. Whether that is due to its actual rareness or its ability to go unnoticed is not known. Because it appears to be rare throughout its range, the coal skink enjoys some level of legal protection in at least five states, but it is not given special protection in Georgia.

Carlos D. Camp

Adult mole skink, Long County (Dirk J. Stevenson)

Mole Skink

Eumeces (Plestiodon) egregius

Description

Adults of this small, slender, elongate lizard are 9–15.2 cm (3.5–5.9 in) in total length and have a maximum body length of 5.7 cm (2.2 in). Adult females are slightly larger than adult males. Unless it has been shortened by injury, the tail length is 165–180 percent of the body length. The background color is gray, brown, or tan. A light stripe runs along each side of the back; these stripes may extend from the head to the base of the tail, but their length is variable. The belly is creamy gray. The relatively heavy tail is pink, reddish, red-orange, or red-brown throughout life. The limbs are small and widely separated when folded against the body. Scales are smooth, shiny, and overlapping. Two rows of enlarged scales extend along the middle of the back. Males have a red-orange wash on the belly and sides that may extend onto the sides of the face during the mating season. Hatchlings measure approximately 20 mm (0.8 in) in body length and 50 mm (2.0 in) in total length. They differ from adults in having a dark chocolate brown background color, more distinct stripes, and a comparatively shorter, more vividly colored tail.

Taxonomy and Nomenclature

The northern mole skink (*E. e. similis*), formerly known as the red-tailed skink, is the only recognized subspecies present in Georgia; the other four are restricted to Florida. (See **Coal Skink** account concerning the genus name *Plestiodon*.)

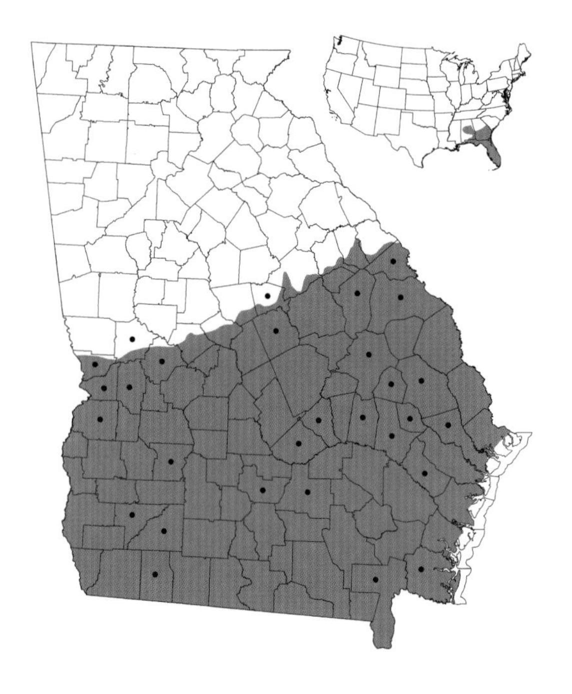

Distribution and Habitat

In Georgia, the mole skink's distribution is limited to the Coastal Plain. This burrowing skink is well suited to arid conditions and is found only in open, well-drained, sandy habitats. Longleaf pine–turkey oak sandhills are preferred, but scrubby oak hammocks and sandy fields may also be acceptable. Extensive areas of unvegetated bare soil maintained by periodic fires or other disturbance are important to this lizard's survival. Shading and litter accumulation have an adverse effect. High, dunelike sand ridges along blackwater rivers and streams (e.g., the Canoochee and Ohoopee rivers) are prime mole skink habitat, as are sandhills supporting populations of the southeastern pocket gopher.

Reproduction and Development

Mating occurs during the fall and winter. Males show a marked response to the scent of females. During copulation, the male bites the sides of the female, creating small V-shaped scars that last for several months. Females lay two to nine (usually three to five) small, elliptical eggs from April to mid-June and brood them until they hatch 1 or 2 months later. Individuals attain sexual maturity in their first or second year.

Habits

Mole skinks spend much of their lives underground. They "swim" through loose, dry sand by plunging the snout into the sand, folding the front limbs back along the body, and undulating the body and tail from side to side. In cooler months they are especially fond of basking within the raised mounds of sand pushed up by southeastern pocket gophers. They may also reside in the small sand piles excavated by geotrupine scarab beetles and the sandy aprons of gopher tortoise (*Gopherus polyphemus*) burrows. The body temperature of individuals basking in these microhabitats may range from 26 to 34° C (79–93° F). Numerous specimens have been collected in March and April in turkey oak sandhills by raking the sand beneath sun-warmed objects such as logs, tin, and even old discarded television sets and rusty cans. Mole skinks are gregarious and, at sites where they are found, may occur in considerable numbers. They are opportunistic predators that feed on a variety of invertebrates, including roaches, spiders, crickets, termites, beetle larvae, and ants. An individual may dash out from its refuge to capture its prey. Like other skinks, mole skinks can break off their tail when roughly handled or grasped by a predator. Known predators include eastern coachwhips (*Masticophis flagellum*), black racers (*Coluber constrictor*), scarlet kingsnakes (*Lampropeltis triangulum elapsoides*), and pigmy rattlesnakes (*Sistrurus miliarius*), and probably also scarlet snakes (*Cemophora coccinea*) and eastern coral snakes (*Micrurus fulvius*).

Conservation Status

This lizard is uncommon in Georgia, though precise evaluation of its status is difficult given its highly secretive lifestyle. Georgia's mole skinks are probably experiencing a slow decline in step with the gradual loss of naturally functioning longleaf pine sandhill ecosystems. The decline, bordering on near extirpation, of the southeastern pocket gopher in the Atlantic Coastal Plain of southeastern Georgia may have negative repercussions for mole skink populations.

Dirk J. Stevenson

Adult five-lined skink, Appling County (Dirk J. Stevenson)

Five-lined Skink

Eumeces (Plestiodon) fasciatus

Description

Adult five-lined skinks are 12.5–21.5 cm (4.9–8.4 in) in total length. Their color is highly variable and depends on age and sex. Scale patterns are probably the best characters to identify this species. Juveniles often exhibit the color pattern that gave this species its name. Five white or yellowish stripes set off by a dark, often black, background run from the head to the base of the bright blue tail, then continue along much of the tail as blue stripes. As females age, the stripes fade and the tail turns gray. Males become nearly uniform brown or tan with age, the tail turns gray, and only very faint stripes remain; a few, presumably old, individuals lose all traces of stripes. The jaws of adult males become bright reddish orange during the breeding season. Similar species include the broadhead skink (*E. laticeps*) and the southeastern five-lined skink (*E. inexpectatus*); the juveniles of both are nearly identical with five-lined skinks in pattern. Five-lined skinks have wider stripes than do the other two species, but stripe width is a relative trait, and juveniles can be positively identified only by looking at scale patterns. Under the base of the tail, the scales in the middle row are enlarged in five-lined and broadhead skinks but not enlarged in southeastern five-lined skinks. Five-lined skinks have two enlarged postlabial

Juvenile five-lined skink, Monroe County (John B. Jensen)

Adult male five-lined skink, Lawrence County, Alabama (William Sutton)

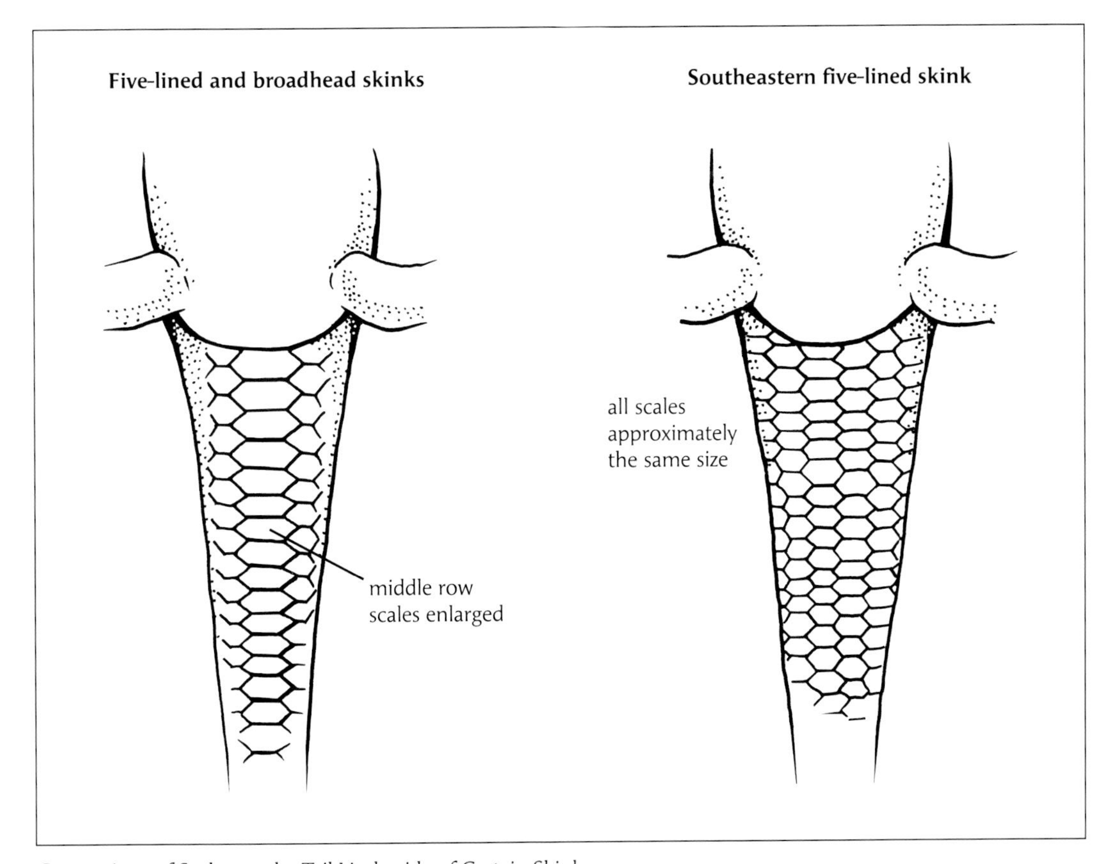

Comparison of Scales on the Tail Underside of Certain Skinks

scales at the back of the jaw and usually four labial scales on the upper lip. Broadhead skinks have no enlarged postlabials and usually five labials. Adults, particularly adult males, are more easily distinguished. Broadhead skinks become much larger, and the males lose the stripes entirely. Male southeastern five-lined skinks lose the middle stripe with age but typically retain the other four stripes; the sides remain dark so that the region between the two side stripes appears black in contrast to the brown of the back. Hatchlings of all three species exhibit strong patterns, often having bright orange stripes on the head. Hatchling five-lined skinks are 51–64 mm (2.0–2.5 in) in total length.

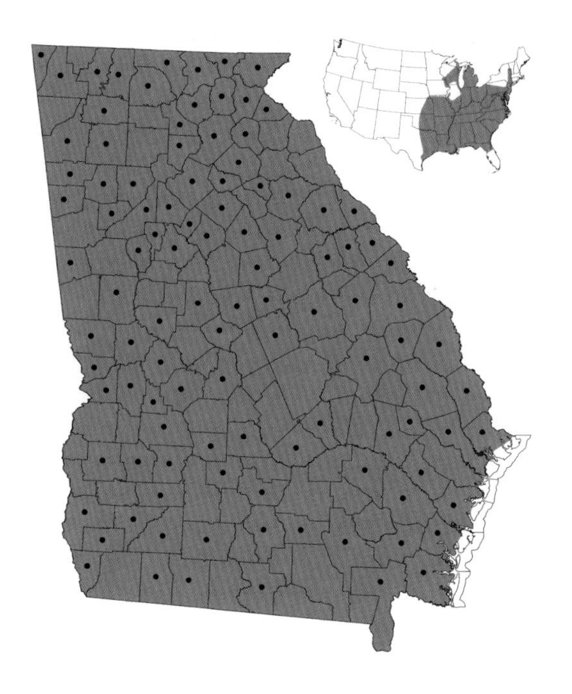

Taxonomy and Nomenclature

No subspecies are recognized. Five-lined skinks are sometimes colloquially known as blue-tailed skinks or scorpions. (See **Coal Skink** account concerning the genus name *Plestiodon*.)

Distribution and Habitat

Five-lined skinks are common in wooded habitats throughout Georgia. They are typically found on the ground in, on, or under logs, leaf piles, or rocks, in moist but not wet microhabitats. Five-lined skinks are also commonly found on and near houses in suburban neighborhoods.

Reproduction and Development

Five-lined skinks breed in the spring. Each female lays 4–14 eggs in late spring or early summer, usually inside a rotten log or stump. Females typically remain with their eggs until hatching and do not actively forage during that 1–2-month period. Some females have been observed to eat one or more eggs at this time.

Habits

Five-lined skinks typically bask on logs or on the ground rather than high on tree trunks as broadhead skinks do, although they may climb short distances to escape danger. On cool days, they are usually inactive and remain in or under rotten logs, under rocks, or under bark piles. A study in Habersham County comparing the ecology of this species with that of coal skinks (*E. anthracinus*) over a 16-month period found that adult activity was concentrated in May while juveniles were active throughout the warm-weather months. Five-lined skinks feed on a variety of small insects, worms, spiders, and, on occasion, young lizards and even mice. They are skittish lizards, often quick to flee from approaching humans, and if captured they are usually just as quick to bite but are incapable of breaking the skin. As is true for other skinks, their tail breaks easily when they are attacked. After detachment, the tail remains twitching on the ground while the skink escapes. If the predator does not consume the tail within a short time, the twitching motion may cause the tail to bury itself in substrate such as leaf litter. Following a successful escape, the lizard may return later and use chemical cues to find and consume its detached tail. Juveniles occasionally wave their

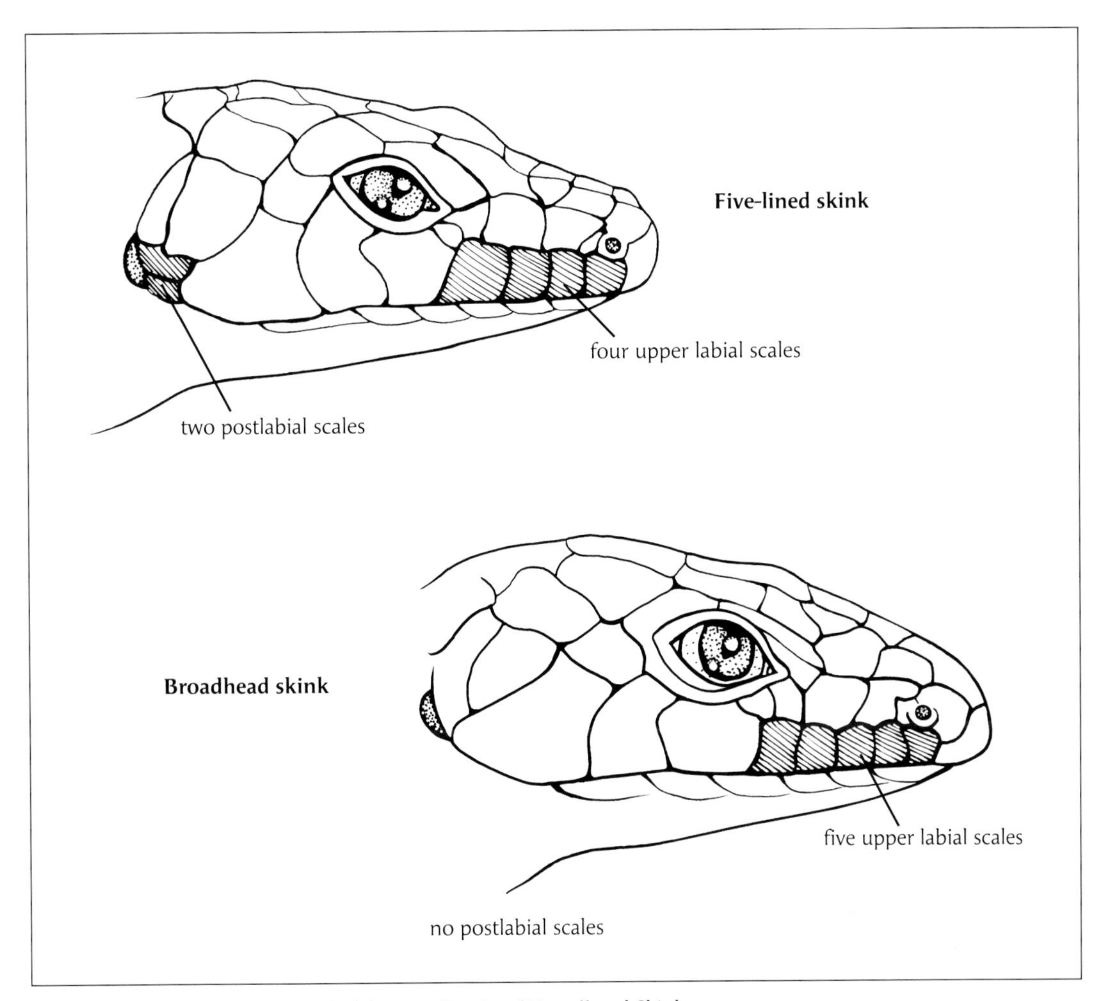

Scale Characteristics on the Head of the Five-lined and Broadhead Skinks

tail when threatened; whether this behavior is an invitation for a predator to attack the tail has not been tested. Skinks, particularly their tails, have been purported to be toxic to household pets, especially cats, but this has not been convincingly demonstrated. Predators include snakes, predatory birds, carnivorous mammals, and the broadhead skink.

Conservation Status

Five-lined skinks are locally abundant within appropriate habitats in Georgia. They can persist in human-altered landscapes such as gardens and forest clear-cuts if insect prey and cover objects remain available.

Christopher T. Winne and
Whit Gibbons

Adult female southeastern five-lined skink, Chatham County (John B. Jensen)

Southeastern Five-lined Skink

Eumeces (Plestiodon) inexpectatus

Description

The southeastern five-lined skink ranges from 14 to 21.6 cm (5.5–8.4 in) in total length. Males grow larger than females. The tail length may be as much as one and a half times the body length. The best way to identify this species is by the uniformity of the scales on the underside of the tail; five-lined (*E. fasciatus*) and broadhead (*E. laticeps*) skinks have an enlarged middle row of scales under the tail (see figure, p. 309). Coloration is highly variable depending on age and sex. Young, averaging approximately 60–70 mm (2.3–2.7 in) in total length, are shiny black with five light stripes; the middle stripe down the center of the back is often thinner than the other stripes. The tail is a striking bright blue in younger animals but fades with age. Adult females retain some juvenile coloration but typically are considerably faded. Adult males typically lose the middle stripe, and the sides remain nearly black; the back becomes brown or bronze. The jaws of adult males enlarge so that the head takes on an arrowhead shape and may be orangish red, especially during the breeding season.

Taxonomy and Nomenclature

No subspecies are recognized. The three medium to large species of southeastern *Eumeces* (southeastern five-lined, five-lined, and broadhead skinks) are very similar in appearance, particularly the young and the females, and the

Adult male southeastern five-lined skink, Monroe County (John B. Jensen)

Juvenile southeastern five-lined skink, Long County (Dirk J. Stevenson)

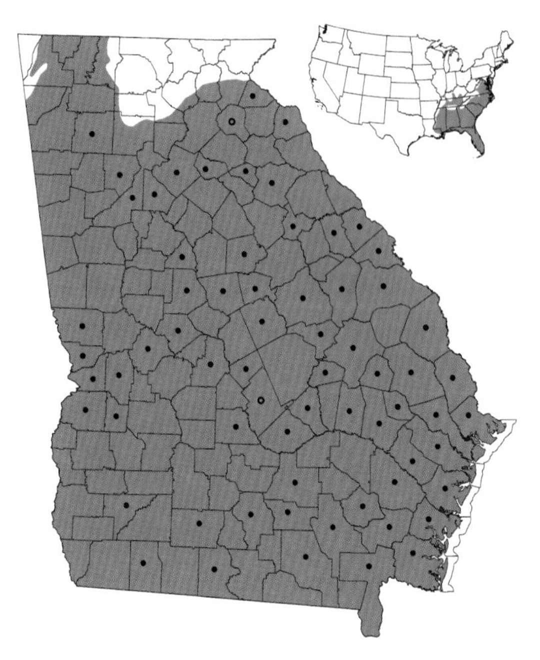

three are believed to be closely related. They are often collectively referred to as blue-tailed skinks. In some areas blue-tailed skinks are locally called scorpion lizards or simply scorpions and are mistakenly thought to be venomous. Historically the southeastern five-lined skink was called the Floridan five-lined skink. (See **Coal Skink** account concerning the genus name *Plestiodon.*)

Distribution and Habitat

The southeastern five-lined skink resides in a wide variety of habitats including natural, urban, and residential environments. Although they often bask on fallen or horizontal logs, fences, or other low structures, these skinks do climb and certainly do so readily to avoid predators or capture. They appear to have a preference for drier habitats than the five-lined and broadhead skinks, and they can be found on many of the state's barrier islands as well as throughout the dry Coastal Plain sandhills. In these areas, although they may be found scurrying through pine needles on the forest floor, they are more likely to be observed on stumps or fallen trees, especially in gaps where sunlight penetrates the forest canopy. In areas of rocky outcrops, common in the Piedmont, these skinks use cracks and crevices as retreats, coming out on the bare rocks to bask.

Reproduction and Development

Males and females can be seen together, especially during the breeding season, and it is not uncommon to find a pair basking in close proximity, often with the male resting his chin on the female's back. Southeastern five-lined skinks breed in early spring and may deposit eggs any time from early June through August. Pregnant females, often distended with eggs, bask frequently, especially just prior to egg-laying.

Females find a secure place, such as a rotting log or vegetation pile, in which to lay their eggs. They attend the eggs throughout incubation, which may take 2–8 weeks. Clutch size typically ranges from 3 to 8 eggs, with a current record of 12 eggs. Sexual maturity is attained at around 21 months of age.

Habits

The main activity periods appear to be in spring and early summer, but these skinks may be found during warmer periods throughout the year. Southeastern five-lined skinks are most often observed while basking; however, during cloudy or cooler weather they can be found under ground debris such as boards and logs as well as in trash piles. Although considerably less arboreal than broadhead skinks, southeastern five-lined skinks ascend high into the trees when pursued or threatened. Specimens are most often observed at ground level and are alert and quick to take cover when disturbed. Invertebrates, including insects, spiders, and arthropods, are probably the preferred foods, and captive specimens readily accept crickets and mealworms. Scarlet kingsnakes (*Lampropeltis triangulum elapsoides*) and scarlet snakes (*Cemophora coccinea*) are known predators. Reports of toxicity, especially to cats, have cropped up in parts of the Southeast, including Georgia, but no scientific tests have confirmed them.

Conservation Status

Although little current information exists on population stability or status, this species is not thought to be vulnerable at this time. In field studies of scrub habitat in central Florida, southeastern five-lined skinks were among the most numerous species encountered.

Gregory C. Greer

Adult male broadhead skink, Grady County (Pierson Hill)

Broadhead Skink

Eumeces (Plestiodon) laticeps

Description

The broadhead skink is the largest of the skinks found in Georgia, reaching a total length of more than 30 cm (11.7 in) and a body length of 14 cm (5.5 in). The scales are smooth, flat, and overlapping. Males have very wide jaws that make their head wider and longer than the head of females of the same length. During the spring breeding season the male's head may get even larger and may become bright orange or red. Color patterns vary considerably with age and sex, although all broadhead skinks have an unmarked belly. Young animals have five or seven distinct, lightly colored stripes extending the length of the back against a black background and a bright blue tail, but these patterns fade with age; the stripes become less striking and the tail turns gray. Males lighten in color, becoming solid brown or olive. Females tend to retain some of the juvenile pattern but also turn brown as their stripes fade. Adult females often retain faint stripes on their sides and sometimes on the center of the back along with a broad, dark band extending along each side. Scale characters are essential to distinguish juvenile broadhead skinks from five-lined (*E. fasciatus*) and southeastern five-lined (*E. inexpectatus*) skinks. The middle row of scales under the tail of broadhead and five-lined skinks is wider than the other scale rows; in southeastern five-lined skinks all scales are the same size. The broadhead skink also has five labial scales and no large postlabial scales at the back of the jaw, in contrast to the five-lined skink, which has four upper labial and two postlabial scales (see figures, pp. 309 and 311). Hatchlings are 57–86 mm (2.2–3.4 in) long.

Juvenile broadhead skink, Monroe County (John B. Jensen)

Adult female broadhead skink, Liberty County (Dirk J. Stevenson)

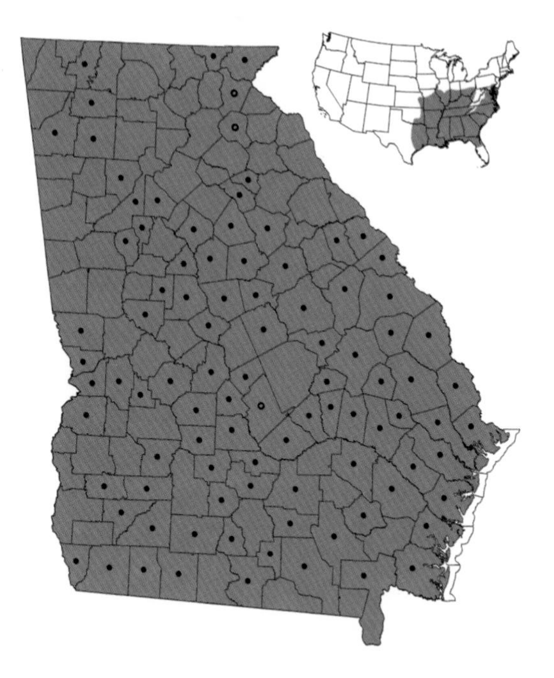

Taxonomy and Nomenclature

No subspecies are recognized. Local names for broadhead skinks include scorpion and red-headed scorpion because they are mistakenly thought to be venomous. This and the other two similar skink species are often collectively known as blue-tailed skinks. (See **Coal Skink** account concerning the genus name *Plestiodon*.)

Distribution and Habitat

Broadhead skinks are present throughout Georgia in habitat types ranging from residential areas to wet forests. They are most often found in moist woodlands in trees or under debris such as decaying logs. Large trees such as live oak and cypress are particularly likely to harbor these skinks, and they are among the most common lizards on Georgia's coastal islands.

Reproduction and Development

Mating occurs from late April to early June in Georgia. Males are larger and have a larger head than females, and they fight one another for mating opportunities. Their bright orange or red head stimulates aggression in other breeding males. The males guard their associated females from intruders, especially when near a favorable nesting site. Larger males enjoy an advantage during these confrontations and have higher mating success than smaller ones. Small males avoid fights with larger males and attempt to court females only in the absence of larger males. Females apparently do not use color in their choice of mates, but do prefer large males with a large head. Females lay clutches of 6–19 soft white eggs in cavities within decomposing logs during summer, usually in June or July. They attend their nests, rarely leaving until the eggs are ready to hatch, and may move the eggs to a new nest site if conditions at the old one become unfavorable. The brooding period is between 1 and 2 months long. The young hatch from late July to September. Sexual maturity is not reached until 21 months of age.

Habits

Broadhead skinks are inactive at low temperatures, and individuals aggregate in hibernation sites during winter. Spring emergence occurs in March or April. These are the most arboreal skinks in Georgia, and they often escape from predators by climbing. Natural tree holes and excavated cavities such as those created by woodpeckers are common refuges. Predators include raptors and snakes, especially eastern coachwhips (*Masticophis flagellum*) and black racers (*Coluber constrictor*), and carnivorous mammals, including house cats. Interestingly, males accompanied by females allow predators to approach closer before fleeing than males alone. Like many lizards, broadhead skinks are likely to lose their tail when harassed. The wiggling tail distracts the enemy and allows the skink to escape. They can regenerate the tail later, but with cartilage rather than bone. Broadhead skinks have a varied diet that includes insects and their larvae, spiders, pill bugs, snails, and small vertebrates, including other skinks. They have powerful jaws for grasping and killing prey. While foraging on the ground they use both visual and chemical cues to locate and identify prey. Many of their victims are inactive during the day when the skinks are foraging, and are probably flushed from retreats before capture. Broadhead skinks feed on wasps and bees without much ill effect, but they usually avoid eating velvet ants (a type of wasp whose females are flightless). They are able to distinguish members of their own species from closely related skinks by their odors and test for these by regularly tasting the substrate and encountered objects with their tongue. These skinks often bite when captured, but the bite is not especially painful and usually does not break the skin.

Conservation Status

Broadhead skinks are common throughout Georgia except in the mountains.

Dean A. Croshaw

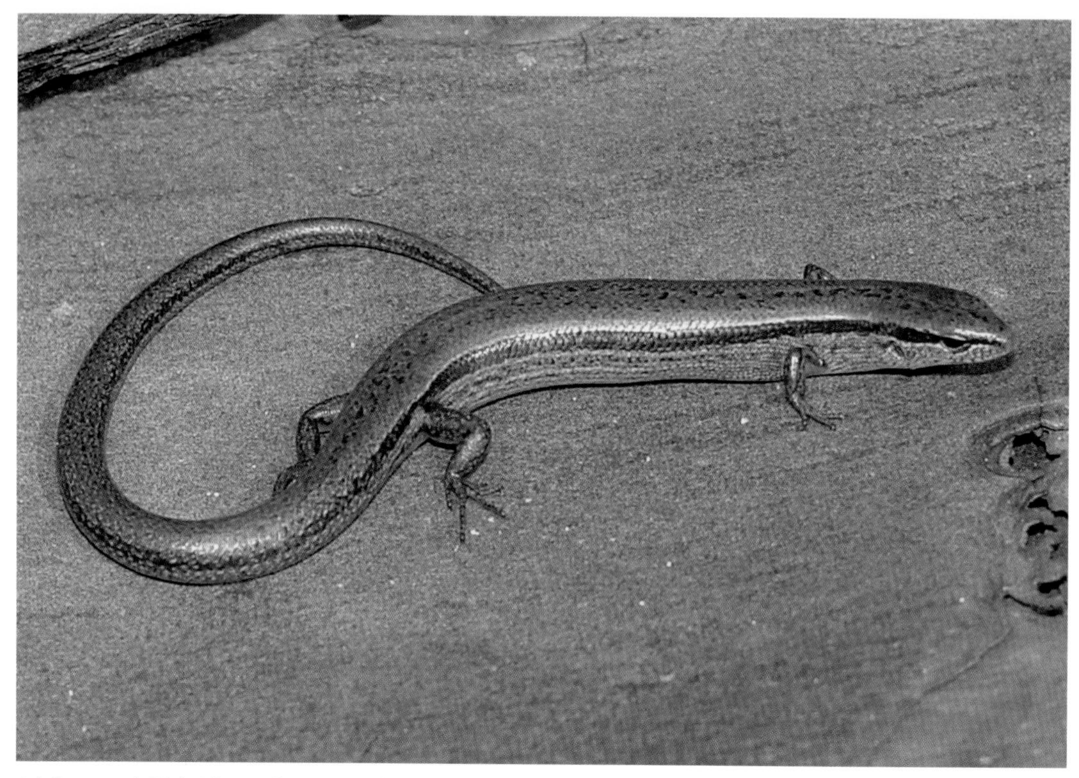

Adult ground skink, Liberty County (Dirk J. Stevenson)

Ground Skink

Scincella lateralis

Description

Ground skinks are small, slender lizards with a long tail, short legs, and smooth, shiny scales. The total length is usually 7.5–14.6 cm (2.9–5.7 in), and the maximum body length recorded is 5.7 cm (2.2 in). The background color ranges from reddish to dark brown, and two dark stripes run down each side of the back. The belly is white or yellowish. Males can be distinguished from females by their yellow belly, particularly during the spring mating period. Hatchlings are about 44 mm (1.7 in) in total length and are identical with adults in appearance. Ground skinks can be distinguished from all of the true skinks (genus *Eumeces*) in Georgia by their lack of light longitudinal stripes on each side of the back.

Taxonomy and Nomenclature

Although the skinks represent a large family of lizards, ground skinks are the only species in the genus *Scincella*. No subspecies are recognized in Georgia or elsewhere. Because of their distinctive coloration, ground skinks are sometimes called brown-backed skinks or little brown skinks.

Distribution and Habitat

Ground skinks are distributed throughout Georgia in forests, grassy woodlands, and even in yards in suburban areas. They prefer habitat with abundant leaf litter and loose soil, especially pine or mixed pine-hardwood forests, and typically do not frequent moist areas. They can, however, be common in dry portions of bottomland forests. They are often found under logs, pine straw, or other debris in wooded areas.

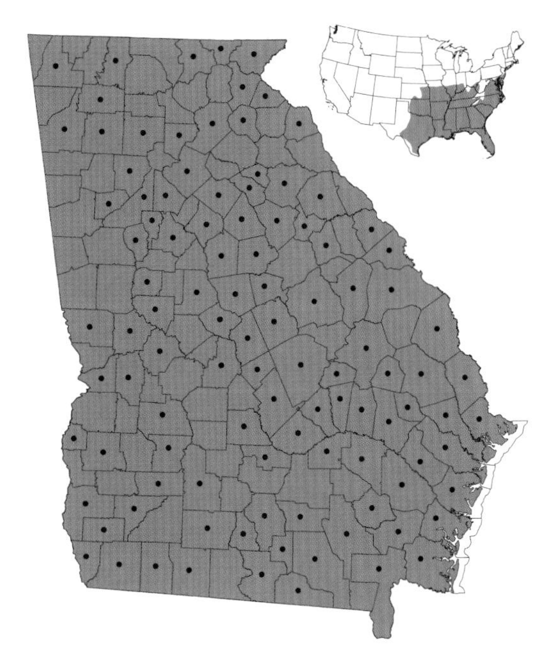

Reproduction and Development

Mating occurs from January through August, but little is known about courtship. Females nest from April to August, laying an average of two or three—up to as many as seven—eggs per clutch. Ground skinks do not guard their nests as some of the true skinks do and may nest several times during the nesting season. When a clutch is deposited, the embryos are already partially developed in the eggs. Females reach maturity and are able to lay eggs the year after they hatch.

Habits

Ground skinks are active during the day. They are most often found during the spring, early summer, and fall but may also be active on cool days during late summer and warm days in winter. They can be seen, or at least heard, scurrying through pine straw and leaf litter, where they are well camouflaged. Their small legs make them appear to "swim" through the leaf litter. Ground skinks seldom climb and spend the majority of their time searching for small insects on the forest floor. They often lose their tail when handled or attacked by a predator. Known predators include eastern bluebirds, black racers (*Coluber constrictor*), and common kingsnakes (*Lampropeltis getula*). Presumably other birds, lizard-eating snakes, and other large skinks also eat ground skinks.

Conservation Status

Ground skinks are common throughout Georgia, and populations appear to be stable.

Brian S. Metts

Racerunners and Whiptails

FAMILY TEIIDAE

Many members of this family, including Georgia's single representative, the six-lined racerunner (*Cnemidophorus sexlineatus*), are less than 40 cm (15.8 in) in total length, but the South American tegus may exceed 100 cm (39 in). Teiid lizards are covered in tiny granular scales everywhere except for the belly, which has distinct rows of large rectangular plates. The 120 or so species in this family are restricted to the New World, ranging over much of both continents as well as the Caribbean islands. Many of them inhabit open, often very dry environments such as deserts or grasslands. Six-lined racerunners are present throughout Georgia. Some non-Georgia species consist only of parthenogenetic females, which lay viable eggs without the benefit of male fertilization.

Adult male six-lined racerunner, Bryan County (Dirk J. Stevenson)

Six-lined Racerunner

Cnemidophorus sexlineatus (Aspidoscelis sexlineata)

Description

Six-lined racerunners are moderate-sized lizards with a long tail. Adults average 15.2–24.1 cm (5.9–9.4 in) in total length. The maximum body length recorded is 7.5 cm (2.9 in). Males and females do not differ in size. Six-lined racerunners have very small, dull, granular scales on the back; large, plated scales on the head; and eight rows of enlarged rectangular scales on the belly. The scales on the tail are rough to the touch. Taken together these scale characteristics distinguish racerunners from skinks (genus *Eumeces*). Six-lined racerunners have six obvious light stripes extending lengthwise down the back and tail on a background of dark brown. Typically the belly is white or, in the case of adult males, bluish. The young are miniature versions of the adults, although the lines extending along the body may be more distinct, and the tail may be washed in light blue. Hatchlings average 28–32 mm (1.1–1.3 in) in body length.

Taxonomy and Nomenclature

Only a single subspecies occurs in Georgia (*C. s. sexlineatus*), and it carries the same common name as the species. The taxonomy of this genus is currently under revision, and many authorities have suggested placing all of the U.S. species in the genus *Aspidoscelis*. Lizards in the genus *Cnemidophorus* are often called whiptail lizards because of the very long, slender tail.

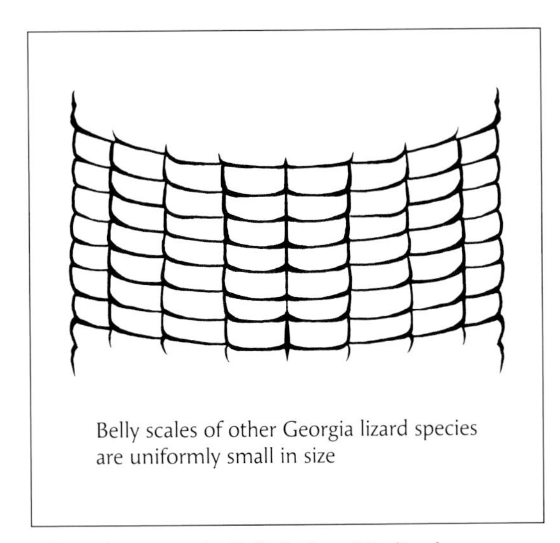

Enlarged Rectangular Belly Scales of Six-lined Racerunners

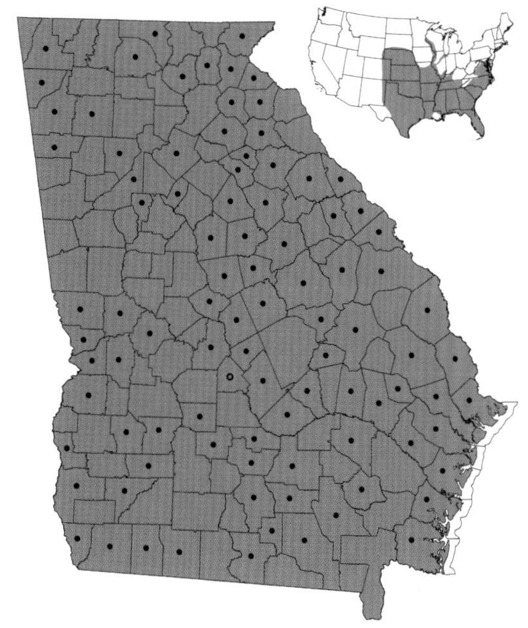

Distribution and Habitat

Six-lined racerunners occur throughout the state. They are most abundant in open habitats such as fields, woodland and roadside edges, and especially sandhills and other sandy habitats. They avoid cool, damp, heavily forested habitats such as those in mountainous areas.

Reproduction and Development

Mating occurs in the spring, and the underside of males becomes bluer at this time. Females lay from one to six eggs in a shallow nest of loose sand or eroded soil during May and June. Several females may lay eggs in colonial nesting sites, which may be near areas where the adults hibernate. The eggs may hatch anytime between late July and September but usually hatch during August.

Habits

As their common name indicates, six-lined racerunners are the fastest lizards in Georgia and are extremely difficult to catch. They are exclusively diurnal and actively forage for a wide variety of insects, usually in a sprint-stop fashion. They sprint to a spot and then stop to scan the ground or low vegetation for prey before sprinting to the next spot. During these foraging events they also shuttle between hot, open areas to warm up and cooler areas of shade to cool down. Racerunners require a higher body temperature for activity than any other lizard species in Georgia and remain inactive in burrows at night and when temperatures are too hot or cool for activity. They rely on their swiftness and burrows to escape predators. Known predators include snakes, birds, small mammals, and eastern glass lizards (*Ophisaurus ventralis*). Six-lined racerunners, like other whiptail lizards, are usually not territorial.

Conservation Status

Racerunners can be locally abundant in appropriate habitats. They fare relatively well in human-altered habitats such as roadsides, rock quarries, piles of riprap, and other disturbed, exposed sunny areas, so long as there is sufficient prey.

Christopher T. Winne

Snakes

SNAKES ARE AMONG the most recognizable components of Georgia's wildlife. Their tubular, limbless bodies evoke more irrational fear than any other animal shape. It is true that some snakes are capable of inflicting venomous, sometimes lethal, bites. But fewer than a dozen people in the United States die from snakebites each year, and most bites result from attempts to catch, handle, kill, or otherwise harass snakes. More people die from bee stings or from hitting deer with their cars than they do from snakebite. Indeed, all but 6 of Georgia's 41 native species are completely harmless, and the vast majority of bites by even our venomous snakes do not result in permanent harm. Only the timber rattlesnake (*Crotalus horridus*), eastern diamondback rattlesnake (*C. adamanteus*), and cottonmouth (*Agkistrodon piscivorus*) pose any realistic threat to human life. When compared with the 400,000 Americans who die annually from tobacco-induced disease, it should be obvious that snakes do not pose a significant threat to humans in our society. Unfortunately, many people consider their fear, no matter how irrational, sufficient justification for killing snakes.

An extreme of irrational behavior toward snakes is the religious practice of "taking up serpents" associated with certain charismatic, historically Appalachian churches, including a few in northern Georgia. Members demonstrate their faith by handling venomous snakes and refusing medical treatment if bitten. The snakes traditionally used in these rituals have been mountain forms such as the relatively docile timber rattlesnake and the nonlethal copperhead (*Agkistrodon contortrix*). One of Georgia's rare snakebite deaths occurred when a snake-handling native of Appalachia moved to Lowndes County and expanded his religious practice to include the more dangerous eastern diamondback rattlesnake (*C. adamanteus*).

People's irrational fear of snakes has spawned a number of ridiculous myths, from the story of the hoop snake that purportedly sports a venomous sting on its tail to the rattlesnake, which, having bitten a farmer's boot leaves its broken fangs to kill future generations of boot wearers. One of the most tenacious tall tales is that of the water skier who falls into a nest of water moccasins in Lake Lanier or Lake Oconee—or whatever the neighborhood reservoir happens to be called.

Biologically speaking, snakes are fascinating, if for no other reason than their ability to cope with the lack of limbs. This seeming handicap has not prevented them from successfully exploiting tropical and temperate environments across the globe. Snakes rival lizards in total numbers, and representatives of their 2,900 or so species occupy every continent except Antarctica; some even occur in the oceans. Snakes fill an amazing variety of ecological niches—burrowers, actively mobile hunters, ambush predators, swimmers, and tree dwellers among them. Several Asian species even glide from tree to tree much like a flying squirrel.

Snakes compensate for their lack of limbs by having many ribs and associated rib muscles. This gives them extreme flexibility and power, a combination that allows some species to kill small prey animals with constricting coils. Enlarged belly scales, or scutes, a characteristic

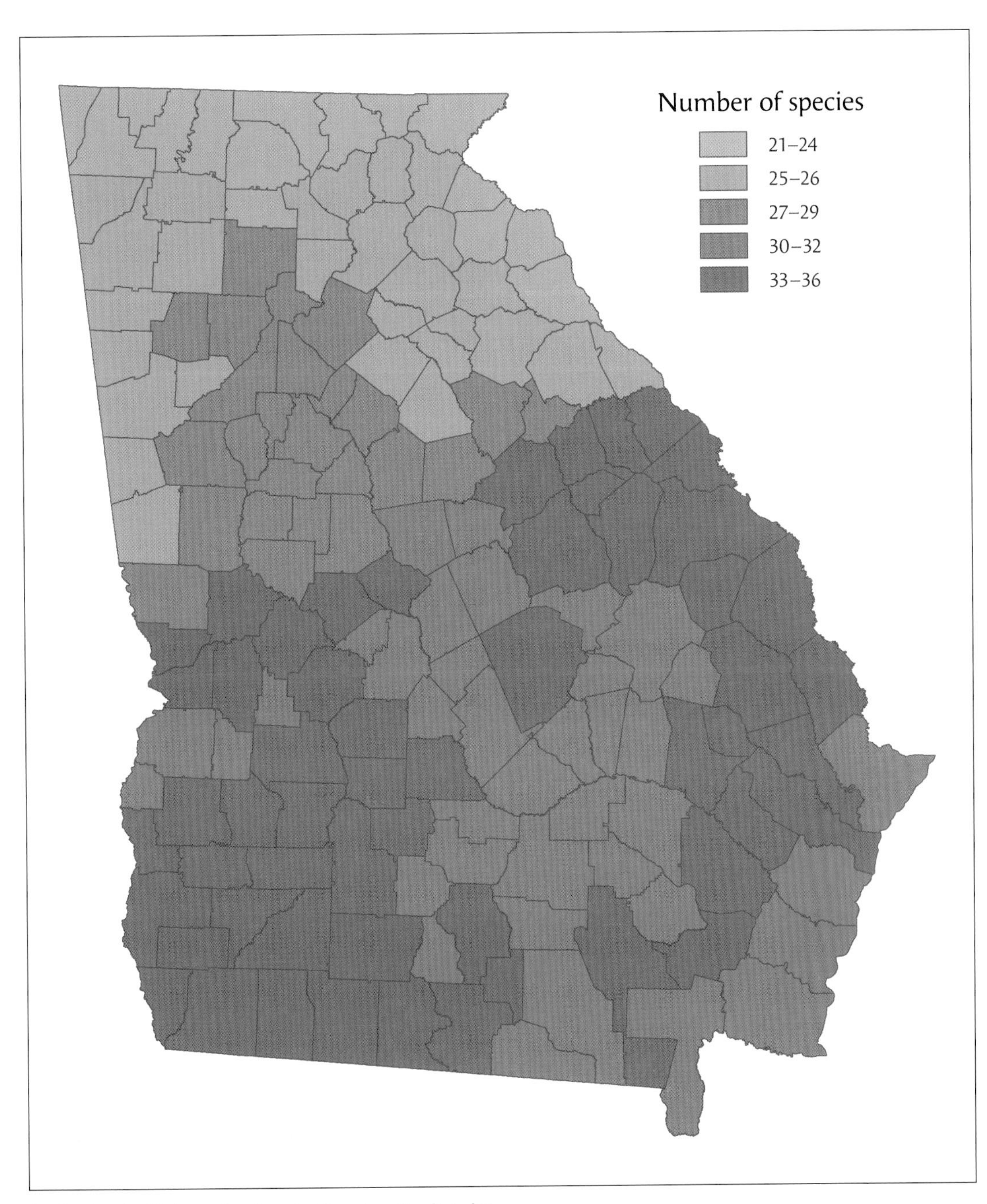

Snake Species Richness by County, Based on Predicted Ranges

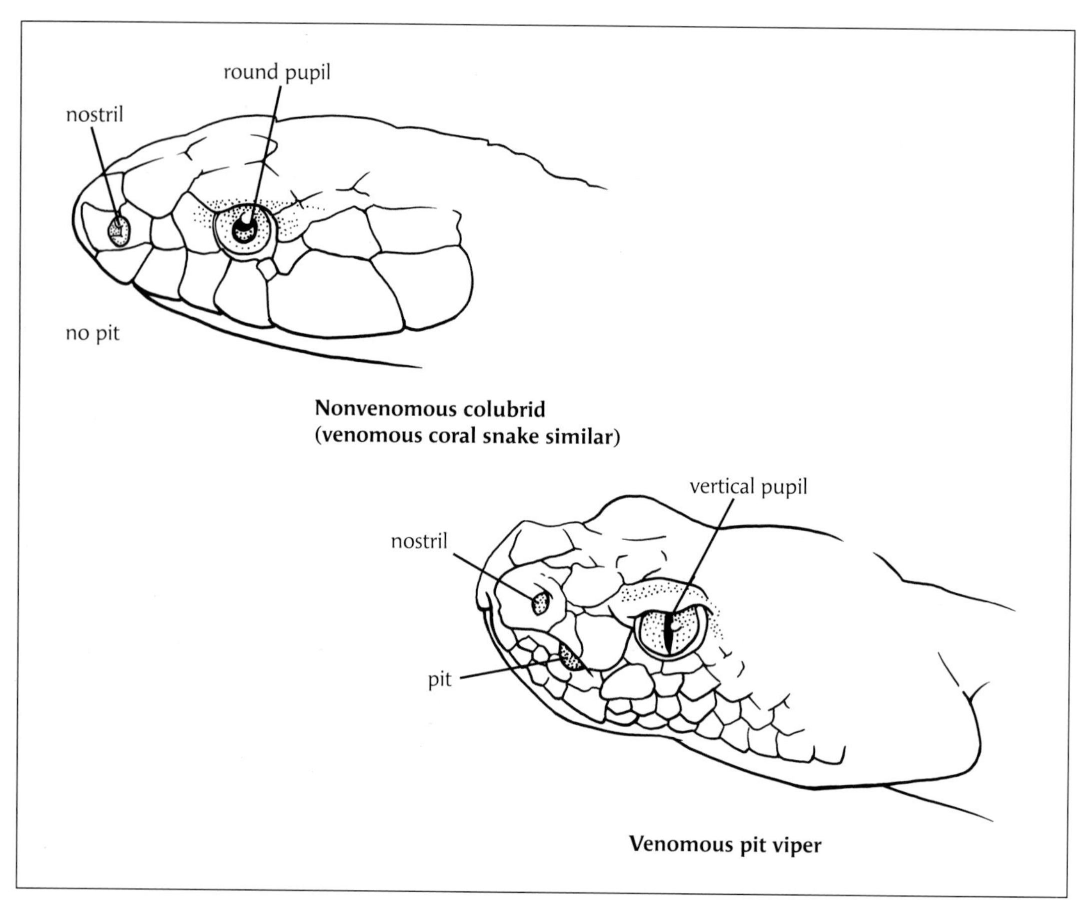

Identifying Features on the Heads of Nonvenomous Colubrid Snakes and Venomous Pit Vipers

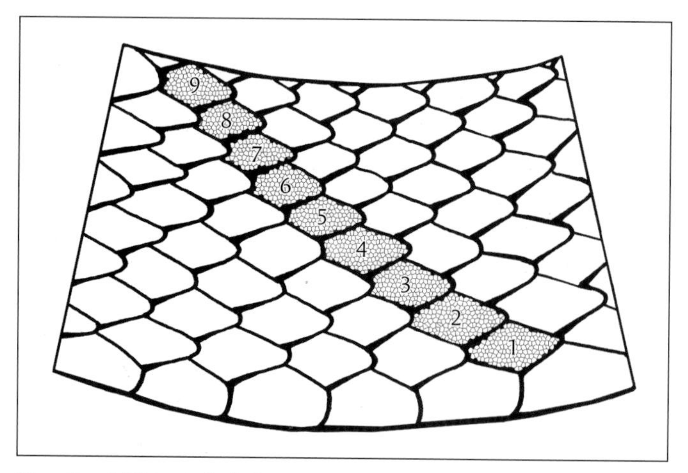

Counting Mid-body Scale Rows

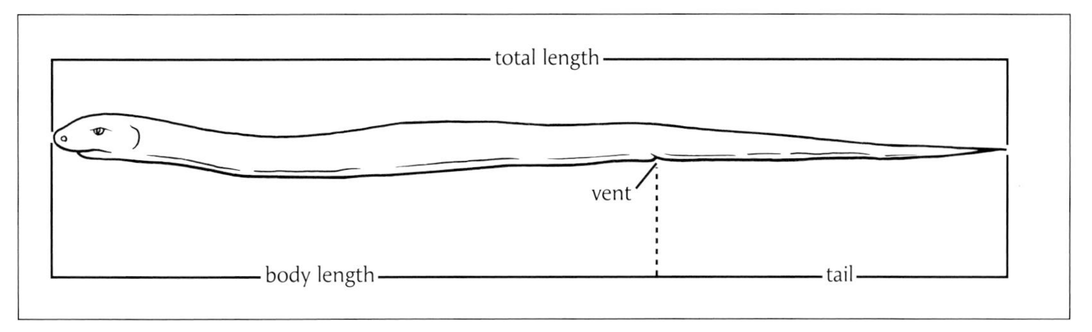

Measurements of Snakes

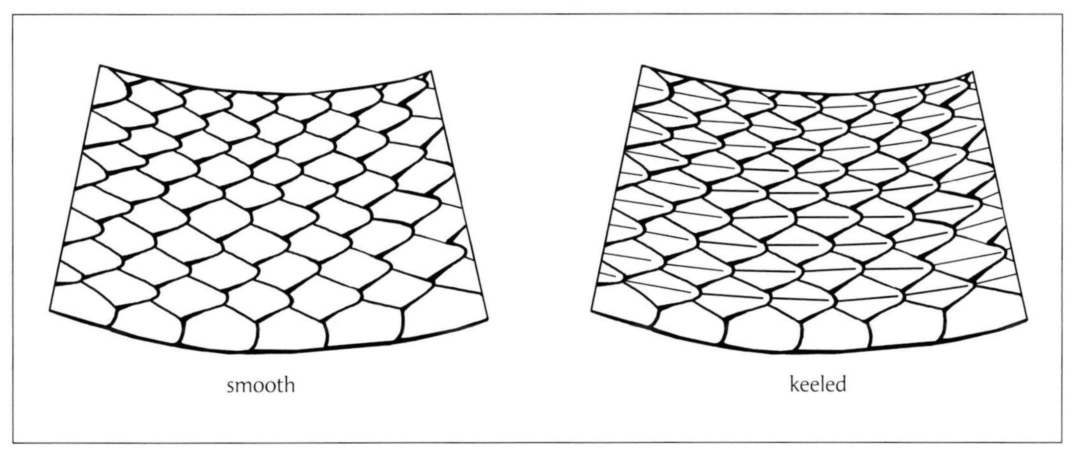

Body Scales of Snakes

that distinguishes native snakes from legless lizards, enable snakes to gain purchase on surfaces so that they appear to move effortlessly across the ground.

All snakes are predators, and most of Georgia's snakes feed on backboned animals. Many species such as the black racer (*Coluber constrictor*) and watersnakes (genus *Nerodia*) simply overpower and swallow live, struggling prey. The rat snakes (*Elaphe*), kingsnakes (*Lampropeltis*), and others constrict their prey to death before eating it. A few like the vipers and elapids have evolved powerful venoms and venom-delivery structures to immobilize and eventually kill prey.

Snake venoms are chemical cocktails of enzymes and related toxins that are produced by modified digestive glands and carried through tubular ducts to the fangs that deliver them. Some of these substances attack various body tissues; others interrupt the transmission of nerve signals. Although there are exceptions, vipers tend to have venoms that are predominantly tissue destroying, and elapids tend to have neurotoxins that disrupt nervous function.

To facilitate swallowing incredibly large food items, snakes have evolved a highly elastic, loosely jointed skull. They grasp their prey, often by the head, with the help of sharp, backwardly curved teeth and proceed to draw it into the throat, pulling first with one side of the head and then the other, very much like a person who alternates left and right hands while pulling on a rope.

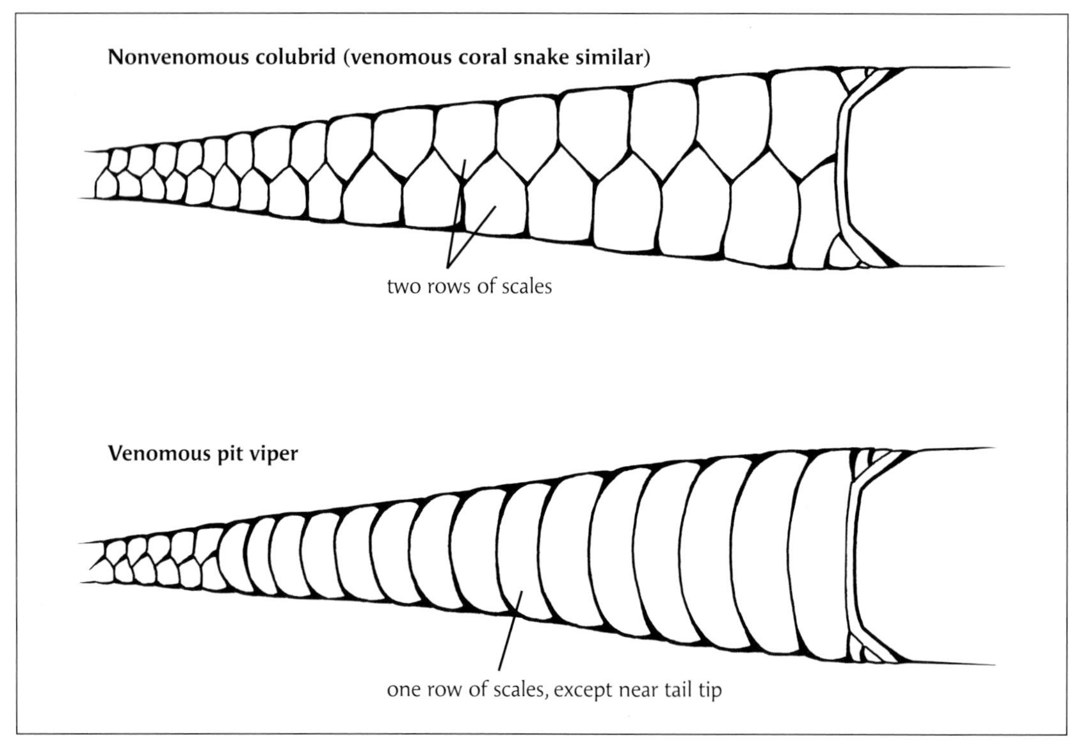

Comparison of Scales on the Underside of the Tails of Nonvenomous Colubrid Snakes and Venomous Pit Vipers

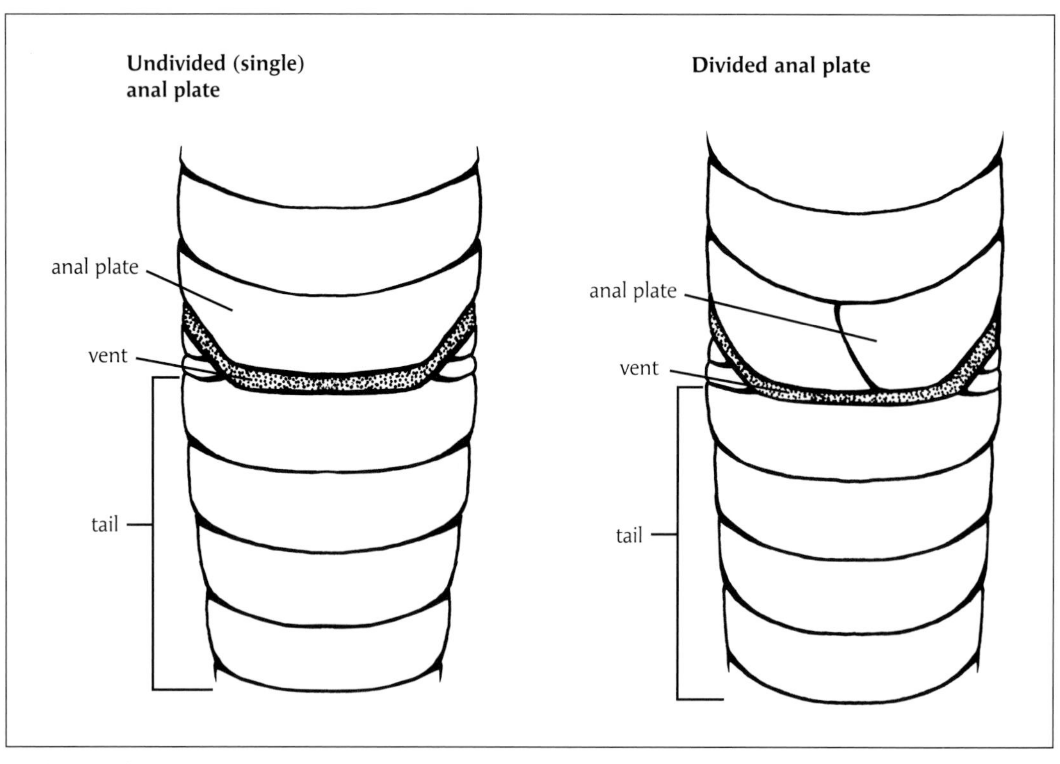

Anal Plates of Snakes

Snakes evolved from lizard ancestors approximately 130 or 140 million years ago, and the two closely related groups share many features, including paired male sex organs (hemipenes) and Jacobson's organ, a sensory device located in the roof of the mouth that samples chemical substances picked up by the tongue. Snakes do not have external ears and cannot hear airborne sounds, but they can detect ground vibrations through their highly sensitive jaw mechanism. Some snakes such as the venomous pit vipers have developed heat-sensing organs that function much like night-vision goggles. Snakes also have a brille—a permanent clear scale covering and protecting each eye—rather than movable eyelids.

Colubrid Snakes

FAMILY COLUBRIDAE

The colubrids are the typical harmless snakes that make up the bulk of Georgia's snake fauna. They are extremely variable in size, and our colubrids range from the crowned snakes (genus *Tantilla*), at less than 33 cm (13 in) maximum length, to the eastern indigo snake (*Drymarchon couperi*), which may exceed 254 cm (100 in). The scales may be smooth and shiny or rough and dull. Most species have a relatively narrow head that is not overly distinct from the neck. This huge family of nearly 1,700 species ranges over most of the world's tropical and temperate regions and exploits a variety of habitats including lakes, streams, forests, grasslands, and deserts. Rodent eaters such as rat snakes (genus *Elaphe*) and kingsnakes (genus *Lampropeltis*) may be common around barns and other outbuildings. Although often mistakenly called moccasins, most of Georgia's aquatic snakes belong to this family. Some taxonomists split the Colubridae into several distinct families. Under this classification system the watersnakes and their relatives (genera *Nerodia*, *Regina*, *Seminatrix*, *Storeria*, *Thamnophis*, and *Virginia*) are placed in the family Natricidae. The harmless rear-fanged snakes, including the genera *Carphophis*, *Diadophis*, *Farancia*, *Heterodon*, and *Rhadinaea*, are in the Xenodontidae. All other harmless Georgia species remain in the Colubridae. In this treatment we retain the old use of the name Colubridae.

Adult eastern worm snake, Paulding County (John B. Jensen)

Eastern Worm Snake

Carphophis amoenus

Description

The eastern worm snake is a small, cylindrical snake with a short tail that terminates in a spine. A few individuals may exceed 33 cm (12.9 in) in total length, and most adults are at least 18 cm (7 in) long. Females are longer than males but have a shorter tail. The head is small and conical and no wider than the neck. The scales are smooth, and the anal plate is divided. The back is solid brown or grayish brown; the belly is lighter, often pinkish. The back and belly colors are sharply delineated on the sides. Freshly shed individuals have an iridescent sheen. Hatchlings are usually 76–102 mm (3–4 in) in total length and a little darker than adults. Other small, unpatterned brownish snakes such as earth snakes (genus *Virginia*) and red-bellied snakes (*Storeria occipitomaculata*) lack iridescence and a spine-tipped tail.

Taxonomy and Nomenclature

Two subspecies have been described: the eastern worm snake (*C. a. amoenus*) and the midwestern worm snake (*C. a. helenae*). The latter form occurs in the far western part of Georgia. With the exception of intergradation zones, particularly in northwestern Georgia, the remainder of the range within the state is occupied by the eastern subspecies.

Distribution and Habitat

Eastern worm snakes occur in the Piedmont and mountains of Georgia in hardwood forests and woodlands up to elevations approaching 1,220 m (4,000 ft). They are most often found under rocks, logs, and other debris. Some herpetologists have suggested that eastern worm snakes are most abundant in moist forests and

Adult eastern worm snake, Bibb County (John B. Jensen)

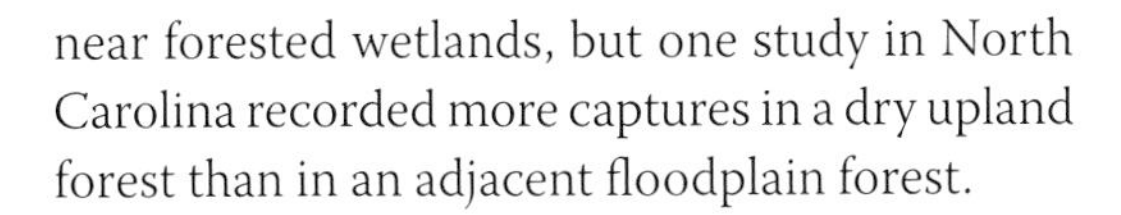

near forested wetlands, but one study in North Carolina recorded more captures in a dry upland forest than in an adjacent floodplain forest.

Reproduction and Development

The life history of eastern worm snakes is not clearly understood because they are very secretive and thus difficult to study. It appears that they have spring and fall breeding seasons, and females have been reported to store sperm from fall matings until after the spring reemergence. Females lay clutches of 1–12 eggs under rocks, logs, or other litter in June and July. The young hatch in August or September. Sexual maturity is apparently reached at around 3 years of age.

Habits

Eastern worm snakes spend most of their lives underground or beneath debris. They burrow by working the small, pointed head into cracks and crevices. They are active from early spring to late fall but are rarely seen during the hottest parts of summer. Peak activity tends to be during the fall months, and they generally remain inactive during extreme temperatures. Activity periods begin mainly in the late afternoon and early evening and rarely last more than 12 hours. Individuals often remain sedentary for as long as 2 weeks even during the season of activity. Eastern worm snakes can have home ranges as large as 723 m^2 (8,033 ft^2), but there is much variation among individuals. Males travel much farther than females. The diet consists primarily of earthworms but may also include other soft-bodied invertebrates such as insect larvae. Documented predators include other snakes such as copperheads (*Agkistrodon contortrix*), black racers (*Coluber constrictor*), and kingsnakes (*Lampropeltis getula* and *L. triangulum*); thrushes, American robins, barn owls, and opossums. Eastern worm snakes release a foul-smelling liquid from the vent when handled, but they are completely harmless to humans and rarely even attempt to bite.

Conservation Status

Eastern worm snake populations are locally dense in favorable habitat, but their secretive nature makes it difficult to assess their overall status.

Dean A. Croshaw

Adult scarlet snake, Baker County (Gabriel J. Miller)

Scarlet Snake

Cemophora coccinea

Description

Scarlet snakes are slender and relatively small; most adults are less than 55 cm (21.5 in) in total length; the maximum length recorded is 83 cm (32.4 in). When viewed from above the body appears to have red, black, and whitish or cream-colored bands. This pattern does not extend to the belly, however, and is more accurately described as black-bordered red saddles separated by a light background color. Small, dark spots or flecks may be apparent in the lighter areas, especially on larger individuals. The belly is immaculate and white or cream colored. The head is red and narrow, and the snout is enlarged and extended, giving the head a noticeably pointed appearance. Body scales are smooth, and the anal plate is undivided. Juveniles are similar to adults. Hatchlings are 113–186 mm (4.4–7.3 in) in total length. Scarlet snakes can be confused with both scarlet kingsnakes (*Lampropeltis triangulum elapsoides*) and coral snakes (*Micrurus fulvius*), but in both of those species the coloration of the sides extends onto the belly. In addition, the coral snake has yellow-bordered black bands rather than black-bordered red saddles.

Taxonomy and Nomenclature

Rangewide, three subspecies are recognized, but only the northern scarlet snake (*C. c. copei*) occurs in Georgia. The genus *Cemophora* contains no other species.

Adult scarlet snake, Monroe County (John B. Jensen)

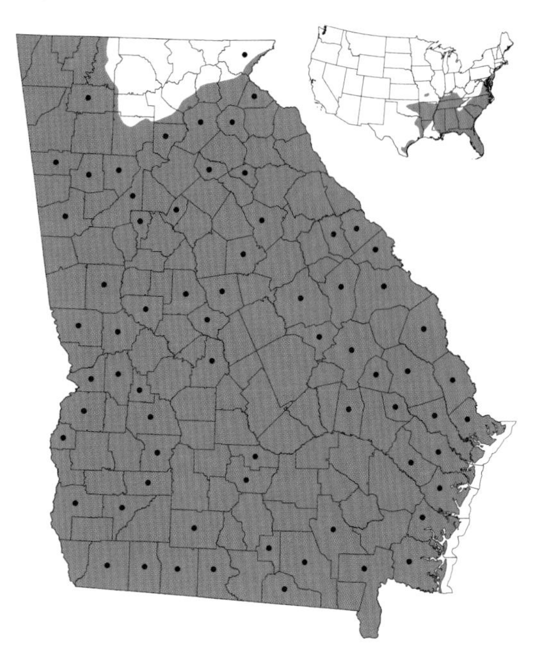

Distribution and Habitat

Scarlet snakes are known from all physiographic provinces in the state except the Cumberland Plateau. They are probably present there as well, since the species has been documented in the Cumberland Plateau of adjacent Alabama. Scarlet snakes appear to be significantly more common and widespread in the Coastal Plain than elsewhere. Pine, hardwood, and mixed pine-hardwood woodlands with sandy or loamy soils are preferred habitats.

Reproduction and Development

Most of what is known about the reproductive biology of scarlet snakes comes from captive animals. Mating is thought to occur in spring, followed by the laying of two to nine (average = five) white, elongated eggs in underground burrows or under objects that provide cover. Females may be able to produce more than one clutch each year. Eggs usually hatch in late summer or fall following an incubation period of 70–80 days, although a March hatching was documented in neighboring South Carolina. Age at maturity has not been reported for this species.

Habits

Aided by their pointed and enlarged snout, scarlet snakes are adept burrowers, which is certainly one reason why they are infrequently encountered. Another reason is that their aboveground activity is almost exclusively nocturnal. Surface movements occur principally during the warmer months. Inactive individuals are occasionally found during the day under logs or rocks. Scarlet snakes hibernate in rotting logs, stumps, or within burrows, and they apparently can tolerate complete flooding of these sites for extended periods. Eggs of other reptiles, especially lizards and snakes, are the primary food, although small lizards, snakes, and amphibians are occasionally killed by constriction. Scarlet snakes also constrict eggs in order to force the contents out. They use sharp enlarged teeth in the back of the mouth to split larger eggs; smaller eggs may be eaten whole. Coral snakes, southern toads (*Bufo terrestris*), and loggerhead shrikes are documented predators, but carnivorous mammals, other predatory birds, and snake-eating snakes such as kingsnakes (genus *Lampropeltis*) and eastern indigo snakes (*Drymarchon couperi*) probably kill and eat scarlet snakes as well. The color and pattern of scarlet snakes may mimic the appearance of venomous coral snakes and thus discourage attacks from predators. Scarlet snakes rarely bite but often release a mild-smelling musk when handled or harassed.

Conservation Status

The status of this very secretive species is difficult to assess.

John B. Jensen and Whit Gibbons

Adult black racer, Floyd County (Bradley Johnston)

Black Racer

Coluber constrictor

Description

Adults are generally 91–152 cm (35.5–59.3 in) in total length; males average slightly larger than females. There is little to no variation in the appearance of adults within Georgia, or between males and females other than the size difference. This smooth-scaled snake is solid black and shiny with a lighter black, dark blue, or grayish belly and a white chin. The head is narrow and the body is slender and relatively round. The anal plate is divided. Juveniles, approximately 250–310 mm (9.8–12.1 in) at hatching, are lighter in color than adults and are patterned with darker red, brown, or gray blotches. These markings fade by the time the snake has achieved a length of approximately 63 cm (24.6 in).

Taxonomy and Nomenclature

Two subspecies are present in Georgia: the northern black racer (*C. c. constrictor*), distributed through the upper half of the state, and the southern black racer (*C. c. priapus*), which is restricted to the lower half. Racers are commonly referred to as black snakes or black runners.

Juvenile black racer, Long County (Dirk J. Stevenson)

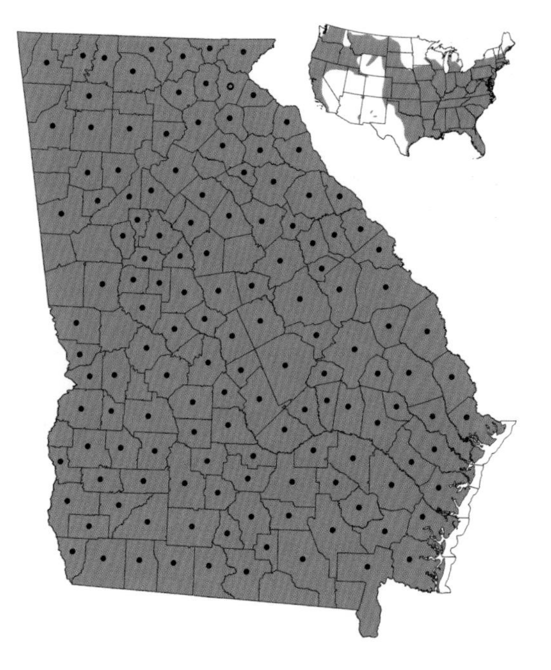

Distribution and Habitat

Black racers are prevalent throughout the state in a variety of habitats but are most common in open areas such as agricultural fields, pine and hardwood forests with thin undergrowth, and around the edges of wetlands and marshes.

Reproduction and Development

Black racers mate in the springtime, usually in April and May. Eggs are laid in the summer, and females produce 4–36 eggs (average = 10–13). Larger females lay more eggs than smaller ones. Eggs are oblong and have a coarse shell whose surface has the textured appearance of grains of salt. Nest sites are moist, covered localities underground or under bark or rocks. Egg clutches have been found buried in abandoned sawdust piles in Georgia. Incubation time is approximately 2 months. Individuals reach sexual maturity in 1 or 2 years, with males maturing faster than females.

Habits

Black racers eat almost any animal they can swallow whole and may have the most diverse diet of any snake in Georgia. They consume insects, small mammals, birds, amphibians, lizards, other snakes, and have even been noted to eat small turtles. They also eat bird eggs and small venomous snakes such as copperheads (*Agkistrodon contortrix*) and pigmy rattlesnakes (*Sistrurus miliarius*). Black racers have an interesting behavior known as periscoping, in which they raise the upper half of their body and visually search for prey. Once the prey is spotted, the racer hunts it down and swallows it alive. Large or difficult-to-handle prey is sometimes chewed before being consumed. Despite its scientific name, this species does not constrict, although individuals will pin down prey with a body loop. Racers are eaten by larger mammals, including raccoons and foxes; birds of prey, such as hawks; and certain other snake species. Black racers are named for their primary defense against predators, their incredible ability to flee quickly. They sometimes retreat up a tree or into a bush and often remain close to a concealed location into which they can quickly escape. When cornered, however, black racers are pugnacious fighters with an impressive array of threats that includes repetitive biting, chewing, and tail vibration.

Individuals sometimes kink their body like an accordion when threatened by a predator or person. Kinking behavior is more commonly associated with rat snakes (genus *Elaphe*) and is believed to be effective camouflage. Black racers are active solely during the daytime and may travel several hundred meters (yards) a day in search of prey. They hibernate for at least a few weeks during the winter months even in southern Georgia and frequently begin to retreat in November into crevices under logs and stumps, into burrows and tunnels, or under leaves and other ground litter. They emerge from hibernation in early spring.

Conservation Status

Although populations should not be assumed to be stable simply because they occur in high numbers, black racers seem more resilient to human alteration of the environment than many other Georgia snake species.

Kimberly M. Andrews and
Whit Gibbons

Adult ringneck snake, Appling County (Dirk J. Stevenson)

Ringneck Snake

Diadophis punctatus

Description

The ringneck snake is a small snake; adults are 25–38 cm (9.8–14.8 in) in total length. Individuals in the northern part of the state are generally larger than those in the south. This species is characterized by smooth black or dark gray scales; a divided anal scale; an orange, yellow, or reddish belly; and an orange, yellow, or cream-colored ring around the neck. A row of black dots (one on each scale) often lines the center of the belly scales but can be absent in some geographic locations or individuals; for instance, the row may be faint and broken or completely absent on ringneck snakes in northern Georgia. The color of the belly and the color and extent of the neck ring are also geographically variable. Individuals from northern Georgia have a yellow belly and complete yellow neck ring. Those in the Coastal Plain have an orange or reddish belly and often have an incomplete ring. The young look like the adults and are 100–140 mm (3.9–5.5 in) at hatching.

Taxonomy and Nomenclature

Two subspecies occur in Georgia: the northern ringneck snake (*D. p. edwardsii*) is confined primarily to the mountainous part of the state, and the southern ringneck snake (*D. p. punctatus*) is found farther south. Ringneck snakes from much of the Piedmont are intermediate between the two.

Adult ringneck snake, Gilmer County (John B. Jensen)

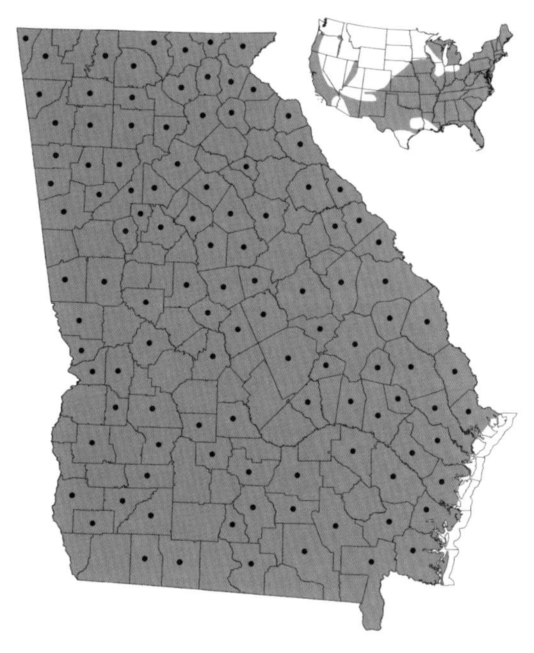

Distribution and Habitat

Ringneck snakes occur throughout the state, often in high densities. A study in Kansas found population densities in some locations that were estimated to be more than 1,800 snakes per hectare (720 per acre). Ringneck snakes use a wide variety of habitats and seek shelter under rocks, logs, and other ground cover.

Reproduction and Development

Mating occurs biannually, in the spring and the fall. Females are capable of storing sperm over the winter for fertilization the following spring. Females lay approximately two to seven eggs in the summer under rocks, in or under decomposing logs, and in other moist, hidden areas. Incubation time is roughly 7 or 8 weeks. While mating, the male uses his head to rub the female's neck and ultimately bites her on the neck ring. The female's response to the rubbing and biting helps the male determine her interest in mating. Individuals normally reach sexual maturity in 3 years.

Habits

Ringneck snakes are active in warm regions year-round, but they hibernate underground or burrow beneath logs during the winter in colder areas, including much of northern Georgia. Aboveground activity occurs primarily at night. For their size, these snakes cover long distances. Individuals often move more than 70 m (230 ft), and some may move more than 1.6 km (1 mi) over the course of a year. Ringneck snakes feed primarily on smaller animals, including invertebrates (e.g., earthworms, insect larvae), amphibians, and reptiles such as small snakes and lizards. The enlarged rear teeth release a mild toxin that mixes with the saliva and paralyzes the prey while the snake is chewing. A wide variety of predators eat ringneck snakes, and a considerable amount is known about these relationships because of the extensive research that has focused on the ecology of this geographically widespread species. Its small size makes the ringneck snake prey for almost anything it encounters while searching for a meal, including snakes, birds, and mammals as well as bullfrogs (*Rana catesbeiana*) and toads (genus *Bufo*). Flight is the first defense against predators. If it cannot escape, the snake may flip over and expose its bright belly. If this sudden change in color distracts or confuses the predator, the ringneck snake turns back over and quickly retreats under decomposing wood or the soil. Ringneck snakes also release a bad-smelling musk when threatened or handled. Individuals sometimes remain motionless (i.e., play dead) or hide the head underneath the body. This species rarely bites in self-defense.

Conservation Status

Ringneck snakes are common and apparently secure in Georgia.

Kimberly M. Andrews and Whit Gibbons

Adult eastern indigo snake, Liberty County (Dirk J. Stevenson)

Eastern Indigo Snake

Drymarchon couperi

Description

This magnificent snake reaches a maximum total length of 263 cm (102.6 in), making it the longest snake in North America. Adult males are 150–260 cm (58.5–101.4 in) in total length; adult females are 150–210 cm (58.5–81.9 in). Large specimens may weigh 2–5 kg (4.4–11.1 lb). Tail length averages about 16 percent of total length in both sexes. Adults are glossy black or bluish black above and below. Recently shed specimens are lustrous and appear iridescent in sunlight. The sides of the head, chin, and throat are pigmented with varying amounts of reddish brown, salmon orange, or cream. The head is indistinct from the neck. The scales are large, smooth, and shiny. Adult males have faint, difficult-to-see keels on the scales along the middle of the back. The anal plate is undivided. Hatchlings, which are 410–615 mm (16–24 in) in total length, have bluish white, poorly defined crossbands or light speckling on the back. Immature specimens have more extensive red on the head and forward half of the belly but otherwise closely resemble adults. The black racer (*Coluber constrictor*) is often confused with the indigo snake, but the slimmer, smaller racer is dull black with a white chin and throat and a divided anal scale. Racers are lithe and built for speed, streaking across roads and paths; the eastern indigo snake moves much more slowly and deliberately.

Adult eastern indigo snake, Evans County (Dirk J. Stevenson)

Juvenile eastern indigo snake, Bryan County (Dirk J. Stevenson)

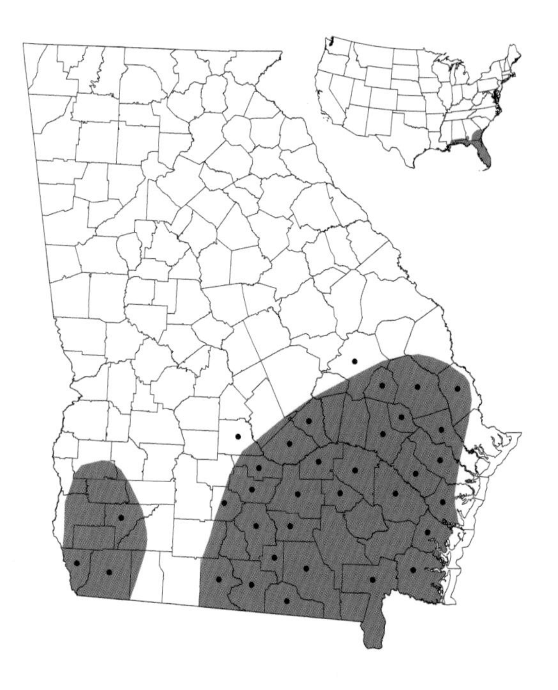

Taxonomy and Nomenclature

Until recently, the eastern indigo was considered a subspecies (*D. corais couperi*) of the wide-ranging *D. corais*, which occurs from southern Texas to Argentina. Color differences, the unique arrangement of the upper lip scales, and geographic isolation distinguish the eastern indigo snake from other species of *Drymarchon*. Other common names include blue indigo snake and blue gopher snake.

Distribution and Habitat

Indigos are closely associated with longleaf pine sandhills of the Coastal Plain that are populated with gopher tortoises (*Gopherus polyphemus*), whose burrows provide shelter for the snakes. Sandhills depend on regular fires to maintain the open-canopied conditions preferred by these reptiles. Undeveloped ridges of coarse white sands lying along the northeastern sides of blackwater streams in the Atlantic Coastal Plain are prime indigo snake country. The deep (3–9 m, or 9.8–29.5 ft), nutrient-poor sands of these barren habitats support a canopy of scattered longleaf pine and stunted turkey oak over a patchy ground cover of saw palmetto, reindeer lichen, and numerous grasses. During the warmer months, eastern indigo snakes often disperse from sandhills and visit a wide variety of habitats, including agricultural fields and successional communities. They also frequent low, shaded, and wet areas such as forested swamps. In habitats lacking gopher tortoises, the snakes shelter in hollow root channels, armadillo burrows, under or inside large logs, and in hummocks of soil at the base of trees.

Reproduction and Development

Eastern indigo snakes mate from October to February. During this time, aggressive males competing for mates are sometimes seen intertwined and fighting. Females lay a single clutch of 4–12 large (75 mm, or 2.9 in, long), oval, granular-surfaced eggs in May or June. Other than two reports of nests associated with gopher tortoise burrows, little is known about the location of egg-laying sites. Eggs hatch from July to September. Males reach sexual maturity in 2 or 3 years; females mature in 3 or 4 years.

Habits

Indigos are diurnal, terrestrial snakes that retire to animal burrows or similar underground retreats during periods of inactivity. In Georgia, gopher tortoise burrows are central to the ecology of this snake, which relies on them for winter dens and may use them for nesting, foraging, and refuge prior to shedding during warm months. Like eastern diamondback rattlesnakes (*Crotalus adamanteus*), indigo snakes frequently bask near tortoise burrows on mild winter days; they are commonly active at temperatures that are rather cool (13–18° C, or 55–65° F) for a snake. Snakes found in the winter or early spring often have small, fluid-filled boils or blisters that ultimately rupture and form crusty lesions. Presumably the high humidity of tortoise burrows combined with cool winter temperatures predispose the snakes to bacterial infections. Adults have the largest home ranges known for any U.S. snake species. A radiotelemetry study in southeastern Georgia found the ranges of several male indigo snakes to exceed 1,200–1,600 ha (3,000–4,000 acres). Individuals may return each winter to the same gopher tortoise colonies, even to the same burrow. This active forager explores the margins of wetlands or probes crevices to locate its prey; although a powerful predator, it is not a constrictor. Eastern indigo snakes eat virtually any vertebrate they can overpower: birds, small mammals, frogs, toads, lizards, turtles—including hatchling gopher tortoises—and snakes, including venomous species. Snakes are seized by the head and chewed vigorously until subdued. When threatened, eastern indigo snakes may flatten the neck vertically, vibrate the tail, hiss, and release a strong and distinctive musk. They seldom bite humans, even when handled.

Conservation Status

The eastern indigo snake was federally listed as Threatened in 1978 because of population declines associated with habitat loss and fragmentation, killing by vehicles and persons, diminishing tortoise numbers, overcollection for the pet trade, and gassing of gopher tortoise burrows by rattlesnake hunters. Except for collection for the pet trade, which is now illegal, all of these factors continue to threaten indigo snake populations. In the 1980s, indigo snakes were released at several Georgia sites (Baker, Decatur, and Dodge counties) in an attempt to recover populations near the edge of the species' range, but these introductions apparently did not establish new populations. Although current population trends are unknown, indigo snakes are encountered with some regularity in rural portions of the lower Coastal Plain and in the Tifton Uplands region of the middle Coastal Plain; there is but a single recent observation for southwestern Georgia west of Interstate 75. The state of Georgia lists this species as Threatened.

Dirk J. Stevenson, Robert A. Moulis, and Natalie L. Hyslop

Adult corn snake, Banks County (John B. Jensen)

Corn Snake

Elaphe guttata (Pantherophis guttatus)

Description

Adult corn snakes are typically 91–122 cm (35.5–47.6 in) in total length but may be up to 183 cm (71.4 in) long. Females are sexually mature at a body length of about 68 cm (26.5 in). Although males reach sexual maturity at a smaller size, adult male corn snakes are usually slightly longer and heavier than adult females. The background color varies from bright orange to russet or gray. A series of black-edged red blotches runs down the middle of the back. The hue and pattern vary geographically. Individuals from coastal areas tend to be brighter orange, while specimens from the inland portions of the state tend to exhibit more gray or brown. The belly of all specimens is patterned with a mix of black, white, and orange that forms a distinct checkerboard pattern. A characteristic physical marking is a spear point or V on top of the head. Just in front of this spear point is a small band of similar color that runs between and then behind the eyes. The scales on the middle of the back have weak keels, but scales elsewhere on the body are smooth. The anal plate is divided. Newly hatched corn snakes are 230–360 mm (9.0–14.1 in) in total length. Although patterned much like their parents, their colors are more subdued, often lacking orange and red.

Taxonomy and Nomenclature

No subspecies are currently recognized. Some taxonomists have suggested that North American members of *Elaphe* should be placed in the genus *Pantherophis*. The belly pattern's

Juvenile corn snake, Bryan County (Dirk J. Stevenson)

Adult corn snake, Glynn County (John B. Jensen)

Adult corn snake in defensive posture, McIntosh County (Dirk J. Stevenson)

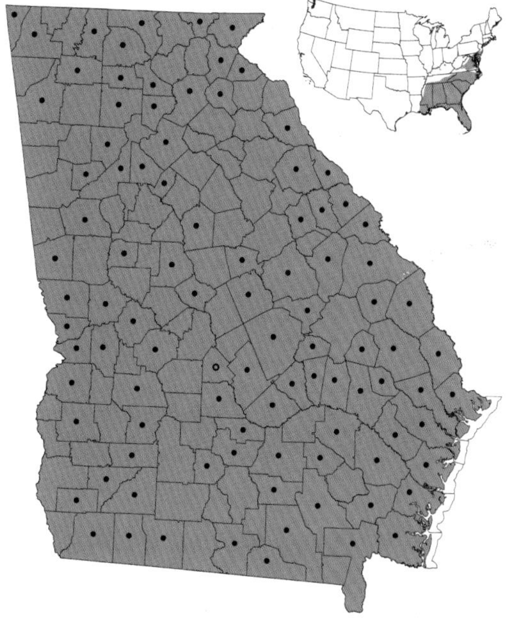

superficial resemblance to maize corn is the likely origin of this snake's common name. Corn snakes are also sometimes called red rat snakes.

Distribution and Habitat

These common snakes are found throughout the state in a variety of habitat types, including but not limited to sandhills, pine flatwoods, maritime forests, mixed pine-hardwood forests, and even suburban settings. Corn snakes seem to be most abundant in habitats with a significant pine component.

Reproduction and Development

In spring, after emerging from the winter dormancy period, male corn snakes begin searching for mates. Breeding typically takes place in April and May, and females lay 5–30 soft eggs in June or July in or under soft rotting logs or in sawdust or debris piles. Young corn snakes hatch in late summer or early fall. Both sexes generally reach adult size in 3 or 4 years.

Habits

Corn snakes spend much of their time burrowing beneath leaf litter and loose soil. They also hide in stump holes and rotting logs and behind the bark of dead trees. They are skilled climbers and are sometimes found in the rafters of old barns and buildings and in the branches and cavities of trees. Climbing corn snakes are often foraging for treefrogs, lizards, bats, and birds and their eggs. On the ground, corn snakes seek out and consume mice and other small rodents, which they kill by constriction. Corn snakes can be found throughout the year in Georgia but are most active from March to November. During the colder months, they may hibernate in protected locations such as stump holes, abandoned buildings, and tree hollows. Corn snakes are usually fairly inoffensive, but they do hiss and may strike, especially if surprised. They may also release a smelly musk. Predators include carnivorous mammals, birds, and other snakes such as common kingsnakes (*Lampropeltis getula*) and eastern indigo snakes (*Drymarchon couperi*).

Conservation Status

Corn snakes are regularly encountered in Georgia, and their populations appear to be stable. Although it is currently illegal to keep native nonvenomous snakes in captivity without a state permit, captive breeding of corn snakes has eliminated the need for wild capture of the species for the pet trade.

Berkeley W. Boone

Adult rat snake, Monroe County (John B. Jensen)

Rat Snake

Elaphe obsoleta (Pantherophis obsoletus)

Description

Male and female rat snakes are similar in appearance, and both reach a total length of 102–188 cm (39.8–73.3 in). Males weigh more than females, averaging 698 g (1.5 lb) to the females' less than 600 g (1.3 lb). Body size can vary greatly within the adult class and even within the sexes. All adults have some dorsal scales that are slightly keeled and a divided anal scale. The basic pattern is distinct in juveniles throughout the state and consists of a series of squarish or H-shaped blotches that run down the back and alternate with blotches on each side. Along the coast and on the barrier islands, rat snakes tend to be greenish, yellowish, or tan. In many, the long edges of the Hs and the side blotches grow together into stripes; in some, the middle of the blotches disappear completely so the animal has stripes rather than blotches. Most individuals in the northern half of the state darken with age, and many become all black with virtually no trace of a pattern. In southwestern Georgia the juvenile pattern remains distinct throughout life and consists of gray blotches on a light gray background. Throughout much of the middle of Georgia most adults are dark gray, brown, or blackish but retain the juvenile pattern. Black individuals are likely to be confused with black racers (*Coluber constrictor*), but rat snakes have

Juvenile rat snake in region where adults are typically black, Clarke County (Jason Norman)

Juvenile rat snake in region where adults are typically black, Floyd County (Bradley Johnston)

Juvenile rat snake in region where adults are typically gray and have a similar pattern, Baker County (David Steen)

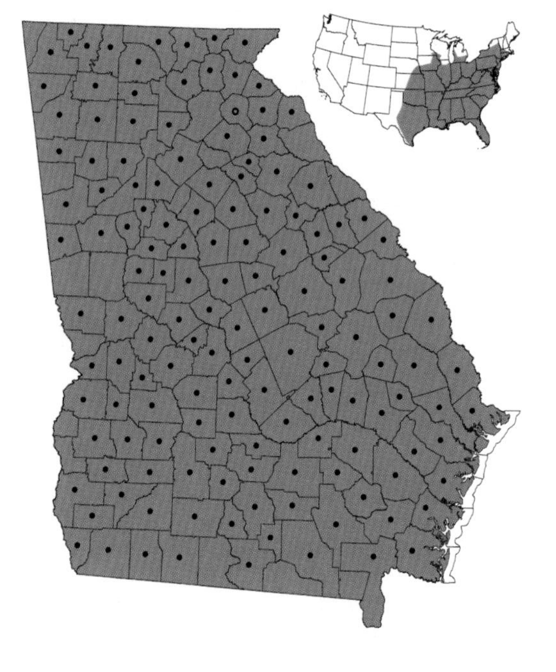

an angular head, are generally a little thicker, and have a distinctly angled belly that gives the body a cross-sectional shape similar to a loaf of bread. Rat snakes are typically 280–400 mm (10.9–15.6 in) long at hatching.

Taxonomy and Nomenclature

Three subspecies of rat snakes occur in Georgia: the black rat snake (*E. o. obsoleta*) in northern Georgia, the gray rat snake (*E. o. spiloides*) in the southwestern part of the state, and the yellow rat snake (*E. o. quadrivittata*) along the coast. Wide zones of intergradation between all these subspecies make it difficult to pin a precise name on the rat snakes found in many areas because they share characteristics of one or more subspecies. Genetic research has shown that color patterns do not reflect evolutionary relationships among rat snake populations. This has led to the suggestion that different lineages of rat snake be recognized as different species. According to this suggested taxonomy, the Chattahoochee River approximates the boundary between the eastern rat snake (*E. alleganiensis*) and the midland rat snake (*E. spiloides*) in Georgia. Some taxonomists suggest that North American members of *Elaphe* should be placed in the genus *Pantherophis*. Rat snakes are locally referred

Adult rat snake, Long County (Dirk J. Stevenson)

to as oak snakes, chicken snakes, black snakes, and pilot black snakes, the latter being based on an old myth that rat snakes lead rattlesnakes to hibernation dens.

Distribution and Habitat

Rat snakes occur statewide in Georgia in a broad range of wooded habitats that contain large trees to climb and rodents or birds to eat, including hardwood, pine, and mixed forests, wetlands, and suburban areas. They also frequent abandoned buildings and barns.

Reproduction and Development

Rat snakes mate in the spring (April, May, and June), and in some areas in the fall as well. During the breeding season males sometimes compete with other males for females by wrestling (male-male combat), a behavior in which each male lifts the front end of his body and attempts to push the other's head to the ground. During mating the male bites the neck of the female. Rat snakes lay an average of 15 eggs, although the clutch size can vary greatly (4–44). Eggs are laid in dark, moist locations such as in stumps, tree holes, or other crevices, and hatch in about 2 months; multiple females may nest together. Females have also been shown to exhibit nest site fidelity, returning to the same site from 1 year to the next. Sexual maturity is reached in approximately 4 years.

Habits

Adult rat snakes eat small mammals, birds, and bird eggs. Juveniles eat amphibians, small lizards, and small rodents. One yellow rat snake was observed climbing approximately 30 m (98 ft) up a loblolly pine to raid a swallow-tailed kite nest. Rat snakes use both vision and smell to locate their prey, which they constrict before swallowing. Eggs are swallowed whole, or the snake breaks them by squeezing them with its trunk muscles while swallowing. Shell fragments are either regurgitated or passed through the digestive tract. Rat snakes are preyed on by a variety of animals, including several snake species, both venomous and nonvenomous; mammals such as raccoons, coyotes, and weasels; and birds of prey, both hawks and owls. Rat snakes commonly defend themselves by kinking, a behavior in which the snake immobilizes and constricts its body lengthwise like an accordion. The kinked animal resembles a stick and may go undetected by the predator. Rat snakes may bite when handled, although many do not if handled gently. They will also wrap around a captor's hand and often release a pungent musk. Rat snakes can be active year-round, but during cold winters they hibernate for 2–4 months in decaying stumps or trees or in abandoned houses. Rat snakes are mostly active during the day, on the ground and in trees or bushes, but also may be active at night during summer months.

Conservation Status

Because of their abundance and excellent climbing abilities, rat snakes are by far the most common snake to enter peoples' houses and other buildings, and they are frequently killed by humans who find them unwelcome in such places. Nevertheless, rat snakes remain very common in both rural and suburban areas of Georgia.

Kimberly M. Andrews and
Whit Gibbons

Adult mud snake, Liberty County (Dirk J. Stevenson)

Mud Snake

Farancia abacura

Description

Mud snakes are large, moderately heavy-bodied snakes that can exceed 206 cm (80.3 in) in total length, but most adults are about 102–137 cm (39.8–53.4 in) long. They are shiny black and red, with the red restricted to triangular or rounded blotches extending up from the red-and-black, checkerboard-like belly pattern. Males are smaller than females and have a shorter tail. Male snakes also have fewer black markings on the underside of the tail and a series of small ridges on the scales on the sides in front of the tail. The chin may be yellowish, and the scales of the upper lip are heavily marked with black blotches that can give the impression of a zipper or large teeth along the mouth. Mud snakes have smooth scales, a divided anal plate, and a sharp tip on the end of the tail. Hatchlings are usually 160–270 mm (6.2–10.5 in) in total length. This species has the largest adult-to-hatchling body length ratio of any of the Georgia snakes; that is, they have the smallest young when compared with the maximum size of the adults. Juveniles have very pronounced red triangles on the sides that sometimes join the red triangles from the other side on the back. Approximately 10 percent (depending on the population and locality) of all mud snakes are anerythristic (i.e., lack all red pigments), and instead have white on their sides and belly. Juvenile mud snakes can be confused with adult black swamp snakes

Adult mud snake with aberrant color and pattern, Wilcox County (H. Bernard Bechtel)

Anerythristic adult mud snake, Echols County (B. W. Smith)

Juvenile mud snake, Lee County (John B. Jensen)

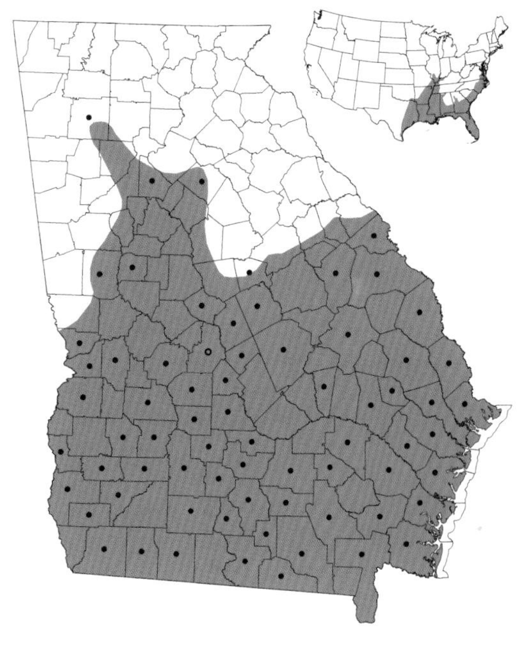

(*Seminatrix pygaea*). The latter do not have any red pigment on the sides of the body or any yellow pigment on the head, however, and the red on their belly is continuous.

Taxonomy and Nomenclature

Two subspecies of mud snakes may be present in Georgia: the western mud snake (*F. a. reinwardtii*) in the extreme western portion of the state and the eastern mud snake (*F. a. abacura*), which occurs throughout most of the species' range in Georgia. Mud snakes in western Georgia are probably intermediate between the western and eastern forms in many characters. Mud snakes have many common names, such as hoop snake, stinging moccasin, and horn snake. The latter two names derive from the spiny tip of the tail; the former comes from this species' tendency to bask in a loose horizontal coil after a large meal.

Distribution and Habitat

Mud snakes are aquatic snakes that occur in most slow-moving, acidic rivers and swamps throughout the upper and lower Coastal Plain of Georgia; a few specimens have been found in the Piedmont. Other habitats include cypress swamps, Carolina bays, river swamps, irrigation ditches, weed-choked ponds, freshwater marshes, and flooded flatwoods. The distribution may be limited by the range of amphiumas (genus *Amphiuma*) and sirens (genera *Pseudobranchus* and *Siren*)—the giant salamanders that are their primary prey.

Reproduction and Development

Little is known about mud snake reproduction. Adults mate from late April through early June, and females lay their eggs on land in sandy nests that they excavate; they sometimes lay their eggs in nests of the American alligator (*Alligator mississippiensis*). Clutch size varies from as few as 4 eggs to more than 100. The female may stay with the eggs until they hatch in August and September. Hatchling mud snakes remain in the nest, or at least on land, until the following spring, when they migrate to water. Mud snakes mature in 2 years, and females probably lay eggs every year.

Habits

Mud snakes may hibernate under logs or other debris near the edge of their aquatic habitat during cold weather, but they can be found any time of the year during warm weather. They become nocturnal during summer. Adults are dietary specialists that eat only salamanders, primarily two-toed amphiumas (*Amphiuma means*) and lesser (*Siren intermedia*) and greater (*S. lacertina*) sirens. Some researchers believe that the spine on the tail is an adaptation for eating these large, slippery animals. Anecdotal evidence indicates that mud snakes use their tail as an anchor when struggling to consume giant salamanders alive. Captured mud snakes sometimes poke the captor's hand with the tail spine. Mud snakes sometimes curl the end of the tail and display the bright red underside when threatened, and juveniles may play dead or go limp if handled excessively.

Conservation Status

The conservation status of mud snakes is poorly known. They are often considered rare, but this may be because of their secretive life history and often inhospitable preferred habitats. Their reproductive strategy of being primarily aquatic but laying eggs on land makes mud snakes vulnerable to upland habitat modifications even when wetlands are protected.

Cameron A. Young and Whit Gibbons

Adult rainbow snake, Baker County (Gabriel J. Miller)

Rainbow Snake

Farancia erytrogramma

Description

This large, heavy-bodied snake is a close relative of the mud snake (*F. abacura*). Adults are typically 91–122 cm (35.5–47.6 in) in total length; the record size is almost 168 cm (65.5 in). Females get considerably larger than males. The head is narrow—about the same width as the neck—and the tongue is very short and reduced. This is a beautiful snake with glossy iridescent scales, a dark background color, and three clearly defined red or pink stripes extending the length of the back. The belly is even more boldly marked with a double row of black spots on a bright yellow to red background. The scales are smooth, and the anal scale is usually divided. Rainbow snakes have a prominent spine on the tip of their tail, presumably to help control slippery food items like eels and large aquatic salamanders. The hatchlings, which are 200–220 mm (7.8–8.6 in) long, are identical in appearance with the adults. Rainbow snakes are unlikely to be confused with any other Georgia snake.

Taxonomy and Nomenclature

Only a single subspecies of the rainbow snake is found in Georgia: the common rainbow snake (*F. e. erytrogramma*). A local name for this snake is eel moccasin.

Distribution and Habitat

This species is found only in the Coastal Plain and is apparently much more abundant in the eastern portion of that province than farther west in Georgia. The rainbow snake is at home in creeks, rivers, and swamps, and may even

Adult rainbow snake, Baker County (David Steen)

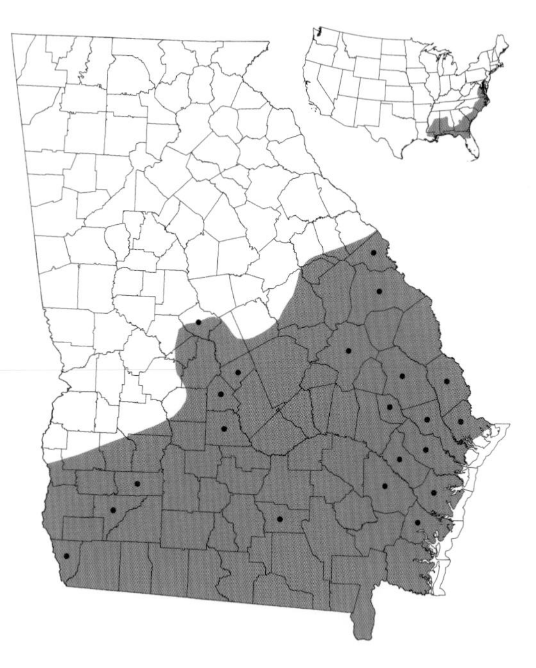

enter brackish marshes on occasion. It seems to prefer clear spring runs and moving water, as opposed to the mud snake, which is fond of mucky swamps and slow-water situations. Although the rainbow snake is strongly aquatic, individuals are occasionally uncovered during excavation activities and plowed up in sandy fields near wetlands, usually while in winter dormancy.

Reproduction and Development

Mating occurs in the spring, and each female lays 10–52 eggs in an underground cavity in sandy soil; females often remain with their eggs until hatching, which takes 60–80 days. In a South Carolina study, hatchlings were believed to have overwintered on land in the vicinity of the nest and then to have moved overland to an isolated wetland in March and April. They remained in the wetland feeding on small amphibians such as tadpoles and larval salamanders and on small fish. There is no direct information on development of rainbow snakes in Georgia, but in neighboring South Carolina, males reach sexual maturity at the end of the second or the beginning of the third year. Females mature at around 3 years.

Habits

Because they are secretive by nature, hiding under logs, mats of vegetation, and other debris, rainbow snakes are rarely encountered in the wild. They are thought to be chiefly nocturnal, although individuals are occasionally encountered basking during the day, usually shortly before shedding. The adults have a definite dietary preference for American eels and are occasionally seen foraging for these fish in streams and rivers. Potential natural predators of adults include large mammals such as raccoons, river otters, and bobcats that inhabit river swamps in Georgia. A red-shouldered hawk was observed feeding on an adult rainbow snake in a lower Savannah River marsh. Wading birds, large fish, and common kingsnakes (*Lampropeltis getula*) probably prey on the young. If captured, this beautiful snake is usually inoffensive, only occasionally poking the collector with its spiny, though harmless, tail.

Conservation Status

Because they are so secretive, little is known about the population status of rainbow snakes. Dams that prevent the up- and downstream movements of catadromous American eels very likely affect upstream populations of these snakes.

Tony Mills

Adult eastern hognose snake, Liberty County (Dirk J. Stevenson)

Eastern Hognose Snake

Heterodon platirhinos

Description

The eastern hognose is a stout, medium-sized snake. Adults are usually 51–84 cm (19.1–32.8 in) in total length, but large individuals may reach 116 cm (45.2 in). Females are generally longer and heavier than males. The scales are keeled, and the anal plate is divided. Eastern hognose snakes are extremely variable in coloration and pattern. The background body color can be yellow, orange, tan, or gray, with brown, rust, or black blotches along the center of the back and smaller alternating spots along either side of the body. In some areas, adults may be almost uniformly black, charcoal, or gray, with the pattern barely visible or entirely absent. The light beige or gray belly is mottled, sometimes heavily, with gray or greenish gray. The underside of the tail generally has less mottling than the belly, giving it a lighter appearance. The young, 130–250 mm (5.1–9.8 in) at hatching, are always conspicuously patterned and can be as variable as the adults, although they are never solid black or gray. The belly of juveniles can be almost solid black. The eastern hognose snake is not venomous but does have enlarged rear teeth in the upper jaw. This species is characterized by its spade-shaped, slightly upturned snout, which is keeled above. Some individuals resemble the southern hognose snake (*H. simus*) in color and pattern, but the latter can be distinguished by its more sharply upturned snout and lack of

Adult eastern hognose snake feigning death, Bryan County (Dirk J. Stevenson)

Juvenile eastern hognose snake, Long County (Dirk J. Stevenson)

Juvenile eastern hognose snake with flared neck, Bryan County (Natalie L. Hyslop)

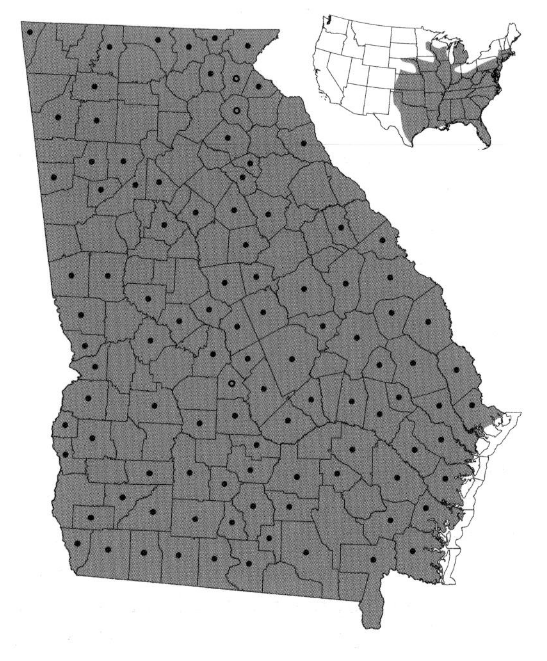

conspicuous mottling on the belly. Black or dark gray forms may be confused with black racers (*Coluber constrictor*), which lack the keeled scales, large head, upturned snout, and stout body of the eastern hognose snake.

Taxonomy and Nomenclature

No subspecies are recognized. Colloquial names for the eastern hognose snake include puff adder, hissing adder, and spreading adder.

Distribution and Habitat

Eastern hognose snakes are burrowers, and they may be found anywhere in Georgia where sandy or loamy soils occur. They inhabit many upland habitat types, including sandhills, mixed oak-pine forests, pine plantations, pine flatwoods, pure hardwood forests, and some disturbed areas such as abandoned fields. They generally avoid densely wooded habitats and wet areas, although they can occur in the terrestrial habitat surrounding wetlands. Eastern hognose snakes are known to use rotting stumps and root holes, rocks, and rubbish piles for overwintering sites.

Reproduction and Development

Mating typically occurs in the spring, although a Baker County pair was found copulating in early February. Clutch size varies widely, with

Adult eastern hognose snake with flared neck, Long County (Dirk J. Stevenson)

females depositing 4–61 thin-shelled eggs in late May–August. Eggs are 25–34 mm (1–1.3 in) in length and take about 60 days to hatch. Hatchlings emerge in July–September. Very little is known about nesting behavior and nest site selection. Sexual maturity is usually reached within 2 years.

Habits

The eastern hognose snake is notorious for its theatrical performances. When threatened, it flares its neck, raises its upper body—perhaps weaving it from side to side like a cobra—expands its body by breathing in deeply, and then expels the air with an impressive hiss. The snake may strike but does not bite. If further harassed, the snake rolls over on its back, writhing vigorously as if in pain. During this stage of the performance the snake may also evert its vent, regurgitate its stomach contents, defecate, or release musk, and may even bleed from its mouth. After a couple of minutes it usually becomes totally inert and limp, sometimes gaping or hanging out its tongue, feigning death. If rolled onto its belly, the snake will promptly roll onto its back again. The eastern hognose snake is strictly diurnal and is most active from April to June and again from September to October, although captures have been reported in each month of the year. These snakes are extremely slow moving and are frequently encountered while attempting to cross roads. Like the southern hognose snake, the eastern hognose is well adapted for consuming toads, its primary prey. The upturned snout helps excavate buried toads, and the rear fangs are used to puncture toads that inflate themselves while being ingested. Less frequently consumed prey items include frogs, salamanders, lizards, small mammals, snails, and arthropods. Juveniles may also consume crickets or other insects. Snake-eating snakes such as the common kingsnake (*Lampropeltis getula*), birds of prey, and carnivorous mammals prey on eastern hognose snakes.

Conservation Status

The eastern hognose snake appears to be relatively common throughout the state, but they are too often killed by humans who mistake their spirited performance as indicative of a dangerous snake.

Tracey D. Tuberville and Kurt A. Buhlmann

Adult southern hognose snake, Baker County (Pierson Hill)

Southern Hognose Snake

Heterodon simus

Description

The maximum total length reported for this small, stout-bodied snake is 61 cm (23.8 in), but adults are typically 33–56 cm (12.9–21.8 in). Females get larger and have a comparatively shorter tail than males. The southern hognose snake is characterized by its prominently keeled scales and sharply upturned snout, which is also keeled above. The tail is often rather blunt, and the anal plate is divided. Dark brown squarish blotches extend along the center of the back with a row of smaller, sometimes more circular blotches offset along either side. The background color is beige or tan. A light orange or tan stripe is usually apparent along the center of the back. The belly is immaculate and cream colored, although faint brown pigment is usually evident near the tail. The coloration and pattern are fairly uniform among individuals, including hatchlings, which are 130–170 mm (5.1–6.6 in) in total length. This snake has enlarged, grooved rear teeth in the upper jaw but does not inject venom. Southern hognose snakes can be easily confused with eastern hognose snakes (*H. platirhinos*) and pigmy rattlesnakes (*Sistrurus miliarius*). Pigmy rattlesnakes are similar in size, although usually somewhat smaller, and have a very similar pattern, including the stripe along the back, but they also have a tiny rattle, vertical pupils, and a small pit between the eye and nostril. Eastern hognose snakes are highly variable in pattern and coloration, and some are remarkably similar to

Adult southern hognose snake, Dodge County (John B. Jensen)

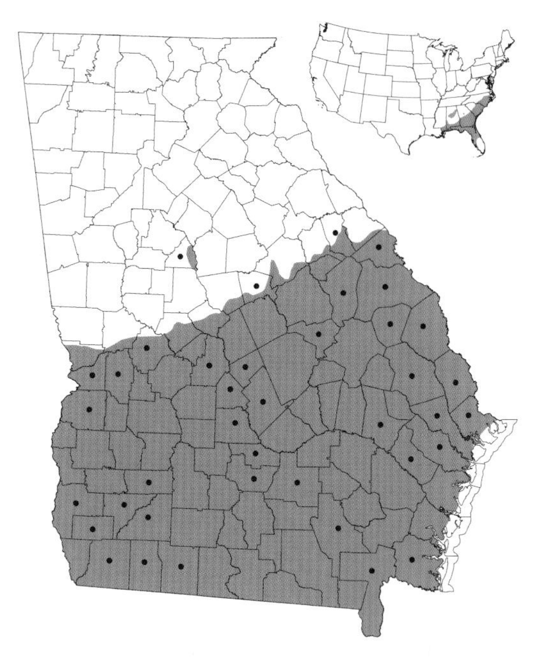

the southern hognose. Eastern hognose snakes are larger, though, and have a less distinctly upturned snout; their belly has gray flecking or mottling that makes it darker than the underside of the tail.

Taxonomy and Nomenclature

No subspecies are recognized. Southern hognose snakes are often colloquially known as puff adders or spreading adders.

Distribution and Habitat

The southern hognose snake is primarily associated with dry, upland habitats with sandy soils such as mixed oak-pine forests, scrub, sandhills, and oak hammocks. Individuals are occasionally reported from agricultural fields, residential areas, and other disturbed habitats. This species is widely distributed in the Coastal Plain of Georgia but tends to occur in small, disjunct populations that are sometimes isolated from one another by several miles. A single specimen from near Lake Jackson represents the only Piedmont record of this species in Georgia.

Reproduction and Development

Most of what is known about reproduction is based on observations of captive animals; no nest has ever been documented in the wild. Data from adult wild-caught females suggests that eggs are laid in late spring or early summer, although some wild-caught females that were subsequently maintained in captivity deposited eggs in late July. Clutch sizes of 6–14 eggs have been reported, and eggs take approximately 65–70 days to hatch. Hatchlings emerge between mid-September and mid-October. Age at maturity is not known.

Habits

This mild-mannered snake rarely tries to bite. Like its more common relative, the eastern hognose snake, the southern hognose sometimes hisses and flares its neck when threatened. Individuals may also roll over and feign death, but are less likely to delight their audience with dramatic displays than are eastern hognose snakes. The southern hognose snake is extremely fossorial. Much of its time is spent underground in burrows and in the stumps and root holes of rotting pines; it rarely occurs under loose cover objects. Individuals found coiled on the surface seldom move when encountered, and their camouflage makes them difficult to spot. Their fossorial habits, cryptic coloration, and secretive behavior ensure that they are relatively infrequently observed, even when specifically targeted by survey efforts. Most individuals are found incidentally on roads, either alive or dead. The species is strictly diurnal in its aboveground activity, and most observations occur from May to June and from September to October. Although southern hognose snakes may occasionally eat lizards, frogs, and small mammals, their diet is composed almost exclusively of toads (*Bufo* and *Scaphiopus*). They have specialized anatomical traits that may help them capture and eat their primary prey; the sharply upturned snout may help them excavate buried toads, and their rear fangs can puncture toads that inflate themselves while being swallowed. Predators include carnivorous mammals, birds of prey, common kingsnakes (*Lampropeltis getula*), and eastern indigo snakes (*Drymarchon couperi*). Eggs and hatchlings may be vulnerable to predation by fire ants.

Conservation Status

Although the southern hognose snake has no formal federal protection, the U.S. Fish and Wildlife Service considers it a Species of Concern. The state of Georgia recognizes it as a Threatened species. The southern hognose snake appears to be declining throughout most of its range and may even be extirpated from some states. Although it is still relatively widespread in Georgia, it appears to have disappeared from some areas. The primary threats are thought to be habitat destruction and degradation, especially intensive or chronic soil disturbance, and road mortality. The spread of introduced fire ants throughout the southern hognose snake's geographic range has also been implicated as a reason for the apparent decline.

Tracey D. Tuberville and John B. Jensen

Adult mole kingsnake, Floyd County (Bradley Johnston)

Mole Kingsnake

Lampropeltis calligaster

Description

Mole kingsnakes are medium-sized snakes. Some adults reach 119 cm (46.4 in) in total length, but most are much smaller. The scales are smooth, and the anal plate is undivided. The dorsal pattern may be light brown to reddish with a series of 35–55 rounded, reddish brown blotches bordered by black. A similar number of smaller blotches or vertical bars on the sides alternate with the back blotches. Older animals may be solid dark brown or greenish gray. The head is small and is not very distinct from the neck. The belly is whitish, yellowish, brownish, or greenish with checkered, alternating, darker squares. Juveniles are brighter in coloration than adults, usually with reddish to gray ground color and darker, bolder blotches. Hatchlings are 152–177 mm (5.9–6.9 in) in body length. This snake is often confused with the corn snake (*Elaphe guttata*) but can be distinguished by its smooth back scales and undivided anal scale; the anal scale is divided in the corn snake.

Taxonomy and Nomenclature

Originally described as *Coluber calligaster*, this species has been known as *L. calligaster* since 1860. Three subspecies are recognized, but the mole kingsnake (*L. c. rhombomaculata*) is the only one found in Georgia. This snake is also commonly known as the mole snake and brown kingsnake.

Juvenile mole kingsnake, Gordon County (Sean Graham)

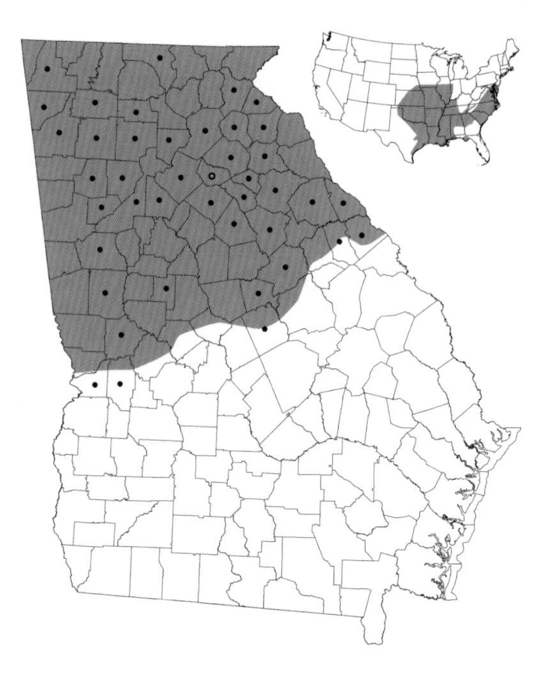

Distribution and Habitat

Most of the Georgia records are from the Piedmont and southern Appalachians, where the mole kingsnake occurs in upland forests, fallow agricultural fields, and pastures. In the Coastal Plain, it is associated with remnants of the longleaf pine savannas that were once prevalent in Georgia. Most records from nearby Florida come from longleaf pine flatwoods.

Reproduction and Development

Little is known about the life history of this highly fossorial and secretive species in Georgia, but the details are believed to be similar throughout its range. Mating takes place in the spring, and 5–16 eggs are laid 32–62 days afterward. Fertile eggs are 35–40 mm (1.4–1.6 in) long and 20–23 mm (0.8–0.9 in) wide. Eggs incubated in the laboratory hatched after 63–65 days at a mean temperature of 26° C (79° F). Sexual maturity is reached in 3 years.

Habits

The ecology of the mole kingsnake is not well known. Individuals are often discovered in fields during plowing, suggesting that they spend much of their life underground and come to the surface to feed. Mole kingsnakes are constrictors that feed on small snakes, lizards, and rodents; hatchlings and juveniles feed primarily on the latter two, and particularly favor skinks (genera *Eumeces* and *Scincella*). Birds, mammals, and snakes, including common kingsnakes (*L. getula*) and black racers (*Coluber constrictor*), are known to prey on mole kingsnakes. Individuals are sometimes found crossing roads—at any hour of the day during spring and after dark in midsummer.

Conservation Status

The mole kingsnake is rarely encountered anywhere in the southern part of its range, and its fossorial behavior makes assessments of its conservation status difficult.

Kenneth L. Krysko and D. Bruce Means

Adult common kingsnake, Monroe County (John B. Jensen)

Common Kingsnake

Lampropeltis getula

Description

The common kingsnake is one of Georgia's larger snakes. Adults are 90–122 cm (35.1–47.6 in) in body length and reach a maximum total length of 208 cm (81.1 in). Males are usually larger than females. Hatchlings are 130–200 mm (5.1–7.8 in) in body length and 230–300 mm (9–11.7 in) total length. The scales are smooth, and the anal plate is undivided. The dorsal pattern of eastern kingsnakes (*L. g. getula*) is solid black to chocolate brown with 17–36 narrow, light (usually yellow, cream, or white) cross bands that widen or divide on the sides, creating a chainlike pattern. The light pigment within each cross-band scale is situated on the forward portion of the scale, although specimens in northwestern Georgia may have the light spot in the center of cross-band scales. Hatchlings may have reddish coloration on the light-colored cross-band scales, and the dark scales of the back between the bands do not usually lighten with age, although interband scales are lighter in some adults from southeastern Georgia. The belly consists of checkered alternating squares of dark and light coloration. The back pattern of snakes living in northwestern Georgia tends to darken with age, and some individuals may become nearly uniformly black.

Adult common kingsnake, Baker County (Gabriel J. Miller)

Adult common kingsnake, Floyd County (John B. Jensen)

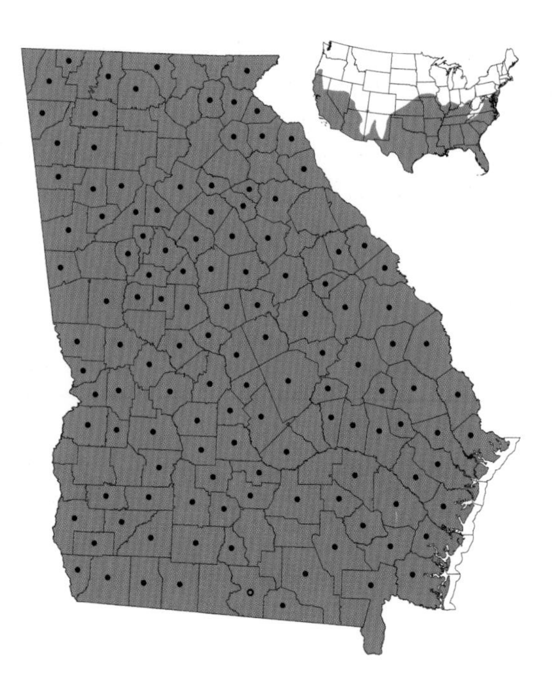

Taxonomy and Nomenclature

Originally described as *Coluber getulus*, this species was known as *L. getulus* for nearly all of the 1900s. The species epithet was changed to *getula* in 1988 to conform with the rules of Latin grammar. Two of the seven recognized subspecies of *L. getula* occur in Georgia: the eastern kingsnake (*L. g. getula*) and the black kingsnake (*L. g. nigra*); the latter is restricted to the northwestern corner of the state. These two taxa are both morphologically and genetically distinct and may warrant recognition as species.

Distribution and Habitat

Common kingsnakes are found throughout Georgia, especially in microhabitats that allow them to burrow and in the vicinity of water. They are also found on the barrier islands. Common kingsnakes occur in a wide variety of habitats but typically shy away from dry sites such as sandhills.

Reproduction and Development

Mating takes place from March through June. Males are encountered more often than females during this time, possibly because they are actively searching for mates. Combat rituals between males have been observed in adjacent Florida populations. The female lays 3–29 eggs in a moist substrate about 45–60 days after mating. Eggs are 31–51 mm (1.2–2.0 in) long and 18–27 mm (0.7–1.1 in) wide. They hatch in about 55–65 days, depending on incubation temperature.

Habits

Common kingsnakes are diurnal and spend much of their life underground, although they appear on the surface as they become active each season. Climatic variables such as temperature and precipitation have major effects on their activity patterns. The diet consists of rabbits, rodents, amphibians, turtle eggs, lizards, and snakes, including members of their own species. These constrictors are renowned for their ability to overcome and eat venomous snakes, hence the name kingsnake. Common kingsnakes usually strike and grasp snake prey near the head before coiling around it and constricting. Once the snake is secured, it is swallowed headfirst. If the snake to be eaten is longer than the common kingsnake, it is folded before being swallowed. Natural predators include birds, mammals, and snakes, including eastern indigo snakes (*Drymarchon couperi*), black racers (*Coluber constrictor*), and other common kingsnakes. Common kingsnakes frequently vibrate the tail and secrete musk when threatened.

Conservation Status

Common kingsnakes are relatively easy to find in many areas of Georgia but are becoming increasingly scarce in parts of the lower Coastal Plain. Their large size, docile disposition, beautiful coloration, and ease of capture have made common kingsnakes a popular target for collectors throughout their range. Because of their well-known appetite for other snakes, especially venomous ones, common kingsnakes are among the very few snakes tolerated by Georgians who otherwise kill snakes on sight.

Kenneth L. Krysko

Adult scarlet kingsnake, Long County (Dirk J. Stevenson)

Scarlet Kingsnake

Lampropeltis triangulum elapsoides

Description

The scarlet kingsnake is a small snake. Adults average 36–51 cm (14–19.9 in) in total length, but some may reach 62 cm (24.1 in). The scales are smooth, and the anal scale is undivided. This snake is arguably one of the most beautiful snakes in the world. Its vibrant colors and delicate build make it a consummate jewel of the Georgia ecosystem. Bright red, black, and cream to yellow bands usually completely encircle the body. The snout is usually red and noticeably pointed. The scarlet kingsnake is easily distinguished from the venomous eastern coral snake (*Micrurus fulvius*) because its brightly colored bands are oriented red-black-yellow instead of red-yellow-black. It differs from the scarlet snake (*Cemophora coccinea*) in that the bands go completely around the body; the scarlet snake has an immaculate white belly. The brightly hued and boldly patterned hatchlings resemble adults and average 130–200 mm (5.1–7.8 in) at hatching.

Taxonomy and Nomenclature

The scarlet kingsnake has historically been treated as a subspecies of the milk snake (*L. triangulum*). It is distinct in both its physical appearance and its ecology, however, and in some areas appears to occur with, or at least in close proximity to, the eastern milk snake (*L. t. triangulum*) without interbreeding. Many

Adult scarlet kingsnake, Liberty County (Dirk J. Stevenson)

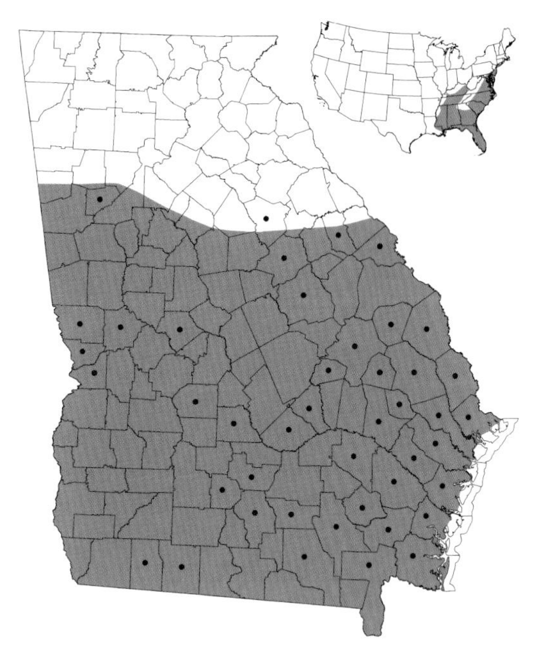

herpetologists thus treat the two as different species. Although further genetic research is needed to resolve this issue, the distinctiveness of this subspecies is sufficient to warrant its treatment here as a separate form.

Distribution and Habitat

The scarlet kingsnake occurs in a variety of wooded habitats but prefers pine flatwoods. Although it is more common in the sandy soils of the Coastal Plain, it can also be found in the clay-based soils of the Piedmont. It may also occur in the Ridge and Valley Province.

Reproduction and Development

Georgia's scarlet kingsnakes presumably mate during spring. Pairs are occasionally collected together under bark, in rotten logs, or under other debris. The female lays two to eight eggs in a rotten log, pile of debris, or underground burrow in April or May. Incubation lasts 45–70 days. The eggs typically hatch in July or August, although an unhatched clutch was discovered in Georgia in September. Age at maturity is not known.

Habits

This snake is rarely encountered during the day but is often found crossing roads at night in late spring or summer. It is more active during rainy weather. Individuals spend their days under debris, underground, or under the bark of decaying stumps and logs. Remaining just under the bark of a vertical or prone log may be an effective way for the snake to regulate its body temperature without the risk of desiccation or predation associated with basking in the open; it may also be a good place to secure lizards and other food items. Although scarlet kingsnakes have been reported to feed on a variety of small vertebrates such as small snakes, mice, and fish, they seem especially fond of lizards, including skinks (genera *Eumeces* and *Scincella*) and anoles (genus *Anolis*). These snakes may be eaten by foxes, skunks, opossums, various bird species, and other snakes. Some scientists believe that the scarlet kingsnake mimics the coral snake, thus protecting itself from predators, but this has never been unequivocally demonstrated.

Conservation Status

Although secretive and rarely seen, this small snake is common in many parts of Georgia.

Tony Mills

Adult eastern milk snake, Walker County (Sean Graham)

Eastern Milk Snake

Lampropeltis triangulum triangulum

Description

Adult eastern milk snakes average 61–90 cm (23.8–35.1 in) in total length; exceptionally large individuals may reach 132 cm (51.5 in). The body is relatively slender. The scales are smooth and usually shiny, and the anal scale is undivided. The back and tail have 24–54 large, irregularly shaped brown, grayish brown, or reddish brown blotches that are bordered in black. Smaller blotches of similar color and shape extend along the lower sides and alternate with those on the back. The background color varies from gray to tan. A light, Y- or V-shaped mark usually appears on the nape or back of the head, but this character is not as reliable for identification as some field guides suggest. The belly is black and white and patterned like a disorganized checkerboard. Hatchlings and young juveniles resemble adults, but their blotches are a deeper reddish brown and may even be true red. Hatchlings may be as long as 267 mm (10.4 in). Corn (*Elaphe guttata*) and juvenile rat (*E. obsoleta*) snakes are the species most likely to be confused with eastern milk snakes, but the scales of both of those species are keeled, and their anal plates are divided.

Taxonomy and Nomenclature

Historically the eastern milk snake has been treated as a subspecies of a wide-ranging species that includes the native scarlet kingsnake (*L. t. elapsoides*) and 23 other subspecies occurring across the Americas. It is distinct in both

Juvenile eastern milk snake, Murray County (Sean Graham)

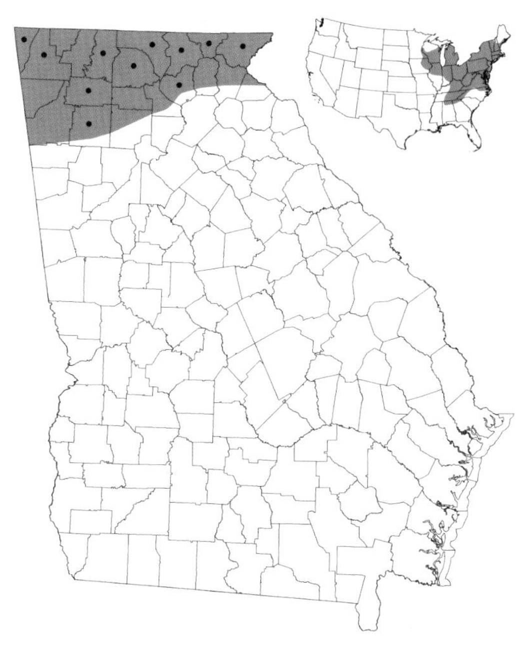

its physical appearance and its ecology, however, and in some areas it appears to occur with, or at least in close proximity to, the scarlet kingsnake without interbreeding. Therefore, many herpetologists believe that eastern milk snakes and scarlet kingsnakes are different species. Although further genetic research is needed to resolve this issue, this subspecies is distinct enough to warrant its treatment here as a separate form. The common name is derived from a myth that these snakes suck milk from cows. Absurd claims that this nonvenomous species poisons the milk while doing so make this tale even more outrageous.

Distribution and Habitat

Eastern milk snakes are found in the mountainous areas of Georgia's Blue Ridge, Cumberland Plateau, and Ridge and Valley provinces. Throughout its U.S. range the eastern milk snake occupies a wide variety of both natural and disturbed habitats; in Georgia, however, most individuals have been found in forested habitats, often in association with rock outcroppings.

Reproduction and Development

Little is known about the reproductive biology of this snake in Georgia, but a considerable amount of information is available from elsewhere in its range. Mating takes place in spring, typically April–May, followed by nesting 30–40 days later. Clutches usually range in size from 9 to 24 eggs and are laid in rotten stumps or logs, under rocks, or in mammal burrows. Eggs typically hatch in late summer after an incubation period of 50–70 days. Age at maturity has not been accurately determined. Captive eastern milk snakes have lived in excess of 20 years, but the longevity of those in the wild surely averages considerably less.

Habits

Eastern milk snakes are shy and secretive, typically staying hidden in rock crevices, under rocks and logs, or, like the related scarlet kingsnake, under the bark of decaying logs and stumps. Studies outside Georgia have shown some seasonal variation in the habitats used, from moist, low-elevation sites during the summer to higher, drier areas presumably associated with hibernation in the fall, winter, and spring. Eastern milk snakes are decidedly nocturnal during the warmer months, but individuals may be active at any time of the day during cooler—but not cold—periods. Instead of basking in direct sunlight, these snakes warm themselves on rocks or other objects heated by the sun's radiation. Eastern milk snakes apparently prefer small mammals such as shrews, mice, and voles but also eat birds, lizards, snakes, frogs, salamanders, fish, worms, caterpillars, and roaches, as well as reptile and bird eggs and carrion. They use constriction to subdue large, active prey. Like other members of the genus, this species can and does eat venomous pit vipers without suffering harmful consequences from envenomation. Indeed, the blood of all *Lampropeltis* species has venom-neutralizing properties. These snakes are not, however, immune to predators such as hawks, owls, raccoons, foxes, skunks, weasels, opossums, and even large bullfrogs (*Rana catesbeiana*). When threatened, eastern milk snakes often vibrate the tail and strike repeatedly; when captured they may release musk.

Conservation Status

Very little is known about eastern milk snake populations in Georgia, but the significant extent of protected public land within their state range would seem to indicate that they are secure.

John B. Jensen

Adult coachwhip, Bryan County (Dirk J. Stevenson)

Coachwhip

Masticophis flagellum

Description

The coachwhip is a long, slender snake with a maximum total length of 260 cm (101.4 in) and average adult body length of 106–152 cm (41.3–59.3 in). The scales are smooth, and the anal plate is divided. The dark chocolate brown or black head sometimes seems to have a velvety texture. The body color fades rearward to caramel or tan at the tail, and the proportion of dark color to light on the body is highly variable. The belly is pale from midbody to tail tip, but the chin, throat, and very front portion of the belly are usually dark. The scales do not form the linear rows typical of other species but are instead arranged in a manner resembling a braided rope or whip. Hatchlings look very different from the adults, being light brown or tan throughout their length with darker bars or markings on the front part of the body. The most similar species in both appearance and behavior is the black racer (*Coluber constrictor*). Both are fast-moving, slender snakes with large eyes and a narrow head. Although some coachwhips are almost totally melanistic like the black racer, some reddish tint is usually evident along the sides of the tail. Coachwhips also get much longer than black racers. The young are easily distinguished from other small brown snakes by their large eyes, smooth body scales, and braided-whip appearance. Hatchlings are 30–44 cm (11.7–17.2 in) in total length.

Adult coachwhip, Baker County (Gabriel J. Miller)

Juvenile coachwhip, Liberty County (Dirk J. Stevenson)

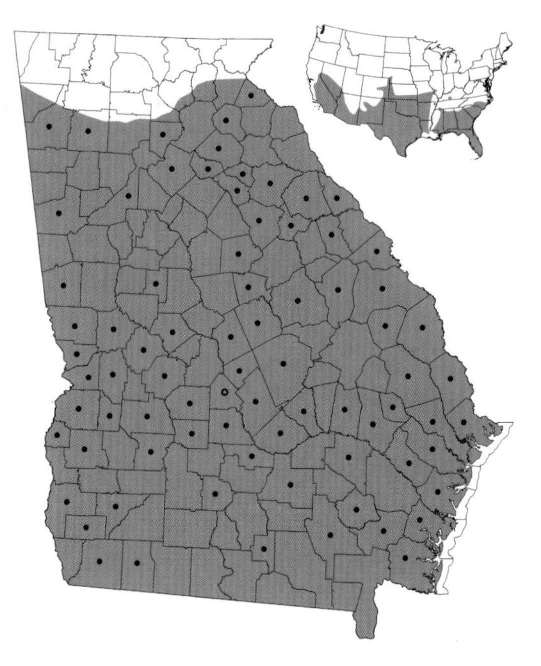

Taxonomy and Nomenclature

Several subspecies are recognized, but only the eastern coachwhip (*M. f. flagellum*) occurs in Georgia. In some areas coachwhips are called whipsnakes.

Distribution and Habitat

The coachwhip is found throughout much of the state but is apparently absent from the Blue Ridge and Cumberland Plateau provinces. It is most often associated with dry upland habitats such as sandhills, longleaf pine woodlands, and scrub, but can persist in some disturbed habitats such as agricultural lands, abandoned old fields, and power line rights-of-way. It is especially common in secondary dune systems of barrier islands.

Reproduction and Development

Mating probably occurs during April and May, shortly after adults emerge from overwintering sites. During the summer nesting season, females lay 4–24 eggs in loose soil; in burrows made by other animals; or in rotting stumps, logs, or root chambers. The eggs, which have distinctive granular nodules that look like grains of salt on the outer surface of the shell, take 45–79 days to hatch, and most hatchlings probably emerge by late September. Information on age at maturity for Georgia coachwhips is not available.

Habits

Although the coachwhip is a relatively conspicuous member of Georgia's wildlife, very little is known about its biology in this state or elsewhere in the eastern United States. Our knowledge of its natural history is largely based on anecdotal observations. The coachwhip is well suited to dry habitats; it is strictly diurnal, can remain active at very high temperatures during the summer, and is extremely resistant to desiccation. The activity period in Georgia usually runs from April or early May through October. During this time, when not active, coachwhips seek refuge in rotting pine stumps, root holes, burrows of other animals such as gopher tortoises (*Gopherus polyphemus*) or mammals, and under a variety of natural and artificial cover objects. Overwintering sites include stumps and root chambers of rotting pines. The eastern coachwhip is a versatile forager in terms of both its diet and its hunting behavior. It hunts actively, constantly moving in search of lizards as well as small mammals, birds, snakes, small turtles, and invertebrates. In addition to hunting on the ground surface, eastern coachwhips probably follow their prey into underground burrows and even into trees and shrubs. Because of their speed and agility, eastern coachwhips are difficult to catch. When threatened or cornered they usually raise the upper body and attempt to strike or bite. If captured they will bite and may also whip their body around vigorously. Predators include birds of prey, carnivorous mammals, and perhaps other snakes.

Conservation Status

Although coachwhips are encountered fairly frequently in some parts of the state, especially the Coastal Plain, the status of this species in Georgia and throughout its range is unknown.

Tracey D. Tuberville and
Whit Gibbons

Adult plain-bellied watersnake, Baker County (John B. Jensen)

Plain-bellied Watersnake

Nerodia erythrogaster

Description

Plain-bellied watersnakes are among the larger semiaquatic snakes in our region; large individuals may be 164 cm (64 in) in total length. Like all watersnakes, plain-bellied watersnakes have a fairly heavy body and keeled scales. The anal plate is usually divided, although an occasional individual has an undivided one. Adults are generally uniform in color and may range from light reddish brown to almost black. True to its common name, this species has an unpatterned belly; its color ranges from light yellowish in northwestern Georgia to bright orange-red elsewhere in the state. No other watersnake in our region has an unpatterned belly, which makes identifying this species easy. Young plain-bellied watersnakes are quite different in appearance from adults. Their markings begin nearest the head as complete cross bands but break up further back into alternating back and side blotches, a pattern similar to that of many northern watersnakes (*N. sipedon*). Although the belly of juvenile plain-bellied watersnakes may be marked with small, wormy, horizontal lines, it lacks the conspicuous markings of the other species. Newborn plain-bellied watersnakes are 180–250 mm (7.0–9.8 in) long and develop the solid coloration of adults at about 300 mm (11.7 in).

Juvenile plain-bellied watersnake, Decatur County (Pierson Hill)

Adult plain-bellied watersnake, Liberty County (Dirk J. Stevenson)

Adult plain-bellied watersnake, Cobb County (Gregory C. Greer)

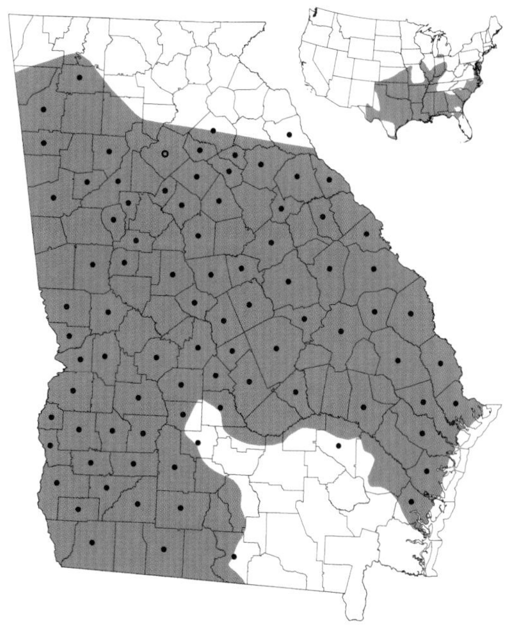

Taxonomy and Nomenclature

Of the six recognized subspecies of *N. erythrogaster*, two are found in Georgia: the red-bellied watersnake (*N. e. erythrogaster*) in the eastern and central parts of the state and the yellow-bellied watersnake (*N. e. flavigaster*) just entering Georgia in the northwest. As their names imply, these subspecies differ in belly color. The two subspecies intergrade in western Georgia, however, and although individuals with lighter bellies can be found, true yellow-bellied snakes are rare in the state. Many Georgians call the plain-bellied watersnake the copper-bellied moccasin.

Distribution and Habitat

Plain-bellied watersnakes are absent from the mountainous regions of northern Georgia and from much of the southeastern Coastal Plain south of the Altamaha River, including the Okefenokee Swamp. They are almost always found in association with some type of aquatic habitat and are particularly common in heavily vegetated temporary wetlands and along slow-moving creeks, rivers, and blackwater cypress swamps. They may also be common around farm ponds and spillways. This species is generally considered to be more terrestrial than other watersnakes and can often be found a fairly long distance from water.

Reproduction and Development

Less is known about reproduction and development of the plain-bellied watersnake than most other watersnake species, but presumably they resemble their congeners in many respects. Mating apparently occurs in spring, and females give birth to relatively large litters (2–55; average = 18) of live young in the late summer or early fall. Other similar-sized watersnakes mature in 2–3 years.

Habits

In or at the edge of water bodies this species may be found at any time of day, but particularly in the morning, evening, or night during the summer and during the day in the cooler months. Like other watersnakes, individuals are frequently observed basking on logs or branches overhanging the water or foraging at night in shallow water. These snakes move overland with particular frequency during rainy or humid summer nights, and many are killed as they try to cross highways. Plain-bellied watersnakes eat a variety of aquatic prey, including fish, amphibians, and crayfish, but several studies have shown that they generally consume a higher proportion of amphibian prey than other watersnakes. This preference for amphibian prey may partially explain the abundance of plain-bellied watersnakes in temporary wetlands, where the lack of fish allows amphibians to thrive. Predators include fish such as largemouth bass; predatory birds such as hawks, herons, and egrets; and other snakes, including common kingsnakes (*Lampropeltis getula*) and cottonmouths (*Agkistrodon piscivorus*). Presumably, plain-bellied watersnakes are also eaten by mammals such as raccoons, river otters, and opossums. When captured, plain-bellied watersnakes display the usual repertoire of watersnake defensive behaviors: they release musk, thrash and twist the body, and bite. Unlike other watersnakes, however, this species is noted for occasionally fleeing onto land when approached rather than diving underwater.

Conservation Status

Plain-bellied watersnake populations in Georgia appear to be stable, and this species continues to be common in many aquatic habitats. Even so, it typically does not reach the high densities observed of other species such as northern and brown (*N. taxispilota*) watersnakes. Elsewhere in this species' range, human activities, particularly habitat degradation, have had a negative impact on populations. And as with other watersnakes, many plain-bellied watersnakes are needlessly killed by humans who mistake them for venomous cottonmouths (water moccasins) or by vehicles as they attempt to cross highways.

John D. Willson and Whit Gibbons

Adult banded watersnake, Liberty County (Dirk J. Stevenson)

Banded Watersnake

Nerodia fasciata

Description

Banded watersnakes are medium-sized watersnakes. Adults are generally 61–102 cm (23.8–39.8 in) in body length; the maximum total length recorded is 159 cm (62.0 in). Females grow considerably larger than males. The pattern consists of complete cross bands along the entire length of the back, squarish or triangular spots on the edges of the belly scales, and a dark stripe from the eye to the angle of the jaw. The scales on the back are heavily keeled, and the anal plate is divided. Color is highly variable among individuals. Cross bands and belly spots can be pale or dark red, brown, or black. Background color can be red, gray, tan, yellow, or, less often, nearly black. As with most watersnakes, the pattern tends to fade with age. Some individuals are completely melanistic with no evidence of markings on the back. Juveniles, which are 164–191 mm (6.4–7.5 in) in body length, have similar banding patterns but more vivid markings and colors than most adults. The geographic range of the banded watersnake partially overlaps that of a closely related species, the northern watersnake (*N. sipedon*). Although the two species are known to hybridize, banded watersnakes can generally be distinguished by the complete cross bands down the entire length of the body; most northern watersnakes have the back blotches offset from the side blotches on the rear half to two-thirds of the

Juvenile banded watersnake, Liberty County (Dirk J. Stevenson)

Adult banded watersnake, Long County (Dirk J. Stevenson)

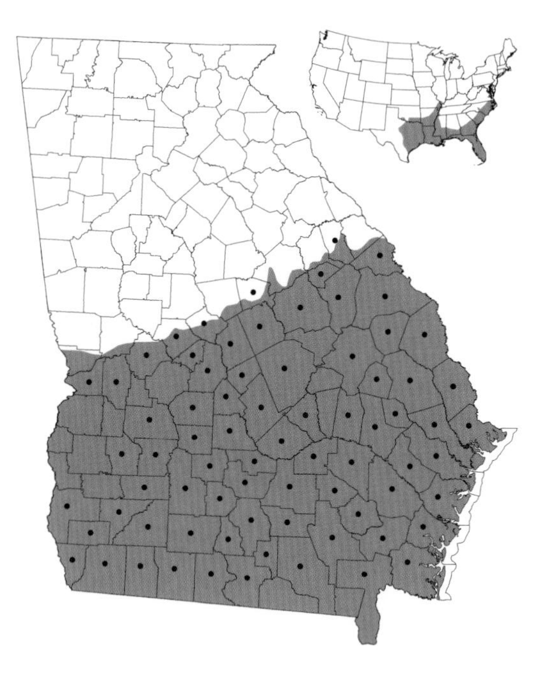

body. Northern watersnakes also tend to have more rounded, half-moon-shaped markings on the belly. Juvenile plain-bellied watersnakes (*N. erythrogaster*) are also superficially similar to banded watersnakes in appearance, but like the northern watersnake their back blotches are offset from the side blotches except near the head.

Taxonomy and Nomenclature

Of the three known subspecies, only the southern banded watersnake (*N. f. fasciata*) is found throughout most of the species' Georgia range. The Florida banded watersnake (*N. f. pictiventris*) may occur in extreme southeastern Georgia. Banded and northern watersnakes may occasionally hybridize in areas where their ranges overlap, such as in southwestern Georgia, where the northern watersnake follows major streams deep into the Coastal Plain.

Distribution and Habitat

Truly a habitat generalist, this snake can be found in any type of freshwater habitat, including isolated wetlands, rivers, streams, lakes, reservoirs, roadside ditches, canals, and swamps throughout most of the Coastal Plain. In the Coastal Plain west of the Flint River, banded and northern watersnakes are segregated by habitat, with banded watersnakes occurring in ponds and lakes and northern watersnakes occurring along major streams.

Reproduction and Development

Several males may aggregate and form a mating ball with one or a few females during courtship, which usually occurs from early April to mid-June. Females give live birth to 2–83 young (average = 22), usually between late June and late August, although sometimes as late as October.

Habits

Banded watersnakes bask on branches, other emergent objects, and shorelines during the day. Generally, they are diurnal in the spring and fall and nocturnal during the summer, although they may be found active during any time of day. During the winter, they hibernate beneath logs or other shelters in swampy areas or in crayfish burrows, although they may come out to bask on warm winter days. The banded watersnake is a generalist predator that feeds on fish, frogs, salamanders, and occasionally crayfish. When threatened and unable to flee, individuals readily bite and exude a foul-smelling musk. They also flatten their head and body to resemble the triangular head and heavy body of the venomous cottonmouth (*Agkistrodon piscivorus*). Common predators likely include cottonmouths, common kingsnakes (*Lampropeltis getula*), large wading birds, raptors, carnivorous mammals, and large fish.

Conservation Status

Banded watersnakes are very common in most bodies of water within their range, even in human-altered landscapes, but many individuals are needlessly killed by humans who mistake them for cottonmouths (water moccasins).

Christopher T. Winne and
Whit Gibbons

Adult eastern green watersnake, Putnam County, Florida (John B. Jensen)

Eastern Green Watersnake

Nerodia floridana

Description

The eastern green watersnake is the largest of the North American watersnakes. Adults reach an average total length of 76–140 cm (29.6–54.6 in); the record total length is 188 cm (73.3 in). Females grow larger than males, and almost all adults over 100 cm (39 in) long are females. This greenish or brownish snake generally lacks distinctive markings; a few individuals have some dark speckles that are more noticeable on the sides than the back. The belly is solid whitish or cream, and the underside of the tail is gray or brown marked with pale spots that are often shaped like half-moons. The head is large and has a row of scales between the eye and the upper lip scales. The scales on the back are keeled, and the anal plate is divided. Juveniles resemble adults except that black or dark brown markings may be present on their sides and back. Their belly is mostly yellow with black markings near and under the tail. Newborns are usually about 224–254 mm (8.7–9.9 in) in total length.

Taxonomy and Nomenclature

This snake was once considered a subspecies of what is now called the western green watersnake (*N. cyclopion*) because of their close resemblance in color and build and because they are the only two watersnakes that have a row of scales between the eye and the upper lip scales. Genetic studies have shown the eastern green watersnake to be a distinct species. No subspecies are recognized. A study of separate populations in Georgia and Florida found no genetic difference between them. This species is also commonly known as the Florida green watersnake.

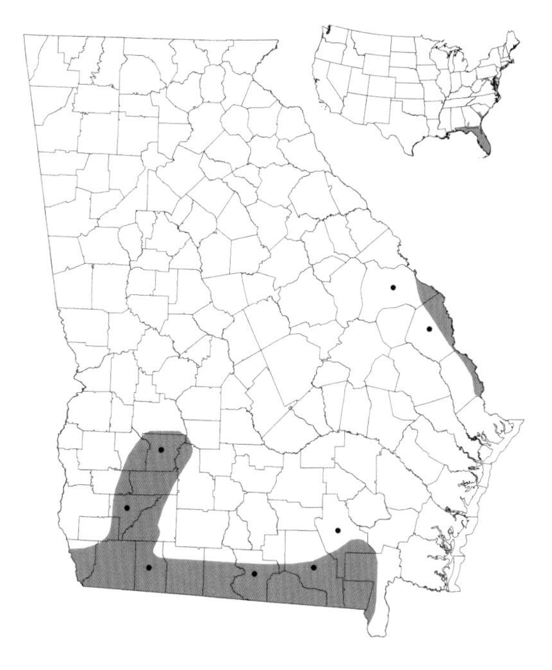

Distribution and Habitat

The eastern green watersnake is commonly found in quiet water with abundant vegetation and minimal tree cover such as marshes, lakes, ponds, swamps, and sloughs of the Coastal Plain. There is a large hiatus in the known range of this species in the eastern Coastal Plain of Georgia. The lack of genetic differences in the populations on either side of this gap may indicate that the species does occur continuously but has gone undetected in this part of the state, or that the two groups were only recently separated.

Reproduction and Development

Breeding takes place as early as February, and females bear litters from late summer into fall. Litters usually contain 20–30 young but can have more than 100; larger females bear larger litters. Gravid females attain an impressive girth and bask conspicuously. Little else is known about the reproduction or development of this species.

Habits

Eastern green watersnakes are often seen basking in the daytime on grassy edges next to water or on logs and branches over water. When alarmed, they may drop or slide into the water to seek cover. These snakes are known to hide in the houses of round-tailed muskrats. Eastern green watersnakes are excellent swimmers, and they obtain most of their food—including larval and adult amphibians, small turtles, fish, and invertebrates—in or near water. They are nocturnal or crepuscular, depending on conditions and time of year, and individuals often travel overland at night. When cornered, they may flatten their body and strike. When captured, they can bite hard and discharge copious quantities of fetid-smelling musk and feces from the vent. Potential predators include carnivorous mammals such as raccoons and river otters, wading birds, American alligators (*Alligator mississippiensis*), cottonmouths (*Agkistrodon piscivorus*), and common snapping turtles (*Chelydra serpentina*). The young may also be eaten by large fish.

Conservation Status

The eastern green watersnake is thought to be secure, although its range is not extensive. In Georgia, it is very localized and seldom encountered.

Sean M. Poppy

Adult northern watersnake, Fulton County (John D. Willson)

Northern Watersnake

Nerodia sipedon

Description

Northern watersnakes are moderate-sized, fairly stout, semiaquatic snakes that reach a maximum length of 150 cm (58.5 in); females grow larger than males. Adults are characterized by dark brown or reddish brown banding on the front part of the body that changes to three rows of alternating dark blotches on the rear of the body. A few individuals have bands extending the entire length of the body. The scales are keeled, and the anal plate is divided. Bands and blotches tend to be reduced in both number and width in Piedmont populations relative to populations in the mountains. Some older individuals become melanistic. The light belly is usually covered by many dark, irregular markings shaped like half-moons. Juveniles, which are 192–308 mm (7.5–12 in) at birth, are similar to the adults, but their patterns are generally bolder. Northern watersnakes can be distinguished from banded watersnakes (*N. fasciata*) by the lack of a stripe from the eye to the corner of the mouth and the lack of complete dark bands down the entire length of the body. Juvenile plain-bellied watersnakes (*N. erythrogaster*) also have cross bands, but the middle portions of their belly scales lack markings. Like other watersnakes, northern watersnakes can be distinguished from cottonmouths (*Agkistrodon piscivorus*) by the double row of scales on the underside of the tail and the lack of a facial sensory pit.

Juvenile northern watersnake, Cobb County (John D. Willson)

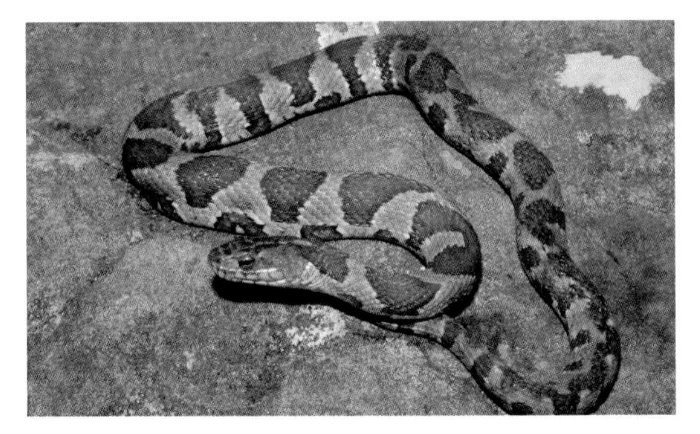

Juvenile northern watersnake, Walker County (John B. Jensen)

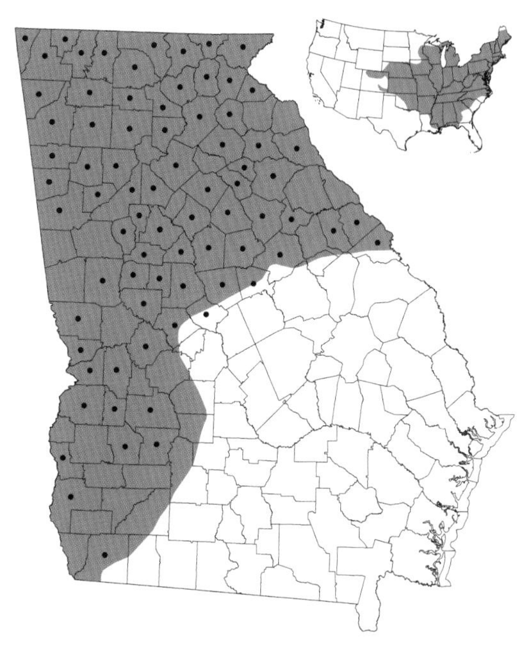

Taxonomy and Nomenclature

The northern watersnake is most closely related to the banded watersnake, and the two may hybridize where their ranges overlap. Two subspecies of the northern watersnake are found in Georgia: the northern watersnake (*N. s. sipedon*) in extreme northeastern Georgia, and the midland watersnake (*N. s. pleuralis*) elsewhere. Other common names for northern watersnakes include banded watersnake, common watersnake, and northern banded watersnake.

Distribution and Habitat

Northern watersnakes can be found in any aquatic habitat within the mountains and Piedmont of Georgia and within the upper Coastal Plain in the western part of the state. They can be especially abundant around the edges of small ponds, where amphibians and small fish abound. In the western part of the Coastal Plain, northern watersnakes are restricted to major streams and banded watersnakes occupy the surrounding lakes and ponds.

Reproduction and Development

Courtship and mating occur in late spring or early summer. Females give live birth to 5–60 (average = 23) young in late summer or early fall. Larger females give birth to more young than smaller females. Most females give birth every year after they reach sexual maturity at 2 or 3 years of age. Most males mature during their second year.

Habits

Northern watersnakes may be active during much of the year in warm areas but hibernate during colder periods in the northern part of the state. They may be active both day and night, but nighttime activity occurs primarily during the warmer months. During the day, northern watersnakes often bask on rocks or on branches overhanging the water. Northern watersnakes feed on a wide variety of aquatic and semiaquatic animals, but primarily on fish and amphibians. They hunt using both sit-and-wait (i.e., ambush) techniques and by active searching. Predators of adults include raccoons and other predatory mammals. Juveniles may be eaten by wading birds and even by largemouth bass. When threatened, northern watersnakes first try to escape; if escape is not possible, they bite repeatedly and release foul musk from their vent.

Conservation Status

Northern watersnakes are common or even abundant in many parts of their range throughout Georgia and are not thought to warrant conservation concern at this time. As is true for many aquatic snakes, they are frequently killed by people who mistake them for cottonmouths.

Michael E. Dorcas

Adult brown watersnake, Emanuel County (Dirk J. Stevenson)

Brown Watersnake

Nerodia taxispilota

Description

This is a medium-to-large, heavy-bodied snake. Females, which grow substantially larger than males, are 73–133 cm (28.5–51.9 in) in body length; males are 50–97 cm (19.5–37.8 in). The largest documented total length is 176.6 cm (68.9 in). The scales are strongly keeled, and the anal plate is divided. The background color is typically light tan to chocolate brown with three alternating rows of dark brown to black, squarish blotches, one row extending along the middle of the back and another row on each side. Large individuals may appear uniformly dark brown from a distance, and specimens from some areas have a rusty color. Occasionally the blotches may be connected, and rare partially striped individuals have been reported. The belly is cream colored to light tan with irregular dark brown spots that may be shaped like half-moons. The head is distinctly triangular, especially in large individuals, and the eyes are positioned high on the head. Because of their large size and dark coloration, brown watersnakes are often confused with the venomous cottonmouth (*Agkistrodon piscivorus*), which can be distinguished from brown and other watersnakes by its vertical pupils, pair of facial sensory pits, and single row of scales on the underside of the tail. Newborns average 240 mm (9.4 in) in body length, with a

Adult brown watersnake eating a bullhead, Burke County (John B. Jensen)

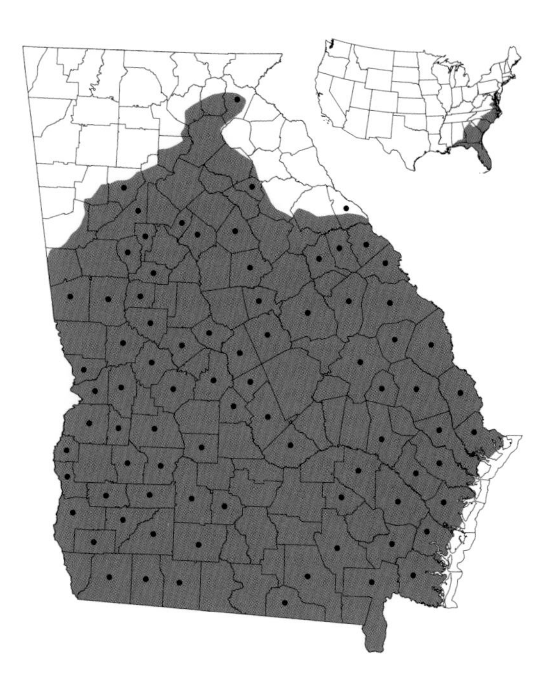

range of 200–280 mm (7.8–10.9 in). The juvenile pattern is like that of the adults, only typically brighter and better defined.

Taxonomy and Nomenclature

The diamondback watersnake (*N. rhombifer*) was at one time considered a subspecies of the brown watersnake, but studies of scale patterns, proteins, and DNA confirmed that they are separate species. No subspecies are currently recognized. The brown watersnake is commonly called water pilot, water rattler, pied-bellied water snake, and moccasin or water moccasin (even by people who understand that it is not a venomous cottonmouth).

Distribution and Habitat

Brown watersnakes are common residents of rivers, streams, swamps, and impoundments throughout the Coastal Plain and along major rivers of the lower Piedmont. In some areas they are the most frequently encountered watersnake, partially because of their fondness for basking on limbs, logs, and stumps over and near the water. In larger rivers they are found most often on the outer cut bank near relatively fast water; however, they can be locally abundant in areas of quiet water in swamps, sloughs, oxbow lakes, beaver ponds, reservoirs, and millponds.

Reproduction and Development

Mating takes place in March–May, earlier in the southern part of the state than in the north. More than one male often courts a female, and mating has been observed on the ground and in trees over water. Young are born from late July through October. The average litter size of females captured in a study along the Savannah River was 18 (range = 4–50), with larger females typically having larger litters. Most of the larger individuals give birth every year. Males reach maturity in approximately 3 years, and females mature in 5–6 years.

Habits

Brown watersnakes of all sizes feed almost exclusively on fish, primarily catfish. Other fish commonly consumed include sunfish, minnows, pirate perch, and suckers. The spines of catfish and sunfish occasionally injure individuals, but they can apparently survive even serious injuries of this type. For example, a female captured on the Savannah River with spines protruding through her body wall was recaptured 5 years later and not only had survived, but had grown more than 40 cm (15.6 in). These snakes forage both during the day and at night and usually take captured fish to the bank or to shallow water before swallowing them, although individuals have been observed swallowing smaller fish while lying on the bottom of quiet pools. Brown watersnakes use a variety of fishing techniques, including wrapping a coil or two around a branch and suspending the rest of the body in the water waiting for an unsuspecting fish to swim by; ambushing passing fish while lying very still on the bottom of a quiet pool; moving along the bottom or the bank probing holes and cavities; and moving the head from side to side with the mouth open ready to grab any fish encountered. Several authors have described brown watersnakes as both the most aquatic and the most arboreal of the watersnakes. Unlike banded (*N. fasciata*) or plain-bellied (*N. erythrogaster*) watersnakes, they seldom venture far from permanent water, although individuals have been seen sunning themselves during late winter around old stump holes on wooded hillsides in the lower Piedmont. These stump holes, presumably overwintering sites, were located 15 m (49.2 ft) or more from the riverbank. These snakes sometimes climb as high as 4–6 m (13.1–19.7 ft), and on Kinchafoonee Creek in southwestern Georgia an individual was observed more than 15 m (49.2 ft) high. It is this fondness for basking on limbs and dropping into the water to escape that often gets this snake into trouble. Many fishermen tell stories of a big moccasin (usually a brown watersnake) dropping into their boat along one of the big rivers, streams, or reservoirs of Georgia. Even though they are not venomous, these snakes bite when handled and release a foul-smelling musk. Both aquatic and terrestrial animals presumably prey on brown watersnakes; predators include American alligators (*Alligator mississippiensis*), large fish, cottonmouths, herons, red-shouldered hawks and other birds of prey, raccoons, and feral pigs.

Conservation Status

Humans often kill brown watersnakes in the mistaken belief that they are venomous. Nevertheless, this species seems to be doing well throughout most of its range in Georgia.

Mark S. Mills and Carlos D. Camp

Adult rough green snake, McIntosh County (Dirk J. Stevenson)

Rough Green Snake

Opheodrys aestivus

Description

This moderate-sized snake averages a total length of 68 cm (26.5 in) and may reach a maximum length greater than 115 cm (44.9 in). Females are larger than males and reach maturity at about 38 cm (14.8 in) in body length; males mature at about 25 cm (9.8 in). The scales are keeled, and the anal plate is divided. This snake's very slender build and vibrant green color are characteristics of an arboreal lifestyle. Its color provides impeccable camouflage in green vegetation, and its build allows it to move on even the smallest branches. Rough green snakes are uniformly green to greenish blue on top with a light yellowish green to white or cream-colored belly. Males have a slightly longer tail than females. Juveniles, 170–230 mm (6.6–9 in) long at hatching, are miniature versions of the adults but are slightly grayer. Interestingly, rough green snakes turn blue after death.

Taxonomy and Nomenclature

No subspecies are currently recognized. These snakes are sometimes called keeled green snakes because their scales are keeled, in contrast to the smooth scales of a northern terrestrial counterpart, the smooth green snake (*O. vernalis*).

Distribution and Habitat

Rough green snakes occur statewide in highly branched vegetation, especially along the edges

Road-killed adult rough green snake, Monroe County (John B. Jensen)

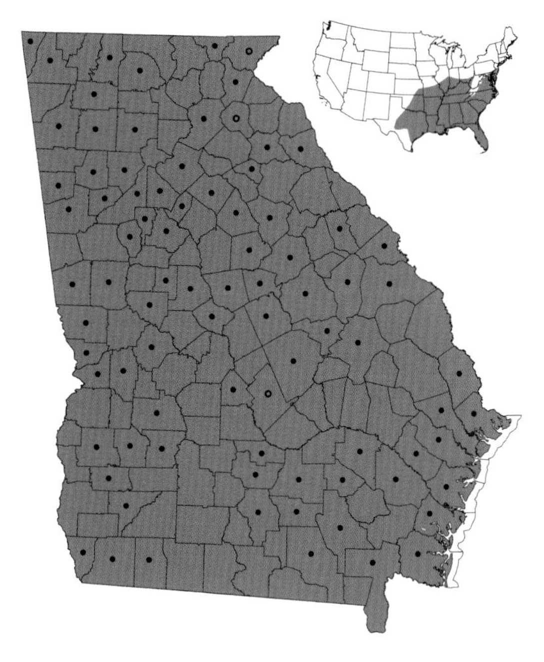

of streams or lakes with alder, willow, and ironwood. They can also be found in the bushy shrubs of forest-edge habitats. Occasional specimens are found on the ground, especially crossing roads.

Reproduction and Development

Females lay one clutch of 3–10 (average = 6) flexible-shelled eggs per season, usually in June or July. The eggs are deposited on the ground beneath rocks or in rotting logs, beneath bark, or in the hollows of dead or living trees, where they incubate for 36–43 days. A female may use the same nest cavity year after year, and nest cavities are sometimes used by more than one female at a time. Males are able to breed in the spring of their second year (21 months old). Some females breed at that age as well; others take a year longer to mature.

Habits

Rough green snakes are diurnal and actively hunt for much of the day. They make small, frequent meals of caterpillars, spiders, grasshoppers, crickets, and dragonflies. This snake usually stays partly covered in the branches of vegetation, where its brilliant camouflage and fluid movements keep it hidden from predators. Birds such as blue jays and swallow-tailed kites are known predators, as are black racers (*Coluber constrictor*) and common kingsnakes (*Lampropeltis getula*). Egg predators include black racers and broadhead skinks (*Eumeces laticeps*). Although rough green snakes are often associated with water-edge habitats, they are rarely observed swimming. They can sometimes be located at night by shining a flashlight up into the bushes alongside a waterway and illuminating the pale underbelly. Some have been found sleeping in a loose coil atop vegetation.

Conservation Status

Although its cryptic coloration makes the rough green snake difficult to detect, this species is thought to be fairly common in Georgia.

Peri A. Mason

Adult pine snake, Baker County (Lora L. Smith)

Pine Snake

Pituophis melanoleucus

Description

Pine snakes are among the largest of Georgia's snakes. Adults reach a total length of 106–183 cm (41.3–71.4 in); the maximum reported length is 229 cm (89.3 in). The ground color ranges from yellowish tan to white with dark brown or reddish markings, although the front of the body may be so dark that no pattern is discernible. From midbody to the tail, the markings usually form saddles on top; on the tail itself, they wrap around the sides to form bands. A row of smaller blotches runs along each side. Snakes from the mountains are usually one of two color phases; one has a background of creamy yellow and the other has a white or light gray background. Most pine snakes from the Coastal Plain are gray or rusty brown and have much less distinct markings than their northern relatives. The belly is creamy white to yellow with little to moderate spotting. Head markings are variable but often include a dark band across the forehead and pigmenting beneath the eyes. The rostral scale is enlarged. The strongly keeled scales and single anal plate set this species apart from lighter forms of the rat snake (*Elaphe obsoleta*), which have weakly keeled scales and a divided anal plate. Newly hatched pine snakes are 340–460 mm (11.7–17.9 in) in total length and resemble adults, although their patterns are often bolder.

Adult pine snake, Evans County (Dirk J. Stevenson)

Adult pine snake, Paulding County (Gregory C. Greer)

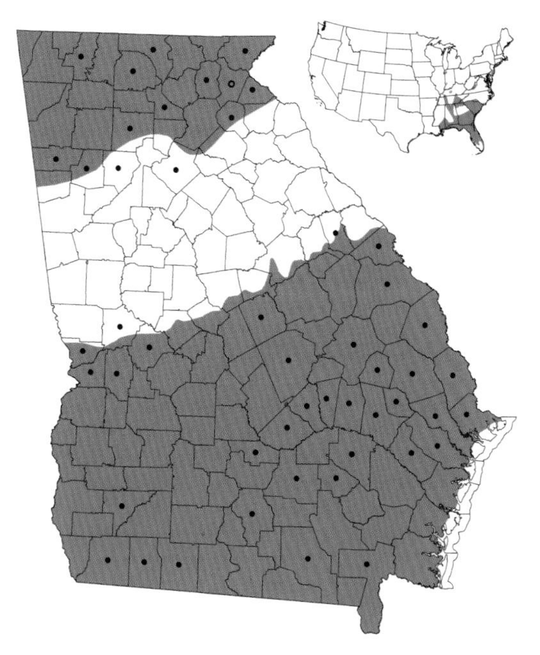

Taxonomy and Nomenclature

The name *Pituophis* is derived from two Latin words meaning "hissing snake," in reference to one of this snake's defensive behaviors. Of the three subspecies of pine snakes, only two occur in Georgia: the northern pine snake (*P. m. melanoleucus*) in the northern part of the state and the Florida pine snake (*P. m. mugitus*) in the Coastal Plain. Pine snakes are sometime referred to colloquially as bull snakes.

Distribution and Habitat

Pine snakes are found primarily in the Coastal Plain and Blue Ridge provinces of Georgia, although a few populations are known from the upper Piedmont and Ridge and Valley provinces. The Coastal Plain populations are associated with xeric habitats with deep sandy soils, including sandhills, scrub, and even agricultural lands and other disturbed habitats. Very little is known about the habitat requirements of the populations outside the Coastal Plain because pine snakes are so rarely encountered there, but individuals have been found in the vicinity of hardwood and mixed oak-pine forests.

Reproduction and Development

Breeding season is April–May. In June and July females lay 3–27 (usually 4–12) eggs underground in open sandy areas with very little canopy cover. In at least some portions of the species' range, females excavate long nesting chambers that may be used by several females and from one year to the next. The eggs are laid in an adherent cluster, and at about 60 mm (2.3 in) in length and 35 mm (1.4 in) in width are among the largest eggs of any North American snake. Hatchlings emerge in September or October after approximately 51–75 days of incubation. Age at maturity has not been published for Georgia's pine snakes.

Habits

Pine snakes have been observed aboveground from April until October but are most commonly encountered during May and June. They are active on the surface only during the day, when they may move as much as 1.5 km (0.9 mi). These snakes are well known for their impressive defensive displays, which include hissing loudly, inflating the body, vibrating the tail noisily against leaf litter, raising the front of the body, and even striking with the mouth closed or partially open. As long as it is not threatened, however, the pine snake is a rather secretive animal that spends much of its time either under rocks, logs, slabs, or artificial debris, or underground in rotting pine stumps and root tunnels, gopher tortoise (*Gopherus polyphemus*) burrows, or burrows of small mammals such as southeastern pocket gophers. Pine snakes sometimes excavate their own burrows. Prey items include small mammals and ground-nesting birds and their eggs. Predators include birds of prey and carnivorous mammals such as foxes and skunks. Scarlet snakes (*Cemophora coccinea*) are known to eat the contents of pine snake eggs, and juvenile pine snakes may fall prey to large snake-eating snakes.

Conservation Status

The seldom-seen pine snake may be declining in portions of its Georgia range due primarily to the fragmentation and loss of habitat and road mortality.

Tracey D. Tuberville and Peri A. Mason

Adult striped crayfish snake, Putnam County, Florida (Pierson Hill)

Striped Crayfish Snake

Regina alleni

Description

Adults average 30–55 cm (11.7–21.5 in) in body length; the maximum reported total length is 71.5 cm (27.9 in). Females are longer than males. The scales are shiny and smooth except for those on or near the tail, which are keeled. The anal plate is divided. The body is dark with three brown stripes, one extending down the middle of the back and one on either side of the body. The lower sides and throat are yellowish, and the belly, which may or may not have a row of spots down the center, is usually buff or straw colored but may be pink or orange in some individuals. Newborns are typically 160–180 mm (6.2–7 in) in total length and resemble adults but often have bolder patterns.

Taxonomy and Nomenclature

Striped crayfish snakes are closely related to glossy crayfish snakes (*R. rigida*). Some authorities consider these two species to be more closely related to black swamp snakes (*Seminatrix pygaea*) than to other members of the genus *Regina* and suggest that the two be placed in a genus by themselves. No subspecies are recognized. The striped crayfish snake is sometimes called the striped swamp snake.

Distribution and Habitat

This species occurs in a very small portion of Georgia's lower Coastal Plain; the vast majority of its range is in Florida. Heavily vegetated and

Adult striped crayfish snake, Lowndes County (John B. Jensen)

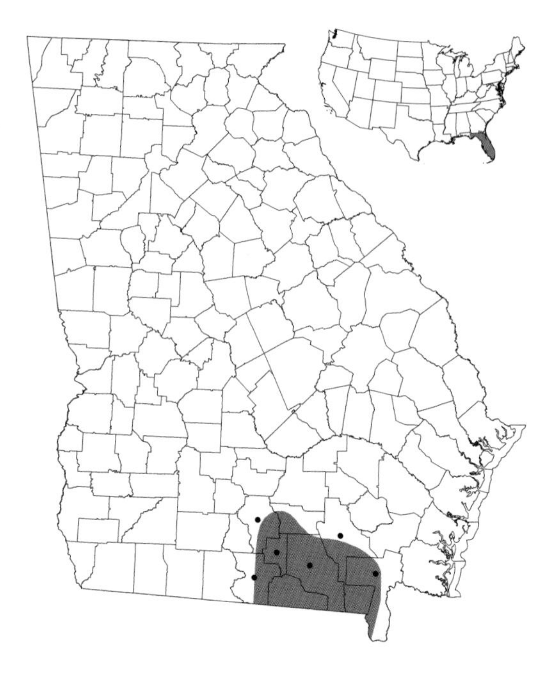

shallow swamps, wet prairies, marshes, ponds, and sluggish streams are the preferred habitats. Individuals are typically found among the roots and submerged stems and leaves of dense, floating herbaceous mats, especially those consisting of frog's bit, pennyworts, and the nonnative water hyacinth.

Reproduction and Development

Very little is known about the breeding biology of this secretive species. The mating season is unknown, but young are born from late spring through late summer. Litter size ranges from 4 to 12.

Habits

Although these snakes are seldom encountered without a dedicated search, striped crayfish snake populations can be quite dense in some areas. One study in Florida revealed a density of nearly 1,300 individuals per hectare (526 per acre). Striped crayfish snakes have specialized chisel-like teeth for subduing and eating crayfish, which make up the bulk of the adults' diet. The snake holds the crayfish in loose, constricting coils and swallows it tail first. Beetles, frogs, and dwarf sirens (*Pseudobranchus striatus*) have also been found in the guts of adults. Juveniles additionally consume grass shrimp and dragonfly larvae. Herons, egrets, river otters, common kingsnakes (*Lampropeltis getula*), black racers (*Coluber constrictor*), and cottonmouths (*Agkistrodon piscivorus*) are documented predators. An odd defensive behavior of unknown purpose and success involves lifting, extending, and arching the front part of the body, followed by swaying and sometimes mouth gaping.

Conservation Status

Very few striped crayfish snakes have been seen in Georgia, but this is likely due to the species' restricted range, secretive nature, and affinity for somewhat inaccessible habitats rather than an indication of true rarity.

John B. Jensen, Michael E. Dorcas, and Whit Gibbons

Adult glossy crayfish snake, Liberty County (John B. Jensen)

Glossy Crayfish Snake

Regina rigida

Description

Glossy crayfish snakes are rather small, somewhat stout snakes that reach a maximum total length of 80 cm (31.2 in) but are typically 46–61 cm (17.9–23.8 in) long. Females tend to grow larger than males. The scales are keeled, and the anal plate is divided. Adults are characterized by a shiny, olive brown to dark brown back, usually with two dark, narrow stripes along either side of the midline. The stripes on some individuals may become obscure with age. Individuals from the eastern portion of the Georgia range also may have dusky stripes along the sides of the throat. The belly is characteristically marked with two rows of half-moons extending the length of the body. The rows converge near the throat into a single row. Young, which vary from 150 to 180 mm (5.9–7.0 in) in total length at birth, are colored like the adults but are usually more boldly marked. The back stripes, which are dark rather than light, and the strong pattern of half-moons on the belly should distinguish glossy crayfish snakes from other species such as queen (*R. septemvittata*) and striped crayfish (*R. alleni*) snakes.

Taxonomy and Nomenclature

Two of the three recognized subspecies are found in Georgia: the glossy crayfish snake (*R. r. rigida*) and the Gulf crayfish snake (*R. r. sinicola*). The purported ranges of these two forms in Georgia roughly correspond to the eastern and western halves of the Coastal Plain, respectively. The closest relative of the glossy crayfish snake is the

Underside of adult glossy crayfish snake, Long County (Dirk J. Stevenson)

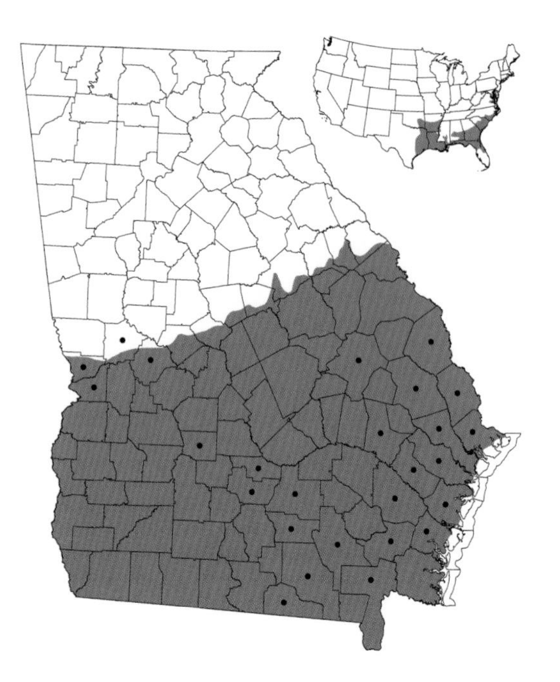

striped crayfish snake, which occurs in Florida and southern Georgia. Recent taxonomic studies indicate that the glossy and striped crayfish snakes may not be as closely related to the queen snake and Graham's crayfish snake (*R. grahamii*) as their current classification in the same genus indicates.

Distribution and Habitat

In Georgia, glossy crayfish snakes are found primarily in bottomland wetlands throughout the lower and upper Coastal Plain, but they can also be found in a variety of other wetland habitats, including isolated ponds, river swamps, canals, and marshes. Their exact distribution is still unknown due to the relatively few records from most areas.

Reproduction and Development

Little is known about the specifics of reproduction. Glossy crayfish snakes apparently mate in early spring to early summer. Females give birth in late summer or early fall to relatively small litters of 6–14 young, and probably reproduce every year if sufficient food is available. Females probably reach sexual maturity during their second or third summer. The age at maturity for males is not known.

Habits

Because of its secretive nature, this snake's ecology is poorly known. The glossy crayfish snake may be active both day and night in warm parts of the year, and even winter basking has been observed on clear, sunny days. Individuals are seldom found far from water, suggesting a highly aquatic existence. As their name implies, they feed primarily on crayfish, hard-shelled as well as newly molted soft-shelled ones, which they subdue by constriction and the use of their specialized chisel-like teeth. Because of their small size, even adult glossy crayfish snakes have many predators, including cottonmouths (*Agkistrodon piscivorus*), common kingsnakes (*Lampropeltis getula*), bullfrogs (*Rana catesbeiana*), and two-toed amphiumas (*Amphiuma means*).

Conservation Status

Glossy crayfish snakes are rarely encountered, and their conservation status is difficult to determine.

Michael E. Dorcas and Whit Gibbons

Adult queen snake, Harris County (John B. Jensen)

Queen Snake

Regina septemvittata

Description

The queen snake is a somewhat slender, medium-sized snake that reaches a maximum total length of 92 cm (35.9 in). The head is not very distinct from the neck. The scales are keeled, and the anal plate is divided. The background color is brown or gray with three darker, usually inconspicuous brown or black stripes on the back and two dull yellow or creamy white stripes along the sides. Two dark brown stripes extend down the center of the lighter colored belly parallel to two wider brown stripes, one along each side of the belly, making the queen snake the only North American water snake with four brown stripes on the underside. The stripes may be obscure on many Georgia specimens, giving them a nearly uniformly dark appearance. Young measure about 200 mm (7.8 in) in total length and resemble adults in coloration.

Taxonomy and Nomenclature

No subspecies are currently recognized. Most herpetologists agree that Graham's crayfish snake (*R. grahamii*) of the central United States is the queen snake's closest living relative. Recent taxonomic studies question the placement of the queen snake, the striped crayfish snake (*R. alleni*), and the glossy crayfish snake (*R. rigida*) in the same genus. Other common names for the queen snake include leather snake, moon snake, North American seven-banded snake, and willow snake.

Adult queen snake, Union County (John B. Jensen)

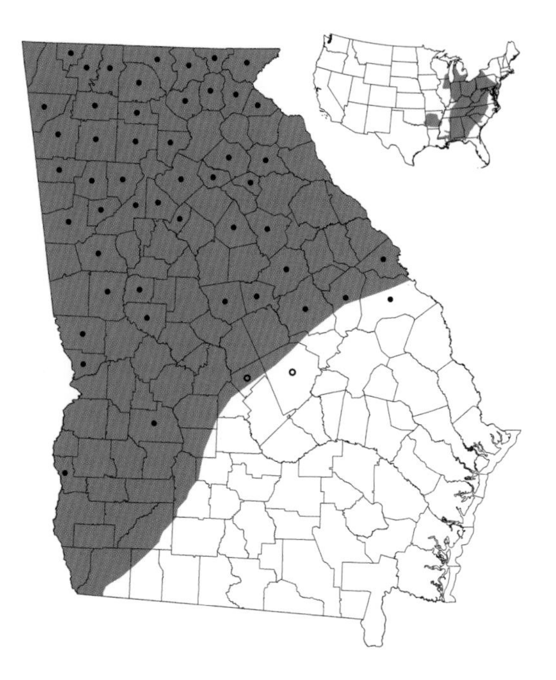

Distribution and Habitat

Queen snakes are found in some localities in the upper Coastal Plain of Georgia but are most common in the Piedmont and mountains. Because they rely on crayfish for food, queen snakes are seldom found far from water. They are particularly common in cool, rocky streams and rivers, where they are commonly found alongside northern watersnakes (*Nerodia sipedon*). Within these habitats queen snakes may be found either in or out of the water. They generally favor relatively open, sunny areas and often seek refuge under flat rocks and undercut banks.

Reproduction and Development

The exact mating season in Georgia is unknown, but queen snakes apparently reach maturity at 2–3 years of age. Like most other aquatic snakes, queen snakes give birth to live young. Litters of 5–23 (average = 8–10) are born in the late summer or early fall.

Habits

Queen snakes can often be found under large rocks or logs near the edges of streams and rivers. They often bask on limbs overhanging water and are considered among the most arboreal of our aquatic snakes. They are presumably most active during the day. Queen snakes feed primarily on crayfish that have recently molted and do not have a hard exoskeleton; soft-shelled crayfish are easier to subdue and pose little danger to the snake. Potential predators include moderate-sized carnivores such as raccoons and foxes, various birds, and other snakes such as the common kingsnake (*Lampropeltis getula*). When threatened, queen snakes first try to escape by dropping into the water. If captured, they usually writhe and release musk, but they rarely bite.

Conservation Status

Queen snakes are fairly common throughout their range in Georgia, but their reliance on stream habitats may put them at risk as human development and urbanization cause further degradation of water quality and subsequent declines of crayfish populations.

Michael E. Dorcas and John D. Willson

Adult pine woods snake, Charlton County (George Gentry)

Pine Woods Snake

Rhadinaea flavilata

Description

The pine woods snake is a small, slender serpent some 25–33 cm (9.8–12.9 in) in total length; females are larger than males. The back is characterized by a velvety iridescence that varies from golden brown to cinnamon brown and fades along the sides. The scales are smooth, and those located on the middle row on the back tend to be slightly darker toward the tail, producing a narrow, faint stripe down the length of the back. The anal plate is divided. The belly ranges from white to very pale yellow but is otherwise unmarked. The top of the head is darker than the back, and a conspicuous brown line extends along either side of the head from the snout to the angle of the jaw, widening as it passes through the eye. The upper lip is white to yellowish. The dark line through the eye and pale lips should distinguish this species from other small snakes. The upper lips are often speckled with tiny dark spots. This spotting is minimal or absent in individuals from Florida but becomes increasingly prominent farther north. Hatchlings average around 120–130 mm (4.7–5.1 in) in total length and resemble adults in color and pattern.

Taxonomy and Nomenclature

The pine woods snake is the sole representative of its genus in the United States; *Rhadinaea* is otherwise chiefly Central American in distribution. No subspecies are recognized. This species was formerly known as the yellow-lipped snake.

Adult pine woods snake, Nassau County, Florida (Dick Bartlett)

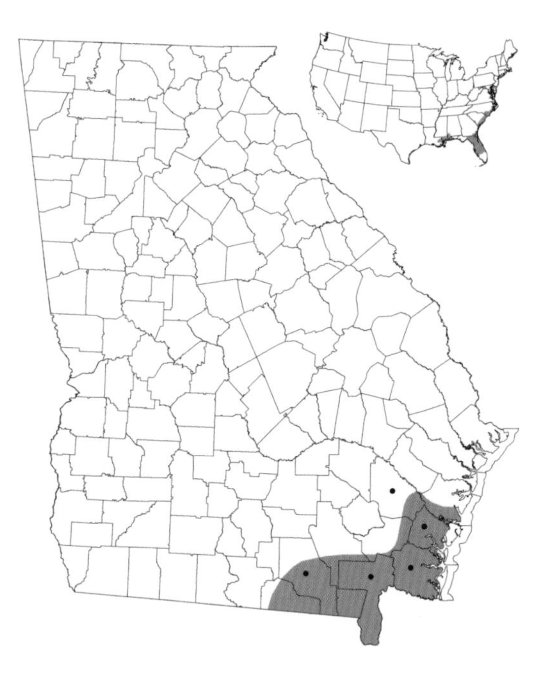

Distribution and Habitat

The pine woods snake exhibits a patchy distribution within the Coastal Plain and in the state of Georgia. All of the state records are confined to the extreme lower Coastal Plain. A disjunct population occurs in South Carolina just below the Fall Line adjacent to the Savannah River, though, and the pine woods snake may inhabit comparable settings in Georgia's upper Coastal Plain. Moist, low woodlands are the preferred habitat, particularly pine flatwoods with an abundance of fallen logs and rotting stumps, which provide ideal cover.

Reproduction and Development

Little is known about the pine woods snake's reproductive behavior. An entwined male and female were found inside a rotten stump during mid-April in North Carolina. Females lay two to four capsule-shaped eggs sometime between late May and early August. Eggs hatch approximately 6 weeks after deposition. Age at maturity is not known.

Habits

Pine woods snakes are highly secretive and rarely encountered. They are typically discovered under loose bark of pine logs or inside decaying pine logs or stumps, most often during the spring. Pine woods snakes feed on small lizards, frogs, and salamanders. Anecdotal accounts add ground skinks (*Scincella lateralis*), southern cricket frogs (*Acris gryllus*), and smaller treefrogs (genus *Hyla*) to the list of prey. All members of the genus *Rhadinaea* are rear-fanged and produce mildly toxic venom that they use to subdue their prey. Smaller prey items are swallowed at once, but larger prey may be held in the mouth for several minutes to allow the venom to take effect. Pine woods snakes do not bite and are harmless to humans. Likely predators of such a small snake include predaceous insects, other snakes, birds, and small mammals.

Conservation Status

Whether the pine woods snake is truly rare or simply rarely encountered is open to question. Its distribution in coastal habitats throughout its geographic range is sporadic and seemingly unpredictable. Likewise, its relatively recent discovery near the Fall Line in bordering South Carolina raises the question of whether the species might also be found in comparable areas of Georgia.

Trip Lamb and Whit Gibbons

Adult black swamp snake, Wakulla County, Florida (Dirk J. Stevenson)

Black Swamp Snake

Seminatrix pygaea

Description

The black swamp snake is the smallest of the North American water snakes. Adults average 25–38 cm (9.8–14.8 in) in body length; the maximum reported body length is 48.5 cm (18.9 in). Females grow considerably larger than males. Pregnant females can be very stout and may weigh in excess of 70 g (2.5 oz). The back is solid black, and the belly is red or reddish orange. The belly scales typically have a black bar on the leading edge covering about one-third of the scale. The scales are shiny and smooth, although some of the scales on the lower sides have light lines that make them appear keeled. The anal plate is divided. Juveniles resemble adults and are 110–150 mm (4.3–5.9 in) in total length at birth.

Taxonomy and Nomenclature

Black swamp snakes are the only members of their genus. Recent evidence suggests that the glossy crayfish snake (*Regina rigida*) and striped crayfish snake (*R. alleni*) may be more closely related to the black swamp snake than to other *Regina* species, but no changes to the current taxonomy have been suggested. Black swamp snakes are sometimes referred to as red-bellied mud snakes, but they are not closely related

Juvenile black swamp snake, Lowndes County (Dirk J. Stevenson)

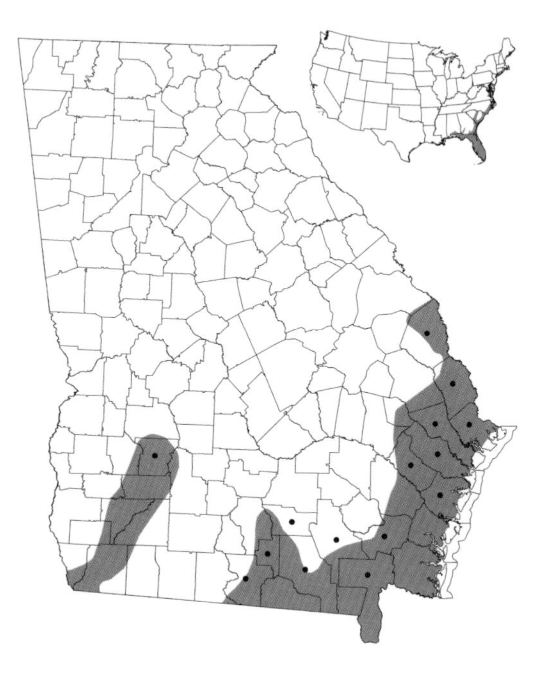

to true mud snakes (*Farancia abacura*). Of the three subspecies, only one (*S. p. pygaea*) is known to occur in Georgia.

Distribution and Habitat

Black swamp snakes are known from widely scattered locations within the Coastal Plain of Georgia. They use a variety of habitats, including swamps, marshes, ponds, lakes, slow-moving streams and rivers, and drainage ditches, but large populations generally occur only in wetlands or canals overgrown with matted aquatic vegetation.

Reproduction and Development

Pregnant females typically give birth annually to 2–14 young in late July–October. Females have larger litters in years following exceptionally high prey availability; the maximum reported litter size is 22 offspring. Both sexes apparently reach sexual maturity within 1 or 2 years.

Habits

These secretive snakes can be found hiding among dense aquatic vegetation such as sphagnum moss, water hyacinths, and other floating and emergent plants. They are particularly susceptible to dehydration and thus do not bask in direct sunlight. They are rarely found far away from water except after an evening rain or under duress when escaping from recently dried wetlands. Some individuals survive extended droughts by burying themselves in damp muck. Black swamp snakes are sometimes found under objects or debris along the edge of water or in crayfish burrows. The diet includes worms, leeches, tadpoles, small frogs, salamanders, and fish. Predators include large fish, other snakes, carnivorous mammals, and birds. This shy snake rarely bites when captured.

Conservation Status

Although black swamp snakes may occur in large numbers at some locations, they appear to be rare or uncommon in most aquatic habitats in Georgia. This perceived rarity may be due to their habit of living within dense aquatic vegetation.

Christopher T. Winne and Sean M. Poppy

Adult brown snake, Dooly County (John B. Jensen)

Brown Snake

Storeria dekayi

Description

The brown snake is a small, cryptically colored snake. The minimum total length of adults is 15 cm (5.9 in); very large individuals may reach 53 cm (20.7 in). The scales are keeled, and the anal plate is divided. The coloration is variable and can be gray, light tan, or brownish red on top. Two rows of darker spots or blotches are present on the first one-third of the body; these fuse in some specimens to form dark stripes for the remaining length. Occasionally the pigment between the stripes is lighter, giving the appearance of one light stripe along the midline. The back stripes take on a checkered or zigzag appearance in some individuals. Less pronounced dark spots are often visible on the sides, alternating with the spots on the back. The tan to pink belly is lighter than the back and has tiny black dots on the sides. The head is rounded and has darker pigmentation on top than on the throat. Patches of dark pigment are often present beneath the eyes. Two relatively large spots at the base of the head sometimes join together to form a band. Newborns are usually dull gray with a distinct neck band of yellow-cream. The length at birth is typically 70–120 mm (2.7–4.7 in). Brown snakes can easily be distinguished from ringneck (*Diadophis punctatus*) and red-bellied (*S. occipitomaculata*) snakes by their drab belly coloration.

Taxonomy and Nomenclature

Most Georgia brown snakes belong to the subspecies known as the midland brown snake (*S. d. wrightorum*). The Florida brown snake (*S. d. victa*), which some authorities consider a

Brown snake hatchling, Floyd County (Bradley L. Johnston)

Adult brown snake, Dekalb County (John D. Willson)

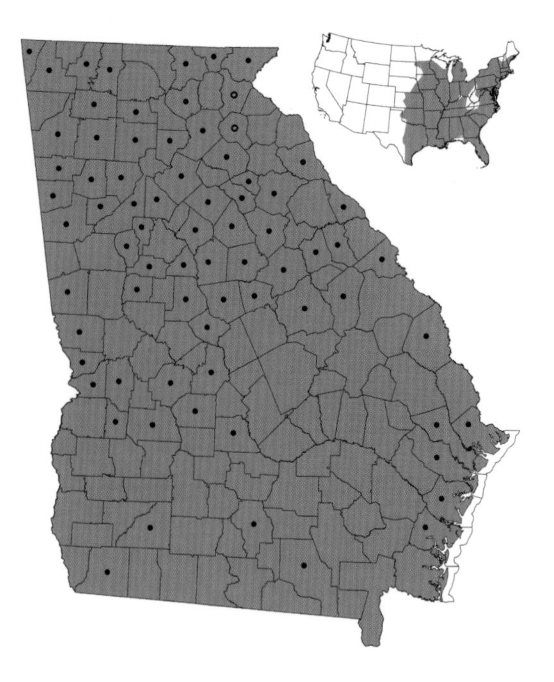

separate species, replaces the midland brown snake in extreme southeastern Georgia. Individuals with characteristics intermediate between those of the northern brown snake (*S. d. dekayi*), whose genetic influence is evident in the Carolinas, and the midland brown snake may be present in northern and eastern Georgia. The species was once named DeKay's snake in honor of the 19th-century New York naturalist James Edward DeKay. "DeKay's snake" is sometimes misinterpreted as "decay snake," owing perhaps to the snake's tendency to hide beneath decaying leaf litter.

Distribution and Habitat

Occurring statewide, this habitat generalist can be found in both hardwood and pine forests and in dry areas adjacent to any freshwater source. Brown snakes seem quite tolerant of development and find sufficient refuge in the gardens and woodpiles of suburban and urban environs.

Reproduction and Development

Brown snakes mate during the spring (March–May) and bear live young in summer to early fall. Those in southerly populations may breed earlier than their northern relatives. While most live-born snakes develop within a membrane inside the mother and draw nutrition from yolk, embryonic brown snakes may actually obtain food from a placenta-like structure. A brood may contain as few as 3 or as many as 41 young, but the average litter size is about 13. Brown snakes are able to breed at the end of their second year.

Habits

Brown snakes may be active day or night. By day they usually move about beneath vegetative cover, but at night they prowl about on the surface for prey. Although brown snakes are terrestrial, they have been observed swimming and climbing. They are active from early spring to late fall and occasionally emerge from hibernation during warm spells in the winter. They may hibernate singly or with other snakes of the same or different species. Their diet includes earthworms, slugs, snails, insects, spiders, small amphibians, and fish. Prey is swallowed whole and alive, without constriction, and snails are pulled from their shells before being swallowed. Like most nonvenomous snakes, brown snakes try to flee when threatened. When they cannot escape, they exhibit defensive behaviors such as playing dead or hiding their head beneath body coils and flattening their body in order to appear larger. When caught, they release a strong-smelling musk from their vent. These small snakes fall prey to a host of animals, including mammals such as shrews, raccoons, opossums, and skunks; birds, including American robins, loggerhead shrikes, brown thrashers, and hawks; toads (genus *Bufo*); spiders; and snake-eating snakes such as black racers (*Coluber constrictor*) and common kingsnakes (*Lampropeltis getula*).

Conservation Status

The brown snake has a large geographic range and seems to do well in suburban and even urban environments. For these reasons, the species is not currently a conservation concern to Georgia herpetologists.

Peri A. Mason and Pierson Hill

Adult red-bellied snake, Bryan County (Dirk J. Stevenson)

Red-bellied Snake

Storeria occipitomaculata

Description

Red-bellied snakes typically range from 20 to 25 cm (7.8–9.8 in) in total length; the maximum length reported is 40.6 cm (16 in). The scales are keeled, and the anal plate is divided. These small, rather drab snakes are colored in grays and browns, although reddish and occasionally black individuals can also be found. The body pattern can also be variable, with some individuals being a solid color and others having stripes that extend the length of the body. The stripes are often faint and may be on the sides; frequently they appear as a single broad line down the center of the back. The head is usually darker than the body, and most individuals have light-colored spots on the back of the neck that in some connect to form a pale neck ring. As the common name implies, the belly is typically reddish, but some individuals have an orange belly, and in a very few it is black. Newborns are typically less than 100 mm (3.9 in) in total length and resemble the adults, although they may be darker. The light coloration on the back of the neck is usually more distinctive on young snakes as well. Red-bellied snakes can be confused with brown snakes (*S. dekayi*) but lack distinct spotting on the back and usually have a brightly colored belly. Their keeled scales distinguish them from ringneck snakes (*Diadophis punctatus*), which always have a dark gray or black body and at least a partial distinctly colored neck ring.

Adult red-bellied snakes, Columbia (brown) and McDuffie (gray) counties (Cameron A. Young)

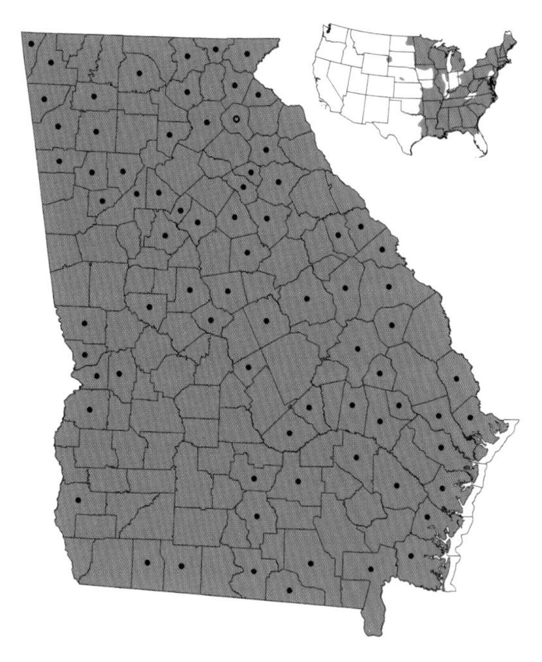

Taxonomy and Nomenclature

Three subspecies are recognized, two of which are in Georgia. The northern red-bellied snake (*S. o. occipitomaculata*) occupies the northernmost counties, and the Florida red-bellied snake (*S. o. obscura*) is found in the southern half of the state. The two subspecies are confusingly similar in appearance and are not easily distinguished, and they intergrade throughout most of central Georgia.

Distribution and Habitat

The geographic range of the red-bellied snake includes all of Georgia, but populations seem to be irregularly distributed; the species is very common in some areas but rare or absent in others. These snakes are prevalent in shaded hardwood and pine forests and may be associated with the terrestrial edges of isolated wetlands and swamps. They seek the shelter of leaf litter, fallen logs, and rocks on the forest floor and are not likely to be found in open-field habitats or in urban areas.

Reproduction and Development

Red-bellied snakes are presumed to mate in the spring but have also been reported to mate in the summer and fall. Most females give birth between July and September, typically earlier in the southern part of the state and later in the north. The young are born live, and a typical litter size is 4–9, although more than 20 young per litter have been reported. Reproductive maturity is achieved in 2 years.

Habits

Red-bellied snakes are most commonly seen in the spring and early fall but may be active in the winter during warm spells, especially in southern Georgia. During the summer, they are active in the early evening and occasionally later in the night. Although these shy and secretive snakes spend most of their time hiding beneath the ground or under debris, they often travel overland. Despite their small size, red-bellied snakes are capable of moving long distances, up to 400 m (1,312 ft). Slugs are among the most common prey, along with earthworms and probably other small invertebrates such as insects and snails. Their small size makes these snakes vulnerable to predators such as wolf spiders, toads (genus *Bufo*), and large terrestrial salamanders in addition to common kingsnakes

(*Lampropeltis getula*), black racers (*Coluber constrictor*), and ground-feeding birds. When threatened, some red-bellied snakes flatten their body, presumably to look larger, and some play dead. Occasional individuals curl their lips and expose black-and-white-striped gums that look like teeth. Red-bellied snakes do not bite defensively and are completely harmless to humans, although they generally release smelly musk when captured.

Conservation Status

These tiny snakes are common and apparently secure in Georgia.

Berkeley W. Boone

Adult southeastern crowned snake, Long County (Dirk J. Stevenson)

Southeastern Crowned Snake

Tantilla coronata

Description

Southeastern crowned snakes are very small, slender snakes. Along with Florida crowned snakes (*T. relicta*), worm snakes (*Carphophis amoenus*), and earth snakes (genus *Virginia*) they are Georgia's smallest serpents. Adults typically are 20–25 cm (7.8–9.8 in) in total length, and males and females are roughly the same size. The maximum recorded length is only 33 cm (12.8 in). The head is relatively flat and pointed. The scales are smooth, and the anal plate is divided. The back is brown, tan, or grayish brown. A light-colored band separates distinct black areas on the head and neck. The belly ranges from white to light pink. Southeastern crowned snakes might be confused with juvenile brown snakes (*Storeria dekayi*), which have keeled scales, or Florida crowned snakes, which at least in Georgia lack the light band separating the black areas of the head and neck. Juvenile southeastern crowned snakes resemble adults and are approximately 76 mm (3 in) long at hatching.

Taxonomy and Nomenclature

No subspecies are recognized. This species has also been called the southeastern black-headed snake.

Distribution and Habitat

Although absent from the Blue Ridge Province and most of the lower Coastal Plain in Georgia,

Adult southeastern crowned snake, Long County (Dirk J. Stevenson)

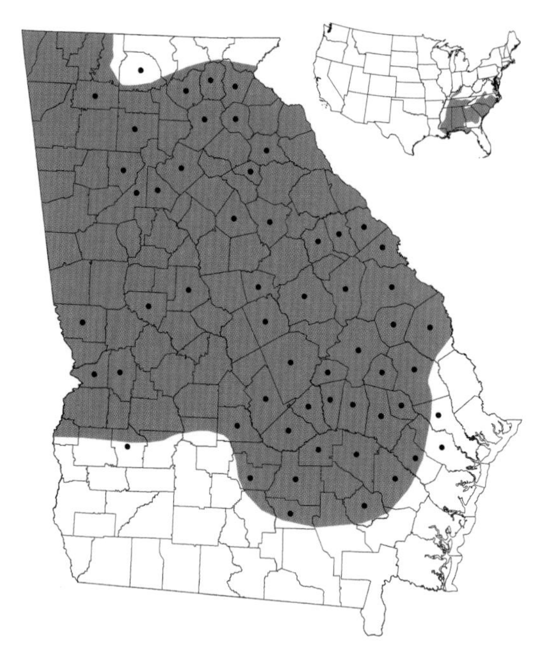

southeastern crowned snakes are common in a variety of situations, including sandy areas and forested habitats. They are almost always found beneath ground litter, rocks, or woody debris, often in fairly high densities.

Reproduction and Development

Southeastern crowned snakes mate throughout the warm parts of the year, from spring to fall. Females lay up to three eggs underground or beneath surface litter in June or July, and the eggs hatch in the fall. Sexual maturity is apparently reached in approximately 2 years. Little else is known about mating and reproduction in this species.

Habits

Southeastern crowned snakes hibernate during winter but become active on warm days in early spring and may be found in Georgia until late fall. They are largely fossorial. Most aboveground travel is during early evening or at night, but individuals are active beneath logs, rocks, and ground litter during the day in the spring, summer, and fall. They are skillful burrowers, especially in sandy soil, and individuals may appear to "swim" into the sand when discovered. Southeastern crowned snakes feed on invertebrates. They have a particular preference for centipedes and insect larvae but also eat earthworms and spiders. Although no distinct venom glands are present, these snakes do have small grooved fangs at the back of the jaw with which they inject venom into insect prey. Their small size makes them vulnerable to most of the carnivorous vertebrates that occur in forested habitats, such as eastern coral snakes (*Micrurus fulvius*), common kingsnakes (*Lampropeltis getula*), and ringneck snakes. Southeastern crowned snakes first react to a predator by trying to burrow into sand or soft soil or by crawling beneath leaves or other ground litter. When captured, they do not bite but may release small quantities of musk from the vent. There are no records of bites or any other indications that these snakes are in any way harmful to humans.

Conservation Status

This species can be found in high numbers throughout much of its range.

Kimberly M. Andrews and
Whit Gibbons

Adult Florida crowned snake, Levy County, Florida (Pierson Hill)

Florida Crowned Snake

Tantilla relicta

Description

Adults are typically 18–23 cm (7–9 in) in total length and 17–19 cm (6.6–7.4 in) in body length. Females are slightly larger than males and have a proportionately shorter tail. The scales are smooth and glossy, and the anal plate is divided. This small, secretive snake is tan to light reddish brown with a cream-colored belly. The somewhat rounded head is dark brown to black with a light patch behind each eye. Juveniles resemble adults but with proportionately larger head and eyes. The Florida crowned snake can be distinguished from other small, smooth-scaled snakes in southern Georgia by the dark head and lack of a pattern on the body. The more common southeastern crowned snake (*T. coronata*), which occurs in northern Georgia, has a light ring between the dark head and dark neck band. Hatchling Florida crowned snakes are 77–95 mm (3.0–3.7 in) in body length.

Taxonomy and Nomenclature

Three subspecies are recognized, but only the central Florida crowned snake (*T. r. neilli*) occurs in Georgia.

Distribution and Habitat

In Georgia, the Florida crowned snake is known only from two Lowndes County specimens.

These represent the northernmost occurrences of the species, which is otherwise restricted to peninsular Florida where it inhabits sandhills and moist hammocks.

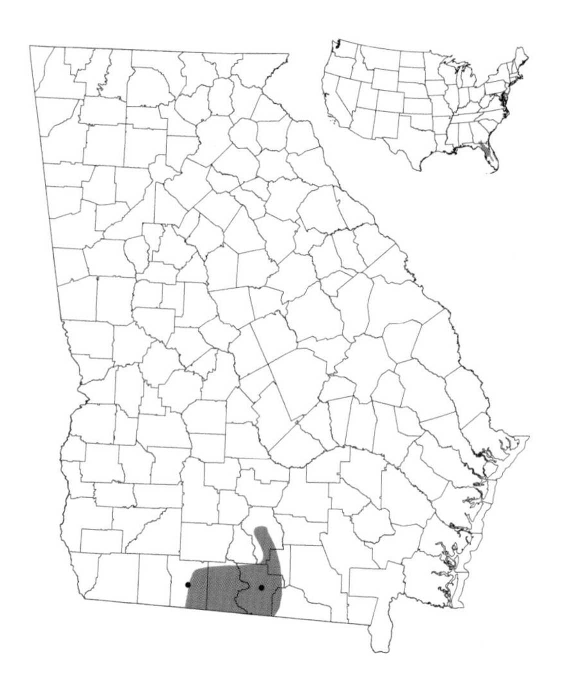

Reproduction and Development

Very little is known of the reproductive biology of the Florida crowned snake. Females with eggs have been found from March through August in western Florida, and recently hatched individuals have been observed in May and June. The extended reproductive period suggests that the species may be able to produce more than one clutch of eggs in a season.

Habits

Our knowledge of the habits of this secretive species is based on data from Florida populations. The Florida crowned snake is a burrower and can be found beneath leaf litter and woody debris or in the sandy mounds made by southeastern pocket gophers. The dark head is thought to be a means of camouflage when the snake is aboveground in thick vegetation. Individuals have been observed basking with only the head exposed above the sand. Peak activity occurs in spring and fall. Florida crowned snakes feed primarily on beetle larvae and to a lesser extent on other soil invertebrates including centipedes and snails. Predators include the other snakes, birds, and possibly mammals.

Conservation Status

Nothing is known of the status of the Florida crowned snake in Georgia, but the existence of only three documented specimens suggests that it is very uncommon.

Lora L. Smith

Adult eastern ribbon snake, Long County (Dirk J. Stevenson)

Eastern Ribbon Snake

Thamnophis sauritus

Description

Eastern ribbon snakes are very slender, medium-sized snakes that reach a total length of 46–102 cm (17.9–39.8 in). The scales are keeled, and the anal plate is single. Two or three yellow stripes, one on each side and usually one along the center of the back, extend the length of the body and contrast with the dark back and sides. The lines are bright and well defined in some individuals, but the middle stripe is often absent or indistinct in specimens from the southeastern part of the state. The belly of all eastern ribbon snakes is solid white or yellowish. The tail is very long in proportion to the body—up to a third of the animal's total length. The lateral stripes are higher up on the sides (on scale rows three and four rather than on rows two and three) than those of the stouter but similar-looking common garter snake (*T. sirtalis*). The labial scales are uniformly white or pale yellow with no markings between them, further distinguishing this species from the common garter snake, which has vertical dark markings between each pair of lip scales. Newborn young are 180–230 mm (7–9 in) in total length and are similar to adults in color and pattern.

Taxonomy and Nomenclature

Three subspecies have been described, two of which occur in Georgia. The common ribbon

Adult eastern ribbon snake, Muscogee County (John B. Jensen)

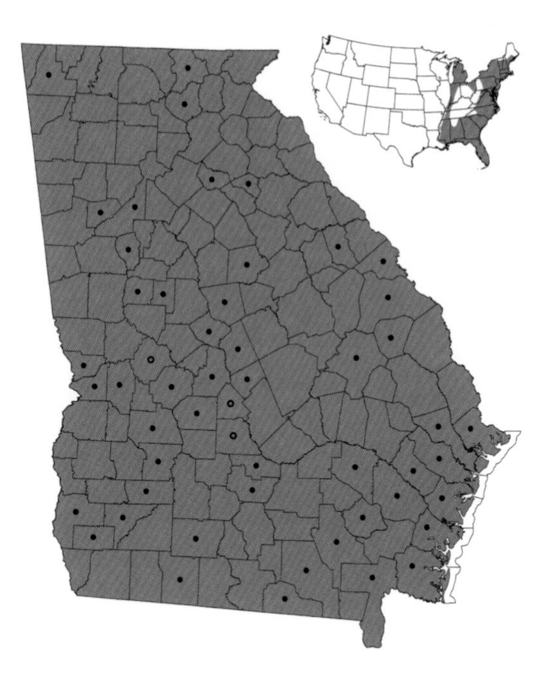

snake (*T. s. sauritus*) is present throughout most of Georgia, and the peninsula ribbon snake (*T. s. sackenii*) occurs in southeastern Georgia.

Distribution and Habitat

Eastern ribbon snakes can be found across the state around most aquatic habitats, including streams, rivers, swamps, wetlands, ponds, and lakes, but are encountered only rarely in the northwestern portion of the state. They are semiaquatic, rarely traveling far from water. They typically shun open water, however, instead preferring grassy shallows and brushy edges.

Reproduction and Development

Eastern ribbon snakes mate in the spring and give birth from July to October. Litter size ranges from 3 to 36 young, and typically averages about a dozen. Individuals reach sexual maturity in 2 or 3 years.

Habits

Eastern ribbon snakes are active predominantly during the daytime and frequently bask in low vegetation. They are often found on the ground along shorelines or in low shrubs and bushes from which they can make a rapid escape by disappearing into the vegetation or into the water. The activity season can be long in Georgia, especially during warm winters, when these snakes may be active year-round. They feed largely on fish and amphibians and may be nocturnal during periods when frogs are breeding. Eastern ribbon snakes also eat spiders and insects but do not appear to feed on earthworms as do common garter snakes. Predators include cottonmouths (*Agkistrodon piscivorus*), common kingsnakes (*Lampropeltis getula*), many birds, large fish, and predatory mammals. The small size and slender build of newborns make them vulnerable to insect predators such as aquatic beetles. Escape is the first form of defense for these snakes, but they may also thrash wildly and release foul-smelling musk from the vent when captured.

Conservation Status

The species is common and apparently secure in Georgia.

Brian D. Todd

Adult common garter snake, Long County (Dirk J. Stevenson)

Common Garter Snake

Thamnophis sirtalis

Description

Common garter snakes can reach a total length of 137 cm (53.4 in), but those found in Georgia are typically less than 76 cm (29.6 in). Females grow larger than males. The tail is usually 20–25 percent of the total length, and it is slightly longer in males than in females. Females tend to have a larger head than males of the same body length. The scales on the back and sides have keels, and the anal plate is undivided. The ground color of the back and sides is greenish gray to black, and the belly is cream colored. A white or yellow stripe runs down each side along scale rows two and three, counting from but not including the wide belly scales. Most individuals in Georgia also have another light stripe along the center of the back. Between the center and side stripes there may be a series of light blotches alternating with the darker background color. The stripes on snakes living along the coast are often bluish. Some individuals have a checkered appearance, with the stripes absent or poorly defined. The head of Georgia individuals is usually uniformly olive to black on top with light-colored lip scales and a cream-colored chin and throat. The eyes are relatively large, and the tongue is reddish. In Georgia, the common garter snake is likely to be confused only with the eastern ribbon snake (*T. sauritus*), which is more slender, has an

Adult common garter snake, Liberty County (Matt O'Connor)

Adult common garter snake, Baker County (Gabriel J. Miller)

Adult common garter snake, Walker County (John B. Jensen)

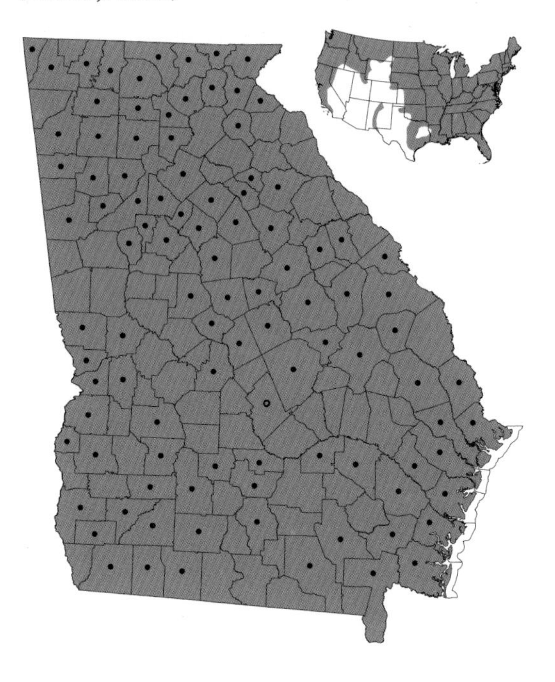

extremely long tail (more than 30 percent of the total body length), and has side stripes higher up on the body (on scale rows three and four). Newborn common garter snakes are tiny (body length = 102–203 mm, or 4–7.9 in), but otherwise they resemble adults.

Taxonomy and Nomenclature

The eastern garter snake (*T. s. sirtalis*) is the only subspecies of common garter snake found in Georgia. The term "garter snake" refers to 19th-century men's garters, which were similarly striped. Being commonly found near human habitations including gardens, they are frequently, but incorrectly, referred to as gardener snakes.

Distribution and Habitat

Statewide in occurrence, common garter snakes are among the most frequently encountered snakes in the United States and can be found in a variety of habitats. Although some individuals wander far from water, the vast majority are found in moist habitats, often on the edge of a wetland. Both temporary and permanent wetlands are used, including those in suburban and urban areas with sufficient prey to maintain snake populations.

Reproduction and Development

Mating takes place primarily from March through June but may also occur in the fall. Multiple males may simultaneously court a female, although this is less common in Georgia than in populations farther north. Females are capable of storing sperm in their reproductive tract for at least a year before using it to fertilize their eggs. Common garter snakes bear live young, and the reproductive tract has a placenta that transmits nourishment to the developing embryos. Litters of 50 or more are common, and large females occasionally produce more than 100 young. Individuals reach sexual maturity in 2 years.

Habits

Common garter snakes can tolerate relatively cool temperatures and may be found aboveground even in the winter months. In most areas, however, individuals hibernate singly or communally underground or in places like rotting logs. These snakes can even survive subzero temperatures for a short period and may hibernate underwater by absorbing oxygen directly through their skin. Common garter snakes eat just about any animal that will fit in their mouth. Juveniles eat larval and small frogs and salamanders as well as fish, insects, and a variety of soft-bodied invertebrates such as slugs and worms. Adults feed primarily on amphibians, fish, and worms as well as the occasional snake, mammal, or bird. Carrion is eaten on occasion, and these snakes will even attempt to swallow road-killed frogs on pavement. The larger head of females may allow them to eat larger prey items than their male counterparts. These snakes have a wide variety of predators, including mammals such as shrews, skunks, and cats; birds such as hawks, crows, and American robins; reptiles such as common kingsnakes (*Lampropeltis getula*), other common garter snakes, and eastern box turtles (*Terrapene carolina*); as well as frogs, toads, large salamanders, a variety of fish, and even spiders. When disturbed, common garter snakes first attempt to flee; their stripes can make them appear to be motionless when in fact they are moving away. If retreat is not an option, they may respond to a perceived attacker by flattening the head and body to resemble a venomous snake. They typically respond to capture by thrashing about, biting, and releasing copious quantities of musk and feces. The musk has a unique malodorous tang and seems to defy removal with soap and water. Although common garter snakes are usually considered harmless and nonvenomous, they actually produce weak venom. They lack enlarged fangs, however, and thus chew the venom into their prey rather than injecting it through a hollow fang. The effects of the venom have not been characterized, although it appears to have a weak myonecrotic (muscle-digesting) capacity. Only a tiny percentage of the thousands of people (including children) bitten by garter snakes each year exhibit minor swelling around a bite, and it can be difficult to determine whether the reaction is due to the venom or is a simple allergic reaction to snake saliva. No common garter snake bite has ever resulted in a life-threatening emergency.

Conservation Status

This snake is very common and probably secure in the state.

Robert N. Reed and Whit Gibbons

Adult rough earth snake, Liberty County (Dirk J. Stevenson)

Rough Earth Snake

Virginia striatula

Description

Rough earth snakes are small snakes; most adults are 18–26 cm (7–10 in) in total length, and the maximum length known is 32.4 cm (12.6 in). The typical color is uniform gray to light brown above and light tan below. The scales on the back and sides are weakly keeled, and the anal plate is usually, but not always, divided. The head is small, and the snout is noticeably pointed. Males, which can reach maturity at a body length of 11.8 cm (4.6 in), may be slightly smaller than females, which do not mature until approximately 18 cm (7 in). Males have a longer tail than females but are otherwise similar in appearance. Newborns are 74–127 mm (2.9–5.0 in) long and may have a light mark immediately behind the head that resembles a neck ring; this mark disappears with age. The related smooth earth snake (*V. valeriae*) can be distinguished by its smooth scales, less pointed snout, and the small black flecks that are usually scattered on the body. Worm snakes (*Carphophis amoenus*) are often similarly drab in appearance but have shiny, smooth scales, and the sides and belly contrast strongly in color.

Taxonomy and Nomenclature

No subspecies are recognized. Because of this snake's secretive nature and the infrequency with which it is encountered, most people refer to it (and several other similarly sized and colored species) simply as a little brown snake.

Distribution and Habitat

Although reported from much of the southern two-thirds of the state, rough earth snakes are apparently absent from the Dougherty Plain in southwestern Georgia. These burrowing snakes occupy a wide range of habitats, including both hardwood and pine forests. This species has adapted to suburban and urban environments and is among the few species that appear to thrive in some of our larger cities.

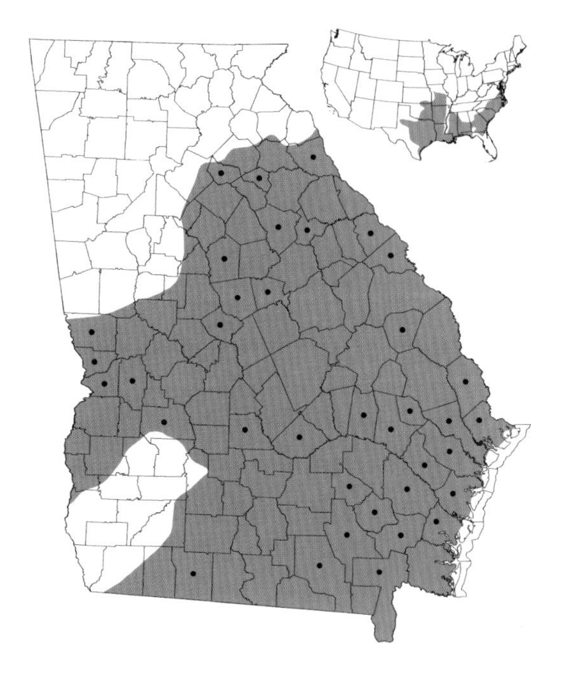

Reproduction and Development

Very little reproductive information is available. Mating occurs in the spring, and females give birth to live young beginning in early to mid-summer and continuing into late summer or early fall. A gestation period of 82 days has been reported. Litter size averages around 9 but can range from 2 to 12, with larger females having more young. Males have been reported to reach maturity as early as 7 months after birth; females do not mature until after more than a year.

Habits

Rough earth snakes are fairly secretive; they spend most of their time underground, in leaf litter, under surface objects such as logs and rocks, or inside deteriorating pine stumps. They are frequently recovered from swimming-pool filters, and house cats sometimes bring individuals inside homes. They feed largely on earthworms but also eat insects (including larvae), slugs, and other invertebrates. The major known predators of this small, inoffensive snake are other snakes, including black racers (*Coluber constrictor*), mole kingsnakes (*Lampropeltis calligaster*), scarlet kingsnakes (*L. triangulum elapsoides*), and eastern coral snakes (*Micrurus fulvius*).

Conservation Status

Rough earth snakes appear to be common in a wide range of habitats within Georgia, including human-altered environments.

Bob Herrington

Adult smooth earth snake, Bibb County (John B. Jensen)

Smooth Earth Snake

Virginia valeriae

Description

Smooth earth snakes are typically 17–25 cm (6.6–9.8 in) in total length; the maximum length known is 39.3 cm (15.3 in). Males have a proportionately longer tail than females and reach sexual maturity at a smaller size (13 cm, or 5.1 in, body length) than females (19 cm, or 7.4 in). These snakes have a narrow head with a somewhat pointed nose, smooth scales, and a divided anal plate. The back is pinkish brown to brownish gray, and the belly is white or yellowish. Tiny dark spots reminiscent of pepper flakes are often scattered on the back, sometimes arranged in rows. Young smooth earth snakes are similar to the adults and are 79–115 mm (3.1–4.5 in) at birth. Rough earth snakes (*V. striatula*), which also occur throughout much of the state, have keeled scales, lack speckling on the back, and sometimes have an indistinct light band across the back of the head. Brown snakes (*Storeria dekayi*) have strongly keeled scales. Worm snakes (*Carphophis amoenus*) are strongly bicolored, often with a pink belly; have a tail tip that terminates as a small spine; and have a much smaller head relative to the size of the body.

Taxonomy and Nomenclature

A single subspecies, the eastern smooth earth snake (*V. v. valeriae*), occurs in Georgia and other southeastern states.

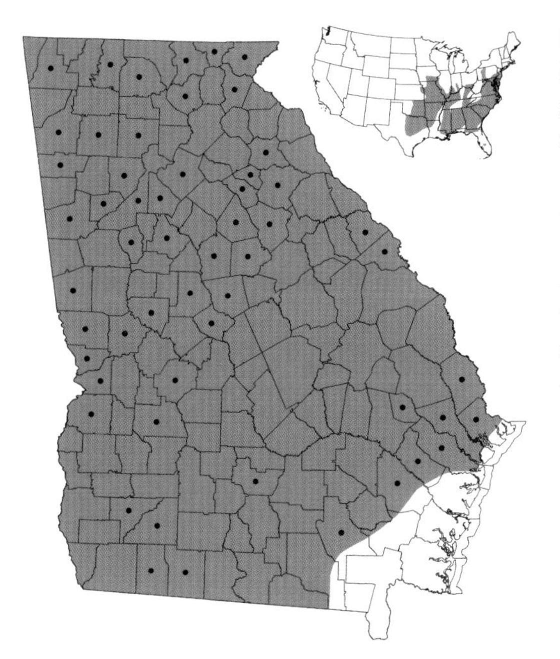

Distribution and Habitat

Found throughout the state save the extreme southeastern corner, smooth earth snakes inhabit pine or hardwood forests, including the margins of swamps and other wetlands. They are also found in fields and rural areas adjacent to woodlands and are common in some suburban areas.

Reproduction and Development

Smooth earth snakes mate in the spring and early summer. Their courtship behavior has not been documented, but mates presumably find one another by following pheromone trails. Females bear 2–14 (usually 6 or 7) live young in mid-to-late summer or early fall. Age at maturity for Georgia populations is unknown.

Habits

Although smooth earth snakes are active nearly year-round, they are rarely seen because they spend most of their time under leaf litter, rocks, or logs. They are active aboveground primarily in early evening or at night during the warmer months. Earthworms make up the bulk of their diet, but they also eat slugs; small snails; and the adults, larvae, and eggs of small, soft-bodied insects. Because of their small size, smooth earth snakes are preyed on by a variety of larger animals, including mammals and other snakes such as common kingsnakes (*Lampropeltis getula*), eastern coral snakes (*Micrurus fulvius*), and black racers (*Coluber constrictor*). Smooth earth snakes may bite when attacked by a predator, but they seldom if ever bite when handled by people. Secrecy and the release of musk are their primary means of defense.

Conservation Status

Smooth earth snakes are common and are presumed to be abundant despite the lack of detailed population studies. Populations in suburban areas may be affected by domestic cat predation.

Betsie B. Rothermel and
Whit Gibbons

Elapid Snakes

FAMILY ELAPIDAE

The elapids come in a variety of lengths, but all are typically slender snakes. Although Georgia's eastern coral snake (*Micrurus fulvius*) rarely exceeds 100 cm (39 in) in length, the largest Asian species, the king cobra, can exceed 5 m (16.4 ft). Elapids often have smooth, shiny scales and a head that is distinct from the body but not particularly large or pronounced. They are venomous and have relatively short, rigid fangs in the front of the upper jaw. The approximately 300 species are distributed across most of the world's tropical and warm temperate regions. Some, like the eastern coral snake, are burrowers; others are active ground dwellers or tree climbers. The sea snakes live in portions of the Indian and Pacific oceans. This family includes some of the world's most notoriously dangerous snakes such as cobras, mambas, kraits, and taipans. Human deaths from Georgia's eastern coral snake, however, are virtually unknown.

Adult eastern coral snake, Long County (Dirk J. Stevenson)

Eastern Coral Snake

Micrurus fulvius

Description

VENOMOUS. Although fairly slender, the eastern coral snake is larger than many people realize. Most adults are 51–76 cm (20–29.6 in) in total length, and the maximum total length recorded is 121 cm (47.2 in). The tail accounts for 8–13 percent of the total length. Females typically have a body length of 56–94 cm (21.8–36.7 in); males are usually 46–69 cm (17.9–26.9 in) in body length and have a slightly longer tail. The anal plate is divided. The smooth, shiny scales give this snake a lacquered appearance, and the alternating red, yellow, and black rings that completely encircle the body are striking in their contrast. The red rings are always in contact with the yellow rings. The rounded snout is black and is followed by a broad yellow band across the head and neck. The body has broad black and red rings, equal in width, that are separated by narrow yellow rings. The red rings are dotted with numerous black flecks that may coalesce on the back into a pair of spots. The tail has three or four broad black rings and two to four narrow yellow rings. The blunt head is scarcely wider than the neck. Young eastern coral snakes, 180–230 mm (7–9 in) in total length at hatching, are patterned like the adults. The scarlet kingsnake (*Lampropeltis triangulum elapsoides*) also has contrasting red, black, and yellow or white rings, but it has a red snout, and the light rings are separated from the red ones by

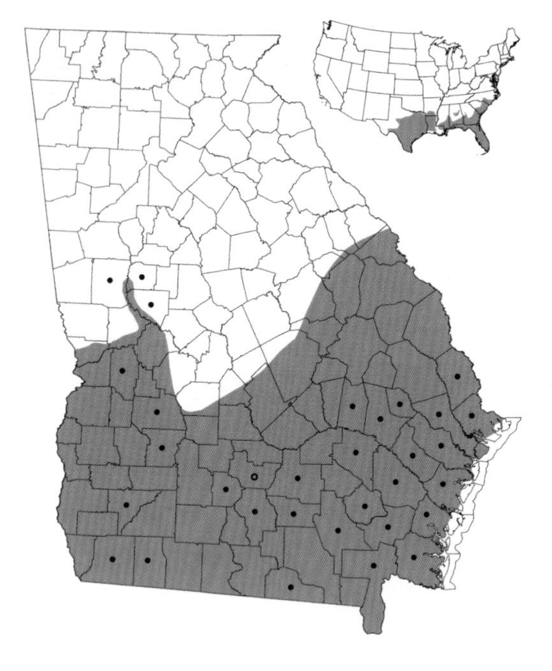

black. The scarlet snake (*Cemophora coccinea*) has a red, pointed snout; red blotches outlined by black on the back; and a white, unmarked belly. A helpful rhyme to distinguish the coral snake from the others goes, "Red touch yellow, harm a fellow; red touch black, friend of Jack."

Taxonomy and Nomenclature

No subspecies are recognized. Until recently the eastern coral snake and the Texas coral snake (*M. tener*) were both considered subspecies of *M. fulvius*, but because their ranges do not overlap, their color patterns differ, and there is no evidence of recent intergradation between the two, they are now recognized as distinct species. Some authorities use the common name harlequin coral snake for the eastern form.

Distribution and Habitat

The distribution of this species in Georgia is imperfectly known. Records are lacking for a large portion of the middle and upper Coastal Plain in the south-central and southeastern parts of the state, although suitable habitat is widespread. A population in the lower Piedmont inhabits both dry upland and sandy floodplain habitats along the Flint River. This species occurs on Ossabaw Island but is not known from any other barrier island. Coastal Plain populations are typically associated with sandy upland habitats such as longleaf pine sandhills and pine–saw palmetto flatwoods. Live oak and other hardwood hammocks on well-drained soils may also support populations. Eastern coral snakes are absent from extensive wetlands and from pine flatwoods underlain by low, poorly drained soils. Individuals lead highly subterranean lives and shelter in virtually any type of underground refuge, including gopher tortoise (*Gopherus polyphemus*) burrows and stump holes.

Reproduction and Development

Coral snakes are the only egg-laying venomous snakes in the United States; the pit vipers are live-bearers. Females lay a single clutch of 3–10 eggs in June or July. The small, smooth eggs are deposited in decaying pine stumps and logs or possibly underground in sandy soil. The young hatch in August or September. Males reach sexual maturity during their second year; females mature at approximately 2 years of age.

Habits

The eastern coral snake is fossorial, secretive, and infrequently encountered. Individuals may be active on the surface during the day or at night, and activity at dawn or dusk is common. Red, yellow, and black snakes found crawling in southern Georgia during daylight hours usually prove to be eastern coral snakes; the scarlet kingsnake and scarlet snake, presumed mimics of the coral snake, are primarily nocturnal. A Florida study found adult male eastern coral snakes to be especially active on the surface during the spring (March–May), when they search for mates, while adult females exhibit pronounced activity in the late summer or fall as they search for prey following egg-laying.

Eastern coral snakes feed exclusively on elongate reptiles. Small snakes, including the juveniles of large species, and slender lizards, particularly skinks (genera *Eumeces* and *Scincella*) and glass lizards (genus *Ophisaurus*), are their primary prey. Eastern coral snakes are active foragers that poke their head into leaf litter and beneath other surface debris. Prey animals are bitten, envenomated, and held until paralysis occurs, and then swallowed head-first. The striking color pattern of the eastern coral snake advertises its toxicity to visually oriented predators. When molested, an agitated eastern coral snake may flatten its body and curl and elevate the tail, waving it as if it were the head. During this display, the head may be hidden under the coils of the body or may aggressively whip from side to side. When grabbed, stepped on, or similarly restrained, individuals will bite. Among the known predators of eastern coral snakes are bullfrogs (*Rana catesbeiana*), eastern indigo snakes (*Drymarchon couperi*), loggerhead shrikes, and red-tailed hawks. Other likely predators include common kingsnakes (*Lampropeltis getula*) and coachwhips (*Masticophis flagellum*). Scarlet snakes may eat the eggs of eastern coral snakes.

Conservation Status

The eastern coral snake is generally distributed and appears to be fairly common in the lower and middle Coastal Plain of southeastern Georgia. Its status in the Piedmont, upper Coastal Plain, and southwestern portion of the state is poorly known.

Dirk J. Stevenson and Robert A. Moulis

Vipers

FAMILY VIPERIDAE

Five of Georgia's six dangerously venomous snake species belong to this family. Vipers are thick-bodied snakes with rough scales. The head is very wide and stands out from the relatively small neck. Georgia's vipers are called pit vipers because they have a heat-sensing depression, or pit, on each side of the face. The pupils are vertically elliptical, and, unique among Georgia snakes, there is a single row of scales underneath the tail (except at the very tip). All vipers are venomous and have relatively long, hollow fangs that fold up against the roof of the mouth when not in use. The more than 200 viper species are located across tropical and temperate regions of Africa, Eurasia, and the Americas. The New World vipers have heat-sensing pits; the majority of Old World vipers lack such structures. Virtually all of the New World vipers, Georgia's included, give birth to fully developed young. Vipers are generally ambush predators that rely on cryptic colors and patterns to make them invisible to unsuspecting prey such as small mammals. Although many viper species are capable of killing humans, death from viper bites is very rare in Georgia.

Adult copperhead, Floyd County (Bradley Johnston)

Copperhead

Agkistrodon contortrix

Description

VENOMOUS. These medium-sized snakes reach a maximum total length of 137 cm (53.4 in). Like most vipers, copperheads are relatively heavy bodied and have vertical pupils and a broad head that is distinct from the neck. Males generally mature at about 50 cm (19.5 in) in total length; females are smaller, maturing at 35–50 cm (13.7–19.5 in). The scales are keeled, and the anal plate is undivided. The background color is usually light brown or gray, but individuals range from rusty orange to pinkish to nearly black. This species is easily identifiable by its pattern of 10–21 dark brown, hourglass- or saddle-shaped cross bands, which are wider at the sides of the body and become narrower on the back. Individuals from mountainous regions in northern Georgia are generally dark with well-defined wide, hourglass-shaped cross bands. Farther south, particularly in the Coastal Plain, copperheads are generally lighter in color and have narrower cross bands that are often broken along the center of the back. Although this handsome pattern stands out when the snake is in the open, it renders copperheads virtually invisible on the leaf litter of the forest floor. The head is unpatterned coppery brown with two small, dark dots on the top just behind the eyes. The belly is generally gray to pinkish with darker blotches. Young copperheads resemble adults but have a bright yellow to greenish tail tip. Newborns average 250 mm (9.8 in) in total length.

Juvenile copperhead, Baker County. Note the bright yellow tail tip. (Gabriel J. Miller)

Adult copperhead, Liberty County (Dirk J. Stevenson)

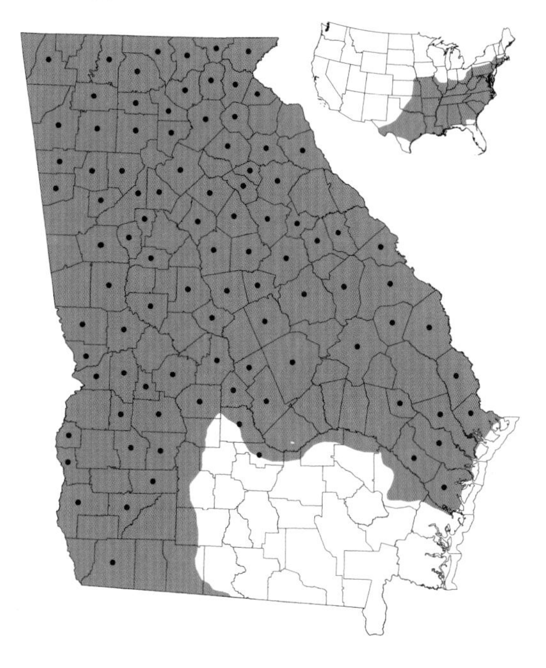

Taxonomy and Nomenclature

Two of the five described subspecies occur in Georgia. The northern copperhead (*A. c. mokasen*) is found in northern Georgia, and the southern copperhead (*A. c. contortrix*) occurs in the lower Coastal Plain. The two subspecies intergrade extensively throughout central Georgia, and specimens resembling either subspecies can be found in most regions. Upland or highland moccasin is another common name that may be used in some regions of the state.

Distribution and Habitat

Copperheads are found throughout much of Georgia but are absent from the southeastern Coastal Plain south of the Altamaha River. They can be found in most forested habitats but are particularly common on rocky, wooded hillsides in the mountains and on swamp and river edges in the Piedmont and Coastal Plain. Habitats with abundant logs, leaf litter, and rocks for cover are preferred, and open habitats such as old fields and agricultural areas are generally avoided. Perhaps due to their remarkable camouflage, copperheads are apparently more tolerant of urban development than many snake species are, and populations often persist in suburban neighborhoods as long as some patches of forest remain.

Reproduction and Development

Copperheads apparently breed in both the spring and fall. During these seasons males actively search out females, locating them by following scent trails, sometimes over long distances. Male copperheads are known to participate in ritualized combat in which they lift 30–40 percent of their bodies vertically off the ground and become entwined. They wrestle to establish dominance but do not bite one another. Although females may breed every year, most reproduce only every other year throughout their lives. Like other vipers, copperheads give birth to relatively few (1–21; average = 7) live young, usually in the late summer or early fall. Copperheads generally mature at 2–3 years of age.

Habits

Copperheads are primarily active from dusk into the middle of the night but are occasionally encountered during the day in cooler times of the year. They are not territorial and have been noted to hibernate communally with other copperheads and timber rattlesnakes (*Crotalus horridus*) in northern parts of Georgia. Copperheads are generalists that will eat almost any animal small enough to be swallowed. Small mammals, especially mice, are common food items, but these vipers also readily consume small birds, snakes, lizards, amphibians, and invertebrates, with a preference for cicadas. Young feed primarily on invertebrates and amphibians. While copperheads often ambush prey, they actively forage for food as well. Juveniles wag the bright yellow tip of their tail to lure prey within striking range, a behavior called caudal luring. Copperheads are potential prey for a variety of predators, including mammals such as opossums, coyotes, and domestic cats; hawks and owls; American alligators (*Alligator mississippiensis*); and other snakes such as black racers (*Coluber constrictor*), eastern indigo snakes (*Drymarchon couperi*), eastern coral snakes (*Micrurus fulvius*), and especially common kingsnakes (*Lampropeltis getula*). Copperheads rely on camouflage to conceal themselves from predators. This species occasionally bites humans, but generally only if picked up, stepped on, or approached too closely. Copperheads have the mildest venom of the venomous snakes in the state, and human deaths from their bites are virtually nonexistent. Copperheads sometimes simply strike the air as a warning to the intruder, and it is not uncommon for them to flee, especially when they are in open areas such as while crossing a road.

Conservation Status

Copperheads are among the most abundant snakes in many regions of Georgia, although their cryptic coloration often hides them from the casual observer.

John D. Willson and Kimberly M. Andrews

Adult cottonmouth gaping, Tattnall County (Dirk J. Stevenson)

Cottonmouth

Agkistrodon piscivorus

Description

VENOMOUS. Cottonmouths are relatively large, heavy-bodied pit vipers. Adults average 76–122 cm (29.6–47.6 in) in total length and reach a maximum length in excess of 183 cm (71.4 in). Males are longer and heavier than females. The scales are keeled, and the anal plate is undivided. These snakes are characterized by wide, dark bands along the body on a lighter brown or olive background. Many adults become very dark with age, and the bands become totally obscured. The belly is tan or gray with some obscure blotches present on most individuals. A well-defined dark stripe runs from the back of each eye to the corner of the jaw. Young cottonmouths, which average 280 mm (10.9 in) at birth, have very bold markings and often closely resemble copperheads (*A. contortrix*). Like copperheads they have a greenish or yellowish tail tip with which they lure prey within striking distance. Watersnakes (genus *Nerodia*) are often mistakenly called cottonmouths because they resemble these vipers, but the dark stripe on the side of the jaw, the heat-sensing pit on the face, and the behavior of gaping and vibrating the tail are distinctive to cottonmouths.

Taxonomy and Nomenclature

Two subspecies are found in Georgia. The Florida cottonmouth (*A. p. conanti*) is found

Juvenile cottonmouth, Long County. Note the bright yellow tail tip. (Dirk J. Stevenson)

Adult cottonmouth, Long County (Dirk J. Stevenson)

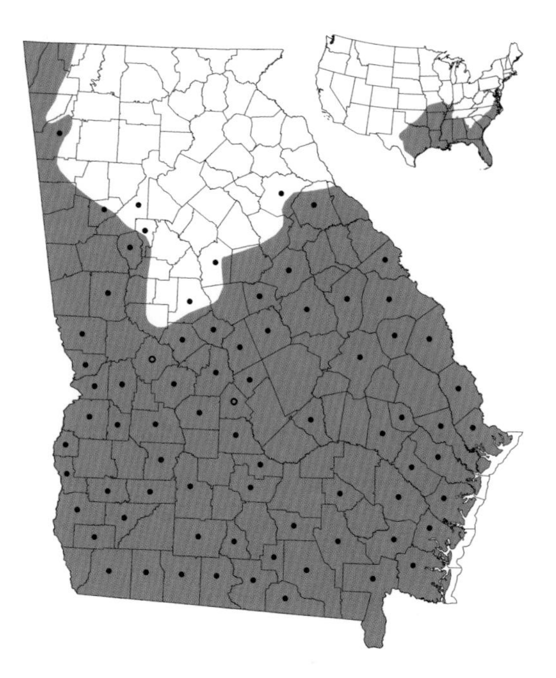

only in the extreme southeastern corner of the state, although much of the Coastal Plain is purported to be a zone of intergradation between it and the eastern cottonmouth (*A. p. piscivorus*). Cottonmouths are generally considered to be close relatives of copperheads. They are called water moccasins in many parts of their range.

Distribution and Habitat

The cottonmouth is found throughout the Coastal Plain in Georgia and in portions of the Piedmont, especially the extreme western part near Alabama. There is only a single Ridge and Valley Province record, but multiple records from this province in extreme eastern Alabama suggest that cottonmouths may be more widespread there than is currently thought. Cottonmouths are semiaquatic snakes that prefer swamplike habitats and can be very abundant in these areas. They can sometimes be found along rivers or in more open areas such as around large lakes. They may move overland long distances and are sometimes found far from any water source. They generally hibernate in dry upland areas, often in stump holes.

Reproduction and Development

Mating may occur anytime during the year but is primarily in the fall. Females can store sperm for long periods before using it to fertilize their eggs. Between 2 and 16 (average = 7) young are born in the fall. Both sexes probably reach sexual maturity at around 3 years of age.

Habits

Cottonmouths are the classic venomous snake of southern river swamps and other wetlands, although people's fears of unprovoked attack are unfounded. One scientific study documented that cottonmouths are highly unlikely to bite a person unless actually picked up. Cottonmouths are active almost year-round in Georgia, even emerging from underground wintering sites on sunny days of winter. They may be active or found coiled on shore or in vegetation at any time of the day or night but are most often active at night during the summer months and during the day during cooler periods in the spring and fall. Cottonmouths are noted for their extremely varied diet, which includes insects, frogs, salamanders, fish, birds, and other snakes, including smaller cottonmouths. They also eat dead animals, including roadkills. Cottonmouths have few predators as adults, although American alligators (*Alligator mississippiensis*) unhesitatingly attack and eat swimming cottonmouths. When alarmed, cottonmouths have a characteristic threat in which they hold open their mouth and display its white lining. Some individuals also vibrate the tail and release a musky odor from the vent. Cottonmouths characteristically swim with the head elevated above the surface of the water.

Conservation Status

Cottonmouths are very common in many areas of Georgia.

Michael E. Dorcas and Whit Gibbons

Adult eastern diamondback rattlesnake, Bryan County (Mark Wallace)

Eastern Diamondback Rattlesnake

Crotalus adamanteus

Description

VENOMOUS. This is Georgia's heaviest-bodied snake; large adults may reach 6 kg (13.2 lbs) in body mass. It is also one of the longest, reaching or possibly exceeding 200 cm (78 in), but more typically measuring 100–150 cm (39–59 in) in total length. Adult males are larger than females. The tip of the tail bears a rattle formed of one or more loose rings of hard keratin that makes a loud whirring noise when shaken. The head is broad and clearly set off from the neck. The scales are keeled, and the anal plate is undivided. The upper surface of the body is patterned by a long row of thin, yellow-bordered diamonds linked together at their tips, each fringing a dark brown center. The belly is uniformly cream colored. The pupils are hidden in the black eyes, which are on the sides of the face and lie within a black or dark brown band bordered by a pair of light yellow stripes that extend down and back to the lips. Juveniles resemble adults and are 310–380 mm (12.1–14.9 in) long at birth.

Taxonomy and Nomenclature

Eastern diamondback rattlesnakes appear similar throughout their geographic range, and no subspecies are recognized.

Distribution and Habitat

In Georgia, eastern diamondbacks are confined to the Coastal Plain, where they occupy upland habitats with an open canopy, especially native longleaf pine forests on sandhills, clay hills, and flatwoods. They are also found in mixed pine-hardwood forests that develop on abandoned

Adult eastern diamondback rattlesnake with aberrant (striped) pattern, Coffee County (John B. Jensen)

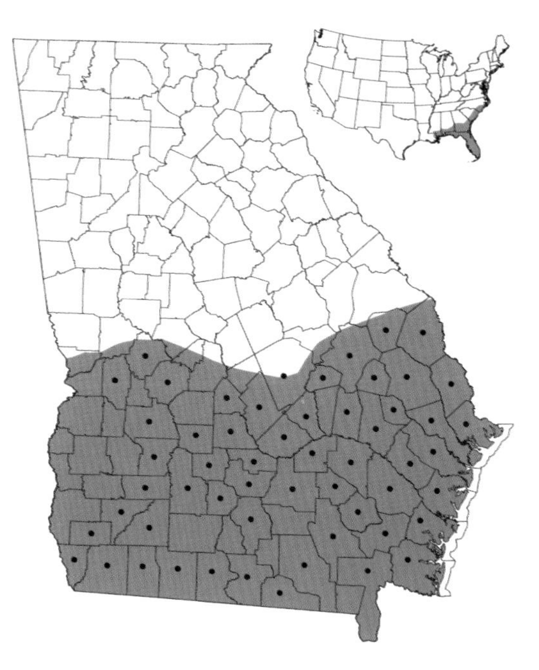

agricultural sites, especially if they are burned frequently and have a relatively open canopy, and in and around open woodlots, brushy pasture borders, and abandoned homesites in suburban and rural areas. Hardwood bottomlands along river and stream courses are used to a lesser degree, especially those adjacent to upland pine forests. The eastern diamondback also inhabits coastal dunes, spits, and barrier islands in interdune meadows containing dense bunchgrasses alternating with shrub thickets and in the dense edges of salt marshes.

Reproduction and Development

Extensive research in southwestern Georgia revealed that courtship and mating take place in August and September, but the females store the sperm over the winter and do not fertilize their ova until April. They give birth to 8–29 live young (average = 14) from mid-August to late September after a gestation period of about 90–120 days. Young snakes grow rapidly, shedding about three times in each of their first two years, then twice a year thereafter. Both sexes reach sexual maturity by the third year, but females first reproduce at 4 years of age. Thereafter they give birth every 2 or 3 years. Litters are born under logs, in stump holes or gopher tortoise (*Gopherus polyphemus*) burrows, and under other cover. Individuals can live up to 20 years, or even longer in captivity.

Habits

From November through March individuals seek underground shelter in stump holes, gopher tortoise burrows, and partially upturned root systems of trees for periods ranging from days to weeks, frequently emerging during warm spells. They remain on the surface almost constantly from April to October, spending most of their time coiled on the ground waiting to strike and envenomate prey. Gravid females, however, often go underground in August and September to give birth. Increased encounters with humans in August and September mostly involve adult males seeking females for mating. Individual snakes often return to the same underground site where they spent the previous winter. Eastern diamondback rattlesnakes are almost exclusively diurnal and rarely move about after dark. Adult males occupy home ranges of up to 240 ha (600 acres); females range over about 80 ha (200 acres) in annually burned mature pinelands. Individuals may have a smaller home range where habitat is damaged and fragmented, and on coastal strands and barrier islands the home range may be as small as 4 ha (10 acres).

The diet consists mainly of rodents, including cotton and wood rats, gray and fox squirrels, and white-footed mice as well as eastern cottontails, marsh rabbits, and assorted other mammals. They eat birds when the opportunity arises. Important predators include red-tailed hawks and other raptors, other snakes such as eastern indigo snakes (*Drymarchon couperi*) and especially common kingsnakes (*Lampropeltis getula*), bobcats, and, when near water, American alligators (*Alligator mississippiensis*), herons, egrets, and wood storks.

Conservation Status

Human impacts have caused the species to decline throughout its range, and it is thought to have disappeared from Louisiana. Habitat loss, fragmentation, and degradation are the most serious threats to eastern diamondback rattlesnakes because their upland habitats are in high demand for agriculture, pine farming, and residential development. Unfortunately, many people kill them on sight. The widespread practice of gassing gopher tortoise burrows to kill rattlesnakes or evict them for use in rattlesnake roundups often harms this species as well as the 300 or so other species of animals that use tortoise burrows. Georgia has historically had three annual rattlesnake roundups, although only two, the ones in Claxton and Whigham, persist; the Fitzgerald Rattlesnake Roundup is no longer held. Data from all three roundups indicates that roundups have a severe negative impact on local rattlesnake populations.

D. Bruce Means

Adult timber rattlesnake, Baker County (David Steen)

Timber Rattlesnake

Crotalus horridus

Description

VENOMOUS. The timber rattlesnake is a large, heavy-bodied snake with a narrow neck, wide head, and short tail ending in a segmented rattle. In the Coastal Plain, where they reach their greatest size, adult females are typically 107–137 cm (41.7–53.4 in) in total length (including the rattle); adult males usually range from 100 to 150 cm (39–58.5 in), but very large individuals may exceed 180 cm (70.2 in). In the mountains, adult females are usually 86–109 cm (33.5–42.5 in) long, males are 84–132 cm (32.7–51.5 in), and the maximum size is probably less than 150 cm (58.5 in). The tail is 6–9 percent of the total length, and males have a longer tail than females. The scales are keeled, and the anal plate is undivided. The background color ranges through various shades of pink, yellow, tan, gray, brown, and olive to velvety black. A series of brown to black chevron-shaped cross bands typically cross the body. The tail is black. Very dark or solid black individuals are common in higher mountains of the northeastern part of the state but are rare elsewhere. Snakes from the higher mountains are typically yellow, yellowish brown, or olive brown, whereas mountain rattlesnakes from lower elevations tend to be dark brown. Those from the Coastal Plain and Piedmont are typically tan, gray, or brown, often with a pinkish tinge. A tan or reddish brown stripe is often present down the center of the back, as is a

Adult timber rattlesnake about to shed, as indicated by cloudy eye, Union County (Dirk J. Stevenson)

Juvenile timber rattlesnake, Upson County (John B. Jensen)

Albino juvenile timber rattlesnake, Lamar County (Jason Clark)

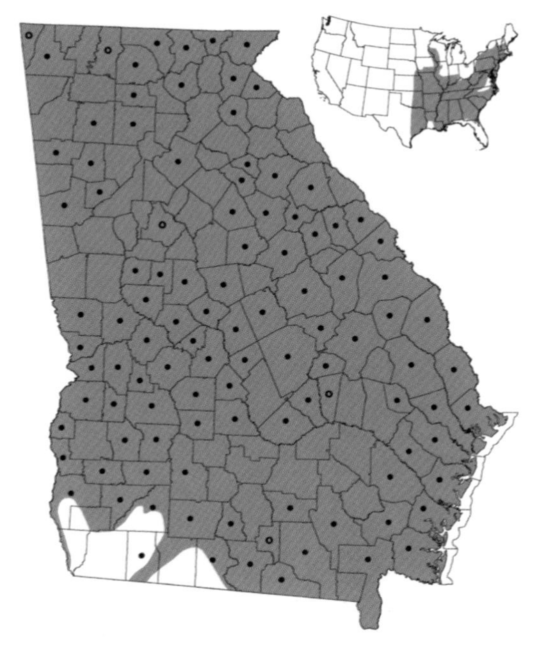

dark stripe behind the eye, especially in Coastal Plain and Piedmont populations. Newborns average about 330 mm (12.9 in) in the Coastal Plain and 280 mm (10.9 in) in the mountains, and are usually grayish and drabber than adults. The cross-banded pattern distinguishes the timber rattlesnake from the eastern diamondback (*Crotalus adamanteus*) and from the much smaller pigmy rattlesnake (*Sistrurus miliarius*).

Taxonomy and Nomenclature

Two subspecies were formerly recognized; snakes from the Coastal Plain and lower Piedmont were called *C. h. atricaudatus* and were often referred to as the canebrake rattlesnake. Recent genetic studies do not support subspecific designation for the geographic variants of this rattlesnake. Nevertheless, specimens from higher elevations, especially above 750 m (2,460 ft), are readily distinguishable from Coastal Plain and Piedmont animals by their somewhat distinctive coloration and smaller size.

Distribution and Habitat

These rattlesnakes are common in much of the heavily wooded country of the Coastal Plain, but in more open areas they are primarily limited to

Adult timber rattlesnake, Lumpkin County (John B. Jensen)

wooded stream corridors. The range is spotty on the barrier islands and along the immediate coast. They may be absent from the Tallahassee Redhills region in the southwestern part of the state. Fire suppression and the concomitant succession of open-canopied longleaf pine habitat to closed-canopy mixed pine-oak forest in the Coastal Plain have probably allowed populations there to expand or increase in abundance. In the Piedmont, the distribution is highly fragmented due to habitat loss and is primarily associated with heavily wooded stream corridors and isolated small mountains. Their distribution in the Georgia mountains is somewhat localized around suitable sites for overwintering and gestating. Denning habitats include root and stump holes, mammal burrows, artificial sites such as stump piles and dilapidated buildings, and, especially in upland regions, rock crevices.

Reproduction and Development

Mating takes place from August to October. Adult males searching for mates are often encountered crossing roads or yards at this time. Females produce young at an average age of 6 years in the Coastal Plain (5–7 years) and 8 years in the mountains (7–9 years). The usual interval between births is probably 2 years (sometimes 3) in the Coastal Plain and typically 3 or 4 years at higher elevations. Females in the Coastal Plain select a secluded place for gestation. In the mountains, gestating females typically use sun-warmed, exposed outcrops where they can maintain the high body temperature necessary for development of the young. The young are usually born from mid-August to mid-September, but births are sometimes delayed by unfavorable weather until early October at higher elevations. Litter size averages about 8 in the mountains and 12 in the Coastal Plain; the maximum recorded is 30. Mothers and young stay together for 1 or 2 weeks until the young shed, at which time all disperse in search of food. Males are mature at a younger age and smaller size than females but continue growing at a faster rate.

Habits

The activity period ranges from about 6 months (mid-April–mid-October) at higher mountain elevations to about 9 months (March–November) in the southern Coastal Plain. No movement studies have been done in Georgia, but elsewhere female timber rattlesnakes typically range 0.8–1.6 km (0.5–1 mi) from the overwintering den, and males move about 1.6–2.4 km (1–1.5 mi). During the summer, most movement is at night. Rattlesnakes overwinter in communal dens in rock crevices in the mountains but tend to den solitarily in ephemeral sites such as stump holes in the Coastal Plain. Timber rattlesnakes are highly cryptic when not moving, making them very effective ambush predators. They prey primarily on rodents such as chipmunks, squirrels, mice, and rats but also consume other appropriately sized mammals as well as birds and lizards. Potential enemies include the black racer (*Coluber constrictor*), eastern indigo snake (*Drymarchon couperi*), coachwhip (*Masticophis flagellum*), common kingsnake (*Lampropeltis getula*), and cottonmouth (*Agkistrodon piscivorus*); and among raptors, the red-tailed hawk and great horned owl. Mammalian predators such as bobcats, opossums, skunks, coyotes, and feral hogs are reported to take timber rattlesnakes as well. Bites from this species may be life threatening; however, individuals are typically reluctant to bite unless thoroughly agitated, and direct physical contact is often necessary to provoke a strike.

Conservation Status

The timber rattlesnake is common in much of the Coastal Plain, but the range is highly fragmented in the Piedmont and is under increasing pressure from residential development. Timber rattlesnakes are hunted to some extent in the mountains, where they are localized around overwintering dens and gestating sites. They are reported to have declined in recent years in most parts of the mountains.

W. H. Martin, Dirk J. Stevenson, and Philip B. Spivey

Adult pigmy rattlesnake, Stephens County (Berkeley W. Boone)

Pigmy Rattlesnake

Sistrurus miliarius

Description

VENOMOUS. This is the smallest of the rattlesnakes. The maximum total length reported is 80 cm (31.2 in), but a typical adult is 40–60 cm (15.6–23.4 in). Males have a longer tail than females but otherwise do not differ in appearance. The scales are keeled, and the anal plate is undivided. The ground color is usually gray or tan, but occasional individuals in some populations can be reddish or almost black. The pattern consists of a series of light-edged dark blotches or spots on the back and one to three rows of dark spots on the sides. There may be a reddish stripe along the center of the back. The belly is usually pale and moderately mottled with dark pigment, although the belly of snakes in southern Georgia is often heavily mottled with black. A dark stripe extends from the back of each eye to the corner of the mouth, and two wavy dark stripes extend from the top of the head to the back of the neck. Pigmy rattlesnakes can be distinguished from nonvenomous snakes by the tiny rattle at the end of their tail and the sensory pit between the eye and the nostril on each side of the head. The interlocking segments of the rattle are poorly notched relative to those of other rattlesnakes, though, and sometimes the rattle is lacking altogether. Pigmy rattlesnakes differ from the other two species of Georgia rattlesnakes in having nine enlarged scales or plates on the head crown instead of many small scales. Newborns have a yellow or green tail and average 150 mm (5.9 in) in total length.

Juvenile pigmy rattlesnake, Stephens County. Note the bright yellow tail tip. (Berkeley W. Boone)

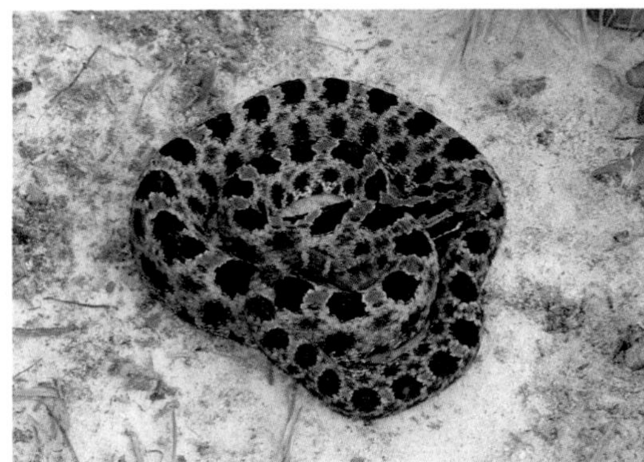

Adult pigmy rattlesnake, Long County (Dirk J. Stevenson)

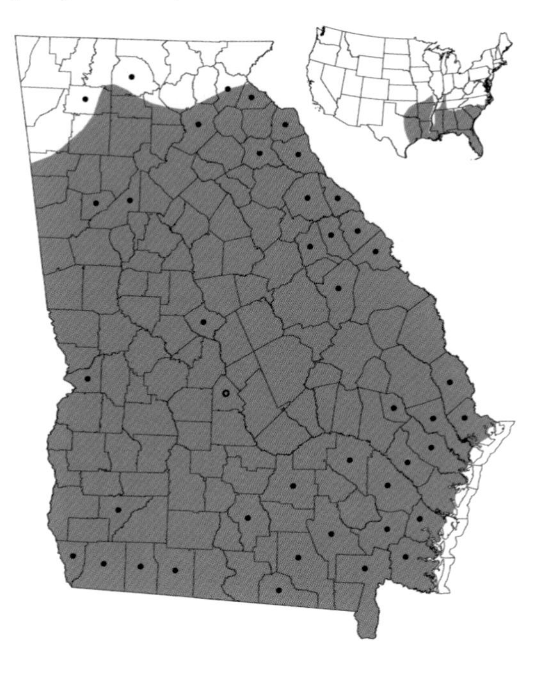

Taxonomy and Nomenclature

Two subspecies are present in Georgia: the dusky pigmy rattlesnake (*S. m. barbouri*) in the southern part and the Carolina pigmy rattlesnake (*S. m. miliarius*) in the rest of the state. Pigmy rattlesnakes are commonly referred to as ground rattlers in many parts of the state.

Distribution and Habitat

Pigmy rattlesnakes are known from the Coastal Plain, Piedmont, and Ridge and Valley provinces. They can be found at elevations ranging from 0 to 500 m (0–1,640 ft) and in habitats ranging from dry sandhills and longleaf pine forests to wet hammocks and seasonally flooded pine flatwoods. In southern Georgia they thrive in saw palmetto thickets, which they share with eastern diamondback rattlesnakes (*Crotalus adamanteus*).

Reproduction and Development

Mating season usually extends from late summer to late fall. Pigmy rattlesnakes start breeding when they are at least 2 or 3 years of age. As is true for all rattlesnakes, the females bear live young; they typically give birth to five to ten newborns in the summer following mating the previous year. Pigmy rattlesnakes produce no more than one litter per year, but they usually reproduce every year, a fairly uncommon phenomenon in rattlesnakes. Males are known to fight with other males during the reproductive season, but this behavior is seldom observed and seems to be absent in certain populations.

Habits

Our knowledge of the biology of pigmy rattlesnakes in Georgia stems primarily from assumptions based on research conducted in Florida. In the southern parts of Georgia they may remain active year-round due to the moderate winter temperatures. In the rest of the state they presumably hibernate in animal burrows or other suitable retreats from October or November to March or April depending on the location and winter temperatures. Activity seems to peak between May and October. This species is active during the day in spring and fall but becomes nocturnal during the hottest part of the year. Pigmy rattlesnakes are sit-and-wait predators. They are opportunistic when it comes to prey choice but seem to show a predilection for amphibians and other reptiles, particularly frogs, lizards, and small snakes. Six of 12 pigmy rattlesnakes from a Georgia study contained reptilian prey in their stomachs. Adults also take small mammals, primarily mice, and the young eat invertebrates such as centipedes, crickets, and spiders. Their small size makes pigmy rattlesnakes vulnerable to a variety of animals that includes carnivorous mammals, raptors, and other snakes such as black racers (*Coluber constrictor*), eastern indigo snakes (*Drymarchon couperi*), and common kingsnakes (*Lampropeltis getula*). Pigmy rattlesnakes have been described as fiery and irritable when disturbed, but recent research on their defensive behavior has shown that they are in fact reluctant to bite. Nonetheless, pigmy rattlesnakes should be left alone because they are still equipped with potent venom.

Conservation Status

In some areas of Georgia these rattlesnakes can be found in very high population densities, but in other places they can be fairly uncommon. They are small and cryptic, however, and many individuals are probably overlooked.

Xavier Glaudas

Turtles

TURTLES ARE NATURE'S tanks, the only four-footed creatures completely protected both above and below by bony armor. The characteristic turtle shell consists of upper (carapace) and lower (plastron) portions that are connected on the sides by a bony bridge. Typically, large, horny scales called scutes cover the surface of the entire shell. Turtles can maximize the shell's protection by withdrawing their head, limbs, and tail inside it. Some, like Georgia's own box turtle (*Terrapene carolina*), have a hinge in the plastron that allows it to shut tightly against the carapace, thereby keeping at bay any predator that cannot actually crush the shell.

The shell itself is a lesson in biophysics that offers insights into evolutionary trade-offs between opposing physical forces. The obvious protective advantage provided by the shell is achieved at the cost of the encumbrance of its weight and the associated loss of agility and speed. Many species compensate for the weight by living in water, whose buoyant effect on the shell mitigates the drawbacks of gravity. For aquatic turtles to achieve any kind of swimming speed, however, the shell must be relatively light and flat to maximize its hydrodynamic function. A flat shell, in turn, sacrifices the strength of the domed shell, which, much like the arch of a Roman bridge, is highly resistant to crushing. Terrestrial turtles, which do not face hydrodynamic problems, develop the most highly domed shells. High domes carry their own price, though, and the tortoises (family Testudinidae), so well adapted for life on land, have developed sturdy, columnar hind legs to help support the extra weight.

Even with its attendant problems, the shell has enabled turtles to be particularly successful, at least as far as persistence goes. The oldest known turtle lived some 220 million years ago, meaning that turtles as a group are at least as old as the earliest dinosaurs. The unique body form of a turtle thus represents one of the oldest surviving body forms among amniotic vertebrates.

As durable as turtles are, both as individuals and as an evolutionary group, they have not developed the kind of species diversity prevalent in some other reptilian groups, most notably the lizards and snakes. Fewer than 300 turtle species exist worldwide, although Georgia has twice as many native turtles (27) as lizards (13).

Though relatively few when species are counted, turtles have been able to take advantage of diverse habitats. Freshwater habitats are the most heavily exploited, and turtles occur in lakes, ponds, streams, and wetlands of various descriptions. Georgia's relatively high number of turtle species is undoubtedly a consequence of the state's diversity of aquatic habitats. The largest turtles have evolved as marine creatures, reminders of a time when the seas supported a variety of large marine reptiles such as the extinct plesiosaurs and ichthyosaurs. Some turtles live exclusively on land, the most notable of these being the tortoises, of which Georgia has a single representative, the gopher tortoise (*Gopherus polyphemus*). The box turtle, which belongs to a family (Emydidae) otherwise dominated by water turtles, is another Georgia turtle that has eschewed an aquatic lifestyle in favor of a terrestrial one.

Regardless of the habitat where they spend

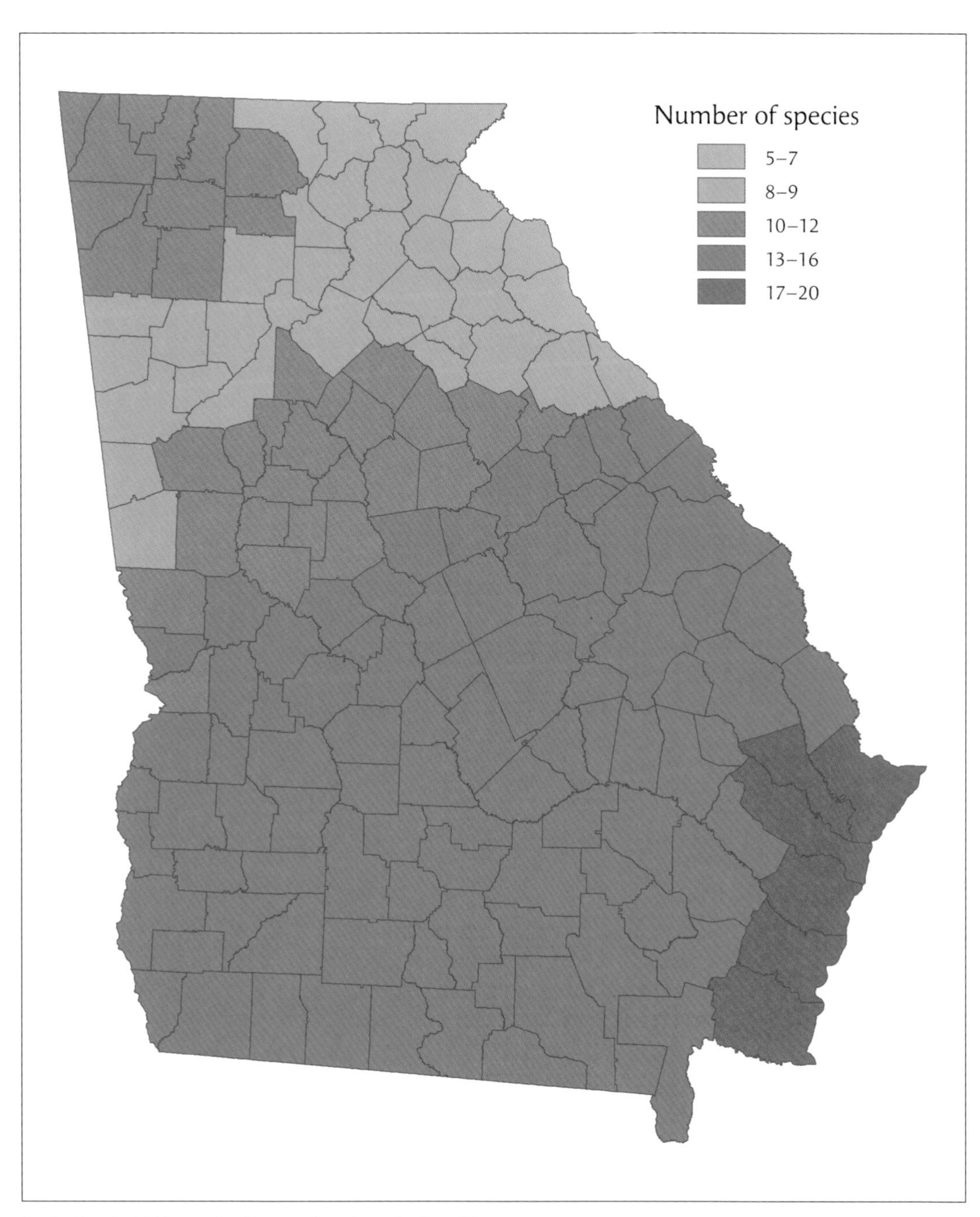

Turtle Species Richness by County, Based on Predicted Ranges

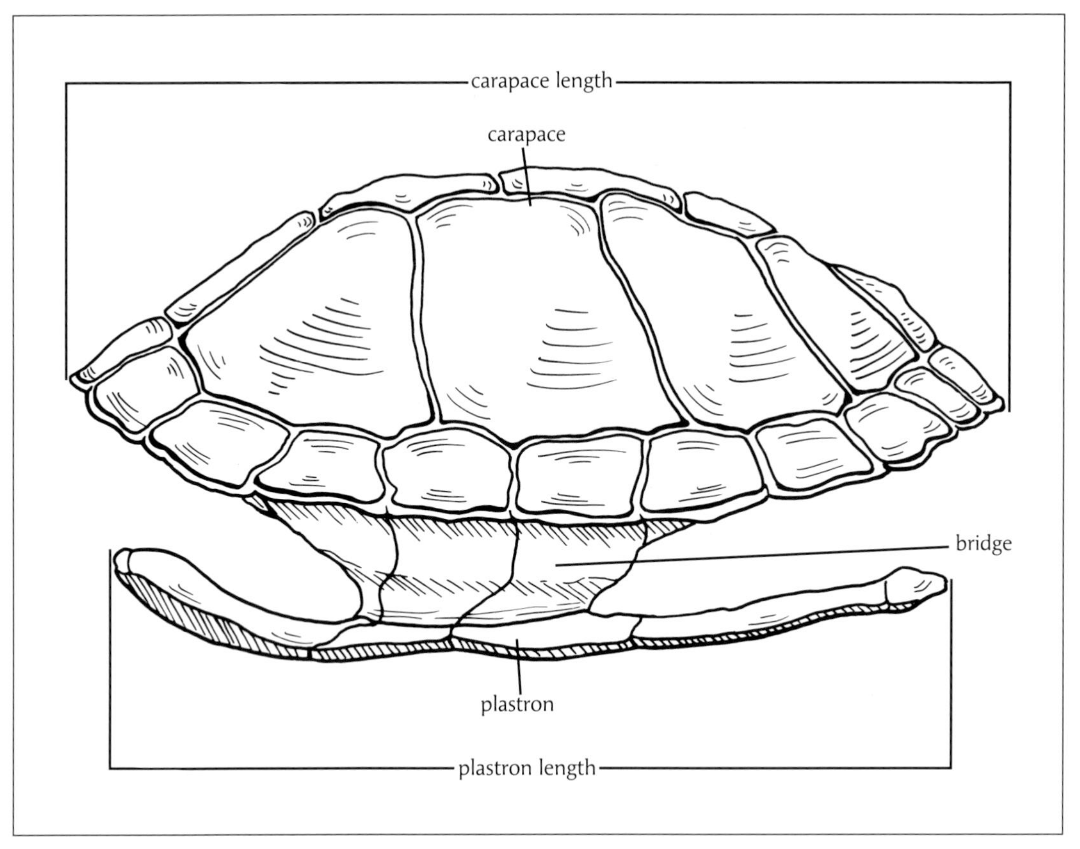

Major Divisions and Measurements of Turtle Shells

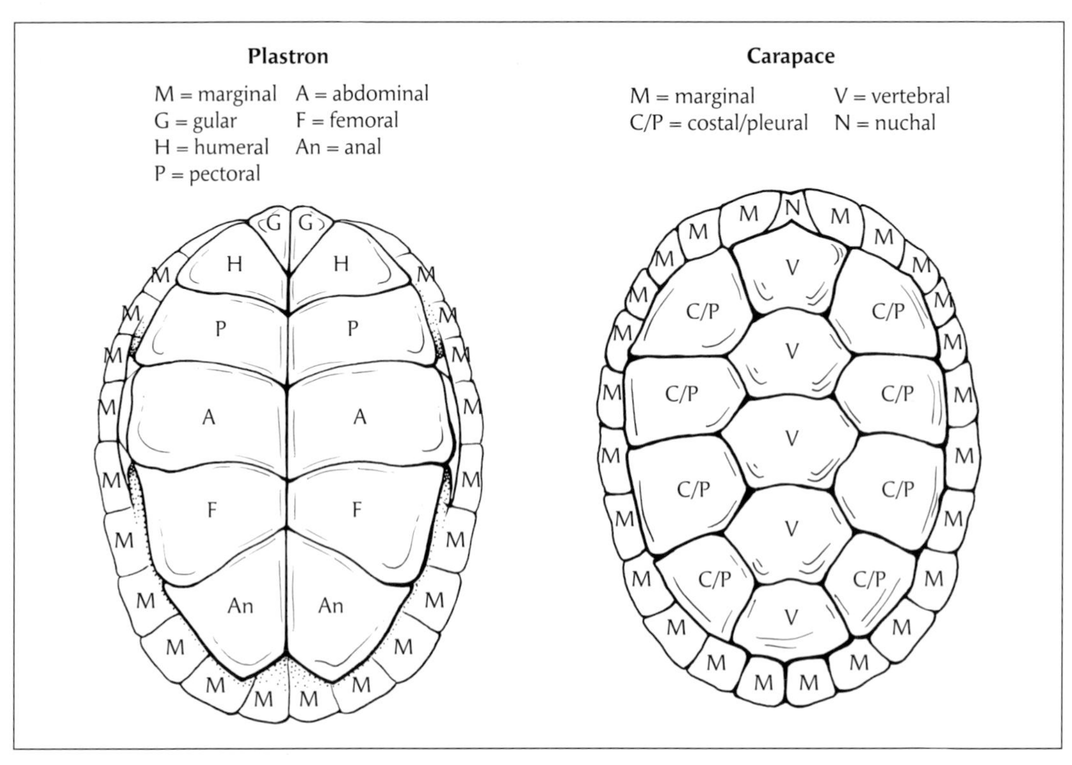

Scutes of Turtle Shells

most of their time, all turtles lay eggs on land. Even the giant marine turtles, whose front limbs are flippers rather than walking legs, waddle up beaches far enough to lay eggs where they will not be flooded by incoming tides. Female turtles typically excavate a flask-shaped nest with their hind limbs, urinating on the ground to soften it for digging. After depositing her eggs, the female covers the nest and abandons them. The eggs then incubate at the temperature of the environment. For many turtle species, the incubation temperature, particularly early in development, determines the sex of the hatchling, with low developmental temperatures typically producing predominantly males and high ones predominantly females. Hatchlings of many Georgia turtles, including painted (*Chrysemys picta*), eastern mud (*Kinosternon subrubrum*), and pond slider (*Trachemys scripta*) turtles, characteristically spend the winter underground in the nest and do not emerge until the following spring.

Hard-shelled Sea Turtles

FAMILY CHELONIIDAE

The marine species are large turtles; some exceed 100 cm (39 in) in carapace length. The shell is relatively flat and is often shaped like a shield. The front limbs are modified into flippers. There are six species in this family, and although they may swim throughout the tropical and temperate oceans of the world, most prefer tropical waters. Georgia's coast is at least occasionally visited by four species, each in a different genus. Female sea turtles nest on beaches around the warm parts of the world, usually producing several clutches in a single season. The loggerhead sea turtle (*Caretta caretta*) is by far the most common of the sea turtles that nest on Georgia's beaches.

Although the juveniles of most species spend their time far out at sea, the adults tend to remain near shore, or at least near the edge of the continental shelf. Each species appears fairly specialized in its diet, some feeding on seaweeds, others on sponges or corals, and still others on crustaceans and mollusks.

Adult female loggerhead sea turtle, Chatham County (Mark Dodd)

Loggerhead Sea Turtle

Caretta caretta

Description

Loggerhead sea turtles are among the world's largest turtles. Adults have a carapace length of 75–112 cm (29.3–43.7 in) and weigh 77–159 kg (170–351 lb). As with other sea turtles, males tend to be smaller than females. The tail of reproductively active males may extend 30 cm (11.7 in) or more beyond the carapace; the tail of females is much shorter. Adults are reddish brown and have a large, thick head, from which they get the name loggerhead; substantial jaws; and two claws on each flipper. These first two characters should distinguish loggerheads from the other sea turtles. Adult loggerheads also house a uniquely large amount of epibiota—small organisms such as barnacles, bivalves, caprellid shrimp, and algae that attach to the carapace. Further, the first costal scute touches the nuchal scute, distinguishing the loggerhead from both Kemp's ridley (*Lepidochelys kempii*) and green (*Chelonia mydas*) sea turtles. Although juvenile loggerheads also have a reddish tone, they can be easily confused with juveniles of other sea turtle species. Hatchlings are 38–50 mm (1.5–2 in) long and weigh approximately 20 g (0.7 oz). They are darker brown than the adults, have a light plastron, and have three keels on the carapace and two on the plastron.

Taxonomy and Nomenclature

Loggerhead sea turtles are the only species within their genus, and the only subspecies occurring in Georgia waters is the Atlantic loggerhead sea turtle (*C. c. caretta*).

Distribution and Habitat

Newborns are pelagic. Those that hatch on Georgia's beaches first seek the rafts of sargassum

Loggerhead sea turtle hatchling, Glynn County (Brad Winn)

Juvenile loggerhead sea turtle, Glynn County (Mark Dodd)

that form along lines where currents converge. When they are approximately 150 mm (5.9 in) long they perform a transatlantic journey that follows the Gulf Stream and the northern loop of the North Atlantic Oceanic Gyre eastward toward the Azores and the eastern Atlantic. They then turn south along western Africa, at which time they are around 30 cm (11.7 in) long, and then head back to the western Atlantic along the southern loop of the gyre. A few individuals move into the Mediterranean Sea. Individuals return to the western Atlantic at 41–46 cm (16–17.9 in) and perhaps 6–10 years of age, and move into shallow coastal waters where they spend the rest of their lives.

Reproduction and Development

Loggerheads nest on Georgia beaches from May to August in open sand dunes. Nesting takes

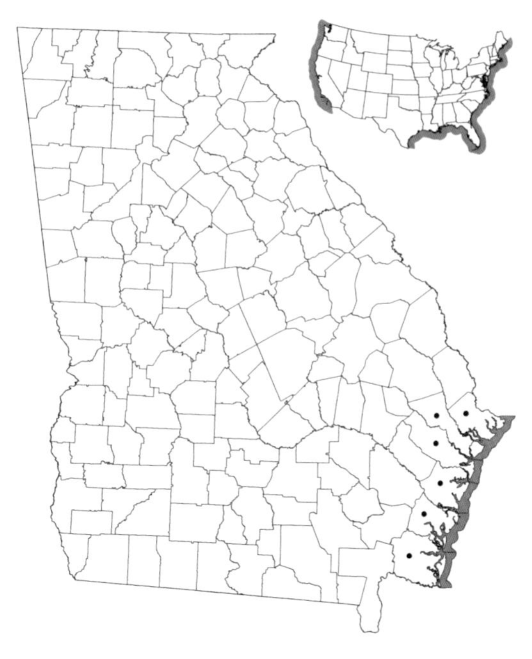

about an hour for an individual female. The average clutch size is 115–125 eggs, and individuals lay from four to six clutches per season at 2-week intervals. Nests incubate for 55–65 days. The incubation temperature determines the sex of the developing embryos. Individual females nest every 2–4 years. They return to the vicinity of their natal beach and are often observed nesting on the same stretch of beach from year to year. Loggerhead sea turtles probably reach sexual maturity at 15–25 years of age.

Habits

Loggerhead sea turtles spend much of the day submerged and sleep at night on the bottom, surfacing periodically to breathe. When in cooler waters they bask while floating on the surface to absorb the sun's warmth. In the winter, loggerhead sea turtles migrate south in the Atlantic to warmer waters to avoid cold stunning, which occurs when the water temperature falls below the threshold at which the turtle can physiologically function. Small fish often accompany them on their travels, feeding on food fragments that trail from the turtles' mouths. Pelagic newborns feed on small invertebrates (e.g., crabs and insects) and fish eggs within the sargassum community. As they get bigger, they also feed on jellyfish and other invertebrates that float in surface waters. After they return to coastal waters, they become benthic (bottom) feeders and retain this foraging strategy as adults. Although they feed opportunistically on dead fish and other accessible prey, their diet at this point consists primarily of crabs and mollusks. Their huge jaws crush the shells, and the fragments are swallowed along with the prey tissue. Native terrestrial predators, primarily raccoons and ghost crabs in Georgia, are mainly a threat to eggs and hatchlings. Nonnative feral hogs and fire ants represent significant sources of predation as well. Hatchlings in the water fall prey to fish and seabirds. They are defenseless at this stage, and survivorship of just a few per clutch is possible only because of their sheer numbers. Adults outgrow most predators except for blue-water sharks.

Conservation Status

Even though loggerhead sea turtles are the most abundant of the sea turtles that nest in Georgia, there are fewer than 3,000 nests in the state each season, and the numbers are decreasing. The species is state listed as Endangered and federally listed as Threatened. Loggerheads are especially vulnerable to drowning when incidentally captured by shrimp trawls, long-lines, gill nets, and other commercial fishing gear. Loggerhead and other sea turtles are attracted to light and become disoriented by artificial lights on beaches. Georgia law requires beachfront residents (i.e., homeowners, businesses, and hotels) to darken lights during the nesting season. Loggerhead sea turtles are further threatened by excessive egg predation by feral hogs and by an overabundance of native predators such as raccoons on some islands. Egg predators can destroy an entire season's reproductive effort. Habitat destruction, both marine and terrestrial, is a serious concern for loggerhead sea turtles nesting on Georgia beaches, as it is for other sea turtles, both locally and globally. One activity of particular concern is coastal armoring with seawalls, riprap, or similar structures, which permanently destroy critical nesting habitat.

Kimberly M. Andrews and
James I. Richardson

Adult green sea turtle with remoras on carapace, U.S. Virgin Islands (Caroline Rogers)

Green Sea Turtle

Chelonia mydas

Description

Green sea turtles reach a carapace length of 88–125 cm (34.3–48.8 in) and weigh 113–204 kg (250–450 lb). Males are smaller than females but have a longer, thicker tail. Adults are characteristically dark brown with a lighter-colored underside and a carapace mottled with light and dark markings. They are most readily distinguished from loggerhead (*Caretta caretta*) and Kemp's ridley (*Lepidochelys kempii*) sea turtles by having four costal scutes on each side rather than at least five. The head has a single pair of prefrontal scales rather than the two pairs found on the hawksbill sea turtle (*Eretmochelys imbricata*). Each flipper has a single claw. Hatchlings are 38–63 mm (1.5–2.5 in) in carapace length, weigh 25 g (0.9 oz), and are black or very dark brown above and light beneath. Their flippers are black trimmed with white. Hatchlings also have a keel on the center of their carapace and a pair of keels on the plastron.

Taxonomy and Nomenclature

The green sea turtle is the only member of its genus. The common name derives from the color of the internal fat bodies rather than the color of the turtle or its herbivorous diet. Taxonomic debate has arisen over whether the Pacific green sea turtle (*C. m. agassizii*) should be considered a separate species. All other green sea turtles, including those in Georgia's coastal waters, are currently recognized as Atlantic green sea turtles (*C. m. mydas*).

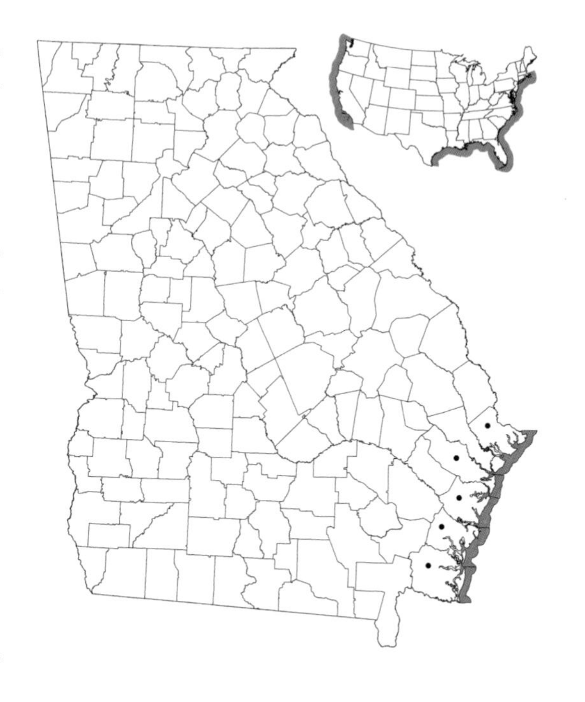

Distribution and Habitat

Green sea turtles can be found in warm marine waters worldwide. They rarely nest on Georgia's coastline, but records exist from Jekyll, Ossabaw, and Wassaw islands. Nonnesting female emergences have been observed at Little Cumberland and Little Saint Simons islands, as have strandings (i.e., dead turtles washed ashore). Green sea turtles are commonly found in nearshore estuarine habitats such as marine grass flats and reefs that contain an abundance of seaweed and algae. Young turtles appear to occupy pelagic environments but move to vegetated areas closer to shore when they reach a carapace length of 25–60 cm (9.8–23.4 in).

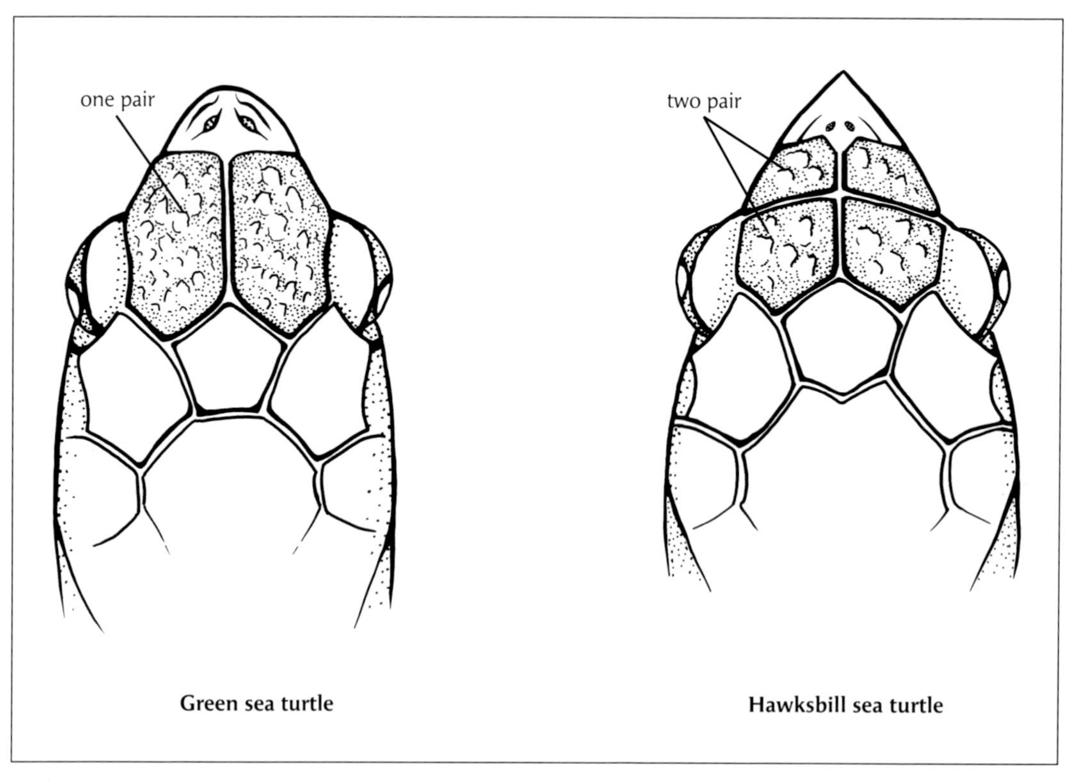

Prefrontal Scutes

Reproduction and Development

Green sea turtles nest throughout the year in the tropics and June–September in the southeastern United States. Nests are usually constructed in open sand; the process, preceded by body pitting, takes 2 or 3 hours. Females of all sea turtle species first clear the nest area and burrow the body into a pit in the sand before digging the nest cavity. While laying, the female uses her rear two flippers to cover the eggs. This behavior, also seen in leatherback sea turtles (*Dermochelys coriacea*), is believed to have evolved to protect the eggs from predators. An average clutch consists of 110 eggs, and females complete one to four nests per season with an internesting interval of approximately 2 weeks. The eggs incubate for approximately 60 days, and the incubation temperature determines the sex of the hatchlings. Most females nest every 2 years, but some wait 4–6 years between nestings. Green sea turtles take longer to reach sexual maturity than any of the other sea turtle species. Females have been reported to be 25–40 years old before they become reproductively active.

Habits

Adults migrate between feeding and nesting grounds, sometimes swimming thousands of miles. Green sea turtles migrate south in the colder months to prevent cold stunning, which occurs when the water temperature is below the threshold at which the turtle can physiologically function. Unlike other sea turtle species, green sea turtles bask terrestrially, most frequently in Hawaii. This is the only instance known for any of the marine turtle species of males and juveniles coming on land. Both adults and juveniles exhibit strong site fidelity for feeding areas, and individuals often return to their natal beach to nest as adults. Juveniles are predominantly carnivorous; adults are primarily herbivorous, although they also consume jellyfish and other marine invertebrates. Juveniles have been found in Georgia waters with marsh grass in their gut. Green sea turtles can pack enormous amounts of vegetation in their gut; even the intestinal tract of young turtles can hold 1–2 liters (1.1–2.1 qt). Hatchlings can be eaten by almost any predator larger than they are, but adults rely on their large size for defense and fall prey mostly to sharks or terrestrial mammals and humans.

Conservation Status

Green sea turtles are state and federally listed as Threatened. Like other sea turtles, they suffer from habitat destruction, including beachfront development and light pollution, which disorients individual turtles. The grassy areas in coastal waters on which they rely for food are being destroyed by water pollution. Green sea turtles have been harvested extensively for their meat and eggs and are thought to be the best tasting of the turtle species. They were so abundant at the time Columbus discovered America that they served as a major food source for the sailors, but extensive and unchecked harvesting has caused the irreparable decline of many populations. Furthermore, green sea turtles suffer from fibropapillomatosis, a disease that is currently widespread. It is thought to occur in situations where stressors have decreased the effectiveness of the turtle's immune system, such as in areas offshore from extensive human development or compromised water quality.

Kimberly M. Andrews

Adult hawksbill sea turtle, Puerto Rico (Michelle Tanya Scharer)

Hawksbill Sea Turtle

Eretmochelys imbricata

Description

Adult hawksbill sea turtles generally have a carapace length of 76–89 cm (29.6–34.7 in) and weigh around 68 kg (150 lb); the record weight is 126 kg (278 lb). These turtles are known for their marbled brown and yellowish shell. Hawksbills have scutes that overlap and are thicker than those of other turtles, although the overlaps are less pronounced in older individuals. Likewise, younger hawksbills have pointed scutes on the rear of the shell that become rounded in older adults due to wear. The keratinous beak has a hawklike overbite. The flippers, head, and tail are all brownish, with the scales themselves being a bit darker than the skin between them. Males have a longer tail than females and use it as an aid in copulation. Females have a distinct claw on each rear flipper. Hatchlings are approximately 38–51 mm (1.5–2 in) in length, dark charcoal to brown, and may have light brown touches on the shell and flippers.

Taxonomy and Nomenclature

Two subspecies are recognized; those in the Atlantic Ocean are Atlantic hawksbill sea turtles (*E. i. imbricata*).

Distribution and Habitat

The hawksbill inhabits the tropical waters and coral reefs of 65 nations and comes ashore only to lay eggs. Hawksbill sightings in Georgia are

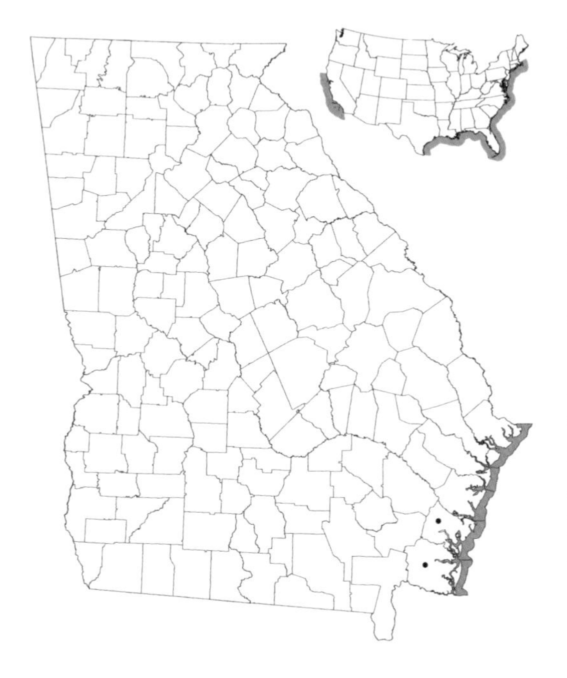

limited to three records, one from 1931 and two from June 1998. The earlier individual was caught in a net off the coast of Savannah, and the later records are of an adult female stranded on Cumberland Island and a juvenile stranded on Jekyll Island a week later. Both had washed ashore and were dead. Hawksbills encountered on the Georgia coast are likely to be Caribbean turtles that have moved north in the Gulf Stream. Unlike other marine turtles, hawksbills prefer to nest in maritime forest.

Reproduction and Development

Nesting is the best-known aspect of the hawksbill's biology because it can be observed on land. North Atlantic hawksbills nest primarily in tropical regions from May to November, but nesting records exist for every month of the year. Like other marine turtles, hawksbills have a keen homing ability, and females return to the region of their birth to nest. They generally migrate to their nesting beach every 2–4, rarely 5 or more, years. Hawksbills typically lay four to six clutches per season, at 2-week intervals, and construct nests in much the same way that the loggerhead sea turtles (*Caretta caretta*) that nest along the Georgia coast do. The eggs, averaging 155 per clutch, incubate for about 2 months before the hatchlings emerge. The sex ratio of a clutch is temperature dependent; individuals that incubate at higher temperatures develop as females, and lower temperatures produce males. Hatchlings stay in the egg chamber for several days before emerging and making a dash for the sea; during that time they stretch and absorb the remaining external yolk sac. Both nesting and hatching are nocturnal events. As is true of most long-lived animals that produce numerous young, hatchling survival in hawksbills is low. Reproductive age is unknown but has been estimated to be 20–30 years.

Habits

This international migrant can travel hundreds of miles between nesting and foraging grounds. Hawksbill sea turtles feed largely on marine sponges but supplement their diet with other marine invertebrates as well. The juveniles use their hawklike beak to pick chicken-liver sponges from crevices in the coral reef; adults feed in deeper waters on larger species such as basket sponges. Like most reptiles, hawksbills are most vulnerable to predators when they are young. On land, hatchlings encounter predators such as fire ants, crabs, and various mammals. When they reach the ocean, fish that inhabit the shallow waters eat them. Birds prey on hatchlings both on land and at sea. As a strategy to defeat their numerous shallow-water predators, hawksbills entering the water for the first time exhibit a behavior known as the swimming frenzy, during which they swim beyond the nearshore zone and subsist on their remaining yolk reserves instead of feeding. Little is known about the habits of young hawksbills after they leave the natal beach. When they reappear at feeding grounds, usually at 20–30 cm (7.8–11.7 in) in length, sharks and humans are their primary predators. Large size and a hard shell are the adult hawksbill's principal defenses.

Hawksbills live a solitary existence; their interactions with other individuals are limited to encounters with other foragers at the feeding grounds. They probably mate during these and other chance encounters and store sperm for the nesting season. Hawksbills sleep underwater, wedged beneath rocks or coral ledges.

Conservation Status

Hawksbills are state and federally listed as Endangered. Their ornate scutes are the source of tortoiseshell, otherwise known as carey or bekko. Tortoiseshell harvest has been the chief reason for the decline of the species in the last century. International trade of hawksbills is now banned, and each country regulates domestic use of eggs, meat, and shells. Additional threats include the degradation of nesting and foraging habitats through beachfront development and the bleaching of coral reefs, and loss from incidental catch (e.g., long-lines) and boat collisions.

Kimberly M. Andrews, Peri A. Mason, and James I. Richardson

Subadult Kemp's ridley sea turtle, Glynn County (Adam Mackinnon)

Kemp's Ridley Sea Turtle

Lepidochelys kempii

Description

The Kemp's ridley is the smallest of the marine turtles. Adults have a carapace length of 52–75 cm (20.3–29.3 in; average = 65 cm, or 25.4 in) and weigh 36–45 kg (80–100 lb). The round, flattened carapace and flippers are gray to olive green on top and lighter on the underside. The foremost costal scute on each side is reduced and touches the nuchal scute. Of the other marine turtles found in Georgia, only the loggerhead (*Caretta caretta*) has a similar characteristic. Kemp's ridleys differ from loggerheads in their greener coloration and four rather than three bridge scutes. Additionally, each bridge scute on a Kemp's ridley bears a conspicuous pore that loggerheads lack. The carapace of adult loggerheads typically has numerous epibionts attached to it, including barnacles, algae, and bivalves; by comparison, Kemp's ridleys have a smoother and cleaner shell. The charcoal-colored hatchlings are 40–50 mm (1.6–2.0 in) in carapace length and weigh approximately 17 g (0.6 oz).

Taxonomy and Nomenclature

Kemp's ridley was first placed in the genus *Thalassochelys* but was later moved into the genus *Lepidochelys* along with the olive ridley (*L. olivacea*) of the Pacific, the only other member of the genus. Until recently, some researchers considered Kemp's ridley to be a subspecies of the olive ridley, but genetic analysis has shown the two to be distinct species. No subspecies are currently recognized. Some say that fishermen took to calling these turtles ridleys because they represented a riddle: did their resemblance to the loggerhead mean that they were hybrids between a loggerhead and another sea turtle species? These turtles are also known as Atlantic ridleys.

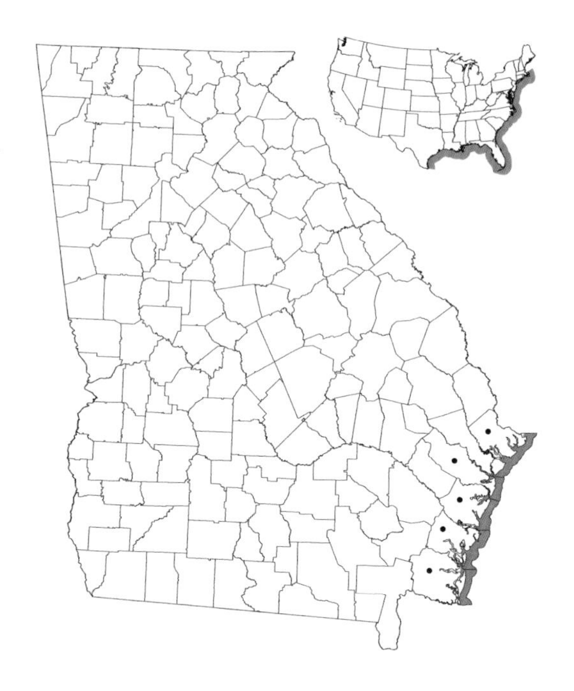

Distribution and Habitat

These turtles inhabit shallow waters, around 200 m (656 ft) deep, in the western Atlantic. Juveniles forage in estuaries, river mouths, bays, and lagoons from Georgia to New England, and adults may venture as far south as Colombia. Unlike most sea turtle species, Kemp's ridleys nest in a restricted area. Although nesting occurs in a few other places on the Gulf coasts of Mexico and Texas, the vast majority of females nest each year on one beach near Rancho Nuevo in Tamaulipas, Mexico. In Georgia, Kemp's ridleys are usually observed as stranded animals that drowned in shrimp trawl nets. Live Kemp's ridleys may be found in the crab-rich muddy or sandy areas around Georgia's barrier islands.

Reproduction and Development

Females make an annual breeding migration to the waters near Rancho Nuevo, where they encounter the typically nonmigratory males. Their reproductive behavior differs from that of other marine turtles in that egg-laying is synchronized; all of the females in the population come ashore to nest within the span of a few days. This phenomenon is known as an *arribada* (Spanish for "arrival"). Nesting is also unusual because it happens during the day; other sea turtles nest at night. Kemp's ridley *arribadas* occur from April to August at intervals of about 28 days. Some researchers have speculated that this relatively long interval allows females to synchronize their next nesting period with other females in the population. The average clutch size is 110 eggs, and females may lay up to four clutches per year. Eggs take 45–58 days to incubate, and incubation temperature determines the sex of the hatchlings. Kemp's ridleys are thought to reach reproductive maturity at 10–16 years of age.

Habits

Juvenile Kemp's ridleys feed largely at the surface on seaweed, pelagic crabs, and mollusks; when they reach a length of about 20 cm (7.8 in) they become bottom feeders. Crabs and mollusks remain their preferred foods, but like most turtles, they feed opportunistically. Satellite-tracking data suggests that Kemp's ridleys may establish residence at a feeding area between nesting seasons. Offspring are produced in large numbers but have many predators early in life, especially birds and fish. Adults rely on their size as defense against sharks, their primary predators.

Conservation Status

Kemp's ridleys are state and federally listed as Endangered. Historically, nesting en masse during daylight hours made the females an easy target for predators, including humans who overexploited the available resource. The estimated 40,000 Kemp's ridleys that nested near Rancho Nuevo in 1947 had dropped into the hundreds by the mid-1980s. However, conservation efforts—including guarding the nesting beach from poachers, protecting eggs in hatcheries, head-starting hatchlings, and translocating eggs to Padre Island, Texas, in hopes that hatchlings would imprint on that beach—have provided hope for the species. Much of this effort continues today, and the population is growing. Despite the use of turtle excluder devices on shrimp trawls, however, the number of Kemp's ridley strandings on the southeastern coast continues to grow.

Peri A. Mason and Kimberly M. Andrews

Snapping Turtles

FAMILY CHELYDRIDAE

The five species of snapping turtles, including the two found in Georgia, have a broad, flattened carapace and a relatively small plastron. The head is large, and the tail is the longest of any living turtle family. Georgia's alligator snapping turtle (*Macrochelys temminckii*), which may reach a carapace length of 80 cm (31.2 in), is one of the world's largest freshwater turtles. Four of the five species occur in the Americas; the fifth lives in southeastern Asia and is sometimes placed in a different family. All are aquatic. Georgia's species live on the bottoms of lakes, ponds, swamps, and relatively quiet sections of streams. Snapping turtles get their common name from what appears to be their perpetual bad mood when they are in contact with humans.

Juvenile common snapping turtle, Tattnall County (Dirk J. Stevenson)

Common Snapping Turtle

Chelydra serpentina

Description

Common snapping turtles are large, aquatic turtles. Adults reach a carapace length of 30–49 cm (11.7–19.1 in) and weigh 4.5–34 kg (10–75 lb). The carapace is medium to very dark brown, sculptured, and serrated on the rear edge. The plastron is yellowish to creamy white and much reduced in size, with more skin and less shell underneath than most other Georgia turtles have. The long, thick, tapering tail is almost as long as the carapace and has a jagged ridge down the center. The wide, sturdy limbs have long, thick claws that deserve respect when these turtles are handled. The feet are noticeably webbed. Males are larger than females and have a thicker tail and a slightly concave plastron. The neck is long, and the large head is usually mottled with dark brown. Small protuberances of skin called tubercles adorn the head and neck; these tubercles may be longer and more extensive on individuals found in extreme south-central Georgia. Common snappers can be distinguished from alligator snapping turtles (*Macrochelys temminckii*) by the latter's row of supramarginal scutes on each side midway around the shell between the marginal and costal scutes (see figure, p. 461). Furthermore, the tail of alligator snappers does not have a jagged ridge, the snout is extended and strongly beaked, and

Adult common snapping turtle, Taylor County (John B. Jensen)

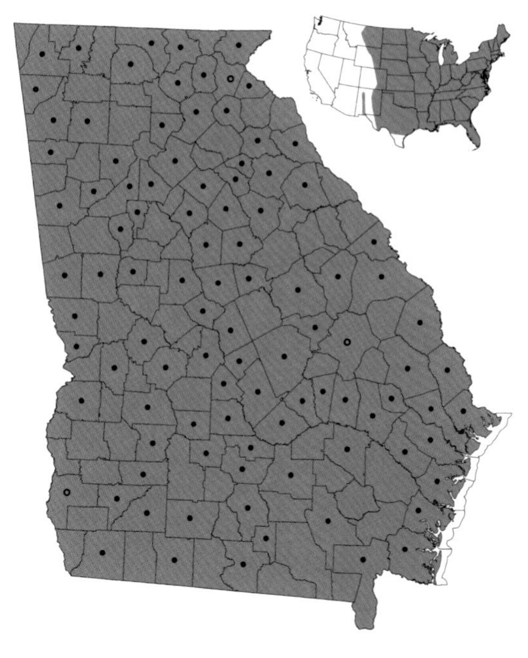

three prominent ridges run the length of the carapace and remain throughout adulthood. Juvenile common snappers, which are approximately 20–30 mm (0.8–1.2 in) at hatching, have three subtle ridges down the carapace that become obscure with age.

Taxonomy and Nomenclature

The Florida snapping turtle (*C. s. osceola*) may occur in extreme south-central Georgia; all other common snappers in the state belong to the subspecies known as the eastern snapping turtle (*C. s. serpentina*). Many local people refer to any snapping turtle as a loggerhead. Large common snapping turtles are sometimes mistakenly referred to as alligator snappers.

Distribution and Habitat

Common snappers can be expected to turn up in almost any permanent or semipermanent aquatic habitat in Georgia, including rivers and streams, swamps, isolated wetlands, large reservoirs, and farm ponds. They can even tolerate brackish water for short periods. Adult females are often encountered on land during the nesting season in late spring, and juveniles and adults of both sexes can be found moving overland during any warm season of the year.

Reproduction and Development

Common snapping turtles mate in the spring in Georgia and usually lay 20–50 or more round, leathery-shelled eggs sometime from late April through mid-June. Females typically excavate their nests in open areas with relatively loose soil. The eggs take 55–95 or more days to develop and hatch, but the hatchlings may overwinter in the nest if weather conditions are unfavorable and emerge between February and May. Like many other turtles, common snappers have temperature-dependent sex determination; that is, the incubation temperature of the eggs determines the sex of the hatchlings. Common snapping turtles may take 10–15 years to reach sexual maturity and may live for more than 40 years.

Habits

Common snappers are often found out of the water and can occasionally be found basking on logs or other structures. They are omnivorous and eat vegetation as well as just about any animal they encounter in their aquatic habitat, including aquatic adult and larval insects, crayfish,

fish, amphibians, snakes, smaller turtles, birds, mammals, and carrion. The eggs and hatchlings often fall prey to raccoons, mink, skunks, and various birds. Adults have few predators other than humans, but American alligators (*Alligator mississippiensis*) and black bears are known to eat them. Common snapping turtles defend themselves with very quick and powerful snaps and can give a dangerous and nasty bite if not approached with proper caution. Males are territorial and are occasionally seen fighting.

Conservation Status

Common snapping turtles are widely distributed in the state and are still common in many areas because they tolerate a broad range of habitats. Humans often kill common snappers in the mistaken belief that they damage game fish and waterfowl populations, and snappers are harvested personally and commercially for food. The degree to which these factors affect populations in Georgia is unknown.

Judith L. Greene

Adult alligator snapping turtle, Decatur County (John B. Jensen)

Alligator Snapping Turtle

Macrochelys temminckii

Description

The alligator snapper is the largest freshwater turtle in the Americas. Males may weigh in excess of 90.5 kg (200 lb) and have carapace lengths up to 80 cm (31.2 in); females are much smaller, rarely exceeding 22.5 kg (50 lb). The broad, rough, brownish carapace has three prominent jagged ridges along its length; the keels of older individuals are often worn and less jagged. The presence of one to five—usually two or three—supramarginal scutes is unique to this species. The cross-shaped plastron is reduced compared with that of most turtles. The enormous head is triangular when viewed from above and has an extended and strongly hooked upper jaw. The eyes are on the sides of the head and are encircled by a row of fleshy projections. Larger, more widely spaced tubercles are present elsewhere on the head, chin, and neck. The skin is typically brown, although many older individuals, especially males, have pale yellowish skin, more so on the head than elsewhere. Very faint stripes are present on the neck of younger turtles, which average 36–42 mm (1.4–1.6 in) in carapace length at hatching. The long tail is about as long as the carapace and has three rows of rounded, hardened tubercles along its length. The tubercles on the tail of the similar common snapping turtle (*Chelydra serpentina*) are distinctly jagged. Common snappers also differ by having eyes positioned higher on the head, a shorter snout, no supramarginal scutes, and much less conspicuous ridges on the carapace.

Underside of adult male alligator snapping turtle, Thomas County (John B. Jensen)

Adult alligator snapping turtle, Quitman County (John B. Jensen)

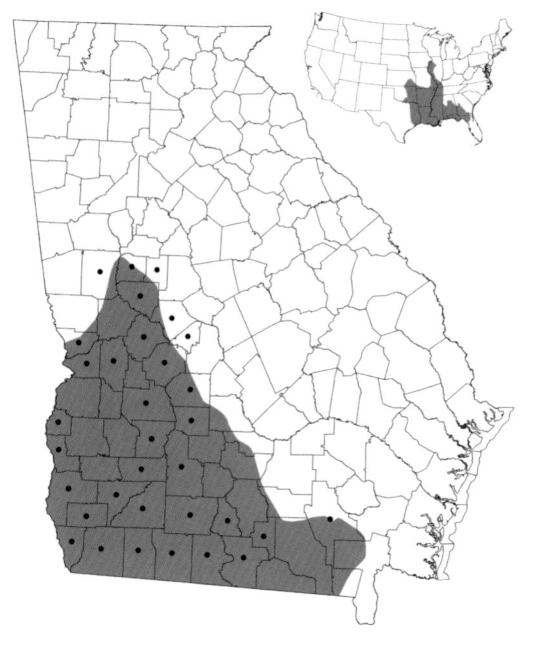

Taxonomy and Nomenclature

The genus was recently changed from *Macroclemys* to *Macrochelys*. Although not recognized as taxonomically distinct, populations inhabiting the Suwannee River drainage are genetically divergent from those found elsewhere. Many people call these turtles loggerheads or alligator turtles.

Distribution and Habitat

Alligator snappers occur in stream and river systems that drain into the Gulf of Mexico, primarily those of the Coastal Plain. Recent Flint River records well upstream into the Piedmont suggest that the full range of the species is still being determined. Large impoundments of these rivers also contain alligator snapper populations. Microhabitats within streams include water beneath or within logjams, undercut banks, rock (primarily limestone) shelters, and deep holes.

Reproduction and Development

Mating takes place in the water during winter or early spring, followed by an April–June nesting season. Nests are often excavated in riverbanks during early morning hours, but two reported from Georgia were found in small agricultural fields, one more than 60 m (197 ft) from the nearest stream. Mature females produce only one clutch of up to 60 round, leathery eggs each year and may nest only every other year. Hatchlings, whose sex is determined by

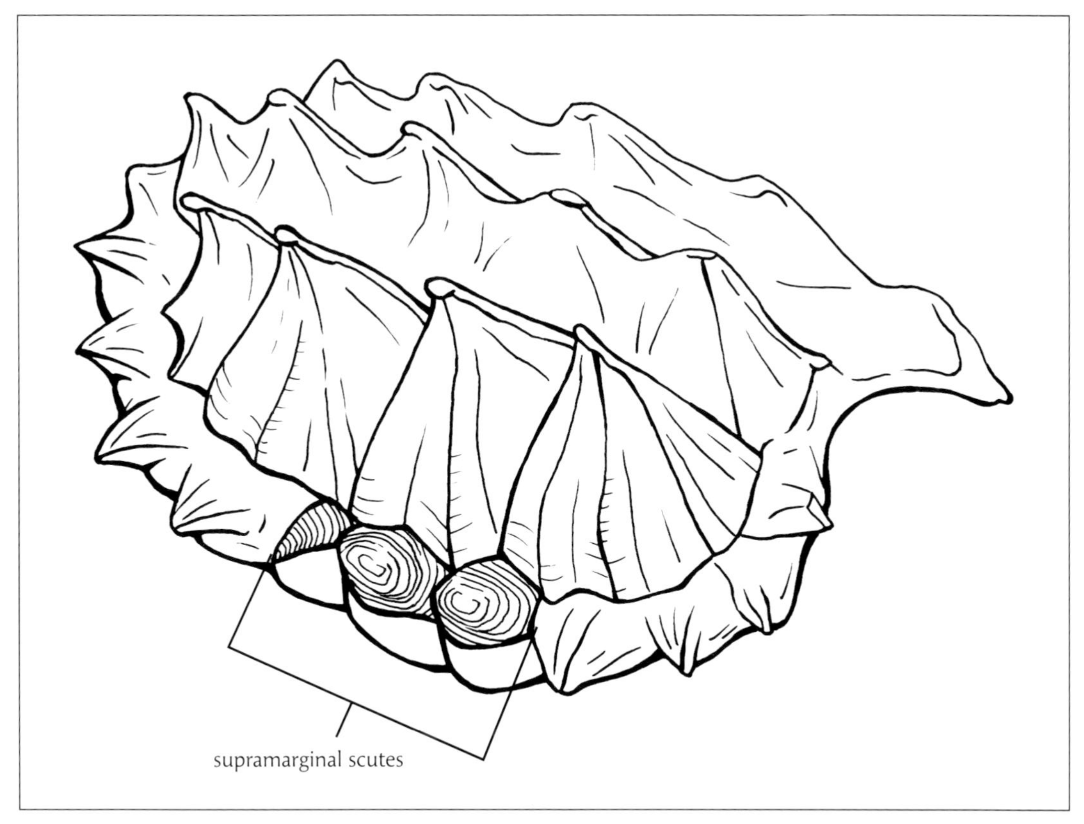

Supramarginal Scutes

the incubation temperature of the eggs, emerge 2.5–3.5 months later. Sexual maturity is reached in approximately 11–13 years.

Habits

Alligator snapping turtles are well known for the wormlike lure inside their bottom jaw that entices unsuspecting fish into the turtle's mouth. The turtle lies otherwise motionless with its mouth wide open and twitches this appendage back and forth. When a fish nips at the "worm" or touches the inside of the mouth, the powerful jaws slam shut with a force that can exceed 453 kg (1,000 lb). Fish are a major component of the diet, but smaller turtles, snakes, young American alligators (*Alligator mississippiensis*), amphibians, waterbirds, crayfish, mollusks, carrion, and even some plant materials that have fallen into the water such as acorns, palmetto berries, and wild grapes are also eaten. Feeding and most other activities are primarily nocturnal. Individuals may wander considerable distances upstream in search of the source of food odors drifting downstream. Adult alligator snappers have no predators other than humans. Raccoons often raid the nests and eat the eggs. Large fish, wading birds, river otters, and other carnivorous animals may take hatchlings. This species is the most aquatic nonmarine turtle in the country, almost never leaving the water except to nest. Individuals forcibly brought to land have a difficult time walking but defend themselves admirably by keeping their jaws wide open and ready to snap.

Conservation Status

Because of their large size and readiness to enter baited traps, alligator snapping turtles were harvested in large numbers by commercial trappers in the 1970s and early 1980s to provide meat to soup canneries. One commercial trapper removed an average of 453 kg (1,000 lb) of alligator snapping turtles per day from the Flint River drainage before the species was given legal protection. As is the case with many other commercially harvested turtle species that exhibit delayed maturity and naturally low reproductive rates, many populations of alligator snapping turtles, including the Flint River one, experienced significant declines. This prompted the legal protection of the species throughout most of its range. Alligator snappers are now considered Threatened in Georgia.

John B. Jensen

Leatherback Sea Turtle

FAMILY DERMOCHELYIDAE

This family has a single distinctive species that is easily recognized by its streamlined shell covered in leathery ridges rather than the horny scutes characteristic of most other turtles. The leatherback is the largest living turtle, averaging approximately 150 cm (58.5 in) in carapace length. As is true of all sea turtles, its front limbs are modified into powerful propulsive flippers. This species, which occurs throughout temperate and tropical oceans of the world, is the most nautical of the sea turtles. Individuals migrate across entire oceans in pursuit of their favorite food, jellyfish.

Nesting adult female leatherback sea turtle, Liberty County (Matt O'Connor)

Leatherback Sea Turtle

Dermochelys coriacea

Description

Adult leatherback sea turtles are the largest turtles in the world. Carapace lengths up to 180 cm (70.2 in) have been recorded. Most of those found along the Georgia coast weigh 400–450 kg (884–995 lb); the species record is 912 kg (2,016 lb). Females presumably get larger than males. Leatherbacks differ from other sea turtles in having a leathery covering of thick skin on the shell with seven distinct longitudinal ridges along the streamlined carapace, which ends in a rounded point at the rear. The shell and body are black with whitish spotting; the chin and undersurfaces of the flippers are lighter, reflecting fat body concentrations. Leatherbacks have flippers on all four feet but differ from other sea turtles in having no claws. Hatchlings resemble adults and are the largest among the sea turtle young with a carapace length of 68–88 mm (2.7–3.4 in) and a body weight of 45 g (1.6 oz). Hatchlings have beaded scales that are later lost.

Taxonomy and Nomenclature

The leatherback sea turtle is the only living species in the genus and the only genus in the family. No subspecies are recognized.

Distribution and Habitat

Leatherback sea turtles have the largest geographic range of any reptile in the world. These pelagic animals migrate through all of the

Leatherback sea turtle hatchling, Glynn County (Brad Winn)

Adult leatherback sea turtle, offshore of Georgia coast (Bruce Hallett)

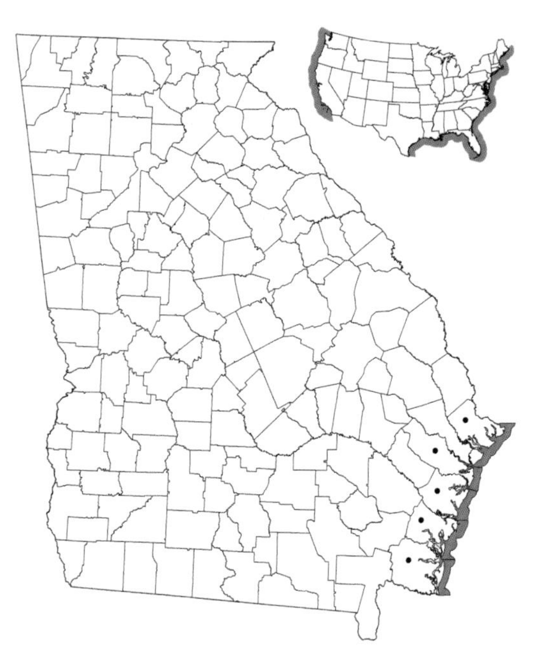

oceans between the Arctic and the Antarctic. After reaching approximately 110 cm (43 in) in carapace length, they are found in upwelling areas where jellyfish are concentrated. Leatherbacks are occasionally sighted swimming off the Georgia coast, and dead ones have washed ashore on Georgia beaches. Most leatherbacks associated with Georgia are migrating north along the coast in the spring and south in the fall.

Reproduction and Development

Leatherback sea turtles generally nest on beaches in the tropics, although limited nesting occurs on temperate beaches in the western Atlantic. The leatherback is such a widely distributed nester that it can be found nesting somewhere in the world almost any time of the year. Between 1996 and 2004, nine leatherback nests were documented on Georgia's beaches. Females generally lay between 50 and 125 eggs, with an average of about 85, and an individual may lay six or seven clutches in a season at intervals of about 1–2 weeks. A female usually does not nest again for 2 or 3 years. The eggs are larger and heavier than those of other sea turtles; most are more than 50 mm (2 in) in diameter and weigh approximately 80 g (2.8 oz). In addition to the viable eggs, the clutch also contains some smaller, round, yolkless eggs whose purpose is not clear. Females in the act of laying cover the eggs with their rear two flippers, a behavior that is believed to have evolved to protect the eggs from predators. The normal incubation period for eggs in the nest is about 60–70 days, but they may take longer to hatch if the nest is shaded and cool. The sex of the hatchlings is determined by the temperature of the eggs during incubation. Leatherbacks are known to mate offshore and may mate en route to nesting grounds from foraging areas. Leatherback males are aggressive maters, and they readily bite females while trying to initiate copulation. The age at maturity has not been determined.

Habits

Leatherbacks migrate thousands of miles—longer distances than the other sea turtle species travel—and store a large amount of fat when in the colder regions of their range. This species is unique among the sea turtles for its ability to thermoregulate; individuals maintain a relatively constant warm internal temperature. Leatherbacks dive to depths up to 1,200 m (3,936 ft). The thick-skinned shell compresses as the turtle goes deeper, thereby adjusting to the increasing pressure of the ocean's depths. Adults are noted for their specialized diet of large jellyfish, but they also eat fish and other sea creatures they encounter. The throat and gut are extremely long and are lined with backward-facing spicules that force large amounts of jellyfish down the throat and prevent them from washing back out as ingested water is expelled. Leatherback sea turtles are so large that they have few marine predators, although they could be prey for sharks or for large terrestrial mammals on nesting beaches where they cannot move quickly. In some parts of the range, humans kill them for eggs, meat, and oil. The leatherback does not accumulate epibionts (e.g., barnacles) on its shell as many of the other sea turtles do. Mild secretions keep the carapace clean, and such organisms cannot attach to the soft shell.

Conservation Status

Leatherbacks are listed as Endangered both by Georgia and by the federal government, with commercial fisheries being the major threat to recovery. In some years leatherbacks are found in high densities in state territorial waters and are incidentally captured and drowned in shrimp trawls. Shrimp fishermen are required to use turtle excluder devices (TEDs) designed to exclude adult leatherbacks, but illegal or improperly installed TEDs remain a problem in some areas. Worldwide, pelagic long-line fisheries are thought to be the primary source of fishery-related mortality. Leatherbacks suffer from widespread environmental destruction due to beachfront development and both terrestrial and marine pollution. Plastic bags floating in the ocean resemble jellyfish and may cause intestinal blockage and death if the turtles ingest them. Other threats include entanglement in plastics, depredation of nests by raccoons and feral hogs, and collisions with boat propellers.

Kimberly M. Andrews and Whit Gibbons

Common Water Turtles

FAMILY EMYDIDAE

The 13 species found in Georgia are small to medium-sized turtles that have a smooth, oval, moderately domed shell and a large plastron. Most have a carapace length less than 20 cm (7.8 in). Largely of northern temperate latitudes, this family includes approximately 40 species distributed primarily across North America and Eurasia. Although a few species, including the eastern box turtle (*Terrapene carolina*), are primarily terrestrial, most emydid turtles live in lakes, ponds, and streams. These are the familiar water turtles often seen basking on fallen trees or floating logs.

Adult painted turtle, Habersham County (John B. Jensen)

Painted Turtle

Chrysemys picta

Description

The painted turtle is a relatively small aquatic turtle with a maximum carapace length of approximately 25 cm (9.8 in). Adult females grow larger than males. The oval, somewhat flat carapace is smooth, keel-less, has unserrated margins, and ranges in color from black to olive. The marginal scutes have prominent red markings. Individuals found in west-central Georgia may have a narrow red line down the center of the carapace. The seams between the large scutes are aligned in most Georgia specimens, forming straight rows across the back, and the scutes are bordered with light olive. Painted turtles in extreme northwestern Georgia may not have the olive border on the scutes, and the seams do not form a straight row across the back. The hingeless plastron is typically yellow but may also be tinged with pink or orange or stained with dark reddish brown pigment. A black or reddish brown spot, which varies in shape and size, is often present in the center of the plastron. Large plastron blotches are more prevalent on painted turtles in northwestern Georgia than in other parts of the state. The skin is black to olive with yellow and red stripes on the neck, legs, and tail. The head and chin have yellow stripes, and a large yellow spot is present behind each eye. The eye is yellow, and a dark stripe extends through the pupil. The upper jaw has two tooth-like notches. Adult males can be differentiated from females by the long claws on their front feet and their long, thick tail with the vent placed beyond the rear edge of the carapace. Females have a smaller tail with the vent inside the rear edge

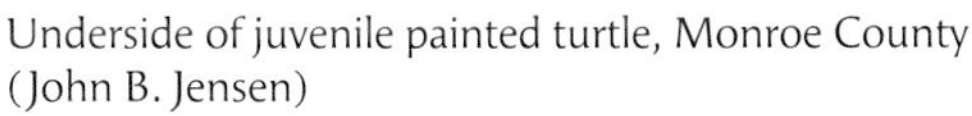

Underside of juvenile painted turtle, Monroe County (John B. Jensen)

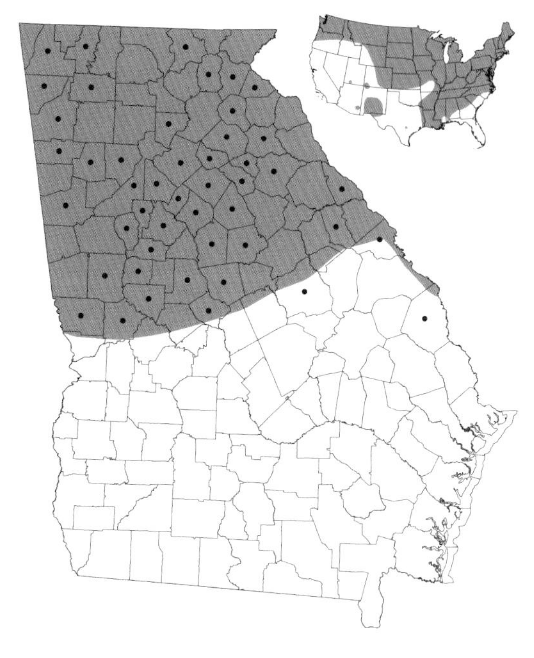

of the carapace. Juvenile painted turtles have a relatively longer head and tail and a deeper and more rounded shell than adults. Hatchlings, approximately 27–30 mm (1.1–1.2 in) in carapace length, often have a slightly keeled carapace and are more brightly colored than adults.

Taxonomy and Nomenclature

Four subspecies are currently recognized; two of these are in Georgia: the eastern painted turtle (*C. p. picta*) is found throughout most of the northern half of Georgia, and the midland painted turtle (*C. p. marginata*) is found in the northwestern corner of the state. Individuals found in areas where the ranges of these subspecies meet may have physical characteristics of both subspecies. Painted turtles from west-central Georgia may share characteristics with a third subspecies, the southern painted turtle (*C. p. dorsalis*), which some herpetologists consider a separate species.

Distribution and Habitat

Painted turtles are found in the Piedmont and mountains of Georgia; only a few records are known from below the Fall Line. They prefer aquatic habitats with soft muddy bottoms, shallow water, and an abundance of aquatic vegetation, such as farm ponds, lakes, freshwater marshes, oxbows, reservoirs, and slow-moving rivers. Seasonal wetlands, roadside ditches, and fast-moving streams and rivers may also be temporarily occupied.

Reproduction and Development

Painted turtles mate in the late winter and spring. Females lay clutches of 2–14 eggs in May, June, and July in open areas, preferably with sandy or loamy soils, relatively close to aquatic habitats. Nesting occurs primarily in the late afternoon and early evening, often following a thunderstorm. Nests are dug with the hind feet and are usually less than 100 mm (3.9 in) deep. Some females lay two or more clutches per nesting season, but not all females nest every year. The elliptical eggs are white, cream, or pinkish; have a smooth, slightly pitted surface; and are somewhat flexible. Eggs hatch in approximately 1.5–2.5 months. The temperature of the nest determines the sex of the hatchlings. Nest temperatures below 27° C (81° F) usually produce all males, and nest temperatures above 30° C (86° F) result in all females. Hatchlings often spend the winter in the nest and emerge the following spring. Young painted turtles grow rapidly, often doubling in size within their first active season. Males reach sexual maturity when they are 2 or 3 years old, but it takes females 6–10 years to mature.

Habits

Painted turtles in Georgia are most active from mid-February to late November, although individuals may be active during warm weather throughout the year. They do not eat until the water temperature reaches approximately 15.5° C (60° F). Painted turtles are primarily diurnal and spend a great deal of time basking. Individuals of all age classes may pile up on logs, rocks, or other favorite basking spots. During the late winter and spring, they bask throughout the day, but summertime basking is generally restricted to the early morning and late afternoon. Painted turtles eat a variety of plants and animals, including algae, duckweed, aquatic invertebrates, fish, and frogs. Hatchlings and juveniles are thought to be more carnivorous than adults. Raccoons are the most important predators of painted turtles, as they raid nests and consume both young and adults. Other predators include foxes, skunks, herons, and birds of prey. When captured, painted turtles usually attempt to escape by scratching or biting.

Conservation Status

Painted turtles are fairly abundant throughout their range in Georgia, even in urban areas. Populations have probably benefited from the construction of farm ponds.

Steven J. Price

Adult spotted turtle, Laurens County (Suzanne L. Collins, CNAH)

Spotted Turtle

Clemmys guttata

Description

Among the smallest freshwater turtles in North America, female spotted turtles rarely exceed 13 cm (5.1 in) in carapace length and 240 g (8.4 oz) in weight. The smooth black carapace has unserrated margins and lacks a keel, and each scute usually has several randomly spaced yellow to cream spots. Each scute on the light yellow to orange plastron has a large black blotch positioned toward its outer edge. The head, neck, and limbs are gray to black with yellow spots on the upper surface and light yellow to salmon red markings below. Females are often slightly longer than males and have a flat plastron, shorter tail, and peculiar color pattern on the head. Their chin and upper jaw are yellow to orange, and their eyes are orange, a trait hatchlings normally have as well. Males have a slightly concave plastron; a long, thick tail; brown eyes; and a tan chin that may become white with age. Hatchlings, which are 28–31 mm (1.1–1.2 in) in carapace length, have a less random arrangement of yellow spots on the carapace. Typically they have only one spot on each scute except for the nuchal scute, which has none.

Taxonomy and Nomenclature

Some taxonomists have suggested subdividing the genus into three genera with only the spotted turtle keeping the name *Clemmys*. No subspecies are currently recognized, and no significantly divergent populations are known.

Adult spotted turtle, Emanuel County (John B. Jensen)

Juvenile spotted turtles, Aiken County, South Carolina (Thomas Luhring)

Adult spotted turtle, McIntosh County (Giff Beaton)

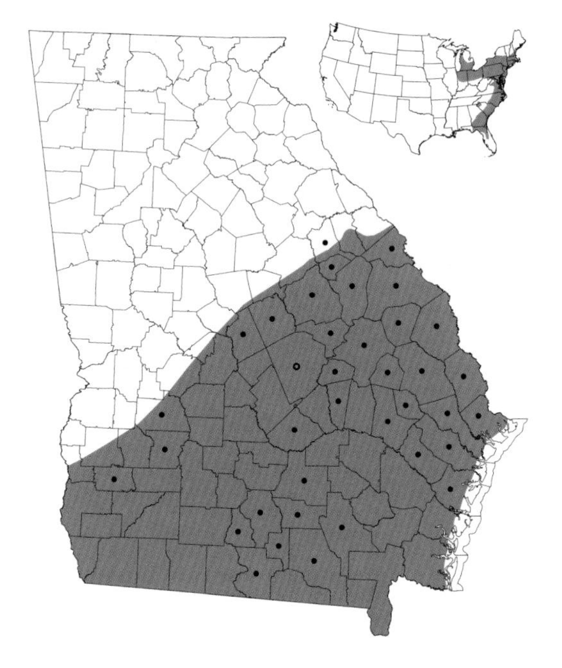

Distribution and Habitat

Spotted turtles are confined to the Coastal Plain in Georgia and are found in a variety of shallow wetlands. Habitats include wooded swamps and streams, marshes, fens, bogs, Carolina bays, and cypress-gum ponds. Slow-moving water with abundant structure including standing trees and shrubs; fallen trees; submerged aquatic vegetation; and emergent sedge, rush, and grass tussocks is preferred.

Reproduction and Development

Spotted turtles often congregate in forested pools to mate immediately following emergence from hibernation in late winter, usually in February–April. Females lay one to three clutches of two to six leathery, elliptical eggs from late April through June. Females often dig their shallow nests in direct sunlight in the sandy loam of wetland edges. Just as often, though, eggs are deposited atop moist sphagnum hummocks, on moss-covered logs, and on grass or sedge tussocks. Hatchlings in Georgia, like those in other southern populations, probably emerge from the nest 60–80 days later. Those from northern populations often overwinter in the nest and emerge the following spring. The sex of hatchlings is determined by the temperature during incubation. Both sexes reach sexual maturity at between 7 and 10 years of age. Captive individuals have lived for more than 30 years.

Habits

Spotted turtles are shy, secretive creatures often seen only as a faint flash of orange or yellow as they drop off a log or swim across the bottom of a blackwater pool. They are among the earliest of the turtles to emerge in the spring, and in the southern portion of the range may be seen basking on warm, sunny days in February. Spotted turtles in Georgia are active primarily during late winter and the cooler parts of the spring, becoming dormant only on the coldest winter days and the warmest days of late spring, summer, and early fall. Feeding and basking can occur during all daylight hours in the cooler months but is generally confined to the morning during warmer periods. Whether inactive or just secretive, spotted turtles are almost never seen during late spring or summer in Georgia. Spotted turtles are primarily carnivorous and eat nearly all types of insects, both aquatic and terrestrial, along with snails and slugs, tadpoles, salamander larvae, and aquatic crustaceans such as crayfish and isopods. Plant material such as algae and aquatic grasses and carrion of all types are also eaten. Known predators include raccoons, skunks, bald eagles, and common snapping turtles (*Chelydra serpentina*), but other carnivorous mammals such as coyotes, foxes, mink, and river otters probably take adults. Hatchlings and small juveniles are vulnerable to these and other predators such as large fish and wading birds. Raccoons, foxes, skunks, other small mammals, and crows raid nests and eat the eggs. Spotted turtles rarely bite and rely on their secretive behavior, cryptic coloration, and ability to hide quickly to avoid predators.

Conservation Status

Because of their striking coloration and mild disposition, spotted turtles have long been popular in the national and international pet trade, and the commercial take appears to have grown in recent years. Unknown numbers of populations are in decline or have been eradicated by road mortality and outright habitat loss as wetlands are bisected by roads or drained for development. Although few states have historically protected this species, some states have halted or now regulate the sale of spotted turtles and protect their habitat. Little is known of the status of the spotted turtle in Georgia, and it has thus been given special protection as Unusual by the state.

Thomas S. B. Akre and Kenneth M. Fahey

Adult bog turtle, Union County (John B. Jensen)

Bog Turtle

Clemmys (Glyptemys) muhlenbergii

Description

The bog turtle is the smallest turtle in North America. Adults average 7.5–9 cm (2.9–3.5 in) in carapace length; the maximum length recorded is 11.4 cm (4.5 in). The orange (sometimes yellowish, reddish, or pinkish) blotch on the head behind each eye is the best character for identifying this species. This blotch usually surrounds the external ear, but the ear itself is often devoid of orange pigment; it may become smaller with age. The head and neck may have small, faint reticulations of orange pigment, and the front legs also have varying amounts of orange, which increases in intensity closer toward the body. The feet are weakly webbed. Bog turtles have a moderately domed carapace, and younger individuals have distinct concentric growth rings within each scute. Each of the larger carapace scutes may have a faint starburst of lighter yellowish to reddish pigment. Both the growth rings and the yellow pigment become indistinct with age and with the accumulation of iron deposits and staining from the sediments through which these turtles constantly burrow. The plastron is unhinged and is usually dark with yellowish blotches. The plastron is flat in females and is strongly concave in males. The tail of adult females is short, with the vent barely extending beyond the rear edge of the plastron. In males, the tail is long and muscular, with the vent well beyond the rear edge of the carapace. The shell length of hatchlings is 23–34 mm (0.9–1.3 in).

Juvenile bog turtle, Union County (Kenneth M. Fahey)

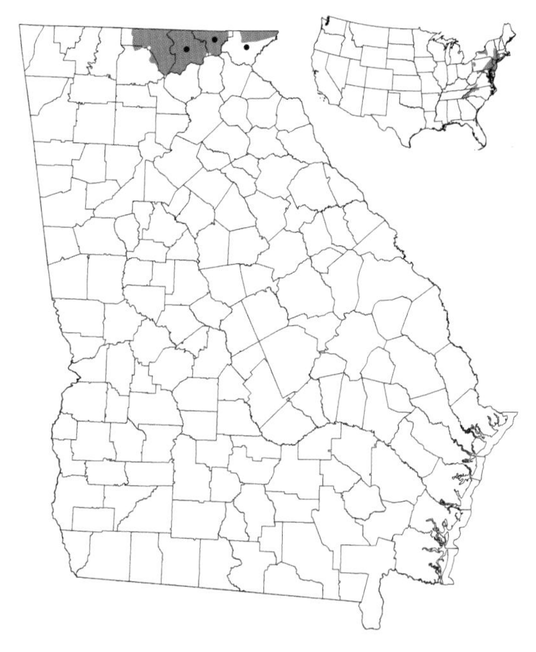

Taxonomy and Nomenclature

Some taxonomists have suggested placing the bog turtle along with the wood turtle (*G. insculpta*) in the genus *Glyptemys*. No subspecies are recognized.

Distribution and Habitat

Bog turtles, first discovered in Georgia in 1979, are confined to Blue Ridge wetlands above 750 m (2,460 ft) in elevation. Only eight populations have been discovered in the state, though more undoubtedly exist. Typical habitats include spring seeps, bogs, fens, wet meadows, marshes, and shallow swamps. Though the habitat type varies, the prerequisites are open wet areas with shallow water and deep, muddy sediments. Habitats capable of supporting a viable bog turtle population may be as small as an acre. Plants associated with these habitats include sedges, grasses, sphagnum, ferns, tag alder, swamp rose, and red maple.

Reproduction and Development

Male turtles roam widely in search of females shortly after becoming active in late March. Mating usually occurs in shallow water from late morning through mid-afternoon. Gravid females have been observed from late May through mid-July in Georgia. Nesting appears to occur predominantly in June during late afternoon and early evening. Nests are usually in high, compact mounds of sphagnum or in the center of sedge tussocks. One Georgia nest was found in a decomposing tree stump. Incubation takes between 52 and 60 days. Hatchlings emerge in late summer or fall and immediately burrow into the substrate. Estimates from scute growth rings indicate that maturity is reached in 7–9 years.

Habits

Bog turtles become active in mid-to-late March in Georgia, depending on weather conditions, and are active through late September to early October. During the active season bog turtles typically are on the surface early in the morning through mid-to-late afternoon. In early spring bog turtles come to the surface to bask as soon as there is sufficient sunlight. Later in the spring and summer, they are most active before the intense midday sun makes it too warm for them. During the warmest parts of the year bog turtles actively burrow below the surface.

They may use the shaded pathways dug by rodents beneath vegetation even during warmer months to remain active at the surface. It was once believed that bog turtles had a bimodal activity pattern and were active predominantly in the early spring and in the fall, but trapping and radiotelemetry studies have shown that the turtles are just difficult to find during late spring and summer when vegetation is dense. Turtles may become inactive if the summer months are extremely hot and dry. Because they are small and dark, bog turtles can soak up enough heat energy through basking to become active more quickly than most other turtles. They then remain active under the mud where they are not easy to see. In the fall, as temperatures become cooler, they spend more and more time underground, coming to the surface only on the warmest days. By mid-October the turtles have returned to a favored hibernation spot where they remain until mid-to-late March. Bog turtles prey on or scavenge a wide variety of invertebrates including insects, earthworms, crayfish, and millipedes; and vertebrates including frogs, salamanders, watersnakes, and mice; they also eat an array of aquatic vegetation, seeds, and berries. Common snapping turtles (*Chelydra serpentina*) as well as large mammalian predators such as raccoons, foxes, and skunks are reported to be predators.

Conservation Status

The bog turtle is currently protected in all states in which it occurs. Northern populations are federally listed as Threatened, and southern populations (including those in Georgia) are listed as Threatened Due to Similarity of Appearance. The state of Georgia currently lists the bog turtle as Endangered. Loss of habitat has been the major factor in bog turtle declines. Many of the wetlands that were prime habitat have been developed as agricultural land or dammed and flooded. Their desirability as pets has also contributed to the bog turtle's decline.

Kenneth M. Fahey

Adult chicken turtle, McIntosh County (Carlos D. Camp)

Chicken Turtle

Deirochelys reticularia

Description

The chicken turtle is a medium-sized, hard-shelled turtle. Females reach approximately 21 cm (8.2 in) in carapace length and grow to be larger than males, which reach 16 cm (6.2 in). Females also have a noticeably shorter, thinner tail. Unlike many other emydid turtles, male chicken turtles do not have longer foreclaws than females. The carapace color ranges from olive to brown to black with a distinct reticulated pattern of yellow lines. The carapace is relatively high domed, especially in females, is longer than wide, and is often pear shaped. The edges of the carapace are smooth and not serrate as those of pond sliders (*Trachemys scripta*) and cooters (genus *Pseudemys*) are. The plastron is pale yellow. Black markings on the bridge between the plastron and carapace, if present, often appear as a black bar or dots; the pattern is unique to each individual. Yellow striping is present on the head and legs. A characteristic broad, yellow stripe on the forelegs helps to distinguish the chicken turtle from similarly colored cooters and pond sliders. Chicken turtles have an extremely long neck; when fully extended the head and neck are approximately as long as the plastron. The head is somewhat flattened relative to that of cooters and pond sliders. Hatchling chicken turtles often have bright yellow carapace edges and vary widely in size from 25 to 34 mm (1.0–1.3 in).

Underside of adult chicken turtle, Long County (Dirk J. Stevenson)

Adult chicken turtle, Liberty County (Brad Winn)

Taxonomy and Nomenclature

The chicken turtle is the only member species of its genus. Of the three recognized subspecies, only the eastern chicken turtle (*D. r. reticularia*) is present in Georgia.

Distribution and Habitat

Chicken turtles are associated with seasonal wetlands such as sinkhole ponds, cypress swales, Carolina bays, and other isolated aquatic habitats of the Coastal Plain that are generally free of large predatory fish. They are only occasionally found in floodplain swamps and borrow pits, although these habitats are probably available in proximity to the preferred seasonal wetlands. Populations rarely exist in farm ponds, lakes,

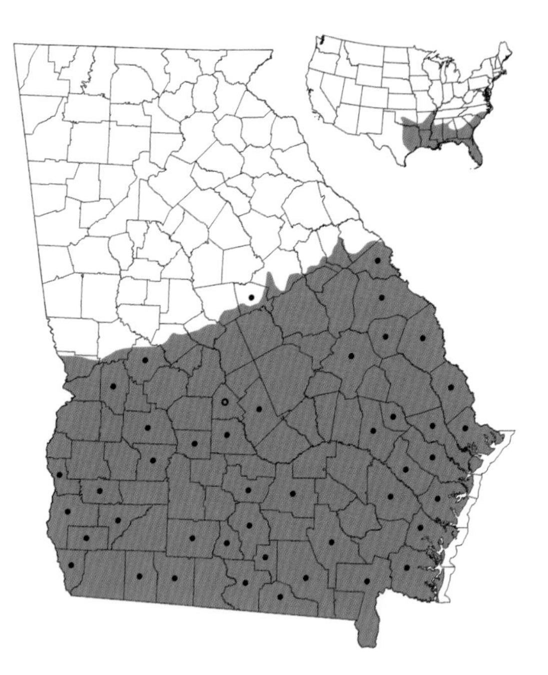

or other permanent wetlands. Ideal habitats include open-canopied seasonal wetlands that hold water from winter through summer, have populations of crayfish and aquatic insects, are surrounded by forest, and are near other similar wetlands.

Reproduction and Development

Chicken turtles have an unusual reproductive pattern. While most co-occurring pond sliders, Florida cooters (*Pseudemys floridana*), eastern mud turtles (*Kinosternon subrubrum*), and common musk turtles (*Sternotherus odoratus*) are nesting during April–June, chicken turtles are foraging on seasonally abundant crayfish and aquatic insects. Chicken turtle females delay nesting until at least early fall and often deposit clutches of eggs on exceptionally warm days between December and early March. Young females produce one clutch; older females generally produce two clutches per year. The number of eggs per clutch usually varies between 5 and 15. Eggs deposited during fall or winter do not begin developing until the following May when ground temperatures are again warm enough for incubation. The eggs hatch 75–90 days later in mid-August, and some hatchlings may emerge from the nest. More commonly, the hatchlings remain in the nest through the next winter, emerging and entering the wetland the following spring. The total time spent in the nest cavity can thus be as long as 18 months. Hatchlings grow rapidly during their first year, and most males are mature by their second year; females mature in 6 or 7 years. Long-term mark-recapture studies have shown that chicken turtles live no more than 15 years, while cooters and sliders may live in excess of 40 years.

Habits

Chicken turtles show strong fidelity to their seasonal wetlands. When these wetlands dry, the turtles bury themselves in loamy, well-drained soil and leaf litter in the surrounding forest. Although they make overland forays to other wetlands, they do not make long migrations to find water when their wetland is dry. They simply wait in the surrounding forest for the wetland to refill. Chicken turtles bask occasionally on logs, but less frequently than cooters and sliders do. They are entirely carnivorous throughout their lives, feeding on crayfish and aquatic insects, notably dragonfly and damselfly larvae. These turtles are more vulnerable to predation by American alligators (*Alligator mississippiensis*) than are cooters and pond sliders, which have thicker shells. River otters also eat chicken turtles.

Conservation Status

The chicken turtle is not listed or protected in Georgia. It is rarely the most abundant turtle in its habitat, and the widespread destruction of seasonal wetlands has most certainly caused a decline in numbers. Populations appear most robust where clusters of seasonal wetlands exist in close proximity in an unfragmented landscape. Timber and logging companies that clear-cut forests to the edges of wetlands eliminate the dry-season habitat that chicken turtles require. The narrow buffers of forest habitat left to protect water quality are insufficient to provide the necessary upland habitat component. Nonnative fire ants may have a significant impact on species such as the chicken turtle whose hatchlings remain in the nest for extended periods.

Kurt A. Buhlmann

Adult male Barbour's map turtle, Decatur County (John B. Jensen)

Barbour's Map Turtle

Graptemys barbouri

Description

Barbour's map turtle is a fairly large riverine turtle. The two sexes exhibit considerable size differences as adults. The carapace of males is generally less than 12 cm (4.7 in) long, while females may reach nearly 33 cm (12.9 in). The smallest adult males have a plastron length of 7 cm (2.7 in); the smallest mature females have a plastron 17.6 cm (6.8 in) long. Adult females also have a greatly enlarged head. Although males tend to retain much of the pattern seen in the young, large females darken considerably and may show little evidence of any pattern or markings. Individuals of all sizes, especially those from environments with clear water, may be partially or completely covered with algae that mask the color, pattern, and even the prominent carapace spines to various degrees. The carapace of some individuals is mostly greenish or olive brown and the pleural scutes have semicircular yellow markings that are often shaped like the letter C. Black-tipped spines or knobs punctuate the keel. As in other map turtles, the patterns and carapace spines are most evident on hatchlings and males. The upper surfaces of the marginal scutes have thin, yellowish, semicircular or C-shaped markings; the lower portions have dark markings on a light background. The plastron is mostly light, although a thin border of dark pigment is usually present along the transverse seams. The skin of the head, legs, and tail is dark olive to nearly black with light yellow or yellowish green markings. The pattern along

Adult female Barbour's map turtle, Decatur County (Andy Day)

Adult male Barbour's map turtle, Stewart County. Note the sharply keeled carapace. (John B. Jensen)

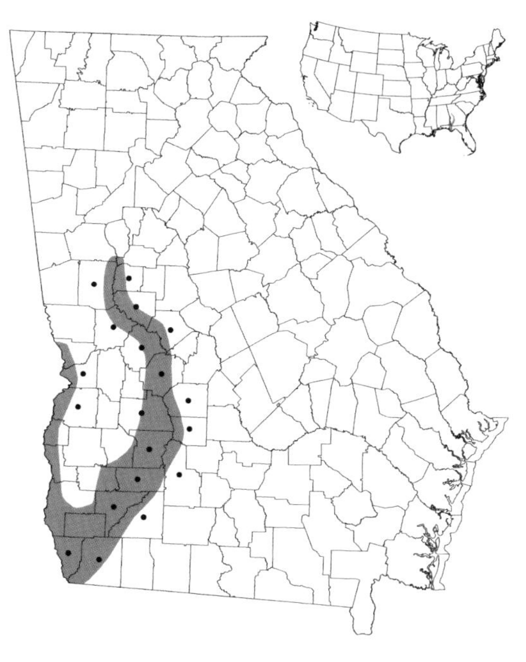

the top of the head consists of a large blotch between the eyes connected to another blotch immediately behind each eye. The feet are webbed. The brightly colored hatchlings are 32–38 mm (1.3–1.5 in) in carapace length. Unlike closely related Alabama map turtles (*G. pulchra*), Barbour's map turtles have a hollow heart-, U-, or Y-shaped area of dark pigment behind the eyes that contains a light concentric pattern. In addition, Barbour's map turtles have a transverse or curved, often comma-shaped, light bar on the underside of the chin rather than the longitudinally directed light bar found on Alabama map turtles.

Taxonomy and Nomenclature

No subspecies are recognized. Map turtles, including Barbour's, that have spiny-keeled shells are often referred to as sawbacks.

Distribution and Habitat

Barbour's map turtles inhabit streams of the Chattahoochee and Flint river drainages that are rich in mollusks and contain ample basking sites in the form of logs, snags, stumps, exposed rocks, and so on. In the Coastal Plain these turtles are characteristically found in clear, limestone-bottomed streams and large rivers that provide similar basking opportunities, but in the Piedmont the limestone is replaced by granite and in certain areas is more reminiscent of mountainous terrain.

Reproduction and Development

There are no reports of mating in wild turtles, although captives have been observed courting during the winter months. The nesting season is June–August in Florida and is probably similar in Georgia. In Florida, these turtles nest on open sandbars. Nest sites reported from Georgia include sandbars, an eroded riverbank, a river bluff, and the base of a man-made rocky embankment. Nests have been found up to 200 m (656 ft) from the water. The incubation temperature determines the sex of hatchlings. Eggs incubated at 25° C (77° F) produce males; those incubated at 30° C (86° F) produce only females. Males reach sexual maturity at 4 years. Females in Florida have been estimated to take from 15 to more than 20 years to reach sexual maturity.

Habits

Barbour's map turtles bask at temperatures as low as 10° C (50° F), and at least one basking Georgia specimen was recorded in late January. Even so, little winter activity takes place in Georgia. Males in a Florida population moved up- and downriver an average distance of 365 m (1,198 ft); females averaged only 273 m (895 ft). During periods of high water, turtles in this population migrated toward the flooded river perimeter where currents were weakest. Large Barbour's map turtles feed primarily on mussels and snails. Smaller individuals, including most males, eat softer-bodied invertebrates, especially caddisfly larvae. Natural predators such as raccoons are probably major contributors to the mortality of eggs and hatchlings.

Conservation Status

Barbour's map turtle is listed as Threatened in Georgia. Large numbers of adult females have been found dead along the Flint River during the past few years, possibly the result of pollution. The illegal so-called sport of plinking (shooting turtles off basking logs or rocks) may also threaten populations in Georgia. Although siltation has devastated native mussel populations in many streams within this turtle's range, nonnative Asian clams have invaded and provided abundant replacement forage.

Robert A. Moulis

Adult male common map turtle, Oregon County, Missouri (John B. Jensen)

Common Map Turtle

Graptemys geographica

Description

The common map turtle is a moderate-sized turtle. Females reach a carapace length of 27 cm (10.5 in); males are generally less than 16 cm (6.2 in) long. This is the only *Graptemys* species in Georgia that lacks prominent spines or knobs on the keel of the carapace. The carapace is olive green with a reticulated pattern of narrow, yellow lines, although dark pigment usually obscures the pattern of adult females. Algae may further obscure the coloration. All four feet are webbed. The skin ranges from dark green to blackish. Lines on the head, neck, and limbs are yellowish. Like other Georgia species in the genus, females have an enlarged head and grow much larger than males. A midline keel on the carapace is apparent on young individuals but becomes less distinctive on adults. Dark lines that border the edges of the plastral scutes of juveniles fade with age. Hatchlings are nearly round, intensely patterned, and are approximately 30 mm (1.2 in) in carapace length.

Taxonomy and Nomenclature

No subspecies are recognized.

Distribution and Habitat

This species is restricted to the extreme northwestern corner of Georgia; the few records in the state are primarily from the Conasauga River drainage. Outside Georgia, the common map turtle generally inhabits large streams and rivers and their impoundments. Areas providing

Adult female common map turtle, Oregon County, Missouri (John B. Jensen)

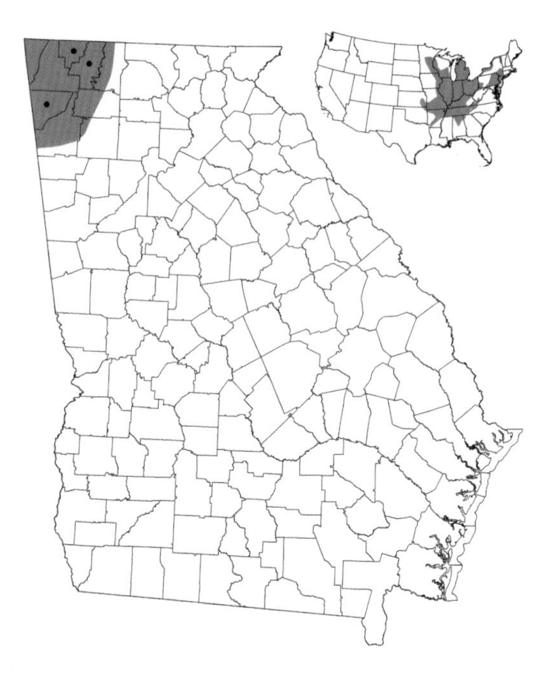

abundant basking surfaces are preferred over other areas, and large adults generally bask farther from shore than medium-sized individuals. Large adults usually avoid areas with emergent vegetation.

Reproduction and Development

Common map turtles probably breed in both the spring and the fall. Most nesting probably occurs from late May to mid-July, often on sandy beaches or sandbars and not in densely wooded areas. Females often begin nesting after dark and may continue until early in the morning. The average clutch size is about 10–12 eggs, and females produce two or more clutches per year. Eggs are 32–35 mm (1.3–1.4 in) long and 21–22 mm (0.8–0.9 in) wide. Sex in developing embryos is determined by incubation temperature, with temperatures warmer than 30° C (86° F) producing mostly females and temperatures as cool as 25° C (77° F) producing mostly males. Incubation takes approximately 75 days. Little information is available on growth and maturity, but males probably reach maturity after 4–6 years; females may take more than 10 years to mature.

Habits

Common map turtles are active primarily during the day, feeding in the morning and late afternoon hours and basking in between. Annual movement patterns include late spring or summer migrations, mostly by adult males, to various locations either upstream or downstream from the overwintering site. Summer retreats generally remain the same year after year. Hatchlings tend to move downstream to establish a summer home range. Known overwintering sites include deep, slow riverine pools and impoundments. The diet is composed of crayfish, mollusks (primarily snails), and insect larvae as well as some fish carrion. Vegetation makes up only a small proportion of the diet and may be ingested incidentally with other foods. Likely predators of adults include raccoons, opossums, skunks, and coyotes. Hatchlings fall prey to gulls, crows, grackles, and red-winged blackbirds. Raccoons, skunks, foxes, and river otters may raid the nests and eat the eggs.

Conservation Status

The common map turtle is listed as Rare in the state of Georgia, primarily because of its restricted range. The accidental introduction of nonnative Asian clams has probably helped this species because much of its native mollusk prey has been decimated by stream degradation.

Robert A. Moulis

Juvenile Alabama map turtle, Elmore County, Alabama (James C. Godwin)

Alabama Map Turtle

Graptemys pulchra

Description

Male Alabama map turtles seldom reach a carapace length of 12 cm (4.7 in); females may exceed 29 cm (11.3 in). The short, broad carapace is more or less olive and has a distinct keel extending along the center with conspicuous spines or knobs. The front portion of the carapace of large females is distinctly humped in profile. Each marginal scute has a concentric yellow semicircle on the upper surface. The seams that separate the lower marginals have wide, dark borders often composed of two or three concentric semicircles. The head and limbs are brown to olive with light yellow or greenish yellow stripes and blotches. A large yellow or light green blotch is present on the head between and behind the eyes. The head of mature females is significantly larger than that of males of similar age. All four feet are webbed. As in other members of this genus, the colors, patterns, and carapace knobs are best developed in the young, although adult males often retain some juvenile characteristics. Hatchlings are approximately 30 mm (1.2 in) in shell length.

Taxonomy and Nomenclature

The Alabama map turtle was for many years considered a highly variable species ranging from Georgia to Louisiana and occurring in several major river drainages. Further research has revealed that it is actually a complex of several species largely separated into distinct river catchments. No subspecies are recognized.

Distribution and Habitat

The Alabama map turtle inhabits waterways in the Ridge and Valley Province of extreme northwestern Georgia. Records are restricted to the Conasauga and Oostanaula rivers. An abundance

Adult female Alabama map turtle, Murray County (John B. Jensen)

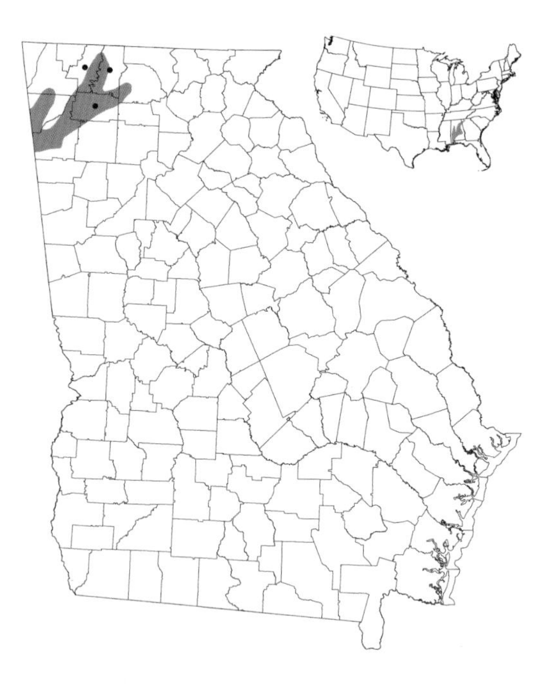

of mollusks (particularly mussels)—an important part of females' diet—seems to be the main requirement for this species. The presence of suitable nesting sites (i.e., expansive sandbars and sandy banks), deep-water pools for shelter, abundant basking sites, and substrates suitable for the attachment and development of mollusks contribute to a stream's suitability for this species.

Reproduction and Development

Alabama map turtles lay as many as seven clutches of four to six eggs each, beginning in late spring and continuing into late summer. Nests are probably placed in sandy soils on beaches or sandbars. Little else is known about the reproductive biology. Females require an average of 14 years to reach sexual maturity; males can apparently reach maturity in 3. Females take more than 20 years to reach their maximum size; males may reach it in 8 years.

Habits

The Alabama map turtle is generally a diurnal turtle that is active from late March through November, although activity may extend throughout the year during warm winters. Like Georgia's other map turtles, this species is extremely wary, and basking individuals quickly retreat to water at the slightest hint of danger. Hatchlings and juveniles prefer basking sites close to shore in thick cover while adults typically bask on large structures in deep water. Turtles less than 100 mm (3.9 in) in carapace length eat primarily insects. Females apparently shift to snails and mussels when they reach a carapace length of 80–100 mm (3.1–3.9 in) and eat insects only occasionally thereafter. Large females commonly feed on the nonnative Asian clam. Fish crows have been implicated as major predators of Alabama map turtle nests during the day, and raccoons—and to a lesser degree skunks—raid them at night. Possible predators of hatchlings include gar and herons. Adults probably have few enemies other than humans.

Conservation Status

Because of its restricted range in Georgia, the Alabama map turtle is listed as Threatened in the state. Environmental alterations that affect water quality often harm mollusks, and thus can ultimately harm turtles that specialize on them. The establishment of nonnative Asian clams may have helped the Alabama map turtle.

Robert A. Moulis

Adult female diamondback terrapin, Glynn County (Dirk J. Stevenson)

Diamondback Terrapin

Malaclemys terrapin

Description

Females of this small to medium-sized turtle range from 15 to 23 cm (5.9–9 in) in carapace length; males are much smaller, averaging 10–14 cm (3.9–5.5 in). Females also have a proportionately larger and wider head. The tail of males is longer and thicker, and the vent is located outside the rear edge of the carapace. Females have a more slender tail, and the vent is in front of or even with the edge of the carapace. Diamondback terrapins show striking variation in color both throughout their range and within local populations. The common name derives from the angular appearance and shape of the carapace scutes. The carapace is usually a combination of gray, green, brown, or nearly black with concentric age rings that sometimes alternate with light and dark coloration. The plastron is usually lighter and the scutes may have dark blotches. The skin is light gray to black with dark spots, speckling, blotches, and rarely stripes. Juveniles are 25–32 mm (1.0–1.3 in) at hatching and are more brilliantly colored than adults; colors of older terrapins tend to fade, and their shells become worn and smooth. The hind feet are webbed for swimming. Few other turtle species can be confused with the diamondback terrapin because of its specific habitat requirements. Other than sea turtles, any turtle found in a brackish or saltwater estuarine habitat is almost certainly a diamondback terrapin. Sea turtles may inhabit the same coastal rivers and creeks but have flippers and are generally much larger.

Two-headed juvenile diamondback terrapin, Glynn County (Adam Mackinnon)

Adult female diamondback terrapin, Glynn County (Brad Winn)

Adult female diamondback terrapin, McIntosh County (Brad Winn)

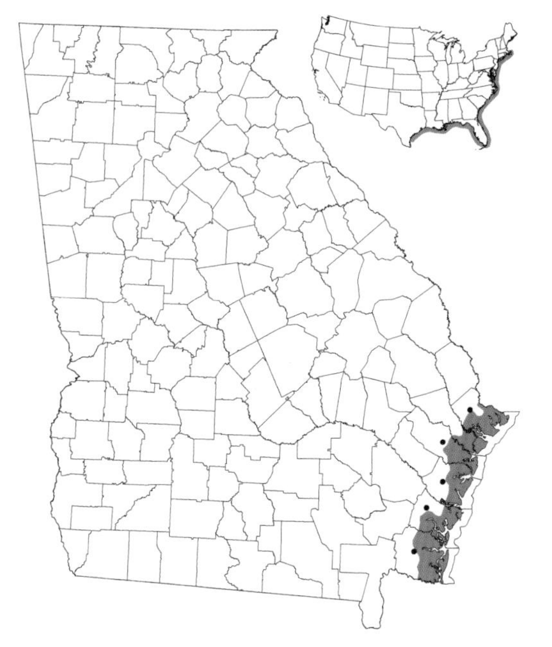

Taxonomy and Nomenclature

Seven subspecies are recognized rangewide, but the Carolina diamondback terrapin (*M. t. centrata*) is the only one found in Georgia. Diamondback terrapins are thought to be more closely related to the map turtles (genus *Graptemys*) than to other turtle groups.

Distribution and Habitat

The diamondback terrapin is a resident of brackish and saltwater estuarine habitats, including tidal creeks, rivers, sounds, and occasionally nearshore ocean waters. This species is closely associated with large expanses of salt marsh such as those in the numerous remote tidal creeks adjacent to Georgia's barrier islands. Juveniles are infrequently seen, and it is assumed that they spend much of their early life hiding under mats of dislodged vegetation along estuarine shorelines. Terrapins burrow in mudflats, shallow channels, and creek banks periodically throughout the year and overwinter in similar habitats.

Reproduction and Development

Mating occurs in the water, and egg-bearing females emerge in late spring or summer to search for suitable nesting sites. Virtually any area of sandy soil may be used, especially sandbars, spits, dunes, and creek banks, as long as the site is above high tide and adjacent to marsh habitat. Unfortunately, edges of parking lots, roadsides, and flower beds often meet that requirement, which brings nesting females in close contact with potential hazards. Clutch size is 7–12 eggs, and females may produce multiple clutches in a single year. As with all turtles, incubation time varies with temperature. Hatching typically occurs after 61–68 days but may take as long as 104 days in cooler periods. Nests deposited later in the season tend to hatch later due to cooler incubating temperatures. Hatchlings may overwinter in the nest and emerge the following spring. Female terrapins mature at around 5 years of age; males mature earlier at 3 years.

Habits

Diamondback terrapins are diurnal and are active for 9 or 10 months of the year in Georgia. They hibernate when water temperatures fall below 10° C (50° F). They are highly aquatic and rarely leave the water except to nest. Their diet varies depending on local habitats and prey availability but includes estuarine mollusks such as snails and bivalves, and crustaceans, mainly small crabs. The periwinkle snail is an important prey item in most areas. These turtles also scavenge on dead fish and other marine animals. Because of their larger head and more powerful jaws, females are capable of feeding on larger and harder-shelled prey items than males are. During high tides, diamondback terrapins have access to the upper reaches of salt marshes and can forage without leaving the water, thus allowing them to evade predators such as raccoons, river otters, and bald eagles. Hatchlings are particularly vulnerable and are eaten by crows, gulls, wading birds, raccoons, rats, mink, foxes, and other mammals.

Conservation Status

Diamondback terrapin soups and stews were wildly popular in the early 1900s, and collectors regularly sold these turtles for $50–60 per dozen. Terrapin populations were heavily stressed during this period and were completely exterminated from many regions. Populations rebounded to a limited degree in the following decades as the popularity of terrapin soup waned. Today, coastal land development is the main threat in Georgia, particularly the increased road mortality it brings. Significant mortality also occurs in the commercial blue crab fishery because many diamondback terrapins drown in crab traps. Recreational crab traps, which people often leave in the water for days on end, and "ghost traps" (detached or abandoned traps) may pose an even greater threat. The diamondback terrapin is listed by the state of Georgia as an Unusual species.

Philip B. Spivey

Adult male river cooter, Murray County (John B. Jensen)

River Cooter

Pseudemys concinna

Description

The river cooter is the largest North American member of its family. The longest carapace length on record is 41.6 cm (16.2 in). Females often exceed 30 cm (11.7 in) at sexual maturity; males may be sexually mature at half that length. The carapace is brown with a reticulate pattern of yellow to yellow-orange lines. The yellow lines on the carapace fade and become thinner and irregular on older individuals. Each marginal scute has a wide, central yellow bar. The lower surfaces of the marginal scutes are marked with rings or blotches centered on their intermarginal seams. The carapace is elongated and often flares out in the rear. Adult females have a higher, more domed carapace than males. The plastron is yellow to yellow-orange with occasional black patterning that is usually confined to areas adjacent to the seams between the scutes; this pattern also fades with age. Yellow to nearly orange stripes are noticeable on the head, chin, and legs. Mature males have greatly elongated claws on the middle three digits of the forelegs and also have a proportionately longer tail than do females. The lower jaws are slightly serrated, especially so in juveniles, but the serrations wear down with age. The hind feet are extensively webbed. Young river cooters are more brightly colored than adults; the carapace is greenish brown, and the reticulation is bright yellow to yellowish orange. The skin is green to

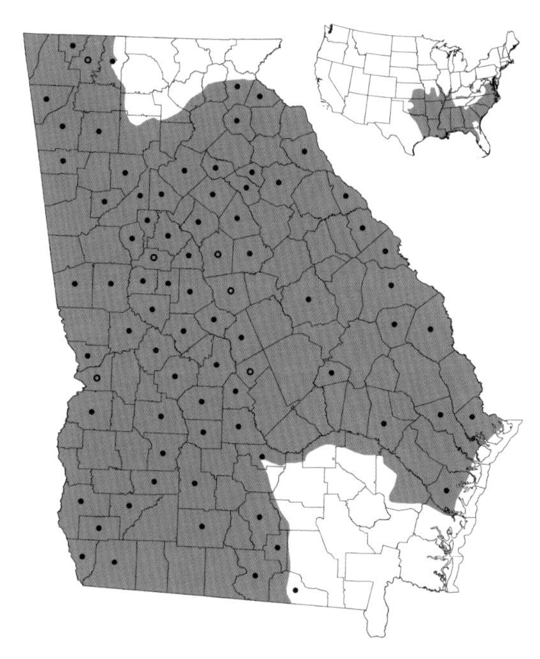

Underside of adult female river cooter, Decatur County (John B. Jensen)

light brown with numerous yellow stripes on the head and neck. Hatchlings are 35–41 mm (1.4–1.6 in) in carapace length. River cooters are sometimes difficult to distinguish from Florida cooters (*P. floridana*), although the carapace of river cooters is relatively flatter and more elongated, and the head stripes are more boldly marked with yellow-orange relative to the thinner yellow-green stripes of Florida cooters. The latter seem never to have orange coloration, and their plastron lacks dark pigmentation.

Taxonomy and Nomenclature

The taxonomy of river cooters and of the genus *Pseudemys* in general has been a source of controversy and remains unresolved. The problem centers on whether the river cooter and the Florida cooter interbreed, and also on what constitute reliable taxonomic characters for identifying the various species and subspecies of *Pseudemys*. In areas where there is distinct separation of habitats, river cooters generally inhabit moving water and Florida cooters inhabit slow-moving or stillwater habitats. In these situations interbreeding rarely occurs. When habitats are disturbed, however, as when rivers are dammed, the two species may occupy the same habitat and interbreeding apparently occurs. River cooters throughout most of Georgia are regarded as belonging to the subspecies *P. c. concinna* (eastern river cooter), although Suwannee cooters (*P. c. suwanniensis*) have been documented in two extreme south-central Georgia streams. Further debate centers on whether the Suwannee cooter should be considered a distinct species. River cooters in northwestern Georgia were at one time considered to belong to the subspecies known as the hieroglyphic river cooter (*P. c. hieroglyphica*), but this subspecies is no longer recognized.

Distribution and Habitat

In Georgia, the river cooter is a species of large creeks, rivers, and the impounded portions of these habitats. River cooters occur throughout the Piedmont and much of both the Coastal Plain and Ridge and Valley provinces. They are found in all major river systems within the state and are particularly abundant in the rocky shoals near the Fall Line. Ideal habitat includes stretches of river with relatively clear water 100 cm (3.3 ft) deep or deeper, abundant basking sites in the form of rocky shoals or fallen logs, and abundant aquatic plant growth.

Reproduction and Development

The nesting season is late May through the end of June. River cooters nest during the day, usually close to the river they inhabit and usually in soils deposited by the river. Females occasionally nest a considerable distance from the river. The incubation period under natural conditions is approximately 60 days. Hatchlings normally remain in the nest until the following spring and appear in rivers in April and May, but some may emerge in the fall if the nest is flooded or disturbed. Hatchlings are usually found in sluggish stretches of the river. Growth rates are most rapid in the first 3–4 years and slow with increasing age. Females mature in 14–15 years in central Alabama and in 10 years in Florida. Males reach sexual maturity in about 6 years.

Habits

River cooters bask extensively during the warm months of the year; most of their daytime activity alternates between feeding and basking. These turtles are active throughout the year in most of Georgia but become inactive when January and February temperatures drop below freezing. Even during this time, however, a period of warm, sunny weather may induce them to come out and bask. The river cooter is primarily herbivorous. Studies in Alabama and West Virginia identified riverweed, eelgrass, pondweed, waterweed, and filamentous algae as important components of the diet. Adult females and their eggs may be vulnerable to raccoons during nesting. Great blue herons, mink, and large predatory fish may eat hatchlings.

Conservation Status

Commercial collection for the Asian food market has been reported in North and South Carolina and likely occurs in Georgia as well. Plinking (shooting aquatic turtles off basking sites for sport) is a not uncommon practice in Georgia. Though river cooter populations in most of Georgia appear large and possibly stable, they cannot support unlimited harvesting or persecution, neither of which is currently regulated or illegal. The Suwannee cooter is currently considered a Species of Concern in Georgia, where it is known only from two tributaries of the Suwannee River.

Kenneth M. Fahey and Kurt A. Buhlmann

Adult male Florida cooter, Liberty County (Dirk J. Stevenson)

Florida Cooter

Pseudemys floridana

Description

With a carapace length up to 39.7 cm (15.5 in), the Florida cooter is one of the largest emydid turtles in North America. Females are larger than males and have a higher-domed carapace. The generally elongated carapace is brown with yellow patterning that often occurs as vertical bars on each pleural scute. The yellow lines on the carapace fade in older individuals and become dull brown to black. The edges of the shell are relatively smooth. The undersides of the marginal scutes often have hollow black circles. The plastron is plain yellow. Yellow to greenish yellow stripes show prominently on the blackish brown head and legs. Mature males have greatly elongated claws on the forelegs and also have a proportionately longer tail than females. The lower jaw is slightly serrated. The hind feet are extensively webbed. Young Florida cooters, 32–35 mm (1.3–1.4 in) at hatching, have a brownish carapace with a reticulate pattern of yellowish lines. A ridge down the center of the carapace is often present but disappears with age. The skin is green with yellow stripes on the head and neck. Florida cooters are sometimes difficult to distinguish from river cooters, although the carapace of river cooters is relatively flatter and more elongated, and the head stripes are a bold yellow-orange compared with the thinner yellow-green stripes on Florida cooters. Although river cooters and Florida cooters often have yellow markings, Florida cooters seem never to have orange coloration. Florida cooters

Florida cooter hatchling, Appling County (Dirk J. Stevenson)

Underside of adult male Florida cooter, Liberty County (Dirk J. Stevenson)

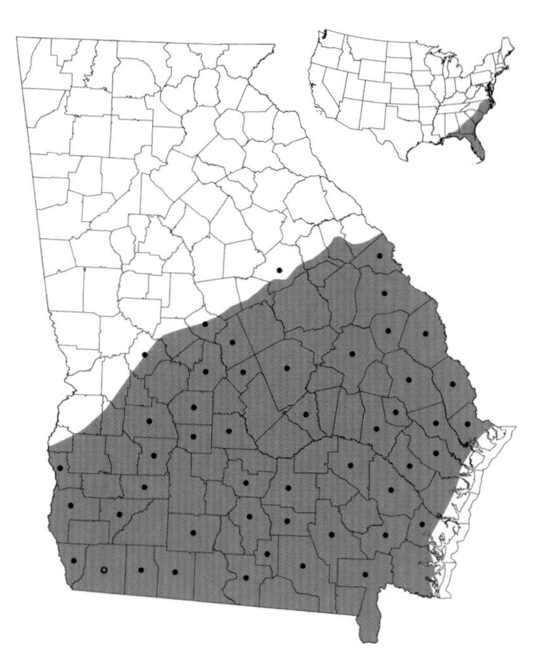

occur together with Florida red-bellied cooters (*P. nelsoni*) in the Okefenokee Swamp, but the latter are easily identified by the broad reddish bands on the carapace.

Taxonomy and Nomenclature

The taxonomy of Florida cooters, and of the genus *Pseudemys* in general, requires further study. Florida cooters and river cooters may interbreed, and the taxonomic characters used to identify the various species and subspecies are unreliable. In areas where there is distinct separation of habitat, Florida cooters inhabit backwaters, isolated wetlands, or permanent wetlands while river cooters primarily inhabit moving water. Florida cooters encounter river cooters in South Carolina when the latter travel downstream from their preferred Piedmont habitats and occupy lower Coastal Plain sections of the same rivers. Florida cooter is also the name of the subspecies (*P. f. floridana*) of the Florida cooter that occurs in Georgia. In peninsular Florida this subspecies is replaced by the peninsula cooter (*P. f. peninsularis*), a taxon that some herpetologists have proposed as a separate species. Florida peninsula cooters closely approach Georgia near the Okefenokee Swamp and may be native here, too.

Distribution and Habitat

Florida cooters inhabit floodplain swamps, isolated wetlands, and the backwaters of Coastal Plain rivers. They prefer semipermanent to permanent wetlands and travel overland to reach them.

Reproduction and Development

Florida cooters nest diurnally during late May or June. The nest complex often consists of as many as three cavities, with a main cavity holding most of the eggs and one or two satellite nests usually holding one egg each. This arrangement is presumably to distract a potential predator from the main nest and thus increase overall hatchling survivorship. Florida cooters often nest multiple times each year. Reported clutch sizes range from 10 to 29 eggs. Females nest in open areas near wetlands and sometimes considerable distances away from them. Nesting on the Savannah River Site in South Carolina, directly across the river from Georgia, often takes place on power line rights-of-way and along road edges. Eggs reportedly take 80–150 days to hatch. Hatchlings often overwinter in the nest and enter the wetlands the following spring. Males reach sexual maturity in 3 or 4 years; females mature at 5–7 years of age.

Habits

Florida cooters commonly bask on floating logs in wetland habitats, often in association with American alligators (*Alligator mississippiensis*). Their high-domed shell may protect the cooters from the crushing power of alligators' jaws. The Florida cooter is primarily herbivorous. Aquatic pond plants and algae are probably the most commonly consumed items. Juveniles are more carnivorous than adults and eat aquatic insects. Raccoons are predators of Florida cooter nests, and hatchlings are likely vulnerable to wading birds and large fish.

Conservation Status

The Florida cooter remains common in suitable habitats in Georgia. Commercial collection for the Asian food market has been reported in North and South Carolina and likely occurs in Georgia as well, and is clearly a cause for potential future concern.

Kurt A. Buhlmann

Adult male Florida red-bellied cooter, Alachua County, Florida (Matt Aresco)

Florida Red-bellied Cooter

Pseudemys nelsoni

Description

Adults of this large freshwater turtle range in length from 20 to 38 cm (7.8–14.8 in), and females are larger than males. This turtle is generally dark with a reddish wash and has a highly arched carapace. The carapace of most adults is black, dark brown, or olive with reddish vertical bars on each marginal scute. Each of the pleural scutes has a wide, light-colored bar that may be red to yellow-orange. This bar forms an inverted Y on the first pleural and an upright Y on the second pleural. Both sexes darken with age, and some older specimens may be very dark with little hint of any red. Although the normally red or orange plastron may rarely be yellow, it almost always has a hint of red, coral, or orange, especially along the edges. The skin is black with bold, narrow yellow lines that are few in number when compared with other members of the genus and pond sliders (*Trachemys scripta*). The arrow-shaped yellow mark on the top of the head and snout is a distinguishing feature. Hatchlings and juveniles are more brightly colored, with an orange to red plastron that is often marked with dark semicircles along the seams of the scutes. Hatchlings are 28–32 mm (1.1–1.3 in) in carapace length, are nearly round, and have a keeled carapace.

Taxonomy and Nomenclature

No subspecies are recognized. The most closely related species are the Alabama red-bellied cooter (*P. alabamensis*) and the red-bellied cooter (*P. rubriventris*).

Underside of adult Florida red-bellied cooter, Alachua County, Florida (Dick Bartlett)

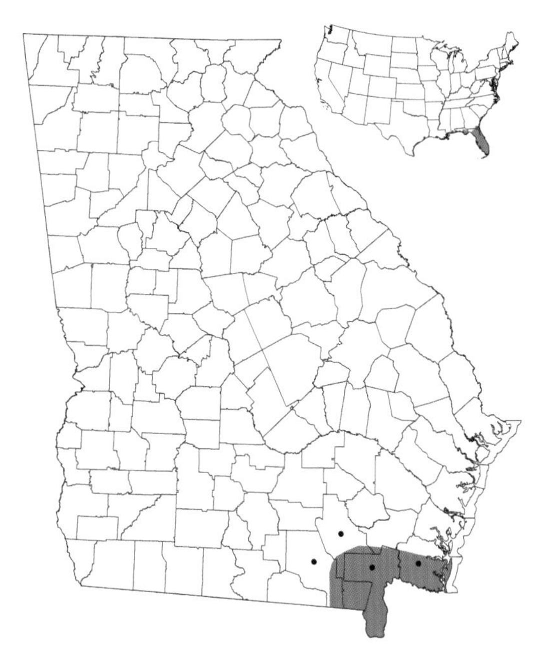

Distribution and Habitat

In Georgia, the Florida red-bellied cooter has been reported only from Cumberland Island, the Saint Marys River, and the Okefenokee Swamp in the extreme southeastern corner of the state. Throughout its range the Florida red-bellied cooter is predominantly an inhabitant of slow-moving fresh water with abundant aquatic plant growth. It may also be found in brackish water.

Reproduction and Development

During courtship the male trails the female, then assumes a position above her in the water and vibrates the long claws on his front toes against the sides of her head. The female may attempt to swim away at first, but a receptive female eventually responds by raising her hind end and stopping all swimming motion. At this point the male drops back, grasps the female's shell, and mates with her. Nesting has been observed mostly in May and June, although one researcher observed a female nesting in October. Nesting may occur year-round in the southern part of the range, but this is not likely in Georgia. Florida red-bellied cooters typically nest in sandy soils at a moderate distance from the water, but they also frequently nest in the sides of American alligator (*Alligator mississippiensis*) nests. This practice may have the advantages of protection by the female alligator, which defends her nest vigorously, and warmth from the rotting nest material, which benefits incubation. Disadvantages may include attack by the female alligator and destruction of the nest when the mother alligator excavates her young. The number of eggs reported for Florida red-bellied cooters ranges from 6 to 31. Incubation probably takes 60–75 days under natural conditions. Hatchlings have been observed from March through December. Male Florida red-bellied cooters mature in 3 or 4 years; females mature in 5–7 years.

Habits

Florida red-bellied cooters are active during the day and spend a considerable amount of time basking, either on logs with other turtles or on mats of floating vegetation. They are active throughout most of the year and appear to have a salt tolerance that is intermediate between that of other freshwater turtles and the diamondback terrapin (*Malaclemys terrapin*). Adults are herbivorous, but hatchlings and juveniles also feed on aquatic insects and other invertebrates. Nest predators include skunks, river otters, raccoons, opossums, and feral hogs. Wading birds, bald eagles, ospreys, watersnakes (genus *Nerodia*), common snapping turtles (*Chelydra serpentina*), American alligators, and various fish may prey on hatchlings and juveniles. Probably the only significant predators of adults are alligators and humans.

Conservation Status

The Florida red-bellied cooter is at the periphery of its range in Georgia and is rarely encountered. The exact status of Georgia populations is not known.

Kenneth M. Fahey and John B. Jensen

Adult box turtle, Floyd County (Bradley Johnston)

Eastern Box Turtle

Terrapene carolina

Description

Adult eastern box turtles typically have a carapace length of about 13–15 cm (5.1–5.9 in) but may reach almost 20 cm (7.8 in). They are highly variable in color and pattern, although most individuals have yellow to orange blotches or streaks against a dark brown or black background on the shell and skin. All eastern box turtles have a high, dome-shaped carapace. The plastron is hinged in the front and can be closed so that the turtle can withdraw completely into its shell. A central keel is present on most individuals, especially juveniles, and can take on the appearance of small round or teardrop-shaped knobs down the middle of the carapace. The large eyes and beaklike upper jaw are best seen in profile. Adult males have a concave indentation in the plastron and short, curved rear claws; females have a flat plastron and long, straight, slender rear claws. Females' carapace tends to have a more pronounced dome. The eyes of males tend to have red irises while the females' are usually browner, but this is not a reliable sex character. Georgia box turtles vary in the number of toes on the hind feet; individuals found in most of the state have four, but those in southern Georgia may have three. Some of the box turtles in southwestern Georgia have three toes on one hind foot and four on the other. Juveniles are more uniform than adults in color and pattern. All have a dark carapace and a light spot or

Underside of adult box turtle fully withdrawn into its shell, Bibb County (John B. Jensen)

Adult box turtle, Baker County (Erin Condon)

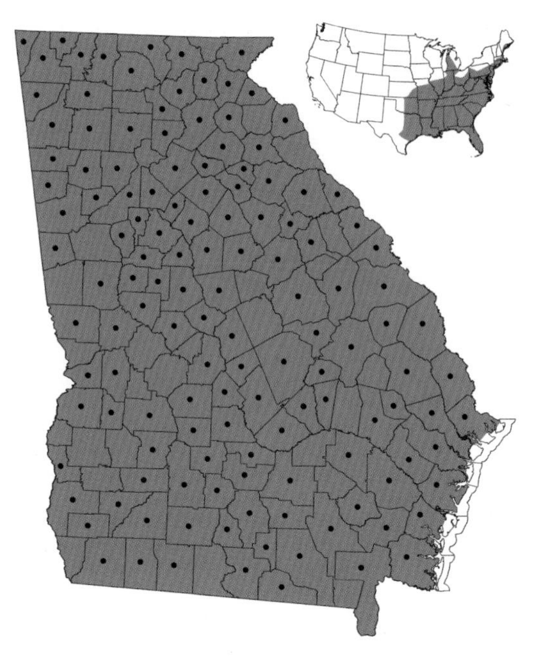

blotch on each scute; the plastron is light with a dark center. The body color changes markedly as box turtles mature. Hatchlings are 30–33 mm (1.2–1.3 in) in carapace length.

Taxonomy and Nomenclature

The subspecies found throughout most of Georgia is *T. c. carolina*, which has the same common name as the species. From extreme western Georgia near the Alabama line and south along the lower Coastal Plain, the eastern box turtle intergrades to varying degrees with three other subspecies: the three-toed box turtle (*T. c. triunguis*) along the Alabama border, the Gulf Coast box turtle (*T. c. major*) in the southwestern corner of Georgia, and the Florida box turtle (*T. c. bauri*) from south-central to southeastern Georgia along the Florida state line. Many rural Georgians refer to eastern box turtles as terrapins.

Distribution and Habitat

Eastern box turtles seem to prefer moist forests with open areas that provide opportunities to bask, and they find such situations in numerous habitats. They are also found in grassy meadows and moderately altered habitats. In some areas they frequent stream sides and floodplains, and they may use aquatic habitats temporarily during drought or hot weather.

Reproduction and Development

Mating occurs from April to June, and females nest in late spring or summer. Fall mating is also common. The male's concave plastron helps him to mount the female and maintain his position during mating. Females lay two to seven eggs in a nest they dig with their hind limbs in relatively loose soil. The eggs are 20–40 mm (0.8–1.6 in) long and 15–25 mm (0.6–1 in) wide and hatch in 60–100 days. Females may lay up to three clutches per year. Hatchlings emerge from the nest in the fall or may overwinter in the nest and then emerge in the spring as temperatures increase. Individuals reach maturity in 5–10 years.

Habits

Eastern box turtles are frequently encountered crossing Georgia roadways during periods of increased activity, especially in the morning hours and after warm rains. They bask in sunny areas during the moderately warm parts of the day. During cold winter weather, they dig a shallow hideaway just below the soil and ground litter and hibernate beneath this insulating layer. Eastern box turtles are omnivorous, and their diet includes plant material (leaf, stem, and fruit), insects and other arthropods, snails, slugs, salamanders, and carrion. There is one documented observation of a Georgia individual consuming a road-killed copperhead (*Agkistrodon contortrix*). Eastern box turtles particularly prefer succulent fruits such as blackberries, may apples, and muscadine grapes; the plants benefit by having their seeds dispersed. In addition, passage through the turtle's gut may actually increase the germination rate of seeds. Mushrooms and other fungi are significant foods in certain populations. Almost all eastern box turtles retreat into their shell and close it tightly when they are threatened. Many a curious child (and adult) who has had a finger caught can speak to the strength of the shell's hinge. This shell-closing behavior provides excellent protection from most predators, especially for larger adult turtles. Even so, adults fall prey to larger mammals such as foxes, raccoons, skunks, and bobcats. Young turtles are taken by kingsnakes (genus *Lampropeltis*), hognose snakes (genus *Heterodon*), copperheads, and cottonmouths (*Agkistrodon piscivorus*); and by predatory birds such as kites, owls, and crows. Introduced species such as hogs and fire ants, as well as domestic dogs and cats, are also major threats to these turtles. Published estimates of eastern box turtle population densities range from 2 to 27 turtles per hectare (1–11 per acre).

Conservation Status

Herpetologists generally agree that the eastern box turtle is in decline over most of its range, with much of the problem attributed to habitat loss. Eastern box turtle populations are seriously jeopardized by habitat fragmentation resulting from urban or suburban development, which is exacerbated by increased road mortality and overcollection for the pet trade. Eastern box turtles are still fairly frequently encountered in Georgia, but state law prohibits their collection.

W. Ben Cash and Whit Gibbons

Adult male pond slider, Baker County (John B. Jensen)

Pond Slider

Trachemys scripta

Description

Adult males typically reach a carapace length of 11–21 cm (4.3–8.2 in); adult females are 17–24 cm (6.6–9.4 in). The carapace varies from brownish to dark green to almost black with yellowish markings extending from the center to the sides. Its surface may have a corrugated appearance, and the seams separating the marginal scutes on the rear are slightly indented. The yellow plastron typically has a pair of black spots on the forward scutes, and spots may also be present on some or all of the other scutes. The skin color ranges from green to brownish. The head, neck, and limbs are striped with yellow, and a large yellow blotch is present on the sides of the head behind the eyes. This blotch may be reddish or orange in individuals from western Georgia. Older males often become melanistic over a period of several years, and the skin, carapace, and plastron, in whole or in part, become dull, smoky black. As males reach sexual maturity, at around 11 cm (4.3 in) in carapace length, the tail and the claws on the front feet begin to grow appreciably longer than those of females of the same body size. Young pond sliders look like the adults except that the shell is greener and the yellow stripes are more vivid. Hatchlings are 25–36 mm (1.0–1.4 in) in carapace length. Pond sliders young enough to have retained their head and body markings are easily distinguished from the other large hard-shelled

Old adult male pond slider, Taylor County (John B. Jensen)

Pond slider hatchling, Long County (Dirk J. Stevenson)

Underside of adult pond slider, Bryan County (Dirk J. Stevenson)

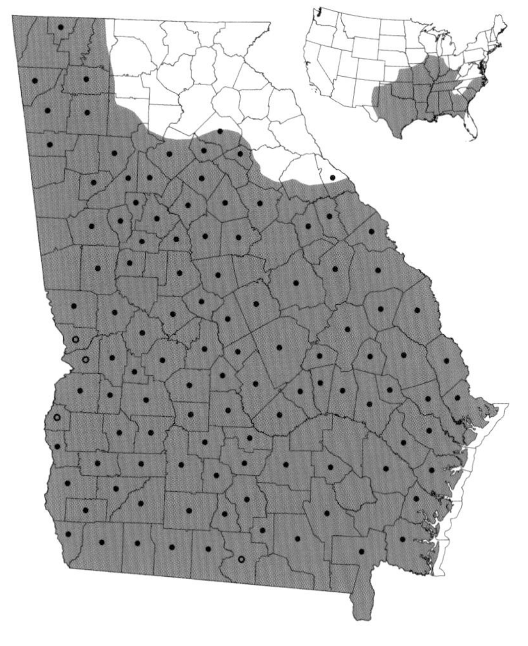

aquatic turtles of Georgia by the large yellow (or reddish) blotch behind the eye and the row of vertical yellow stripes on the back of the legs. No other moderate-sized hard-shelled turtles in Georgia are dull black, so older melanistic males are unlikely to be confused with any other species.

Taxonomy and Nomenclature

Of the three subspecies found in the United States, the yellow-bellied slider (*T. s. scripta*) is the most prevalent in Georgia. This subspecies intergrades with the red-eared slider (*T. s. elegans*) along the southwestern and northwestern borders of the state. Influence from red-eared sliders may also be apparent at locations (e.g., Fort Yargo State Park in Barrow County) where individuals of this subspecies, once sold as pets in dime stores throughout the state, have been released. Populations near the Tennessee border are possibly intergrades between the Cumberland slider (*T. s. troostii*) and either of the other two subspecies. All sliders were once assigned to the genus *Chrysemys.*

Distribution and Habitat

Pond sliders live in almost any nonmoving or slow-flowing aquatic habitat. They are among the most common turtles in many Coastal Plain and Piedmont wetlands, both seasonal and permanent. They are noticeably absent from mountainous stream habitats in the northeastern corner of the state.

Reproduction and Development

Pond sliders mate from late fall into spring, including during the winter on sunny days. Mating takes place in the water and involves an elaborate courtship that includes the male vibrating his long foreclaws in front of the female's face. Females may nest from April to July, but most nesting occurs in May and June. Some nest two or, rarely, three times in a year. The average clutch size is about 6, but the number of eggs varies from 2 to 15. Large females generally lay more eggs than smaller individuals. The eggs hatch in mid-to-late summer after an incubation period of 2–3 months. The hatchlings characteristically remain underground in the nest through fall and winter and then dig out in early spring and travel to the nearest wetland. Males mature at an age of approximately 4 years, and females mature in about 8 years. Individuals have been documented to live more than 20 years in the wild.

Habits

Pond sliders are conspicuous turtles in Coastal Plain and Piedmont habitats because of their ubiquity and abundance, their tendency to bask openly on logs, and the propensity of both males and females to travel overland to mate. These turtles may move more than 4 km (2.5 mi) between wetlands. Juveniles and adults of both sexes may move overland from one aquatic habitat to another to escape drought conditions. Females leave aquatic habitats to find suitable nesting sites on land. Juveniles eat aquatic insects, dead fish, and other animals. Adults are typically vegetarians, consuming algae, aquatic plants, and even shoreline vegetation. Given the opportunity, however, they scavenge or eat live animal food. During cold weather in late fall and winter, they cease feeding and become inactive except for mating activities. Adults can frequently be seen swimming or basking on cool but sunny days during winter and have even been observed at night swimming beneath ice in lakes. Numerous animals prey on pond sliders, especially on juveniles and eggs. American alligators (*Alligator mississippiensis*) eat the adults in the water, although their strong, thick shell protects the largest of these turtles. On land they are attacked by raccoons and coyotes, which can reach the soft parts of the turtle's body with teeth or claws. Predators on juveniles also include gar and catfish, cottonmouths (*Agkistrodon piscivorus*), crows and wading birds, and river otters. When alarmed, these turtles withdraw their head and limbs inside the shell. Pond sliders of any age often bite when captured.

Conservation Status

Pond sliders are very common in Georgia. This species, and more specifically the red-eared slider, is one of the few North American animals that have become a threat to native species in other countries.

Whit Gibbons

Mud and Musk Turtles

FAMILY KINOSTERNIDAE

All four of Georgia's species of mud and musk turtles have carapace lengths less than 15 cm (5.9 in). The shell is oblong and moderately domed, and the plastron is often hinged. The more than 20 species that make up the family are principally aquatic and live in lakes, streams, and wetlands in temperate and tropical areas of North, Central, and South America.

Adult striped mud turtle, Emanuel County (John B. Jensen)

Striped Mud Turtle

Kinosternon baurii

Description

Adults typically have a carapace length of 7.5–10 cm (3–4 in) and may reach 12.7 cm (5 in). The smooth, rounded carapace is black, brown, or olive and may have three cream-colored or yellowish lengthwise stripes, one along the center and one along each side. Georgia specimens often have faint stripes or lack them altogether. The plastron is often a rich mahogany but may also be yellow. Each plastral scute is outlined by dark pigment. The plastron is almost as long as the carapace and has a horizontal hinge on both the forward and rear halves. The head and soft parts of the body are dark gray or black. The feet are webbed. The small, rounded head has two yellowish stripes, which may be continuous or broken, on each side of the head and yellow stripes on the chin. There are also yellow flecks on the top and sides of the head. Females average slightly larger than males and have a short, stocky tail that extends only slightly beyond the rear edge of the carapace. The longer, thicker tail of males ends in a spiny tip, and the vent is located well beyond the rear margin of the carapace. Males also have a smaller and slightly concave plastron and a patch of rough scales on the inside of each rear leg. Hatchlings have three distinct lengthwise keels on the carapace, including a well-defined center keel and less evident keels along each side. The plastron is yellow or yellow-orange with dark blotches throughout its center and black spots along the marginal scutes. Carapace length in hatchlings

Underside of adult striped mud turtle, Clayton County (John B. Jensen)

Adult striped mud turtle, Coffee County (Dirk J. Stevenson)

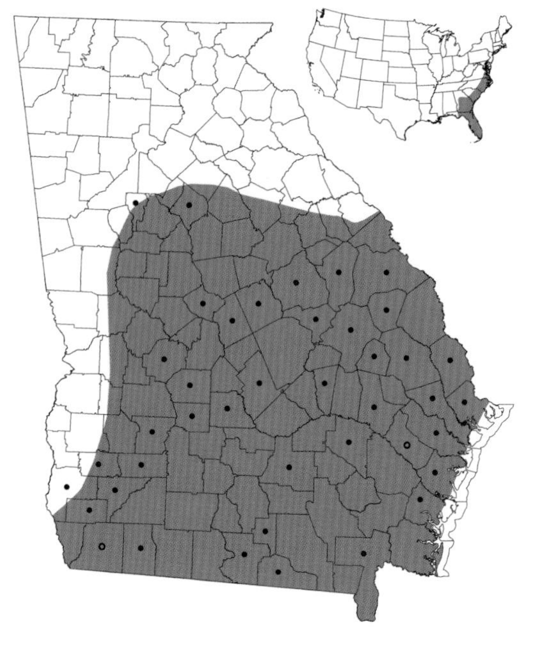

averages 15–25 mm (0.6–1.0 in). This species is easily confused with both the eastern mud turtle (*K. subrubrum*) and the common musk turtle (*Sternotherus odoratus*). The fairly prominent head stripes of striped mud turtles should distinguish them from eastern mud turtles, whose occasional head stripes are in the form of weak mottling. The dark gray or blackish skin and, if present, the three lengthwise stripes on the carapace further distinguish the striped mud turtle from the eastern mud turtle, which has brown or yellowish skin and never has shell stripes. The common musk turtle has a small plastron with a single hinge.

Taxonomy and Nomenclature

No subspecies are recognized.

Distribution and Habitat

Striped mud turtles are known from the Flint River drainage basin and from tributaries of the Chattahoochee River in extreme southwestern Georgia and throughout most of the Coastal Plain west of the coast. There are very few coastal records for this species, and it is known from only a single barrier island (Cumberland Island). Recent occurrences have also been documented along the Flint and Ocmulgee river drainages in the lower Piedmont. This freshwater turtle often moves overland and shelters on land when its aquatic habitats dry up. In Georgia, it is associated with quiet, shallow, blackwater swamps, ponds, and pools situated within the floodplains of rivers and permanent streams. Forest trees typical of these sites include bald cypress, water tupelo, and overcup oak, and a common shrub is buttonbush. Aquatic vegetation may be present but is often sparse or absent. In coastal areas, slightly brackish habitats may support populations. This turtle is absent from isolated wetlands where its close relative, the eastern mud turtle, is often abundant.

Reproduction and Development

Females lay one to six small, elliptical eggs and may nest up to three times per year. Clutch size is positively correlated with female size. The eggs may hatch in 3 or 4 months, but winter may interrupt the development of eggs laid during the fall. Observations of captives and wild specimens indicate that striped mud turtles reach sexual maturity in 3–6 years. This is apparently a long-lived turtle; a female estimated to be 10 years old when captured lived an additional 49 years in captivity.

Habits

Striped mud turtles are active in the spring and fall and are often encountered in shallow, litter-choked pools where they prowl the bottom feeding on plant debris, a variety of small animals, and carrion. Georgia specimens have been observed preying on crayfish. Striped mud turtles migrate to and from aquatic habitats as these fill and dry. When active terrestrially, individuals often shelter for extended periods in shallow burrows dug in sandy soil. These turtles show fidelity to their home ponds and regularly use the same terrestrial retreats. Raccoons, foxes, and common kingsnakes (*Lampropeltis getula*) are likely egg predators. American alligators (*Alligator mississippiensis*) and bald eagles are known predators of adults in Florida. Unlike other members of its family, the striped mud turtle is very mild tempered and seldom bites in self-defense. A newly captured individual typically withdraws its head into the shell.

Conservation Status

Recent museum records and field observations by herpetologists indicate that this turtle is widespread and fairly common in river floodplains and blackwater creek swamps throughout most of the Coastal Plain of southern Georgia.

Robert A. Moulis and Dirk J. Stevenson

Adult eastern mud turtle, Baker County (Gabriel J. Miller)

Eastern Mud Turtle

Kinosternon subrubrum

Description

Eastern mud turtles are small, nondescript, semiaquatic turtles with a smooth, oval carapace that is some shade of dark brown to black. Adult males and females reach similar carapace lengths of about 7.5–10.0 cm (2.9–3.9 in). Males have an enlarged tail and a slightly concave plastron. The plastron of both sexes is yellowish to reddish brown and has both front and rear horizontal hinges. All four feet are webbed. The head is medium sized, brown, and marked with mottling and/or partial stripes. The striped mud turtle (*K. baurii*) is similar in appearance but has distinct stripes on the head. The hatchlings are less than 25 mm (1.0 in) in carapace length and have a keel down the middle of the carapace; the carapace is dark brown and the plastron is a mottled orange to red.

Taxonomy and Nomenclature

Three subspecies are currently recognized, but only the eastern mud turtle (*K. s. subrubrum*), with the same common name as the species, is present in Georgia.

Distribution and Habitat

Found throughout Georgia except in the Blue Ridge Province, eastern mud turtles spend much of their lives in shallow, quiet, or slow-moving waters such as isolated wetlands, farm ponds, and drainage ditches. They are also common in swamps and the backwaters of rivers and large streams. These turtles have a limited tolerance for brackish water in coastal areas. They prefer habitats with dense aquatic vegetation.

Eastern mud turtle hatchling, Long County. Note scar where yolk sac was attached. (Dirk J. Stevenson)

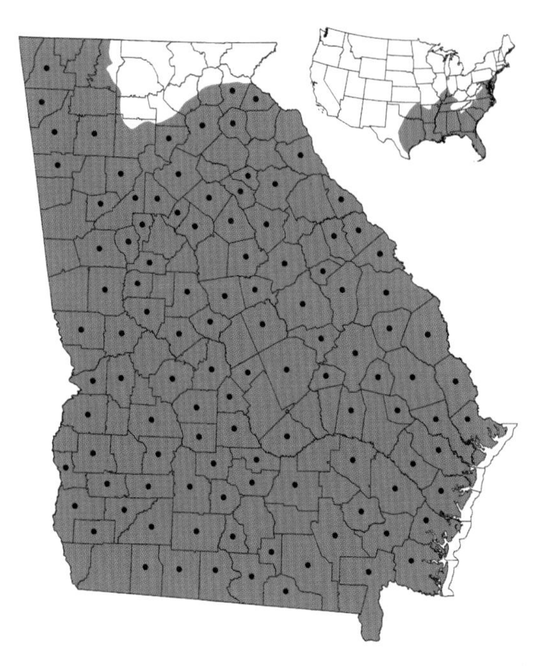

Reproduction and Development

Eastern mud turtles mate in the spring in Georgia. The male's concave plastron facilitates his mounting of the female. Females typically lay two to five (maximum = eight) brittle, hard-shelled eggs in shallow nests from late April through mid-June. Some females lay multiple clutches (up to three) within a single season. Eggs may take 90 to more than 100 days to develop and hatch, but hatchlings usually do not leave the nest in late summer or fall. Instead, they overwinter in the nest and emerge between February and May of the next year. Eastern mud turtles reach sexual maturity in 4–6 years and may live more than 25 years.

Habits

Eastern mud turtles often inhabit aquatic sites that dry periodically. When their wetland dries, they may move to a more permanent aquatic site but are more likely to burrow beneath the soil on land around the wetland habitat. Most eastern mud turtles hibernate or aestivate on land and have been known to survive for several months on land during droughts. Like many other turtles, however, they are subject to desiccation when exposed and usually choose to move about on land after warm rains. Eastern mud turtles are most commonly seen in Georgia as they cross highways while traveling overland. They are seldom seen basking. These omnivorous turtles eat, among other things, aquatic insects and their larvae, crayfish, amphibian larvae, small snakes, aquatic plants, algae, and sometimes carrion such as dead fish. They feed while walking along the bottom in shallow water and occasionally feed on land, where both sexes may spend a considerable amount of time. Adults have few predators because of the strong shell and their ability to close up completely so that no soft parts of the body are exposed, but a wide variety of predators eat the eggs and juveniles. Known nest predators include common kingsnakes (*Lampropeltis getula*), fish crows, raccoons, gray foxes, opossums, weasels, and skunks. Documented predators of small individuals include blue crabs, gar, watersnakes (genus *Nerodia*), and cottonmouths (*Agkistrodon piscivorus*).

Conservation Status

Eastern mud turtles are widely distributed in the state and are still common in many areas because they tolerate a broad range of habitat types. They are less attractive than other turtles and thus have largely avoided commercial exploitation for the pet trade.

Judith L. Greene

Adult loggerhead musk turtle, Murray County (John B. Jensen)

Loggerhead Musk Turtle

Sternotherus minor

Description

The typical adult carapace length is 7.5–11.5 cm (2.9–4.5 in); the largest individuals reach 14.5 cm (5.7 in). Juveniles have a prominent central keel down the length of the carapace and sometimes another keel along each side, but these become less pronounced in adults. Turtles in the northwestern corner of the state rarely retain the keels on the sides of the carapace. The tan to dark brown carapace has overlapping vertebral scutes that are bordered with dark seams and are often marked with radiating streaks or dark spots. The pink or yellowish plastron is small, not completely protecting the legs, and has a single horizontal hinge. The head is light brown with dark spots or, in northwestern Georgia, dark, wormy stripes, and two fleshy protuberances called barbels adorn the chin. Mature males have a much larger head than females have, and females have a shorter tail that barely reaches the edge of the carapace. Hatchlings are 22–27 mm (0.9–1.1 in) long. Loggerhead musk turtles can be readily distinguished from common musk turtles because the latter have light-colored stripes on the head and barbels on the neck as well as the chin. Mud turtles (genus *Kinosternon*) have a smoother carapace at all sizes and a plastron with two hinges that completely covers the limbs and tail when closed against the carapace.

Taxonomy and Nomenclature

Two subspecies have been described: *S. m. minor* (the loggerhead musk turtle) occurs throughout

Underside of juvenile loggerhead musk turtle, Fannin County (Roger Birkhead)

Juvenile loggerhead musk turtle, Baker County (Aubrey M. Heupel)

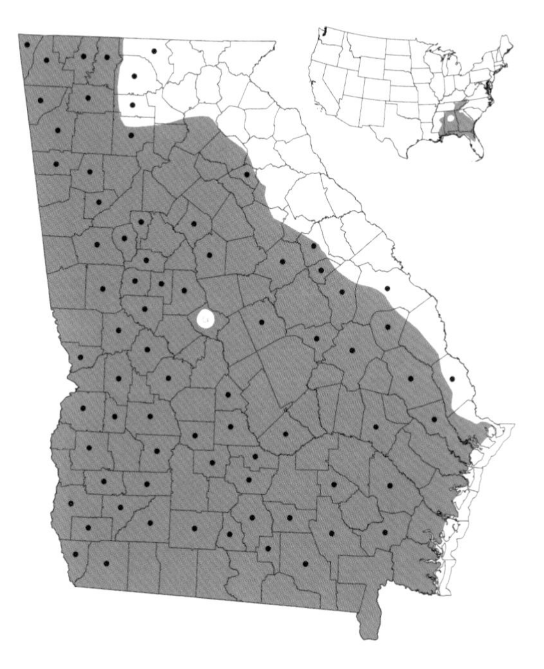

most of central and southern Georgia, and the stripe-necked musk turtle (*S. m. peltifer*) is found only in the northwestern corner of the state.

Distribution and Habitat

In Georgia this species frequents large creeks, rivers, oxbows, swamps, and impoundments. It is not found in the Savannah River drainage or in the Blue Ridge Province.

Reproduction and Development

Females lay one to four clutches of two or three elliptical, hard-shelled eggs each year. The egg-laying season begins in the spring and lasts through the summer. The female digs a nest and deposits the eggs in organic debris along a stream bank or in a sandy area nearby. Eggs are 21–33 mm (0.8–1.3 in) long. The young hatch after approximately 60–120 days of incubation. Males mature in an average of 4 years, but females may be as old as 8 when they reach sexual maturity.

Habits

Like other musk turtles, loggerhead musk turtles are highly aquatic and spend most of their day submerged in water, crawling along the bottom in search of prey or mates. They bask infrequently and generally do not move overland except to nest. Most activity occurs at dawn and dusk or at night. These turtles are highly carnivorous but may occasionally consume plant material. Juveniles feed primarily on insects and small aquatic crustaceans; adults feed mostly on snails and clams that they crush with their large jaws. The primary predator of adults is probably the American alligator (*Alligator mississippiensis*), at least where their ranges overlap in the southern half of Georgia. Alligator snapping turtles

(*Macrochelys temminckii*) and large common snappers (*Chelydra serpentina*) also eat adult loggerhead musk turtles. Egg predators include common kingsnakes (*Lampropeltis getula*), crows, opossums, skunks, and raccoons. Adults are capable of delivering a powerful bite when handled. Both adults and juveniles may release a foul-smelling musk when handled or threatened.

Conservation Status

Loggerhead musk turtles appear to be common in suitable habitats across much of the species' range in Georgia. Increased water pollution and siltation of streams affects prey abundance, however, and may ultimately affect certain loggerhead musk turtle populations.

Brian D. Todd

Adult common musk turtle, Bryan County (Dirk J. Stevenson)

Common Musk Turtle

Sternotherus odoratus

Description

Common musk turtles are only 5–13.7 cm (2–5.3 in) in carapace length and have an elongate, dark, nondescript carapace. A central ridge often extends along the top of the carapace of younger individuals, but the top of the shell typically becomes smoother on older adults. The relatively small plastron has a single horizontal hinge in the front and a shallow notch in the back. Common musk turtles have two yellow to white stripes on each side of the head and fleshy barbels on both the neck and chin. The male's tail is longer and thicker than the female's. Males also have a smaller plastron than females and consequently have more skin exposed on the underside. Hatchlings are 18–23 mm (0.7–0.9 in) in carapace length. Their carapace is nearly black, and the plastron is mottled with black and dull yellow. The head stripes and central ridge on the carapace are very distinctive on young specimens, and no plastral hinge is evident. Three turtle species found in Georgia may superficially resemble the common musk turtle. Loggerhead musk turtles (*S. minor*) have a larger head, lack head stripes, and have barbels on the chin only. Eastern (*Kinosternon subrubrum*) and striped (*K. baurii*) mud turtles have two hinges on the plastron—one forward and one toward the rear—and a smoother, more rounded carapace.

Underside of adult male common musk turtle, Evans County (Dirk J. Stevenson)

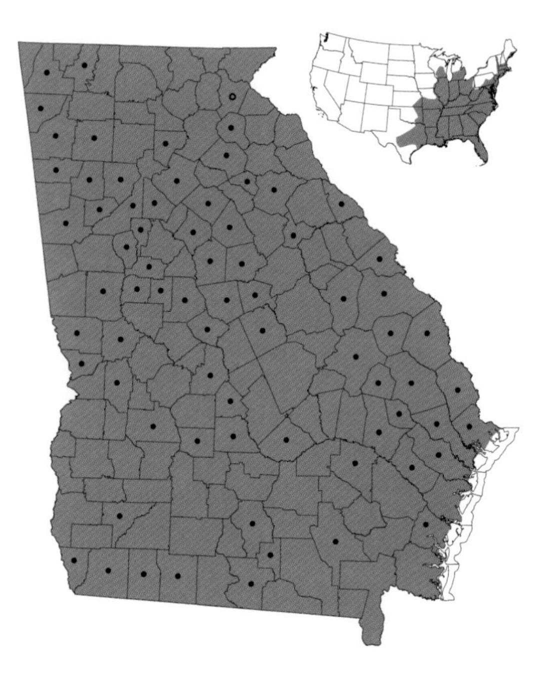

Taxonomy and Nomenclature

No subspecies are recognized. Some authors place this species in the genus *Kinosternon*, and some confusion has existed in the spelling of the generic name *Sternotherus* (e.g., *Sternothaerus*). Common musk turtles are often called stinkpots or stinking jims.

Distribution and Habitat

Common musk turtles can be found in almost any freshwater habitat in Georgia, including rivers and streams, oxbow lakes, and reservoirs, as long as shallow areas and abundant vegetation are available. Although they prefer permanent bodies of water, they are also common in many temporary wetland habitats such as river swamps, some isolated wetlands, and small ponds.

Reproduction and Development

Mating may take place throughout the year but typically occurs in April and May and then again in September and October. Females usually nest every year, producing one to six clutches with an average clutch size of about five eggs (range = one to nine). Common musk turtles sometimes lay their eggs in rotting vegetation such as that found in stump holes and American alligator (*Alligator mississippiensis*) nests rather than digging an underground nest as most other turtles do. The white, hard-shelled eggs are about 25 mm (1 in) long and 15 mm (0.6 in) wide. Hatchlings may emerge in the late summer and fall, but some overwinter in the nest and move to aquatic habitats in February or March. Throughout the range maturity has been estimated to occur anywhere between 2 and 7 years of age in males and between 2 and 11 in females; most Georgia turtles are probably mature at about 2–4 years.

Habits

Common musk turtles are highly aquatic and seldom leave the water except to nest or to seek refuge on land when a wetland is drying. They typically do not bask in open sun as much as other turtles do but are often found on branches overhanging the water in shady areas. Attesting to their strongly aquatic tendencies, many individuals, especially adults, have algae growing on the rear or on the entire carapace, and small leeches are frequently present on the plastron and exposed skin. These turtles eat mostly carrion and snails, but the diet includes a wide variety of plant and animal foods, both living and dead, such as algae and aquatic vegetation, worms, leeches, aquatic insects and their larvae, and fish eggs. Adults, juveniles, and eggs have a diverse array of predators, including largemouth bass, bullfrogs (*Rana catesbeiana*), common kingsnakes (*Lampropeltis getula*), cottonmouths (*Agkistrodon piscivorus*), red-shouldered hawks, bald eagles, crows, herons, raccoons, and skunks. Musk turtles get their name from the distinctive odor released from the vent when they are captured. This pungent musk probably deters some predators, and these turtles will bite if given the opportunity.

Conservation Status

Common musk turtles are abundant throughout most of their range, including Georgia.

Brian S. Metts

Tortoises

FAMILY TESTUDINIDAE

Tortoises have a highly domed shell that does not shed the surface scutes; instead, these accumulate as concentric growth rings over the course of the tortoise's life. The hind limbs are elephantine. These turtles live on land, where they browse on flowers, fruits, and foliage. Although Georgia's sole tortoise (gopher tortoise, *Gopherus polyphemus*) does not exceed 38 cm (14.8 in) in carapace length, the tortoises in the genus *Geochelone* may reach 130 cm (51.2 in). Distributed across tropical and warm temperate regions, tortoises occur principally in relatively dry environments in the Americas, Africa, Eurasia, and a number of island groups, the most famous being the Galápagos. The large island species were once highly prized by sailors because tortoises could survive at sea for many months with virtually no care and thereby provided a long-term source of fresh meat. Tortoises were so frequently taken aboard ships that the American and British navies actually used the word "tortoise" as a verb to describe this activity.

Adult male gopher tortoise, Long County (Andy Day)

Gopher Tortoise

Gopherus polyphemus

Description

This large terrestrial turtle reaches a maximum carapace length of about 38 cm (14.8 in). Adults typically have a uniformly dark brown to grayish black carapace and a lighter-colored plastron that on some individuals has darker mottling or a raylike pattern. The hatchlings and juveniles are a bit more colorful in that the center of each scute is often bright yellow or orange. Growth rings are conspicuous on younger individuals, but the carapace of older individuals is often polished smooth from wear. The plastron is not hinged but has a projection under the chin and a deep notch at the tail. The shell of hatchlings is quite soft, and remains so until they reach about 150 mm (5.9 in) in carapace length. The front feet are wide, flat, and covered with hard scales, and each has large, flat nails. The elephantine hind feet are smaller and round. Males differ generally from females by having a longer plastral projection, a more concave plastron, and a smaller space between the plastron and carapace where the short, stubby tail emerges. Adult males are generally smaller than females as well. Hatchlings are approximately 42 mm (1.6 in) long.

Taxonomy and Nomenclature

No subspecies are currently recognized. Gopher tortoises are sometimes colloquially referred to as gophers or gopher turtles. This is the official state reptile of Georgia.

Adult female gopher tortoise, Pierce County (Jim Flynn)

Gopher tortoise hatchling, Bryan County (Natalie L. Hyslop)

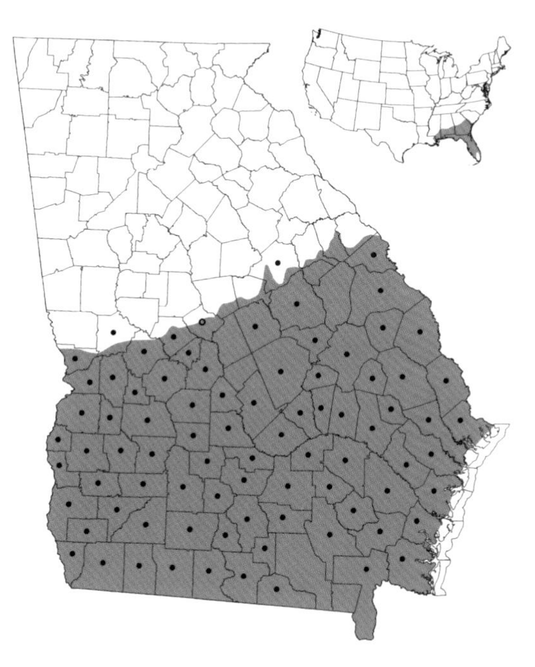

Distribution and Habitat

In Georgia, gopher tortoises are generally confined to areas of deep, sandy soils south of the Fall Line. Open habitats with abundant groundcover vegetation, such as frequently burned longleaf pine and scrub oak woodlands (especially sandhills), are preferred, although some disturbed habitats such as road and utility rights-of-way, field edges, and fence rows are also used. Areas with thick, shrubby vegetation are avoided or abandoned as they hinder movement, shade out potential forage plants, and limit potential nest sites.

Reproduction and Development

Mating occurs in spring, followed by nesting from April to July and hatchling emergence from August to October. Some courtship and mating have been observed in the fall. Eggs are hard-shelled, spherical, and 38–51 mm (1.5–2 in) in diameter. A single clutch of five to seven eggs is typically laid in a flask-shaped nest 15–25 cm (5.9–9.8 in) deep outside the burrow entrance. Females occasionally nest in open, sunny areas away from the burrow. Males reach sexual maturity in 16–18 years; females may take as long as 21 years.

Habits

The most conspicuous sign of gopher tortoises in the landscape is the presence of their burrows, which they dig using their shovel-like front feet. These sloping holes, which average 4.6 m (15 ft) in length and 1.8 m (6 ft) in depth, are where tortoises spend most of their time. Burrows provide stable, moderate temperatures and humidity throughout the year. They also provide refuge from the frequent fires that maintain the habitat these animals require. Many invertebrates and other vertebrates also find food and refuge in gopher burrows; some are completely dependent on them. Eastern indigo snakes (*Drymarchon couperi*), coachwhips (*Masticophis flagellum*), southern toads (*Bufo terrestris*), and gopher frogs (*Rana capito*) are among the many reptiles and amphibians often found in gopher tortoise burrows. Gopher tortoises use multiple burrows throughout the year, and occasionally more than one tortoise occupies a single burrow. Gopher tortoises are most active from May to September but emerge from their burrows whenever warm weather allows. The diet consists almost entirely of grasses and low-growing, herbaceous plants. Succulent fruit is relished when encountered. Gopher tortoises are important seed dispersers for many of their forage plants. Major predators reported to prey on eggs and hatchlings include mammals such as raccoons, foxes, and skunks; native and nonnative fire ants; and snakes such as coachwhips and eastern indigo snakes. Nest predation is high, and recruitment of juvenile tortoises into the population is low. Adults have few natural predators, although raccoons and coyotes may be able to kill a tortoise by breaking the edge of the shell or biting into soft tissue on the underside. Gopher tortoises do not vocalize but when disturbed can produce a loud hissing sound when the head and limbs are retracted.

Conservation Status

The gopher tortoise is state listed as Threatened in Georgia and is considered to be declining throughout its range. Although the U.S. government considers the gopher tortoise Threatened in the western portion of its range, the species is not federally protected in Georgia. Primary threats include the destruction, fragmentation, and poor management of its preferred habitats. Intensive pine silviculture has replaced much of the historic sandhills habitat in Georgia and left it too heavily shaded to support tortoises. Disease, invasive species, historical predation by humans, and gassing of burrows for rattlesnake roundups have also been implicated as contributing to the species' decline.

Roger Birkhead and Tracey D. Tuberville

Softshell Turtles

FAMILY TRIONYCHIDAE

The softshells are very flat, pie-shaped, highly aquatic turtles with a long, thin, tubular snout. Rather than being covered by the typical bone and horny scutes, the shell surface consists of thick, leathery skin and soft, pliable tissue. Georgia's two species generally do not exceed 45 cm (17.7 in) in carapace length. Softshells live in North America, Africa, Asia, and the East Indies. They are very fast swimmers because their flat body reduces water resistance. The flatness of the shell also allows individuals to burrow into muddy or sandy bottoms, from which they periodically reach their long neck and tubular snout to the surface for air.

Adult Florida softshell, Liberty County (Dirk J. Stevenson)

Florida Softshell

Apalone ferox

Description

The Florida softshell is the largest softshell turtle in North America. Females, which can reach up to approximately 63 cm (24.6 in) in carapace length, are larger than males, which reach only 32 cm (12.5 in). Males have a longer, thicker tail than females. The soft, leathery shell has reduced bone structure relative to that of the hard-shelled turtles. All softshell turtles have a long, soft, pointed snout and a long, flexible neck. Compared with the spiny softshell (*A. spinifera*), the carapace of Florida softshells is distinctly longer than wide. The adult's carapace is usually dark brown with darker mottling, and its front edge has numerous flattened, small bumps rather than the pointed, spiny projections found on spiny softshells. The plastron is generally gray to white. The head and legs of adults are typically mottled gray and brown. The feet are extensively webbed and clawed. Hatchlings are brightly colored, with a reddish, orange, or yellow stripe that extends diagonally along each side of the head from the eye to the base of the lower jaw and gradually fades with age. The carapace is olive with brownish spots and has a bright yellow to orange border. The plastron is dark. Hatchlings are usually longer than wide, and the shell is generally 38–39 mm (1.5 in) in length.

Juvenile Florida softshell, Baker County (Aubrey M. Heupel)

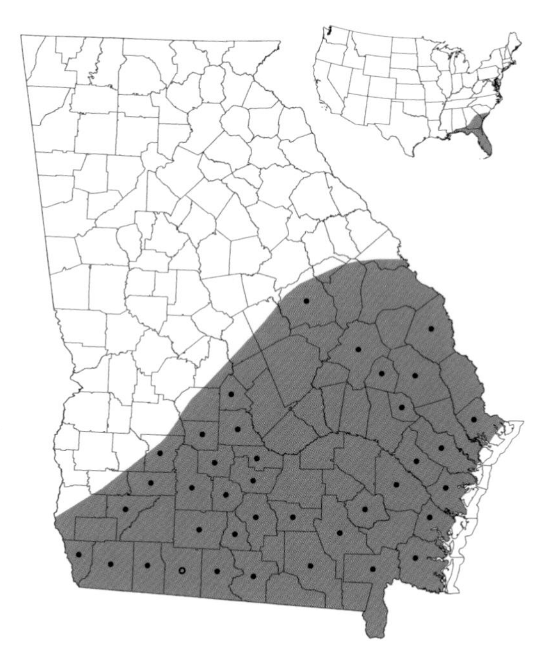

Taxonomy and Nomenclature

The genus *Apalone* contains three species. All are found in the southeastern United States, but only two are found in Georgia. Older field guides refer to this taxonomic group as the genus *Trionyx*; prior to that, the genus was called *Amyda*. The Florida softshell has no recognized subspecies.

Distribution and Habitat

Florida softshells have the smallest overall range of all the North American softshells. In Georgia, they are present throughout much of the Coastal Plain in lakes, canals, sloughs, large springs, and the slow-moving portions of rivers. They sometimes occupy ephemeral cypress ponds and ponds on golf courses.

Reproduction and Development

Little is known about the nesting behavior of Florida softshells, but they presumably nest during June and July in Georgia, often in nest mounds constructed by American alligators (*Alligator mississippiensis*). Females lay their eggs during the day, often in the morning. Clutch sizes reportedly may reach 24 eggs, and more than two clutches may be produced in a given year. Incubation takes 60–70 days, depending on the ambient temperature. Hatchlings are not known to overwinter in the nest. Age at maturity has not been reported for Georgia populations. Captive individuals have lived for 25 years.

Habits

Like all softshells, the Florida softshell is a strong, fast swimmer. Also like other softshells, Florida softshells are assumed to engage in underwater cutaneous breathing; that is, they exchange oxygen and carbon dioxide through their skin rather than their lungs. They are frequently seen basking at the surface and can be distinguished from Florida cooters (*Pseudemys floridana*) and Florida red-bellied cooters (*P. nelsoni*) by the two separate bumps—the eyes and the tip of the snout—that appear on the water's surface rather than one. This species is by far the most terrestrial of the softshell turtles, and individuals are often spotted moving overland between lakes and ponds. Florida softshells reportedly consume aquatic insects, snails, fish, and some plant material, and are not averse to scavenging on dead fish, frogs, and birds. In the few studies conducted, insects and snails made up the bulk of the food items consumed. Their neck gives them a long reach, and Florida softshells may bite a careless or inexperienced human handler. Predators include American alligators, with which they regularly co-occur.

Conservation Status

Although abundant and frequently observed in Florida, little is known about the population status of Florida softshells in Georgia. Large numbers have been harvested in Florida in recent years for food, and large numbers of both hatchlings and adults have been sold to Asian markets for use as pets and food. The degree to which this is practiced in Georgia is unknown, but it is certainly cause for concern.

Kurt A. Buhlmann

Adult male spiny softshell, Muscogee County (John B. Jensen)

Spiny Softshell

Apalone spinifera

Description

Spiny softshells are moderate-sized turtles. Females, which reach approximately 35 cm (13.7 in) in carapace length, are larger than males, which reach only 18 cm (7 in). Males have a longer, thicker tail than females. The flattened shell is soft and leathery, and it has reduced bone structure relative to that of the hard-shelled turtles. All softshell turtles have a long, soft, pointed snout. The carapace is usually brown to tan or olive gray. Juveniles have large black or brown spots and rings on the carapace that remain visible on mature males but become less obvious on aging females. Adult females have an unorganized mottled pattern on the carapace. Two or more rows of curved brown lines are often present on the rear of the carapace. The carapace is only slightly longer than wide, and may appear nearly round. The shell differs in this regard from that of the Florida softshell (*A. ferox*), whose carapace is quite obviously longer than wide. The front edge of the spiny softshell's carapace has numerous spiny projections, and the entire carapace may have the texture of sandpaper. The plastron is generally cream colored. Two white or yellowish lines on the sides of the head are not parallel and often meet near the back of the jaw. The feet are extensively webbed and clawed. Hatchlings are 30–40 mm (1.2–1.6 in) in shell length and resemble adult males in coloration and pattern.

Juvenile spiny softshell, Appling County (Dirk J. Stevenson)

Adult female spiny softshell, Lanier County (John B. Jensen)

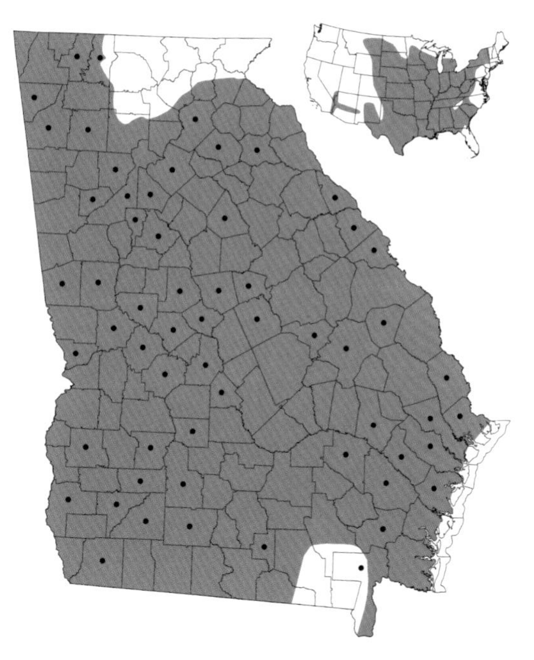

Taxonomy and Nomenclature

Spiny softshells formerly belonged to the genus *Amyda*, and more recently to *Trionyx*. Only one subspecies is documented from Georgia: the Gulf Coast spiny softshell (*A. s. aspera*). Tennessee River tributaries in the extreme northwestern corner of the state may be inhabited by the eastern spiny softshell (*A. s. spinifera*), but no confirmed records exist at this time.

Distribution and Habitat

Spiny softshells occur in all the major physiographic provinces of Georgia, but in the Blue Ridge they are confined to the foothills. They prefer clean, clear, sandy-bottomed streams but are also present in the sediment-laden rivers of the lower Coastal Plain. Spiny softshells are never associated with isolated seasonal wetlands but are commonly found in impoundments, including farm ponds and large reservoirs. Large downed trees and rock ledges are favored basking sites. The downstream, submerged tails of sandbars are frequent burrowing and shallow-water basking sites.

Reproduction and Development

Males search for receptive females at basking sites; mating occurs in deep water with the pair swimming in place, as males do not grasp the female with the forelimbs. Nesting follows during May and June, with females seeking out sandbars and sandy riverbanks that are less likely to become flooded during summer storms. A large female may produce a clutch of up to 40 eggs, but the average clutch size is 12–18 eggs. Incubation takes 60–90 days. The eggs hatch in late summer and the hatchlings immediately enter the water rather than remaining in the nest over the winter as some Georgia turtles do. Unlike most turtles, the sex of hatchlings is determined genetically, not by incubation temperature. Information on age at maturity for Georgia softshells is not available. Spiny softshells have lived more than 20 years in captivity.

Habits

Spiny softshells are thoroughly aquatic and are strong, fast swimmers. They frequently bask on logs, rock outcrops, and sandbars, often remaining close to, or even partially submerged in, the water. They are wary and quickly return to the water if disturbed. Spiny softshells pump water in and out of the mouth, possibly to aid in detection of prey. Some gas exchange takes place across the skin. They are primarily carnivorous and consume a wide variety of aquatic insects, crayfish, and fish; most researchers have found crayfish to be the most abundantly consumed food item. Able to strike quickly, spiny softshells can sometimes capture small fish, but large fish are most likely to be scavenged after they have died from other causes. Enemies include fishermen, who often capture them on hook and line. Nesting females are vulnerable to raccoons. An unwary spiny softshell could be consumed by an alligator (*Alligator mississippiensis*), although speed in the water is this turtle's defense.

Conservation Status

The spiny softshell is widely distributed in Georgia and is locally abundant. Fishermen often harvest these turtles for food. In North Carolina, commercial harvest of turtles—mostly cooters, sliders, and softshells—is increasing to fuel human demands for food in Asia. No laws currently regulate softshell harvesting in Georgia.

Kurt A. Buhlmann

Species of Possible Georgia Occurrence

Whether resulting from range expansion, introduction, or the discovery of previously cryptic populations, amphibians and reptiles currently unknown from Georgia (at least as extant, well-established populations) may be revealed in the future. Based on current ranges and known colonization abilities, the amphibians and reptiles noted below appear to have the greatest chance of becoming a part of Georgia's known fauna.

Pine Barrens Treefrog (*Hyla andersonii*)

Calling adult male pine barrens treefrog, Okaloosa County, Florida (Aubrey M. Heupel)

A Richmond County sighting reported in the literature is of questionable validity. Subsequent searches for this frog in and around the reported observation site have failed to reveal its presence or the presence of suitable habitat. Pine barrens treefrogs are known from the surrounding states of South Carolina, Florida, and Alabama, however, and because seemingly suitable habitat exists in the Coastal Plain of Georgia, cryptic, yet-to-be-discovered populations may be present. (4_Pine_barrens_treefrog.mp3)

Gray Treefrog (*Hyla versicolor*)

Calling adult male gray treefrog, Butler County, Iowa (Aubrey M. Heupel)

Cope's gray treefrog (*Hyla chrysoscelis*), of statewide occurrence in Georgia, is indistinguishable from *H. versicolor* except for differences in the trill rate of their calls and genetic characteristics. Adding to the identification difficulties, trill rates can vary in response to changing air temperatures, and the calls of hybrids between Cope's gray treefrogs and bird-voiced treefrogs (*H. avivoca*) are almost identical with those of the gray treefrog. It is possible that gray treefrogs occur in the state undetected. Populations have been documented in neighboring North Carolina and Tennessee. (35_Gray_treefrog.mp3)

Cuban Treefrog
(*Osteopilus septentrionalis*)

Adult Cuban treefrog, Chatham County (John B. Jensen)

Individuals have been found in Savannah and Brunswick, but no breeding population is known. However, this nonnative invader has become established in Jacksonville, Florida, after quickly marching north through the Florida peninsula, where large populations now occur. It may eventually colonize parts of coastal and southern Georgia. (36_Cuban_treefrog.mp3)

Mabee's Salamander
(*Ambystoma mabeei*)

Adult Mabee's salamander, Berkeley County, South Carolina (Pierson Hill)

A questionable record exists from a single Burke County site. This species does occur on the South Carolina side of the Savannah River, but not opposite Burke County. Otherwise, this salamander is not known to occur on the Georgia side of the river, which may be a significant dispersal barrier.

Pigmy Salamander
(*Desmognathus wrighti*)

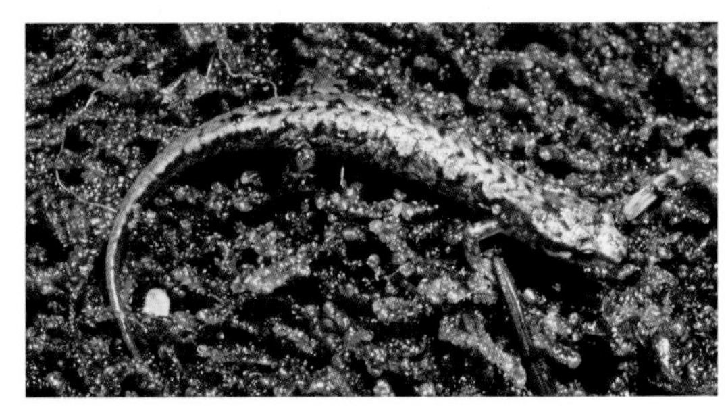

Adult pigmy salamander, Grayson County, Virginia (Steve Roble)

This high-elevation species occurs in the Nantahala National Forest of North Carolina near the Georgia state line. If it occurs in Georgia, the Nantahala Mountains in northern Rabun and Towns counties would be the likely areas, especially considering that a population occurs on Standing Indian Mountain just across the line in North Carolina.

Indo-Pacific Gecko
(*Hemidactylus garnotii*)

Like the Cuban treefrog, this invasive exotic is rapidly expanding its range northward through

Adult Indo-Pacific gecko, Glynn County (John B. Jensen)

Underside of adult Indo-Pacific gecko, Glynn County (John B. Jensen)

peninsular Florida and may become established in Georgia in the not-so-distant future. All Indo-Pacific geckos are females that hatch from unfertilized eggs, a rare reproductive condition known as parthenogenesis. Thus, arrival of a single individual might be sufficient to initiate a population.

Texas Horned Lizard (*Phrynosoma cornutum*)

Adult Texas horned lizard, Payne County, Oklahoma (Kevin Enge)

An introduced population once existed at a single site in Chatham County but has apparently died out. A single individual was captured in Camden County in 2001. Because they are interesting to look at and easy to capture, many Texas horned lizards have been brought to Georgia by people returning from the southwestern United States. Releases have led to the establishment of populations in several coastal areas of Florida and the Carolinas.

Brahminy Blind Snake (*Ramphotyphlops braminus*)

A few individuals of this unisexual (parthenogenic), wormlike snake have been found in Wayne and Dougherty counties. These Asia natives are well established in many parts of Florida, where they continue to spread through shipments of soil and potted plants. Because only a single successful female is necessary to

Adult Brahminy blind snakes, Pinellas County, Florida (Dick Bartlett)

start a population, this tiny exotic snake seems likely to become a part of Georgia's fauna if it is not one already.

Short-tailed Snake (*Stilosoma extenuatum*)

The Florida crowned snake (*Tantilla relicta*) and the Florida worm lizard (*Rhineura floridana*), both known from just a few specimens taken in extreme south-central Georgia, share very similar habitats and range with the very rare short-tailed snake, which is known to occur fairly close to the Georgia line in Florida. It would not, therefore, be surprising to find short-tailed snakes in the region of Georgia occupied by these two associates.

Adult short-tailed snake, Marion County, Florida (Kevin Enge)

Olive Ridley Sea Turtle (*Lepidochelys olivacea*)

Nesting adult female olive ridley sea turtles, Costa Rica (Michael Jensen)

Better known from the Pacific Ocean, this species is considered rare in the western North Atlantic outside the tropics. Olive ridleys have been documented from the Caribbean Sea, along the coast of Florida, and far off the coast of eastern Canada. Georgia waters could be inhabited as well, but no evidence for that exists at this time.

Ouachita Map Turtle (*Graptemys ouachitensis*)

Adult Ouachita map turtle, Lauderdale County, Alabama (Dick Bartlett)

The Tennessee River, which is inhabited by a robust population of Ouachita map turtles, meanders to within 3 miles of northwestern Georgia at several points. Future sampling in the larger Georgia tributaries of the Tennessee River such as Lookout Creek may reveal the presence of this species in the state.

Glossary

Advertisement call: Vocalization made by male frogs to attract potential mates.

Aestivation: State of dormancy during hot, dry conditions.

Aggression call: Vocalization made by male frogs to deter competing males from establishing a calling area nearby.

Algae: Single-celled plantlike organisms that often form colonies.

Ambient: Surrounding.

Amniotic egg/membrane: Fluid-filled sac or shell that surrounds a developing embryo.

Amplexus: Mating grip used by male amphibians.

Anthropogenic: Made or caused by humans.

Aposematic: Referring to a warning signal that may be audible or visible, as a color.

Arboreal: Tree dwelling.

Arthropod: Taxonomic group of invertebrates that includes insects, crustaceans, arachnids, and others.

Barbels: Fleshy, whiskerlike protuberances such as the chin adornments of stinkpots (*Sternotherus odoratus*).

Biomass: The total mass of organisms or a select group of organisms in a given area.

Blue Ridge Divide: The land divide that separates streams that drain into the Tennessee River system from those that drain into the Chattahoochee or Savannah River.

Body length: Standardized length measured from the tip of the snout to the rear of the vent.

Bridge: The section of a turtle shell that connects the lower part (plastron) to the upper part (carapace).

Canopy: The top spread of forest branches and leaves; the treetops. Variation in this forest structure component determines several microhabitat variables such as light penetration.

Canthus rostralis: A bony ridge running from the eye to the nostril, as in spring salamanders (*Gyrinophilus porphyriticus*).

Carapace: The upper shell of a turtle.

Carolina bay: Geologically unique depressional wetland of the Coastal Plain oriented on a northwest-southeast axis and usually oval or elliptical.

Carrion: Dead and decaying flesh.

Catadromous: Migrating from freshwater streams to the ocean for breeding purposes, as the American eel.

Character displacement: Phenomenon in which closely related species differ more in areas where they co-occur so as to distinguish one another or as a result of competition.

Chromosome: Cellular structures that contain the cell's genetic material.

Cladoceran: A type of minute aquatic crustacean commonly referred to as a water flea; it is an abundant member of the freshwater zooplankton.

Class: A higher, inclusive taxonomic group below the level of phylum, such as the class Amphibia.

Cocoon: The self-constructed aestivation chamber of certain salamanders formed of mucous secretions.

Competitive exclusion: The exclusion of one or more species from habitats or areas occupied by a similar species due to competitive disadvantages for food, shelter, and/or other resources.

Copepod: A type of minute, aquatic crustacean; it is an abundant member of the freshwater zooplankton.

Costal grooves/folds: Vertical grooves on the sides of salamanders corresponding to the attachment of body wall musculature to the ribs and vertebrae; the number present is often used for species identification.

Cranial crests: Bony ridges on the head of true toads (genus *Bufo*); often used as diagnostic features for species identification.

Crepuscular: Active at dawn and/or dusk.

Cryptic species: Two related species that share indistinguishable, or nearly so, external features.

Deciduous: Type of vegetation that loses its foliage at the end of the growing season.

Detritus: Particles of decomposing organic material.

Devonian period: Geologic time period occurring approximately 350–400 million years before the present.

Dewlap: An often colorful skin flap, well developed in males, that extends along the throat of lizards; it is used for territorial displays and mate attraction. Also referred to as a throat fan.

Diurnal: Active during the day.

Dorsolateral ridge/fold: Fold of skin on frogs that extends along the edges of the back and is helpful in identifying many species.

DNA: Deoxyribonucleic acid; major component of chromosomes. DNA analysis is often useful in determining relationships of organisms to one another.

Dougherty Plain: Distinct region of southwestern Georgia consisting of limestone features such as caves, springs, and sinkholes.

Eft: The sexually immature terrestrial life stage of newts (genus *Notophthalmus*).

Elephantine: Elephant-like, as in the rear feet of gopher tortoises (*Gopherus polyphemus*).

Endemic: Occurring only in a particular region.

Estuarine: Associated with the coastal regions where fresh and salt water mix.

Eutrophication: The process (either natural or human caused) of sediment and nutrient deposition in wetlands leading to changes in the overall community structure.

Excrescence: Black, keratinized, and hardened pads on the inner thighs and toe tips of breeding male newts (genus *Notophthalmus*).

Explosive breeding: Breeding characterized by intense activity over a short duration and involving a large number of participants; characteristic of certain species such as the wood frog (*Rana sylvatica*).

Extirpation: Regional disappearance of a population or populations of a species.

Fall Line: Abrupt transition between the Piedmont and Coastal Plain physiographic provinces that often determines the geographic range boundary of organisms in the southeastern United States. It approximately coincides with what is colloquially referred to as the Gnat Line. In Georgia, the Fall Line is roughly oriented along a line between the cities of Columbus, Macon, and Augusta.

Family: Taxonomic group lower and less inclusive than class and order. Often includes more than one genus and many species, such as the lungless salamander family, Plethodontidae.

Fauna: Animal component of an ecosystem.

Flatwoods: Coastal Plain habitat characterized by flat topography and an open pine canopy.

Floodplain: The low-lying area adjacent to rivers and streams that is periodically flooded during seasonal or heavy rainfall.

Floridan Aquifer: A water-bearing rock formation underlying southern Alabama, southeastern Georgia, southern South Carolina, and all of Florida. It provides water for many municipalities and habitat for the subterranean Georgia blind salamander (*Haideotriton wallacei*).

Food web: The complex interconnections between predators and prey in an ecosystem.

Fossorial: Soil-dwelling; burrowing.

Friable soil: Loose soil.

Froglet: A small, recently metamorphosed frog.

Gassing: The unethical and illegal injection of toxic fumes into gopher tortoise (*Gopherus polyphemus*) burrows to collect rattlesnakes; it often results in the death of other burrow occupants.

Genus (pl. genera): A lower and less inclusive taxonomic group under family. Usually consists of several closely related species, such as the woodland salamander genus, *Plethodon.*

Gigging: The practice of impaling frogs on barbed spears to collect them for food.

Gravid: Containing well-developed eggs.

Guano: Bat or bird dung.

Hemipene: One of two paired sex organs of male lizards and snakes.

Herbaceous: Referring to nonwoody plants or plant material.

Herpetofauna: The amphibian and reptile component of an ecosystem.

Herpetology: The scientific study of amphibians and reptiles.

Herringbone: Pattern of bent vertical lines, or chevrons, reminiscent of the appearance of a fish's skeleton.

Hybridization: The interbreeding of two distinct species.

Ichthyosaur: An ancient sharklike reptile.

Intergradation: The interbreeding of distinct subspecies.

Invertebrate: Type of animal without a backbone.

Iris: The pigment membrane of the eye.

Isolated wetland/pond: Depressional wetland that is not connected to any stream drainage and usually fills with rainwater.

Jacobson's organ: A chemical-sensing organ in the roof of the mouth of snakes and lizards.

Karst: A landscape composed of irregular limestone bedrock and characterized by fissures, caves, sinkholes, and underground streams.

Keratinized: Converted and hardened into the protein keratin; keratinized structures include hair, fingernails, claws, feathers, reptilian scales, and the rattle of rattlesnakes (genera *Crotalus* and *Sistrurus*).

Larva (pl. larvae): Immature stage of amphibians, usually possessing gills and tail fins; equivalent to tadpole in frogs.

Lateral line: A sensory system found on the sides and head of larval and some paedomorphic adult salamanders that detects vibrations caused by movement in the water.

Life history: The history of changes undergone by an organism over the course of its life, such as age at first reproduction.

Longitudinal: Lengthwise.

Melanistic: Completely dark or dominated by dark pigment.

Mental gland: A gland on the chin of certain male salamanders that produces secretions used in courtship.

Mesozoic era: Geological time period occurring approximately 65–250 million years before the present.

Metamorphosis: The dramatic change in form that occurs in many amphibians as they transform from an aquatic larva into a terrestrial juvenile.

Müllerian mimicry: A physical resemblance in which two or more unrelated noxious organisms exhibit similar warning colors or patterns.

Natural history: The study of the biology of organisms and their origins, evolution, behavior, and interrelationships with other species.

Nocturnal: Active during the night.

Nomenclature: A system of names that, in the case of scientific nomenclature, organizes organisms based on their interrelationships.

Noxious: Harmful to health, painful, or irritating in some way.

Nymph: Immature stage of certain insects.

Obligate: Able to survive only in a specific environment.

Omnivorous: Eating both animal and plant material.

Opalescent: Having a milky, iridescent appearance.

Order: A taxonomic group below the level of class and above the level of family, as in the order Caudata.

Osteoderm: A bony scale in the skin of some reptiles.

Ostracod: A type of minute aquatic crustacean commonly referred to as a seed shrimp.

Ova: Undeveloped eggs.

Oxbow: A historic meander or bend in a stream that, due to a change in the stream's course, was isolated and became a bow-shaped lake.

Paedomorphosis: Sexual maturity reached in the larval body form; i.e., without undergoing metamorphosis.

Parotoid gland: A poisonous gland found on the head, behind each eye, of toads and some salamanders.

Parthenogenesis: Reproduction without a male gamete; that is, using an unfertilized egg.

Pelagic: Living in the open ocean.

Pelham Escarpment: A slope in southwestern Georgia that divides waters that flow to the Flint River drainage from those that flow to the Ochlockonee River drainage. Most of the Flint River drainage waters flow underground along this slope.

Perennial: Lasting from one year to the next.

Photosynthesis: Process in which plant cells use sunlight to produce sugar.

Physiographic province: A region of the landscape characterized by unique physical attributes and geologic history.

Pit: A heat-sensing depression found on the side of the head of certain snakes (family Viperidae) that is used to detect and precisely aim strikes at warm-blooded prey.

Placenta: A membranous organ housing a developing embryo and which is attached to the uterus of the mother by an umbilical cord.

Plastron: The lower shell of a turtle.

Pleistocene epoch: Geologic time period occurring approximately 1.7–0.1 million years before the present.

Plesiosaur: A member of an extinct group of carnivorous marine reptiles.

Protuberance: A bulging or knoblike anatomical feature.

Pterosaur: A member of an extinct group of winged, flying reptiles.

Rain call: Daytime call of uncertain purpose issued by male frogs of certain species, often in response to rainfall.

Reticulation: A netlike pattern.

River system/drainage: The network of all seeps, creeks, and streams that comprise the water source of a certain river, including the river itself.

Rostral scale: Scale at the tip of the snout that is often enlarged or elongated in burrowing reptiles.

Sagpond: A typically round, isolated wetland type in northwestern Georgia that is formed when subsurface limestone bedrock dissolves and the ground above sags. Sagponds are similar in formation, structure, and ecology to sinkhole ponds or limesinks of the Coastal Plain.

Sandhill: A dry, sandy, upland habitat type found in the Coastal Plain; often characterized by scattered longleaf pine and abundant scrub oaks, especially turkey oak.

Savanna: A flat, treeless or nearly treeless grassland habitat.

Scute: An enlarged, platelike scale such as the plates on a turtle's shell or the scales on a snake's belly.

Seep/seepage: A habitat characterized by water trickling out of the ground, usually on a slope.

Siltation: The process (either natural or human caused) of sediment deposition in streams and wetlands.

Silviculture: Tree farming.

Sinkhole pond: A typically round, isolated wetland type in the Coastal Plain that is formed when subsurface limestone bedrock dissolves and the ground above sinks into this space.

Spatulate: Shaped like a spatula; flattened.

Species: A taxonomic level below genus and consisting of interbreeding organisms that are reproductively isolated from other such groups. The scientific name consists of the genus and the specific epithet, such as *Plethodon petraeus* (Pigeon Mountain salamander).

Species complex/group: Two or more species that are more closely related to one another than to other members of the genus.

Sphagnum: A type of loose moss found in moist, acidic habitats; often called peat moss.

Spicules: Needlelike structures found in the throat of leatherback sea turtles (*Dermochelys coriacea*).

Spiracle: The tubular vent that exits the gill chamber of tadpoles.

Stygobitic: Living in underground waters.

Subspecies: A distinct race of a species that, although able to interbreed with other races, shows unique characteristics and occupies a definable geographic region. A subspecies is named using a trinomial, with a third term tacked on to its species name, such as *Pseudotriton ruber schencki* (black-chinned red salamander).

Successional community/habitat: An ecological community or habitat that changes through time due either to natural disturbances, such as lightning-caused fires, or to an absence of them.

Sympatry: The occurrence of two or more species in the same area.

Tail musculature: In tadpoles, the central portion of the tail that contains the muscles; the tail excluding the fins.

Taxonomy: The science of naming organisms.

Temperate (region): The geographic portion of the earth's surface that lies between the Arctic and tropical regions.

Terrestrial: Referring to land or living on land.

Territory: A specific area occupied and defended by an organism.

Tetrapod: A four-legged vertebrate.

Toadlet: A very young toad that has recently metamorphosed from the tadpole stage.

Topographic: Relating to physical relief of the landscape.

Total length: Standardized length measured from the tip of the snout to the tip of the tail.

Transformation: In amphibians, the change from a larval body form into an adult body form through metamorphosis.

Transverse: Spanning from one side to the other.

Tympanum: Eardrum or ear membrane.

Venomous: An animal capable of injecting poisonous

saliva (venom) to subdue or kill and begin digestion of prey.

Vent: The common opening of the reproductive and excretory systems; also referred to as the cloaca.

Vermiculation: A wormlike pattern.

Vertebrate: Type of animal with a backbone.

Vocal sac: The inflatable part of a male frog's throat that amplifies its voice (call).

Wet prairie: An open Coastal Plain wetland type characterized by an abundance of emergent and floating herbaceous plants, especially grasses, sedges, and water lilies.

Xeric: Dry.

Zooplankton: Minute, often microscopic animals that live in the water column.

Selected References for Further Study

To encourage further reading on individual species, groups, or topics, we offer a sample of the peer-reviewed scientific literature available for each species as well as for each of the subjects of the introductory sections. This is not meant to be a comprehensive review, but the references cited here do give an idea of the types of research that have been or, in the case of poorly studied forms, have not been done. Anyone specifically interested in conducting research on any of Georgia's amphibians or reptiles should find in these admittedly abbreviated lists of references species for which a better scientific understanding is sorely needed.

Taxonomy

Collins, J. T., and T. W. Taggart. 2002. Standard Common and Current Scientific Names for North American Amphibians, Turtles, Reptiles, and Crocodilians. 5th ed. Lawrence, Kans.: Center for North American Herpetology.

Committee on Standard English and Scientific Names. 2000. Scientific and Standard English Names of Amphibians and Reptiles of North America North of Mexico, with Comments Regarding Confidence in Our Understanding. Society for the Study of Amphibians and Reptiles, Herpetological Circular no. 29.

Frost, D. R., and D. M. Hillis. 1990. Species in concept and practice: herpetological considerations. Herpetologica 46:87–104.

Conservation

Gibbons, J. W., D. E. Scott, T. J. Ryan, K. A. Buhlmann, T. D. Tuberville, B. S. Metts, J. L. Greene, T. Mills, Y. Leiden, S. Poppy, and C. T. Winne. 2000. The global decline of reptiles: déjà vu amphibians. BioScience 50:653–666.

Heyer, W. R., M. A. Donnelly, R. W. McDiarmid, L. C. Hayek, and M. S. Foster, eds. 1994. Measuring and Monitoring Biological Diversity: Standard Methods for Amphibians. Washington, D.C.: Smithsonian Institution Press.

Lannoo, M., ed. 2005. Amphibian Declines: The Conservation Status of United States Species. Berkeley: University of California Press.

Semlitsch, R. D. 2003. Amphibian Conservation. Washington, D.C.: Smithsonian Institution Press.

Amphibians and Reptiles

Martof, B. S. 1956. Amphibians and Reptiles of Georgia. Athens: University of Georgia Press.

Mount, R. H. 1975. The Reptiles and Amphibians of Alabama. Auburn, Ala: Auburn University Agricultural Experiment Station.

Pough, F. H., R. M. Andrews, J. E. Cadle, M. L. Crump, A. H. Savitzky, and K. D. Wells. 1998. Herpetology. Upper Saddle River, N.J.: Prentice Hall.

Zug, G. R., L. J. Vitt, and J. P. Caldwell. 2001. Herpetology. 2nd ed. New York: Academic Press.

Amphibians

Duellman, W. E., ed. 1999. Patterns of Distribution of Amphibians: A Global Perspective. Baltimore: Johns Hopkins University Press.

Duellman, W. E., and L. Trueb. 1986. Biology of Amphibians. New York: McGraw-Hill.

Heatwole, H., and B. K. Sullivan, eds. 1995. Amphibian Biology, vol. 2: Social Behavior. Chipping Norton, New South Wales: Surrey Beatty and Sons.

Frogs

Altig, R. 1970. A key to the tadpoles of the continental United States and Canada. Herpetologica 26:180–207.

Vial, J. L. 1973. Evolutionary Biology of the Anurans. Columbia: University of Missouri Press.

Wright, A. H. 2002. Life-Histories of the Frogs of the Okefinokee Swamp, Georgia. Ithaca, N.Y.: Cornell University Press.

Wright, A. H., and A. A. Wright. 1949. Handbook of Frogs and Toads. 3rd ed. Ithaca, N.Y.: Comstock.

BUFONIDAE

***Bufo americanus*, American Toad**

Beiswenger, R. E. 1977. Diel patterns of aggregative behavior in tadpoles of *Bufo americanus* in relation to light and temperature. Ecology 58:98–108.

Green, D. M. 2005. *Bufo americanus.* Pp. 386–390 *in* M. J. Lannoo, ed., Declining Amphibians: The Conservation Status of United States Species. Berkeley: University of California Press.

Masta, S. E., B. K. Sullivan, T. Lamb, and E. J. Routman. 2002. Molecular systematics, hybridization, and phylogeography of the *Bufo americanus* complex in eastern North America. Molecular Phylogenetics and Evolution 24:302–314.

Neill, W. T. 1949. Hybrid toads in Georgia. Herpetologica 30:30–32.

Wilbur, H. M. 1977. Density-dependent aspects of growth and metamorphosis in *Bufo americanus.* Ecology 58:196–200.

***Bufo fowleri*, Fowler's Toad**

Allard, H. A. 1908. *Bufo fowleri* (Putnam) in northern Georgia. Science 28:655–656.

Clarke, R. D. 1977. Postmetamorphic survivorship of Fowler's toad, *Bufo woodhousei fowleri.* Copeia 1977:594–597.

Green, D. M. 2005. *Bufo fowleri.* Pp. 408–412 *in* M. J. Lannoo, ed., Declining Amphibians: The Conservation Status of United States Species. Berkeley: University of California Press.

Masta, S. E., B. K. Sullivan, T. Lamb, and E. J. Routman. 2002. Molecular systematics, hybridization, and phylogeography of the *Bufo americanus* complex in eastern North America. Molecular Phylogenetics and Evolution 24:302–314.

Neill, W. T. 1949. Hybrid toads in Georgia. Herpetologica 30:30–32.

***Bufo quercicus*, Oak Toad**

Ashton, R. E. 1979. *Bufo quercicus* oak toad. Catalogue of American Amphibians and Reptiles 222:1–2.

Greenberg, C. H., and G. W. Tanner. 2005. Spatial and temporal ecology of oak toads (*Bufo quercicus*) on a Florida landscape. Herpetologica 61:422–434.

Punzo, F. 2005. *Bufo quercicus.* Pp. 432–433 *in* M. J. Lannoo, ed., Declining Amphibians: The Conservation Status of United States Species. Berkeley: University of California Press.

***Bufo terrestris*, Southern Toad**

Brower, L. P., J. V. Z. Brower, and P. W. Westcott. 1960. Experimental studies of mimicry 5. The reactions of toads (*Bufo terrestris*) to bumblebees (*Bombus americanorum*) and their robberfly mimics (*Mallophora bomboides*), with a discussion of aggressive mimicry. American Naturalist 94:343–356.

Jensen, J. B. 2005. *Bufo terrestris.* Pp. 436–438 *in* M. J. Lannoo, ed., Declining Amphibians: The Conservation Status of United States Species. Berkeley: University of California Press.

Smith, G. C. 1976. Ecological energetics of three species of ectothermic vertebrates. Ecology 57:252–264.

HYLIDAE

***Acris crepitans*, Northern Cricket Frog**

Caldwell, J. P. 1982. Disruptive selection: a tail color polymorphism in *Acris* tadpoles in response to differential predation. Canadian Journal of Zoology 60:2817–2818.

Gray, R. H., L. E. Brown, and L. Blackburn. 2005. *Acris crepitans.* Pp. 441–443 *in* M. J. Lannoo, ed., Declining Amphibians: The Conservation Status of United States Species. Berkeley: University of California Press.

Irwin, J. T., J. P. Costanzo, and R. E. Lee Jr. 1999. Terrestrial hibernation in the northern cricket frog, *Acris crepitans.* Canadian Journal of Zoology 77:1240–1246.

Johnson, L. M. 1991. Growth and development of larval northern cricket frogs (*Acris crepitans*) in relation to phytoplankton abundance. Freshwater Biology 25:51–59.

Smith, G. R., A. Todd, J. E. Retting, and F. Nelson. 2003. Microhabitat selection by northern cricket frogs (*Acris crepitans*) along a west-central Missouri creek: field and experimental observations. Journal of Herpetology 37:383–385.

***Acris gryllus*, Southern Cricket Frog**

Jensen, J. B. 2005. *Acris gryllus.* Pp. 443–445 *in* M. J. Lannoo, ed., Declining Amphibians: The Conservation Status of United States Species. Berkeley: University of California Press.

Mecham, J. S. 1964. Ecological and genetic relationships of the two cricket frogs, genus *Acris*, in Alabama. Herpetologica 20:84–91.

Neill, W. T. 1954. Taxonomy, nomenclature, and distribution of southeastern cricket frogs, genus *Acris*. American Midland Naturalist 43:152–156.

Hyla avivoca, Bird-voiced Treefrog

Jamieson, D. H., S. E. Trauth, and C. T. McAllister. 1993. Food-habits of male bird-voiced treefrogs, *Hyla avivoca* (Anura, Hylidae), in Arkansas. Texas Journal of Science 45:45–49.

Redmer, M. 2005. *Hyla avivoca*. Pp. 448–449 *in* M. J. Lannoo, ed., Declining Amphibians: The Conservation Status of United States Species. Berkeley: University of California Press.

Redmer, M., D. H. Jamieson, and S. E. Trauth. 1999. Notes on the diet of female bird-voiced treefrogs (*Hyla avivoca*) in southern Illinois. Transactions of the Illinois State Academy of Science 92:271–275.

Smith, P. W. 1966. *Hyla avivoca*. Catalogue of American Amphibians and Reptiles 28:1–2.

Hyla chrysoscelis, Cope's Gray Treefrog

Costanzo, J. P., R. E. Lee, and M. F. Wright. 1992. Freeze tolerance as an overwintering adaptation in Cope's gray treefrog (*Hyla chrysoscelis*). Copeia 1992:565–569.

Gerhardt, H. C., M. B. Pacek, L. Barnett, and K. G. Torke. 1994. Hybridization in the diploid-tetraploid treefrogs *Hyla chrysoscelis* and *Hyla versicolor*. Copeia 1994:51–59.

Ptacek, M. B. 1996. Interspecific similarity in life-history traits in sympatric populations of gray treefrogs, *Hyla chrysoscelis* and *Hyla versicolor*. Herpetologica 52:323–332.

Ralin, D. B. 1977. Evolutionary aspects of mating call variation in a diploid-tetraploid cryptic species complex of treefrogs (Anura). Evolution 31:721–736.

Ritke, M. E., and C. A. Lessman. 1994. Longitudinal study of ovarian dynamics in female gray treefrogs (*Hyla chrysoscelis*). Copeia 1994:1014–1022.

Hyla cinerea, Green Treefrog

Gerhardt, H. C., R. E. Daniel, S. A. Perrill, and S. Schramm. 1987. Mating behaviour and male mating success in the green treefrog. Animal Behaviour 35:1490–1503.

Höbel, G., and H. C. Gerhardt. 2003. Reproductive character displacement in the acoustic communication system of green tree frogs (*Hyla cinerea*). Evolution 57:894–904.

Perrill, S. A., and R. E. Daniel. 1983. Multiple egg clutches in *Hyla regilla, Hyla cinerea* and *Hyla gratiosa*. Copeia 1983:513–516.

Redmer, M., and R. A. Brandon. 2005. *Hyla cinerea*. Pp. 452–454 *in* M. J. Lannoo, ed., Declining Amphibians: The Conservation Status of United States Species. Berkeley: University of California Press.

Hyla femoralis, Pine Woods Treefrog

Hoffman, R. L. 1988. *Hyla femoralis*. Catalogue of American Amphibians and Reptiles 436:1–3.

Mitchell, J. C. 2005. *Hyla femoralis*. Pp. 454–455 *in* M. J. Lannoo, ed., Declining Amphibians: The Conservation Status of United States Species. Berkeley: University of California Press.

Wilbur, H. M. 1982. Competition between tadpoles of *Hyla femoralis* and *Hyla gratiosa* in laboratory experiments. Ecology 63:278–282.

Hyla gratiosa, Barking Treefrog

Caldwell, J. P. 1982. *Hyla gratiosa*. Catalogue of American Amphibians and Reptiles 298:1–2.

Mecham, J. S. 1960. Introgressive hybridization between two southeastern treefrogs. Evolution 14:445–457.

Mitchell, J. C. 2005. *Hyla gratiosa*. Pp. 455–456 *in* M. J. Lannoo, ed., Declining Amphibians: The Conservation Status of United States Species. Berkeley: University of California Press.

Travis, J. 1983. Variation in development patterns of larval anurans in temporary ponds. I. Persistent variation within a *Hyla gratiosa* population. Evolution 37:496–512.

Hyla squirella, Squirrel Treefrog

Babbitt, K. J., and G. W. Tanner. 1997. Effects of cover and predator identity on predation of *Hyla squirella* tadpoles. Journal of Herpetology 31:128–130.

Binckley, C. A., and W. J. Resetarits Jr. 2001. Reproductive decisions under threat of predation: squirrel treefrog (*Hyla squirella*) responses to banded sunfish (*Enneacanthus obesus*). Oecologia 130:157–161.

Mitchell, J. C., and M. J. Lannoo. 2005. *Hyla squirella*. Pp. 456–458 *in* M. J. Lannoo, ed., Declining Amphibians: The Conservation Status of United States Species. Berkeley: University of California Press.

Pseudacris brachyphona, Mountain Chorus Frog

Forester, D. C., S. Knoedler, and R. Sanders. 2003. Life history and status of the mountain chorus frog (*Pseudacris brachyphona*) in Maryland. Maryland Naturalist 46:1–15.

Green, N. B. 1938. The breeding habits of *Pseudacris brachyphona* (Cope) with a description of the eggs and tadpole. Copeia 1938:79–82.

Green, N. B. 1964. Postmetamorphic growth in the mountain chorus frog, *Pseudacris brachyphona* Cope. Proceedings of the West Virginia Academy of Science 36:34–38.

Mitchell, J. C., and T. K. Pauley. 2005. *Pseudacris brachyphona.* Pp. 465–466 *in* M. J. Lannoo, ed., Declining Amphibians: The Conservation Status of United States Species. Berkeley: University of California Press.

Moriarty, E. C., and D. C. Cannatella. 2004. Phylogenetic relationships of the North American chorus frogs (*Pseudacris:* Hylidae). Molecular Phylogenetics and Evolution 30:409–420.

Pseudacris brimleyi, Brimley's Chorus Frog

Brandt, B. B., and C. F. Walker. 1933. A new species of *Pseudacris* from the southeastern United States. Occasional Papers of the Museum of Zoology, University of Michigan 272:1–7.

Gosner, K. L., and I. H. Black. 1958. Notes on the life history of Brimley's chorus frog. Herpetologica 13:249–254.

Hoffman, R. L. 1983. *Pseudacris brimleyi.* Catalogue of American Amphibians and Reptiles 311:1–2.

Mitchell, J. C. 2005. *Pseudacris brimleyi.* Pp. 466–467 *in* M. J. Lannoo, ed., Declining Amphibians: The Conservation Status of United States Species. Berkeley: University of California Press.

Moriarty, E. C., and D. C. Cannatella. 2004. Phylogenetic relationships of the North American chorus frogs (*Pseudacris:* Hylidae). Molecular Phylogenetics and Evolution 30:409–420.

Pseudacris crucifer, Spring Peeper

Austin, J. D., S. C. Lougheed, L. Neidrauer, A. A. Chek, and P. T. Boag. 2002. Cryptic lineages in a small frog: the post-glacial history of the spring peeper, *Pseudacris crucifer* (Anura: Hylidae). Molecular Phylogenetics and Evolution 25:316–329.

Crump, M. L. 1984. Intraclutch egg size variability in *Hyla crucifer* (Anura: Hylidae). Copeia 1984:302–308.

Marshall, V. T., S. C. Humfeld, and M. A. Bee. 2003. Plasticity of aggressive signaling and its evolution in male spring peepers, *Pseudacris crucifer.* Animal Behaviour 65:1223–1234.

Moriarty, E. C., and D. C. Cannatella. 2004. Phylogenetic relationships of the North American chorus frogs (*Pseudacris:* Hylidae). Molecular Phylogenetics and Evolution 30:409–420.

Smith, D. C., and J. Van Buskirk. 1995. Phenotypic design, plasticity, and ecological performance in two tadpole species. American Naturalist 145:211–233.

Pseudacris feriarum, Upland Chorus Frog

Fouquette, M. J. 1975. Speciation in chorus frogs I. Reproductive character displacement in the *Pseudacris nigrita* complex. Systematic Zoology 24:16–22.

Moriarty, E. C., and D. C. Cannatella. 2004. Phylogenetic relationships of the North American chorus frogs (*Pseudacris:* Hylidae). Molecular Phylogenetics and Evolution 30:409–420.

Moriarty, E., and M. J. Lannoo. 2005. *Pseudacris triseriata* complex. Pp. 485–488 *in* M. J. Lannoo, ed., Declining Amphibians: The Conservation Status of United States Species. Berkeley: University of California Press.

Smith, D. C. 1987. Adult recruitment in chorus frogs: effects of size and date at metamorphosis. Ecology 68:344–350.

Pseudacris nigrita, Southern Chorus Frog

Caldwell, J. P. 1987. Demography and life history of two species of chorus frogs (Anura: Hylidae) in South Carolina. Copeia 1987:114–127.

Leja, W. T. 2005. *Pseudacris nigrita.* Pp. 474–475 *in* M. J. Lannoo, ed., Declining Amphibians: The Conservation Status of United States Species. Berkeley: University of California Press.

Pechmann, J. H. K., and R. D. Semlitsch. 1986. Diel activity patterns in the breeding migrations of winter-breeding anurans. Canadian Journal of Zoology 64:1116–1120.

Pseudacris ocularis, Little Grass Frog

Harper, F. 1939. Distribution, taxonomy, nomenclature, and habits of the little treefrog (*Hyla ocularis*). American Midland Naturalist 22:134–149.

Jensen, J. B. 2004. Little grass frog *Pseudacris ocularis.* P. 27 *in* R. E. Mirachi, M. A. Bailey, T. M. Haggerty, and T. L. Best, eds., Alabama Wildlife, vol. 3: Imperiled Amphibians, Reptiles, Birds, and Mammals. Tuscaloosa: University of Alabama Press.

Jensen, J. B. 2005. *Pseudacris ocularis.* Pp. 475–477 *in* M. J. Lannoo, ed., Declining Amphibians: The Conservation Status of United States Species. Berkeley: University of California Press.

Marshall, J. L., and C. D. Camp. 1995. Aspects of the feeding ecology of the little grass frog, *Pseudacris ocularis* (Anura: Hylidae). Brimleyana 22:1–7.

Pseudacris ornata, Ornate Chorus Frog

Brown, L. E., and D. B. Means. 1984. Fossorial behavior and ecology of the chorus frog *Pseudacris ornata.* Amphibia-Reptilia 5:261–273.

Caldwell, J. P. 1987. Demography and life history of two species of chorus frogs (Anura: Hylidae) in South Carolina. Copeia 1987:114–127.

Harper, F. 1937. A season with Holbrook's chorus frog (*Pseudacris ornata*). American Midland Naturalist 18:260–272.

Jensen, J. B. 2005. *Pseudacris ornata.* Pp. 477–478 *in* M. J. Lannoo, ed., Declining Amphibians: The Conservation Status of United States Species. Berkeley: University of California Press.

LEPTODACTYLIDAE

Eleutherodactylus planirostris, Greenhouse Frog

Goin, C. J. 1947. Studies on the life history of *Eleutherodactylus ricordii planirostris* (Cope) in Florida. University of Florida Studies, Biological Sciences Series 4:1–66.

Meshaka, W. E. Jr. 2005. *Eleutherodactylus planirostris.* Pp. 499–500 *in* M. J. Lannoo, ed., Declining Amphibians: The Conservation Status of United States Species. Berkeley: University of California Press.

MICROHYLIDAE

Gastrophryne carolinensis, Eastern Narrow-mouthed Toad

Anderson, P. K. 1954. Studies in the ecology of the narrow-mouthed toad, *Microhyla carolinensis carolinensis.* Tulane Studies in Zoology 2:15–46.

Garton, J. S., and H. R. Mushinsky. 1979. Integumentary toxicity and unpalatability as a defensive mechanism in *Gastrophryne carolinensis.* Canadian Journal of Zoology 57:1965–1973.

Mitchell, J. C., and M. J. Lannoo. 2005. *Gastrophryne carolinensis.* Pp. 501–503 *in* M. J. Lannoo, ed., Declining Amphibians: The Conservation Status of United States Species. Berkeley: University of California Press.

Nelson, C. E. 1972. *Gastrophryne carolinensis.* Eastern narrow-mouthed toad. Catalogue of American Amphibians and Reptiles 120:1–4.

Walls, S. C., D. G. Taylor, and C. M. Wilson. 2002. Interspecific differences in susceptibility to competition and predation in a species-pair of larval amphibians. Herpetologica 58:104–118.

PELOBATIDAE

Scaphiopus holbrookii, Eastern Spadefoot

Hansen, K. L. 1958. Breeding pattern of the eastern spadefoot toad. Herpetologica 14:57–67.

Jansen, K. P., A. P. Summers, and P. R. Delis. 2001. Spadefoot toads (*Scaphiopus holbrookii holbrookii*) in an urban landscape: effects of nonnatural substrates on burrowing in adults and juveniles. Journal of Herpetology 35:141–145.

Neill, W. T. 1957. Notes on metamorphic and breeding aggregations of the eastern spadefoot, *Scaphiopus holbrooki* (Harlan). Herpetologica 13:185–187.

Palis, J. G. 2005. *Scaphiopus holbrookii.* Pp. 511–513 *in* M. J. Lannoo, ed., Declining Amphibians: The Conservation Status of United States Species. Berkeley: University of California Press.

Pearson, P. G. 1955. Population ecology of the spadefoot toad, *Scaphiopus h. holbrooki* (Harlan). Ecological Monographs 25:233–267.

RANIDAE

Rana capito, Gopher Frog

Greenberg, C. H. 2001. Spatio-temporal dynamics of pond use and recruitment in Florida gopher frogs (*Rana capito aesopus*). Journal of Herpetology 35:74–85.

Jensen, J. B., M. A. Bailey, E. L. Blankenship, and C. D. Camp. 2003. The relationship between breeding by the gopher frog, *Rana capito* (Amphibia: Ranidae), and rainfall. American Midland Naturalist 150:185–190.

Jensen, J. B., and S. C. Richter. 2005. *Rana capito.* Pp. 536–538 *in* M. J. Lannoo, ed., Declining Amphibians: The Conservation Status of United States Species. Berkeley: University of California Press.

Palis, J. G. 1998. Breeding biology of the gopher frog, *Rana capito,* in western Florida. Journal of Herpetology 32:217–233.

Semlitsch, R. D., J. W. Gibbons, and T. D. Tuberville. 1995. Timing of reproduction and metamorphosis in the Carolina gopher frog (*Rana capito capito*) in South Carolina. Journal of Herpetology 29:612–614.

***Rana catesbeiana,* Bullfrog**

Casper, G. S., and R. Hendricks. 2005. *Rana catesbeiana.* Pp. 540–546 *in* M. J. Lannoo, ed., Declining Amphibians: The Conservation Status of United States Species. Berkeley: University of California Press.

Howard, R. H. 1978. The evolution of mating strategies in bullfrogs, *Rana catesbeiana.* Evolution 32:850–871.

Justis, C. S., and D. H. Taylor. 1976. Extraocular photoreception and compass orientation in larval bullfrogs, *Rana catesbeiana.* Copeia 1976:98–105.

Smith, A. K. 1977. Attraction of bullfrogs (Amphibia, Anura, Ranidae) to distress calls of immature frogs. Journal of Herpetology 11:234–235.

Stinner, J., N. Zarlinga, and S. Orcutt. 1994. Overwintering behavior of adult bullfrogs, *Rana catesbeiana,* in northeastern Ohio. Ohio Journal of Science 94:8–13.

***Rana clamitans,* Green Frog**

Berven, K. A., D. E. Gill, and S. J. Smith-Gill. 1979. Countergradient selection in the green frog, *Rana clamitans.* Evolution 33:609–623.

Martof, B. S. 1953. Home range and movements of the green frog, *Rana clamitans.* Ecology 34:529–543.

Martof, B. S. 1953. Territoriality in the green frog, *Rana clamitans.* Ecology 34:165–174.

Martof, B. S. 1956. Factors influencing the size and composition of populations of *Rana clamitans.* American Midland Naturalist 56:224–245.

Martof, B. S. 1956. Growth and development of the green frog, *Rana clamitans,* under natural conditions. American Midland Naturalist 55:101–117.

***Rana grylio,* Pig Frog**

Lamb, T. 1984. The influence of sex and breeding condition on microhabitat selection and diet in the pig frog, *Rana grylio.* American Midland Naturalist 111:311–318.

Richter, S. C. 2005. *Rana grylio.* Pp. 555–557 *in* M. J. Lannoo, ed., Declining Amphibians: The Conservation Status of United States Species. Berkeley: University of California Press.

Wood, K. V., J. D. Nichols, H. P. Percival, and J. B. Hines. 1998. Size-sex variation in survival rates and abundance of pig frogs, *Rana grylio,* in northern Florida wetlands. Journal of Herpetology 32:527–535.

***Rana heckscheri,* River Frog**

Altig, R., and M. T. Christensen. 1981. Behavioral characteristics of the tadpoles of *Rana heckscheri.* Journal of Herpetology 15:151–154.

Butterfield, B. P., and M. J. Lannoo. 2005. *Rana heckscheri.* Pp. 557–558 *in* M. J. Lannoo, ed., Declining Amphibians: The Conservation Status of United States Species. Berkeley: University of California Press.

Hansen, K. L. 1957. Movements, area of activity, and growth of *Rana heckscheri.* Copeia 1957:274–277.

***Rana palustris,* Pickerel Frog**

Redmer, M. 2005. *Rana palustris.* Pp. 568–570 *in* M. J. Lannoo, ed., Declining Amphibians: The Conservation Status of United States Species. Berkeley: University of California Press.

Resetarits, W. J. Jr. 1986. Ecology of cave use by the frog, *Rana palustris.* American Midland Naturalist 116:256–266.

Schaaf, R. T. Jr., and P. W. Smith. 1971. *Rana palustris.* Catalogue of American Amphibians and Reptiles 117:1–3.

***Rana sphenocephala,* Southern Leopard Frog**

Butterfield, B. P., M. J. Lannoo, and P. Nanjappa. 2005. *Rana sphenocephala.* Pp. 586–587 *in* M. J. Lannoo, ed., Declining Amphibians: The Conservation Status of United States Species. Berkeley: University of California Press.

Caldwell, J. P. 1986. Selection of egg deposition sites: a seasonal shift in the southern leopard frog, *Rana sphenocephala.* Copeia 1986:249–253.

McCallum, M. L., S. E. Trauth, M. N. Mary, C. McDowell, and B. A. Wheeler. 2004. Fall breeding of the southern leopard frog (*Rana sphenocephala*) in northeastern Arkansas. Southeastern Naturalist 3:401–408.

Ryan, T. J., and C. T. Winne. 2001. Effects of hydroperiod on metamorphosis in *Rana sphenocephala.* American Midland Naturalist 145:46–53.

Saenz, D., J. B. Johnson, C. K. Adams, and G. H. Dayton. 2003. Accelerated hatching of southern leopard frog (*Rana sphenocephala*) eggs in response to the presence of a crayfish (*Procambarus nigrocinctus*) predator. Copeia 2003:646–649.

***Rana sylvatica*, Wood Frog**

Berven, K. A. 1990. Factors affecting population fluctuations in larval and adult stages of the wood frog (*Rana sylvatica*). Ecology 71:1599–1608.

Camp, C. D., C. W. Condee, and D. G. Lovell. 1990. Oviposition, larval development, and metamorphosis in the wood frog, *Rana sylvatica* (Anura: Ranidae), in Georgia. Brimleyana 16:17–21.

Davis, M. S., and G. W. Folkerts. 1986. The life history of the wood frog, *Rana sylvatica* LeConte (Amphibia: Ranidae), in Alabama. Brimleyana 12:29–50.

Howard, R. D. 1980. Mating behavior and mating success in wood frogs, *Rana sylvatica*. Animal Behaviour 28:705–716.

Meeks, D. E., and J. W. Nagel. 1977. Reproduction and development of the wood frog, *Rana sylvatica*, in eastern Tennessee. Herpetologica 29:188–191.

***Rana virgatipes*, Carpenter Frog**

Given, M. F. 1988. Growth rate and cost of calling activity in male carpenter frogs, *Rana virgatipes*. Behavioral Ecology and Sociobiology 22:153–160.

Gosner, K. L., and I. H. Black. 1968. *Rana virgatipes*. Catalogue of American Amphibians and Reptiles 67:1–2.

Means, D. B., and S. P. Christman. 1992. Carpenter frog: *Rana virgatipes* Cope. Pp. 26–29 *in* P. E. Moler, ed., Rare and Endangered Biota of Florida, vol. 3: Amphibians and Reptiles. Gainesville: University Press of Florida.

Salamanders

Altig, R., and P. H. Ireland. 1984. A key to salamander larvae and larviform adults of the United States and Canada. Herpetologica 40:212–218.

Bishop, S. C. 1947. Handbook of Salamanders. The Salamanders of the United States, of Canada, and of Lower California. Ithaca, N.Y.: Comstock.

Bruce, R. C., R. G. Jaeger, and L. D. Houck, eds. 2000. The Biology of Plethodontid Salamanders. New York: Klewer Academic/Plenum.

Petranka, J. W. 1998. Salamanders of the United States and Canada. Washington, D.C.: Smithsonian Institution Press.

Sever, D. M., ed. 2003. Reproductive Biology and Phylogeny of Urodela. Enfield, N.H.: Science Publishers.

AMBYSTOMATIDAE

***Ambystoma cingulatum*, Flatwoods Salamander**

Anderson, J. D., and G. K. Williamson. 1976. Terrestrial mode of reproduction in *Ambystoma cingulatum*. Herpetologica 32:214–221.

Means, D. B., J. G. Palis, and M. Baggett. 1996. Effects of slash pine silviculture on a Florida population of flatwoods salamander (*Ambystoma cingulatum*). Conservation Biology 10:426–437.

Palis, J. G. 1997. Breeding migration of *Ambystoma cingulatum* in Florida. Journal of Herpetology 31:71–78.

Pauly, G. B., O. Piskurek, and H. B. Shaffer. 2007. Phylogeographic concordance in the southeastern United States: the flatwoods salamander, *Ambystoma cingulatum*, as a test case. Molecular Ecology 16:415–429.

Whiles, M. R., J. B. Jensen, J. G. Palis, and W. G. Dyer. 2004. Diets of larval flatwoods salamanders, *Ambystoma cingulatum*, from Florida and South Carolina. Journal of Herpetology 38:208–214.

***Ambystoma maculatum*, Spotted Salamander**

Bachmann, M. D., R. G. Carlton, J. M. Burkholder, and R. G. Wetzel. 1986. Symbiosis between salamander eggs and green algae: microelectrode measurements inside eggs demonstrate effect of photosynthesis on oxygen concentration. Canadian Journal of Zoology 64:1586–1588.

Ducey, P. K., and P. Ritsema. 1988. Intraspecific aggression and responses to marked substrates in *Ambystoma maculatum* (Caudata: Ambystomatidae). Copeia 1988:1008–1013.

Husting, E. L. 1965. Survival and breeding structure in a population of *Ambystoma maculatum*. Copeia 1965:352–362.

Phillips, C. A. 1994. Geographic distribution of mitochondrial DNA variants and the historical biogeography of the spotted salamander, *Ambystoma maculatum*. Evolution 48:597–607.

Pough, F. H. 1976. Acid precipitation and embryonic mortality in spotted salamanders, *Ambystoma maculatum*. Science 192:68–70.

***Ambystoma opacum*, Marbled Salamander**

Jackson, M. E., D. E. Scott, and R. A. Estes. 1989. Determinants of nest success in the marbled salamander (*Ambystoma opacum*). Canadian Journal of Zoology 67:2277–2281.

Kaplan, R. H., and M. L. Crump. 1978. The noncost of brooding in *Ambystoma opacum*. Copeia 1978:99–103.

Krenz, J. D., and D. E. Scott. 1994. Terrestrial courtship affects mating locations in *Ambystoma opacum*. Herpetologica 50:46–50.

Scott, D. E. 1998. A breeding congress. Natural History 107:26–28.

Scott, D. E. 2005. *Ambystoma opacum*. Pp. 627–632 *in* M. J. Lannoo, ed., Declining Amphibians: The Conservation Status of United States Species. Berkeley: University of California Press.

***Ambystoma talpoideum*, Mole Salamander**

Raymond, L. R. and L. M. Hardy. 1991. Effects of a clearcut on a population of the mole salamander, *Ambystoma talpoideum*, in an adjacent unaltered forest. Journal of Herpetology 25:509–512.

Semlitsch, R. D. 1981. Terrestrial activity and summer home range of the mole salamander (*Ambystoma talpoideum*). Canadian Journal of Zoology 59:315–322.

Semlitsch, R. D. 1987. Relationship of pond drying to the reproductive success of the salamander *Ambystoma talpoideum*. Copeia 1987:61–69.

Semlitsch, R. D. 1988. Allotropic distribution of two salamanders: effects of fish predation and competitive interactions. Copeia 1988:290–298.

Semlitsch, R. D., and J. W. Gibbons. 1985. Phenotypic variation in metamorphosis and paedomorphosis in the salamander *Ambystoma talpoideum*. Ecology 66:1123–1130.

***Ambystoma tigrinum*, Tiger Salamander**

Church, S. A., J. N. Kraus, J. C. Mitchell, D. R. Church, and D. R. Taylor. 2003. Evidence for multiple Pleistocene refugia in the postglacial expansion of the eastern tiger salamander, *Ambystoma tigrinum tigrinum*. Evolution 57:372–383.

Davidson, E. W., M. Parris, J. P. Collins, J. E. Longcore, A. P. Pessier, and J. Brunner. 2003. Pathogenicity and transmission of chytridiomycosis in tiger salamanders (*Ambystoma tigrinum*). Copeia 2003:601–607.

Gehlbach, F. R. 1967. *Ambystoma tigrinum* (Green). Catalogue of American Amphibians and Reptiles 52:1–4.

Semlitsch, R. D. 1983. Structure and dynamics of two breeding populations of the eastern tiger salamander, *Ambystoma tigrinum*. Copeia 1983:608–616.

Whiteman, H. H., J. P. Sheen, E. B. Johnson, A. van Deusen, R. Cargille, and T. W. Sacco. 2003. Heterospecific prey and trophic polyphenism in larval tiger salamanders. Copeia 2003:56–67.

AMPHIUMIDAE

***Amphiuma means*, Two-toed Amphiuma**

Baker, C. L. 1945. The natural history and morphology of Amphiumae. Journal of the Tennessee Academy of Science 20:55–91.

Gunzburger, M. S. 2003. Evaluation of the hatching trigger and larval ecology of the salamander *Amphiuma means*. Herpetologica 59:459–468.

Snodgrass, J. W., J. W. Ackerman, A. L. Bryan Jr., and J. Burger. 1999. Influence of hydroperiod, isolation, and heterospecifics on the distribution of aquatic salamanders (*Siren* and *Amphiuma*) among depression wetlands. Copeia 1999:107–113.

Sorensen, K. 2004. Population characteristics of *Siren lacertina* and *Amphiuma means* in north Florida. Southeastern Naturalist 3:249–258.

Weber, J. A. 1944. Observations on the life history of *Amphiuma means*. Copeia 1944:61–62.

***Amphiuma pholeter*, One-toed Amphiuma**

Jensen, J. B. 1999. One-toed amphiuma: *Amphiuma pholeter*. Pp. 94–95 *in* T. W. Johnson, J. C. Ozier, J. L. Bohannon, J. B. Jensen, and C. Skelton, eds., Protected Animals of Georgia. Social Circle, Ga.: Georgia Department of Natural Resources, Wildlife Resources Division.

Means, D. B. 1992. One-toed amphiuma: *Amphiuma pholeter* Neill. Pp. 34–38 *in* P. E. Moler, ed., Rare and Endangered Biota of Florida, vol. 3: Amphibians and Reptiles. Gainesville: University Press of Florida.

Means, D. B. 1996. *Amphiuma pholeter* Neill. Catalogue of American Amphibians and Reptiles 622:1–2.

Means, D. B. 2005. *Amphiuma pholeter*. Pp. 645–646 *in* M. J. Lannoo, ed., Declining Amphibians: The Conservation Status of United States Species. Berkeley: University of California Press.

Neill, W. T. 1964. A new species of salamander, genus *Amphiuma*, from Florida. Herpetologica 20:62–66.

CRYPTOBRANCHIDAE

***Cryptobranchus alleganiensis*, Hellbender**

Humphries, W. J., and T. K. Pauley. 2000. Seasonal changes in nocturnal activity of the hellbender, *Cryptobranchus alleganiensis*, in West Virginia. Journal of Herpetology 34:604–607.

Jensen, J. B. 1999. Hellbender: *Cryptobranchus alleganiensis*. Pp. 98–99 *in* T. W. Johnson, J. C. Ozier, J. L. Bohannon, J. B. Jensen, and C. Skelton, eds., Protected Animals of Georgia. Social Circle, Ga.: Georgia Department of Natural Resources, Wildlife Resources Division.

Nickerson, M. A., K. L. Krysko, and R. D. Owen. 2003. Habitat differences affecting age class distributions of the hellbender salamander, *Cryptobranchus alleganiensis*. Southeastern Naturalist 2:619–629.

Phillips, C. A., and W. J. Humphries. 2005. *Cryptobranchus alleganiensis*. Pp. 648–651 *in* M. J. Lannoo, ed., Declining Amphibians: The Conservation Status of United States Species. Berkeley: University of California Press.

Wheeler, B. A., E. Prosen, A. Mathis, and R. F. Wilkinson. 2003. Population declines of a long-lived salamander: a 20+-year study of hellbenders, *Cryptobranchus alleganiensis*. Biological Conservation 109:151–156.

PLETHODONTIDAE

***Aneides aeneus*, Green Salamander**

Bruce, R. C. 1968. The role of the Blue Ridge Embayment in the zoogeography of the green salamander. Herpetologica 24:185–194.

Corser, J. D. 2001. Decline of disjunct green salamander (*Aneides aeneus*) populations in the southern Appalachians. Biological Conservation 97:119–126.

Gordon, R. E. 1952. A contribution to the ecology and life history of the plethodontid salamander, *Aneides aeneus*. American Midland Naturalist 47:666–701.

Jensen, J. B. 1999. Green salamander: *Aneides aeneus*. Pp. 96–97 *in* T. W. Johnson, J. C. Ozier, J. L. Bohannon, J. B. Jensen, and C. Skelton, eds., Protected Animals of Georgia. Social Circle, Ga.: Georgia Department of Natural Resources, Wildlife Resources Division.

Waldron, J. L., and W. J. Humphries. 2005. Arboreal habitat use by the green salamander, *Aneides aeneus*, in South Carolina. Journal of Herpetology 39:486–492.

***Desmognathus aeneus*, Seepage Salamander**

Donovan, L. A., and G. W. Folkerts. 1972. Foods of the seepage salamander *Desmognathus aeneus* Brown and Bishop. Herpetologica 28:35–37.

Harrison, J. R. 1967. Observations on the life history, ecology and distribution of *Desmognathus aeneus aeneus* Brown and Bishop. American Midland Naturalist 77:356–370.

Harrison, J. R. 2005. *Desmognathus aeneus*. Pp. 696–698 *in* M. J. Lannoo, ed., Declining Amphibians: The Conservation Status of United States Species. Berkeley: University of California Press.

Promislow, D. E. L. 1987. Courtship behavior of a plethodontid salamander, *Desmognathus aeneus*. Journal of Herpetology 21:298–306.

***Desmognathus apalachicolae*, Apalachicola Dusky Salamander**

Means, D. B. 1993. *Desmognathus apalachicolae* Means and Karlin. Catalogue of American Amphibians and Reptiles 556:1–2.

Means, D. B. 2005. *Desmognathus apalachicolae*. Pp. 698–699 *in* M. J. Lannoo, ed., Declining Amphibians: The Conservation Status of United States Species. Berkeley: University of California Press.

Means, D. B., and A. A. Karlin. 1989. A new species of *Desmognathus* from the eastern Gulf Coastal Plain. Herpetologica 45:37–46.

***Desmognathus auriculatus*, Southern Dusky Salamander**

Dodd, C. K. Jr. 1998. *Desmognathus auriculatus* at Devil's Millhopper State Geological Site, Alachua County, Florida. Florida Scientist 61:38–45.

Means, D. B. 1974. The status of *Desmognathus brimleyorum* Stejneger and an analysis of the genus *Desmognathus* (Amphibia: Urodela) in Florida. Bulletin of the Florida State Museum, Biological Sciences 18:1–100.

Means, D. B. 1999. *Desmognathus auriculatus*. Catalogue of American Amphibians and Reptiles 681:1–6.

Means, D. B. 2004. Southern dusky salamander, *Desmognathus auriculatus* (Holbrook). Pp. 22–23 *in* R. E. Mirarchi, M. A. Bailey, T. M. Haggerty, and T. L. Best, eds., Alabama Wildlife, vol. 3: Imperiled Amphibians, Reptiles, Birds, and Mammals. Tuscaloosa: University of Alabama Press.

Means, D. B. 2005. *Desmognathus auriculatus*. Pp. 700–701 *in* M. J. Lannoo, ed., Declining

Amphibians: The Conservation Status of United States Species. Berkeley: University of California Press.

***Desmognathus conanti*,**
Spotted Dusky Salamander

Bonett, R. M. 2002. Analysis of the contact zone between the dusky salamanders *Desmognathus fuscus fuscus* and *Desmognathus fuscus conanti* (Caudata: Plethodontidae). Copeia 2002:344–355.
Jones, R. L. 1986. Reproductive biology of *Desmognathus fuscus* and *Desmognathus santeetlah* in the Unicoi Mountains. Herpetologica 42:323–334.
Means, D. B., and R. M. Bonett. 2005. *Desmognathus conanti.* Pp. 705–706 *in* M. J. Lannoo, ed., Declining Amphibians: The Conservation Status of United States Species. Berkeley: University of California Press.
Sites, J. W. Jr. 1978. The foraging strategy of the dusky salamander, *Desmognathus fuscus* (Amphibia, Urodela, Plethodontidae): an empirical approach to predation theory. Journal of Herpetology 12:373–383.
Verrell, P. A. 1995. The courtship behavior of the spotted dusky salamander, *Desmognathus fuscus conanti* (Amphibia: Caudata: Plethodontidae). Journal of Zoology (London) 235:515–523.

***Desmognathus folkertsi*,**
Dwarf Black-bellied Salamander

Camp, C. D. 2004. *Desmognathus folkertsi.* Catalogue of American Amphibians and Reptiles 782:1–3.
Camp, C. D., and J. L. Marshall. 2007. Reproductive life history of the dwarf black-bellied salamander, *Desmognathus folkertsi.* Southeastern Naturalist 7:669–684.
Camp, C. D., and S. G. Tilley. 2005. *Desmognathus folkertsi.* Pp. 706–708 *in* M. J. Lannoo, ed., Declining Amphibians: The Conservation Status of United States Species. Berkeley: University of California Press.
Camp, C. D., S. G. Tilley, R. M. Austin Jr., and J. L. Marshall. 2002. A new species of black-bellied salamander (genus *Desmognathus*) from the Appalachian Mountains of northern Georgia. Herpetologica 58:471–484.

***Desmognathus marmoratus*,**
Shovel-nosed Salamander

Bruce, R. C. 1985. Larval periods, population structure and the effects of stream drift in larvae of the salamanders *Desmognathus quadramaculatus* and *Leurognathus marmoratus* in a southern Appalachian stream. Copeia 1985:847–854.
Camp, C. D., and S. G. Tilley. 2005. *Desmognathus marmoratus.* Pp. 711–713 *in* M. J. Lannoo, ed., Declining Amphibians: The Conservation Status of United States Species. Berkeley: University of California Press.
Jones, M. T., S. R. Voss, M. B. Ptacek, D. W. Weisrock, and D. W. Tonkyn. 2005. River drainages and phylogeography: an evolutionarily significant lineage of shovel-nosed salamander (*Desmognathus marmoratus*) in the southern Appalachians. Molecular Phylogenetics and Evolution 38:280–287.
Martof, B. S. 1962. Some aspects of the life history and ecology of the salamander *Leurognathus.* American Midland Naturalist 67:1–35.
Martof, B. S., and D. C. Scott. 1957. The food of the salamander *Leurognathus.* Ecology 38:494–501.

***Desmognathus monticola*, Seal Salamander**

Brock, J., and P. Verrell. 1994. Courtship behavior of the seal salamander, *Desmognathus monticola* (Amphibia: Caudata: Plethodontidae). Journal of Herpetology 28:411–415.
Bruce, R. C. 1989. Life history of the salamander *Desmognathus monticola,* with a comparison of the larval periods of *D. monticola* and *D. ochrophaeus.* Herpetologica 45:144–155.
Bruce, R. C., and N. Hairston Sr. 1990. Life-history correlates of body-size differences between two populations of the salamander, *Desmognathus monticola.* Journal of Herpetology 24:124–134.
Camp, C. D., and S. G. Tilley. 2005. *Desmognathus monticola.* Pp. 713–716 *in* M. J. Lannoo, ed., Declining Amphibians: The Conservation Status of United States Species. Berkeley: University of California Press.

***Desmognathus ocoee*, Ocoee Salamander**

Camp, C. D., and S. G. Tilley. 2005. *Desmognathus ocoee.* Pp. 719–721 *in* M. J. Lannoo, ed., Declining Amphibians: The Conservation Status of United States Species. Berkeley: University of California Press.
Martof, B. S., and F. L. Rose. 1963. Geographic variation in southern populations of *Desmognathus ochrophaeus.* American Midland Naturalist 69:376–425.
Tilley, S. G. 1980. Life histories and comparative

demography of two salamander populations. Copeia 1980:806–821.

Tilley, S. G., and M. J. Mahoney. 1996. Patterns of genetic differentiation in salamanders of the *Desmognathus ochrophaeus* complex (Amphibia: Plethodontidae). Herpetological Monographs 10:1–42.

Desmognathus quadramaculatus, Black-bellied Salamander

Austin, R. M. Jr., and C. D. Camp. 1992. Larval development of black-bellied salamanders, *Desmognathus quadramaculatus,* in northeastern Georgia. Herpetologica 48:313–317.

Camp, C. D. 1996. Bite scar patterns in the black-bellied salamander, *Desmognathus quadramaculatus.* Journal of Herpetology 30:543–546.

Camp, C. D. 1997. The status of the black-bellied salamander (*Desmognathus quadramaculatus*) as a predator of heterospecific salamanders in Appalachian streams. Journal of Herpetology 31:613–616.

Camp, C. D., and T. P. Lee. 1996. Intraspecific spacing and interaction within a population of *Desmognathus quadramaculatus.* Copeia 1996:78–84.

Camp, C. D., J. L. Marshall, and R. M. Austin Jr. 2000. The evolution of adult body size in black-bellied salamanders (*Desmognathus quadramaculatus* complex). Canadian Journal of Zoology 78:1712–1722.

Eurycea bislineata complex, Two-lined Salamanders

Camp, C. D., J. L. Marshall, K. R. Landau, R. M. Austin Jr., and S. G. Tilley. 2000. Sympatric occurrence of two species of the two-lined salamander (*Eurycea bislineata*) complex. Copeia 2000:572–578.

Guy, C. J., R. E. Ratajczak, and G. D. Grossman. 2004. Nest-site selection by southern two-lined salamanders (*Eurycea cirrigera*) in the Georgia Piedmont. Southeastern Naturalist 3:75–88.

Kozak, K. H., and R. R. Montanucci. 2001. Genetic variation across a contact zone between montane and lowland forms of the two-lined salamander (*Eurycea bislineata*) species complex: a test of species limits. Copeia 2001:25–34.

Rose, F. L., and F. M. Bush. 1963. A new species of *Eurycea* (Amphibia: Caudata) from the southeastern United States. Tulane Studies in Zoology and Botany 10:121–128.

Smith, S., and G. Grossman. 2003. Stream microhabitat use by larval two-lined salamanders (*Eurycea cirrigera*) in the Georgia Piedmont. Copeia 2003:531–543.

Eurycea guttolineata, Three-lined Salamander

Carlin, J. L. 1997. Genetic and morphological differentiation between *Eurycea longicauda longicauda* and *E. guttolineata* (Caudata: Plethodontidae). Herpetologica 53:206–217.

Freeman, S. L., and R. C. Bruce. 2001. Larval period and metamorphosis of the three-lined salamander, *Eurycea guttolineata* (Amphibia: Plethodontidae), in the Chattooga River watershed. American Midland Naturalist 145:194–200.

Marshall, J. L. 1999. The life-history traits of *Eurycea guttolineata* (Caudata, Plethodontidae), with implications for life-history evolution. Alytes 16:97–110.

Ryan, T. J., and B. A. Douthitt. 2005. *Eurycea guttolineata.* Pp. 743–745 *in* M. J. Lannoo, ed., Declining Amphibians: The Conservation Status of United States Species. Berkeley: University of California Press.

Eurycea longicauda, Long-tailed Salamander

Anderson, J. D., and P. J. Martino. 1966. The life history of *Eurycea l. longicauda* associated with ponds. American Midland Naturalist 75:257–279.

Carlin, J. L. 1997. Genetic and morphological differentiation between *Eurycea longicauda longicauda* and *E. guttolineata* (Caudata: Plethodontidae). Herpetologica 53:206–217.

Ireland, P. H. 1979. *Eurycea longicauda.* Catalogue of American Amphibians and Reptiles 221:1–4.

Ryan, T. J., and C. Conner. 2005. *Eurycea longicauda.* Pp. 747–750 *in* M. J. Lannoo, ed., Declining Amphibians: The Conservation Status of United States Species. Berkeley: University of California Press.

Eurycea lucifuga, Cave Salamander

Hutchison, V. H. 1958. The distribution and ecology of the cave salamander, *Eurycea lucifuga.* Ecological Monographs 28:1–20.

Hutchison, V. H. 1966. *Eurycea lucifuga.* Catalogue of American Amphibians and Reptiles 24:1–2.

Juterbock, J. E. 2005. *Eurycea lucifuga.* Pp. 750–753 *in* M. J. Lannoo, ed., Declining Amphibians: The Conservation Status of United States Species. Berkeley: University of California Press.

Peck, S. B. 1974. The food of the salamanders *Eurycea lucifuga* and *Plethodon glutinosus* in caves. National Speleological Bulletin 36:7–10.

***Eurycea quadridigitata* complex, Dwarf Salamanders**

Harrison, J. R. III, and S. I. Guttman. 2003. A new species of *Eurycea* (Caudata: Plethodontidae) from North and South Carolina. Southeastern Naturalist 2:159–178.

McMillan, M. A., and R. D. Semlitsch. 1980. Prey of the dwarf salamander, *Eurycea quadridigitata*, in South Carolina. Journal of Herpetology 14:422–424.

Powders, V. N., and R. Cate. 1980. Food of the dwarf salamander, *Eurycea quadridigitata*, in Georgia. Journal of Herpetology 14:81–82.

Semlitsch, R. D. 1980. Growth and metamorphosis of larval dwarf salamanders (*Eurycea quadridigitata*). Herpetologica 36:138–140.

Trauth, S. E. 1983. Reproductive biology and spermathecal anatomy of the dwarf salamander (*Eurycea quadridigitata*) in Alabama. Herpetologica 39:9–15.

***Gyrinophilus palleucus*, Tennessee Cave Salamander**

Beachy, C. K. 2005. *Gyrinophilus palleucus*. Pp. 775–776 *in* M. J. Lannoo, ed., Declining Amphibians: The Conservation Status of United States Species. Berkeley: University of California Press.

Brandon, R. 1965. A new race of the neotenic salamander *Gyrinophilus palleucus*. Copeia 1965:346–352.

Brandon, R. 1967. *Gyrinophilus palleucus*. Catalogue of American Amphibians and Reptiles 32:1–2.

Cooper, J. E. 1968. The salamander *Gyrinophilus palleucus* in Georgia, with notes on Alabama and Tennessee populations. Journal of the Alabama Academy of Science 39:182–185.

Cooper, J., and M. Cooper. 1968. Cave associated herpetozoa II. Salamanders of the genus *Gyrinophilus* in Alabama caves. Bulletin of the National Speleological Society 30:9–24.

***Gyrinophilus porphyriticus*, Spring Salamander**

Adams, D. C., and C. K. Beachy. 2001. Historical explanations of phenotypic variation in the plethodontid salamander *Gyrinophilus porphyriticus*. Herpetologica 57:353–364.

Bruce, R. C. 1972. Variation in the life cycle of the salamander *Gyrinophilus porphyriticus*. Herpetologica 28:230–245.

Bruce, R. C. 2003. Ecological distribution of the salamanders *Gyrinophilus* and *Pseudotriton* in a southern Appalachian watershed. Herpetologica 59:301–310.

Lowe, W. H. 2005. Factors affecting stage-specific distribution in the stream salamander *Gyrinophilus porphyriticus*. Herpetologica 61:135–144.

***Haideotriton wallacei*, Georgia Blind Salamander**

Brandon, R. A. 1967. *Haideotriton* and *H. wallacei*. Catalogue of American Amphibians and Reptiles 39:1–2.

Carr, A. F. Jr. 1939. *Haideotriton wallacei*, a new subterranean salamander from Georgia. Occasional Papers of the Boston Society of Natural History 8:333–336.

Jensen, J. B. 1999. Georgia blind salamander: *Haideotriton wallacei*. Pp. 100–101 *in* T. W. Johnson, J. C. Ozier, J. L. Bohannon, J. B. Jensen, and C. Skelton, eds., Protected Animals of Georgia. Social Circle, Ga.: Georgia Department of Natural Resources, Wildlife Resources Division.

Means, D. B. 1992. Rare Georgia blind salamander: *Haideotriton wallacei* Carr. Pp. 49–53 *in* P. E. Moler, ed., Rare and Endangered Biota of Florida, vol. 3: Amphibians and Reptiles. Gainesville: University Press of Florida.

Means, D. B. 2005. *Haideotriton wallacei*. Pp. 779–780 *in* M. J. Lannoo, ed., Declining Amphibians: The Conservation Status of United States Species. Berkeley: University of California Press.

***Hemidactylium scutatum*, Four-toed Salamander**

Babcock, S. K. 2001. Caudal vertebral development and morphology in three salamanders with complex life cycles (*Ambystoma jeffersonianum, Hemidactylium scutatum,* and *Desmognathus ocoee*). Journal of Morphology 247:142–159.

Corser, J. D., and C. K. Dodd Jr. 2004. Fluctuations in a metapopulation of nesting four-toed salamanders, *Hemidactylium scutatum*, in the Great Smoky Mountains National Park, USA, 1999–2003. Natural Areas Journal 24:135–140.

Harris, R. N. 2005. *Hemidactylium scutatum*. Pp. 780–781 *in* M. J. Lannoo, ed., Declining Amphibians: The Conservation Status of United States Species. Berkeley: University of California Press.

Hess, Z. J., and R. N. Harris. 2000. Eggs of *Hemidactylium scutatum* (Caudata: Plethodontidae) are unpalatable to insect predators. Copeia 2000:597–600.

O'Laughlin, B. E., and R. N. Harris. 2000. Models of metamorphic timing: an experimental evaluation with the pond-dwelling salamander *Hemidactylium scutatum* (Caudata: Plethodontidae). Oecologia 124:343–350.

***Plethodon glutinosus* complex, Slimy Salamanders**

Beamer, D. A., and M. J. Lannoo. 2005. *Plethodon glutinosus.* Pp. 808–811 *in* M. J. Lannoo, ed., Declining Amphibians: The Conservation Status of United States Species. Berkeley: University of California Press.

Camp, C. D., and J. L. Marshall. 2000. The role of thermal environment in determining the life history of a terrestrial salamander. Canadian Journal of Zoology 78:1702–1711.

Highton, R. 1989. Biochemical Evolution in the Slimy Salamanders of the *Plethodon glutinosus* Complex in the Eastern United States. Part I: Geographic Protein Variation. Illinois Biological Monographs 57:1–78.

Jensen, J. B., and M. R. Whiles. 2000. Diets of sympatric *Plethodon petraeus* and *Plethodon glutinosus.* Journal of the Elisha Mitchell Scientific Society 116:245–250.

Marshall, J. L., C. D. Camp, and R. G. Jaeger. 2004. Potential interference competition between a patchily distributed salamander (*Plethodon petraeus*) and a sympatric congener (*Plethodon glutinosus*). Copeia 2004:488–495.

***Plethodon jordani* complex, Jordan's Salamanders**

Ash, A. N., R. C. Bruce, J. Castanet, and H. Francillon-Vieillot. 2003. Population parameters of *Plethodon metcalfi* on a 10-year-old clearcut and in nearby forest in the southern Blue Ridge Mountains. Journal of Herpetology 37:445–452.

Hairston, N. G. 1980. The experimental test of an analysis of field distributions: competition in terrestrial salamanders. Ecology 61:817–826.

Hairston, N. G. 1983. Growth, survival, and reproduction of *Plethodon jordani:* trade-offs between selective pressures. Copeia 1983:1024–1035.

Highton, R., and R. B. Peabody. 2000. Geographic protein variation and speciation in salamanders of the *Plethodon jordani* and *Plethodon glutinosus* complexes in the southern Appalachian Mountains with the description of four new species. Pp. 31–93 *in* R. C. Bruce, R. G. Jaeger, and L. D. Houck, eds., The Biology of Plethodontid Salamanders. New York: Kluwer Academic/Plenum.

Nishikawa, K. C. 1985. Competition and the evolution of aggressive behavior in two species of terrestrial salamanders. Evolution 39:1282–1294.

***Plethodon petraeus,* Pigeon Mountain Salamander**

Jensen, J. B. 1999. Pigeon Mountain salamander: *Plethodon petraeus.* Pp. 104–105 *in* T. W. Johnson, J. C. Ozier, J. L. Bohannon, J. B. Jensen, and C. Skelton, eds., Protected Animals of Georgia. Social Circle, Ga.: Georgia Department of Natural Resources, Wildlife Resources Division.

Jensen, J. B., C. D. Camp, and J. L. Marshall. 2002. Ecology and life history of the Pigeon Mountain salamander. Southeastern Naturalist 1:3–16.

Jensen, J. B., and M. R. Whiles. 2000. Diets of sympatric *Plethodon petraeus* and *Plethodon glutinosus.* Journal of the Elisha Mitchell Scientific Society 116:245–250.

Marshall, J. L., C. D. Camp, and R. G. Jaeger. 2004. Potential interference competition between a patchily distributed salamander (*Plethodon petraeus*) and a sympatric congener (*Plethodon glutinosus*). Copeia 2004:488–495.

Wynn, A. H., R. Highton, and J. F. Jacobs. 1988. A new species of rock-crevice dwelling *Plethodon* from Pigeon Mountain, Georgia. Herpetologica 44:135–143.

***Plethodon serratus,* Southern Red-backed Salamander**

Camp, C. D. 1986. Distribution and habitat of the southern red-back salamander, *Plethodon serratus* Grobman (Amphibia: Plethodontidae), in Georgia. Georgia Journal of Science 44:136–146.

Camp, C. D. 1988. Aspects of the life history of the southern red-back salamander *Plethodon serratus* Grobman in the southeastern United States. American Midland Naturalist 119:93–100.

Camp, C. D. 1999. Intraspecific aggressive behavior in southeastern small species of *Plethodon:* inferences for the evolution of aggression in terrestrial salamanders. Herpetologica 55:248–254.

Camp, C. D., and L. L. Bozeman. 1981. Foods of two species of *Plethodon* (Caudata: Plethodontidae) from Georgia and Alabama. Brimleyana 6:163–166.

Highton, R., and T. P. Webster. 1976. Geographic protein variation and divergence in populations of the salamander *Plethodon cinereus.* Evolution 30:33–45.

***Plethodon ventralis,* Southern Zigzag Salamander**

Beamer, D. A., and M. J. Lannoo. 2005. *Plethodon ventralis.* Pp. 849–850 *in* M. J. Lannoo, ed., Declining Amphibians: The Conservation Status of United States Species. Berkeley: University of California Press.

Highton, R. 1997. Geographic protein variation and speciation in the *Plethodon dorsalis* complex. Herpetologica 53:345–356.

***Plethodon websteri,* Webster's Salamander**

Beamer, D. A., and M. J. Lannoo. 2005. *Plethodon websteri.* Pp. 852–853 *in* M. J. Lannoo, ed., Declining Amphibians: The Conservation Status of United States Species. Berkeley: University of California Press.

Camp, C. D. 1999. Intraspecific aggressive behavior in southeastern small species of *Plethodon:* inferences for the evolution of aggression in terrestrial salamanders. Herpetologica 55:248–254.

Camp, C. D., and L. L. Bozeman. 1981. Foods of two species of *Plethodon* (Caudata: Plethodontidae) from Georgia and Alabama. Brimleyana 6:163–166.

Highton, R. 1979. A new cryptic species of salamander of the genus *Plethodon* from the southeastern United States (Amphibia: Plethodontidae). Brimleyana 1:31–36.

Semlitsch, R. D., and C. A. West. 1983. Aspects of the life history and ecology of Webster's salamander, *Plethodon websteri.* Copeia 1983:339–346.

***Pseudotriton montanus,* Mud Salamander**

Bruce, R. C. 1969. Fecundity in primitive plethodontid salamanders. Evolution 23:50–54.

Bruce, R. C. 1974. Larval development of the salamanders *Pseudotriton montanus* and *P. ruber.* American Midland Naturalist 92:173–190.

Bruce, R. C. 1975. Reproductive biology of the mud salamander, *Pseudotriton montanus,* in western South Carolina. Copeia 1975:129–137.

Hunsinger, T. W. 2005. *Pseudotriton montanus.* Pp. 858–860 *in* M. J. Lannoo, ed., Declining Amphibians: The Conservation Status of United States Species. Berkeley: University of California Press.

Neill, W. T. 1948. Salamanders of the genus *Pseudotriton* from Georgia and South Carolina. Copeia 1948:134–136.

***Pseudotriton ruber,* Red Salamander**

Bruce, R. C. 1978. Reproductive biology of the salamander *Pseudotriton ruber* in the southern Blue Ridge Mountains. Copeia 1978:417–423.

Hunsinger, T. W. 2005. *Pseudotriton ruber.* Pp. 860–862 *in* M. J. Lannoo, ed., Declining Amphibians: The Conservation Status of United States Species. Berkeley: University of California Press.

Neill, W. T. 1948. Salamanders from the genus *Pseudotriton* from Georgia and South Carolina. Copeia 1948:134–135.

Organ, J. A., and D. J. Organ. 1968. Courtship behavior of the red salamander, *Pseudotriton ruber.* Copeia 1968:217–223.

Semlitsch, R. D. 1983. Growth and metamorphosis of larval red salamanders (*Pseudotriton ruber*) on the Coastal Plain of South Carolina. Herpetologica 39:48–52.

***Stereochilus marginatus,* Many-lined Salamander**

Bruce, R. C. 1971. Life cycle and population structure of the salamander *Stereochilus marginatus* in North Carolina. Copeia 1971:234–246.

Christman, S. P., and H. I. Kochman. 1975. The southern distribution of the many-lined salamander, *Stereochilus marginatus.* Florida Scientist 38:139–141.

Foard, T., and D. L. Auth. 1990. Food habits and gut parasites of the salamander *Stereochilus marginatus.* Journal of Herpetology 24:428–431.

Ryan, T. J. 2005. *Stereochilus marginatus.* Pp. 862–863 *in* M. J. Lannoo, ed., Declining Amphibians: The Conservation Status of United States Species. Berkeley: University of California Press.

Wood, J. T., and R. H. Rageot. 1963. The nesting of the many-lined salamander in the Dismal Swamp. Virginia Journal of Science 14:121–145.

PROTEIDAE

***Necturus* cf. *beyeri,* Alabama Waterdog**

Guyer, C. 2005. *Necturus* cf. *beyeri.* P. 873 *in* M. J. Lannoo, ed., Declining Amphibians: The Conservation Status of United States Species. Berkeley: University of California Press.

Neill, W. T. 1963. Notes on the Alabama waterdog, *Necturus alabamensis.* Herpetologica 19:166–174.

***Necturus maculosus,* Mudpuppy**

Bishop, S. C. 1926. Notes on the habits and development of the mudpuppy *Necturus maculosus*

(Rafinesque). New York State Museum Bulletin 268:5–60.

Matson, T. O. 2005. *Necturus maculosus.* Pp. 870–871 *in* M. J. Lannoo, ed., Declining Amphibians: The Conservation Status of United States Species. Berkeley: University of California Press.

Nickerson, M. A., K. L. Krysko, and R. D. Owen. 2002. Ecological status of the hellbender (*Cryptobranchus alleganiensis*) and the mudpuppy (*Necturus maculosus*) salamanders in the Great Smoky Mountains National Park. Journal of the North Carolina Academy of Science 118:27–34.

Viosca, P. Jr. 1937. A tentative revision of the genus *Necturus* with descriptions of three new species from the southern Gulf drainage area. Copeia 1937:120–138.

Necturus punctatus, Dwarf Waterdog

Ashton, R. E. Jr., A. L. Braswell, and S. I. Guttman. 1980. Electrophoretic analysis of three species of *Necturus* (Amphibia: Proteidae), and the taxonomic status of *Necturus lewisi* (Brimley). Brimleyana 4:43–46.

Braswell, A. L., and R. E. Ashton Jr. 1985. Distribution, ecology, and feeding habits of *Necturus lewisi* (Brimley). Brimleyana 10:13–35.

Dundee, H. A. 2005. *Necturus punctatus.* Pp. 871–873 *in* M. J. Lannoo, ed., Declining Amphibians: The Conservation Status of United States Species. Berkeley: University of California Press.

Meffe, G. K., and A. L. Sheldon. 1987. Habitat use by dwarf waterdogs in South Carolina streams, with life history notes. Herpetologica 43:490–496.

SALAMANDRIDAE

Notophthalmus perstriatus, Striped Newt

Dodd, C. K. Jr. 1993. Cost of living in an unpredictable environment: the ecology of striped newts *Notophthalmus perstriatus* during a prolonged drought. Copeia 1993:605–614.

Dodd, C. K. Jr., and L. V. LaClaire. 1995. Biogeography and status of the striped newt (*Notophthalmus perstriatus*) in Georgia, USA. Herpetological Natural History 3:37–46.

Jensen, J. B. 1999. Striped newt: *Notophthamlus perstriatus.* Pp. 102–103 *in* T. W. Johnson, J. C. Ozier, J. L. Bohannon, J. B. Jensen, and C. Skelton, eds., Protected Animals of Georgia. Social Circle, Ga.: Georgia Department of Natural Resources, Wildlife Resources Division.

Johnson, S. A. 2002. Life history of the striped newt at a north-central Florida breeding pond. Southeastern Naturalist 1:381–402.

Johnson, S. A. 2003. Orientation and migration distances of a pond-breeding salamander (*Notophthalmus perstriatus,* Salamandridae). Alytes 21:3–22.

Notophthalmus viridescens, Eastern Newt

Gabor, C. R., and C. C. Nice. 2004. Genetic variation among populations of eastern newts, *Notophthalmus viridescens:* a preliminary analysis based on allozymes. Herpetologica 60:373–386.

Gill, D. E. 1978. The metapopulation ecology of the red-spotted newt, *Notophthalmus viridescens* (Rafinesque). Ecological Monographs 48:145–166.

Healy, W. R. 1974. Population consequences of alternative life histories in *Notophthalmus v. viridescens.* Copeia 1974:221–229.

Hunsinger, T. W., and M. J. Lannoo. 2005. *Notophthalmus viridescens.* Pp. 889–894 *in* M. J. Lannoo, ed., Declining Amphibians: The Conservation Status of United States Species. Berkeley: University of California Press.

Reilly, S. M. 1990. Biochemical systematics and evolution of the eastern North American newts, genus *Notophthalmus* (Caudata: Salamandridae). Herpetologica 46:51–59.

SIRENIDAE

Pseudobranchus striatus, Northern Dwarf Siren

Liu, F. R., P. E. Moler, H. P. Whidden, and M. M. Miyamoto. 2004. Allozyme variation in the salamander genus *Pseudobranchus:* phylogeographic and taxonomic significance. Copeia 2004:136–144.

Moler, P. E. 2005. *Pseudobranchus striatus.* Pp. 909–910 *in* M. J. Lannoo, ed., Declining Amphibians: The Conservation Status of United States Species. Berkeley: University of California Press.

Moler, P. E., and J. Kezer. 1993. Karyology and systematics of the salamander genus *Pseudobranchus* (Sirenidae). Copeia 1993:39–47.

Siren intermedia, Lesser Siren

Collette, B. B., and F. R. Gehlbach. 1961. The salamander *Siren intermedia intermedia* in North Carolina. Herpetologica 17:203–204.

Fauth, J. E., and W. J. Resetarits Jr. 1999. Biting in the

salamander *Siren intermedia intermedia:* courtship component or agonistic behavior. Journal of Herpetology 33:493–496.
Godley, J. S. 1983. Observations on the courtship, nests and young of *Siren intermedia* in southern Florida. American Midland Naturalist 110:215–219.
Noble, G. K., and B. C. Marshall. 1932. The validity of *Siren intermedia* LeConte, with observations on its life history. American Museum Novitates 532:1–17.
Parmley, D., and G. Gaddis. 1999. Occurrence of *Siren intermedia* Le Conte (Amphibia: Caudata) in central Georgia. Georgia Journal of Science 57:192–198.

***Siren lacertina,* Greater Siren**

Etheridge, K. 1990. Water balance in estivating sirenid salamanders (*Siren lacertina*). Herpetologica 46:400–406.
Hanlin, H. G., and R. H. Mount. 1978. Reproduction and activity of the greater siren, *Siren lacertina* (Amphibia: Sirenidae), in Alabama. Journal of the Alabama Academy of Science 49:31–39.
Snodgrass, J. W., J. W. Ackerman, A. L. Bryan Jr., and J. Burger. 1999. Influence of hydroperiod, isolation, and heterospecifics on the distribution of aquatic salamanders (*Siren* and *Amphiuma*) among depression wetlands. Copeia 1999:107–113.
Sorensen, K. 2004. Population characteristics of *Siren lacertina* and *Amphiuma means* in north Florida. Southeastern Naturalist 3:249–258.
Ultsch, G. R. 1973. Observations on the life history of *Siren lacertina.* Herpetologica 29:304–305.

Reptiles

Gans, C., and D. W. Tinkle. 1977. Biology of the Reptilia, vol. 7: Ecology and Behaviour A. London: Academic Press.
Gans, C., and R. B. Huey. 1988. Biology of the Reptilia, vol. 16: Ecology B. Defense and Life History. New York: Alan R. Liss.

Crocodilians

Lockwood, C. C. 2002. The Alligator Book. Baton Rouge: Louisiana State University Press.
Thorbjarnarson, J. 1996. Reproductive characteristics of order Crocodylia. Herpetologica 52:8–24.

ALLIGATORIDAE

***Alligator mississippiensis,* American Alligator**

Brisbin, I. L., C. A. Ross, M. C. Downes, M. A. Staton, and B. Gammon. 1986. A Bibliography of the American Alligator. Aiken, S.C.: Savannah River National Environmental Research Park Program.
Hunt, R. H., and J. J. Ogden. 1991. Selected aspects of the nesting ecology of American alligators in the Okefenokee Swamp. Journal of Herpetology 25:448–453.
Metzen, W. D. 1977. Nesting ecology of alligators on the Okefenokee National Wildlife Refuge. Proceedings of the Annual Conference of the Southeastern Association of Fish and Wildlife Agencies 31:29–32.
Ruckel, S. W., and G. W. Steele. 1984. Alligator nesting ecology in two habitats in southern Georgia. Proceedings of the Annual Conference of the Southeastern Association of Fish and Wildlife Agencies 38:212–221.

Worm Lizards

Gans, C. 1978. The characteristics and affinities of the Amphisbaenia. Transactions of the Zoological Society of London 34:347–416.

RHINEURIDAE

***Rhineura floridana,* Florida Worm Lizard**

Carr, A. F. Jr. 1949. Notes on the eggs and young of the lizard *Rhineura floridana.* Copeia 1949:77.
Gans, C. 1967. *Rhineura floridana.* Catalogue of American Amphibians and Reptiles 43:1–2.
Malvaney, A., T. A. Castoe, K. G. Aston, K. L. Krysko, and C. L. Parkinson. 2005. Evidence of population genetic structure within the Florida worm lizard, *Rhineura floridana* (Amphisbaenia: Rhineuridae). Journal of Herpetology 39:118–124.
Meylan, P. A. 1984. The northwestern limit of distribution of *Rhineura floridana* with comments on the dispersal of amphisbaenians. Herpetological Review 15:23–24.
Neill, W. T. 1951. The eyes of the worm lizard, and notes on the habits of the species. Copeia 1951:177–178.

Lizards

Cooper, W. E. Jr. 1995. Foraging mode, prey chemical discrimination, and phylogeny in lizards. Animal Behaviour 50:971–985.

Pianka, E. R. 1973. The structure of lizard communities. Annual Review of Ecology and Systematics 4:53–74.

Smith, H. M. 1946. Handbook of Lizards. Ithaca, N.Y.: Comstock.

Vitt, L. J., and E. R. Pianka, eds. 1994. Lizard Ecology: Historical and Experimental Perspectives. Princeton: Princeton University Press.

ANGUIDAE

Ophisaurus attenuatus, Slender Glass Lizard

Holman, J. A. 1971. *Ophisaurus attenuatus.* Catalogue of American Amphibians and Reptiles 111:1–3.

McConkey, E. H. 1954. A systematic study of the North American lizards of the genus *Ophisaurus.* American Midland Naturalist 51:133–171.

Ophisaurus compressus, Island Glass Lizard

Bartlett, R. D. 1985. Notes on the natural history and reproductive strategy of the island glass lizard, *Ophisaurus compressus.* British Herpetological Society Bulletin 11:19–21.

Holman, J. A. 1971. *Ophisaurus compressus.* Catalogue of American Amphibians and Reptiles 113:1–2.

Ophisaurus mimicus, Mimic Glass Lizard

Jensen, J. B. 2004. Mimic glass lizard *Ophisaurus mimicus.* Pp. 57–58 *in* R. E. Mirarchi, M. A. Bailey, T. M. Haggerty, and T. L. Best, eds., Alabama Wildlife, vol. 3: Imperiled Amphibians, Reptiles, Birds, and Mammals. Tuscaloosa: University of Alabama Press.

Moler, P. E. 1992. Mimic glass lizard: *Ophisaurus mimicus.* Pp. 247–250 *in* P. E. Moler, ed., Rare and Endangered Biota of Florida, vol. 3: Amphibians and Reptiles. Gainesville: University Press of Florida.

Palmer, W. M. 1987. A new species of glass lizard (Anguidae: *Ophisaurus*) from the southeastern United States. Herpetologica 43:415–423.

Palmer, W. M. 1992. *Ophisaurus mimicus.* Catalogue of American Amphibians and Reptiles 543:1–2.

Ophisaurus ventralis, Eastern Glass Lizard

Holman, J. A. 1971. *Ophisaurus ventralis* (Linnaeus). Eastern glass lizard. Catalogue of American Amphibians and Reptiles 115:1–2.

Witz, B. W., and D. S. Wilson. 1993. Clutch size, egg mass, and incubation period in an eastern glass lizard, *Ophisaurus ventralis.* Florida Field Naturalist 21:36–37.

GEKKONIDAE

Hemidactylus turcicus, Mediterranean Gecko

Klawinski, P. D., R. K. Vaughan, D. Saenz, and W. Godwin. 1994. Comparison of dietary overlap between allopatric and sympatric geckos. Journal of Herpetology 28:225–230.

Meshaka, W. E. 1995. Reproductive cycle and colonization ability of the Mediterranean gecko (*Hemidactylus turcicus*) in south-central Florida. Florida Scientist 58:10–15.

Punzo, F. 2001. The Mediterranean gecko, *Hemidactylus turcicus:* life in an urban landscape. Florida Scientist 64:56–66.

Selcer, K. W. 1986. Life history of a successful colonizer: the Mediterranean gecko, *Hemidactylus turcicus,* in southern Texas. Copeia 1986:956–962.

PHRYNOSOMATIDAE

Sceloporus undulatus, Eastern Fence Lizard

Crenshaw, J. W. Jr. 1955. The life-history of the southern spiny lizard, *Sceloporus undulatus undulatus* Latreille. American Midland Naturalist 54:257–298.

McGovern, G. M., C. B. Knisley, and J. C. Mitchell. 1986. Prey selection experiments and predator-prey size relationships in eastern fence lizards, *Sceloporus undulatus,* from Virginia. Virginia Journal of Sciences 37:9–15.

Miles, D. B., R. Noecker, W. M. Roosenburg, and M. M. White. 2002. Genetic relationships among populations of *Sceloporus undulatus* fail to support present subspecific designations. Herpetologica 58:277–292.

Niewiarowski, P. H. 1995. Effects of supplemental feeding and thermal environment on growth rates of eastern fence lizards (*Sceloporus undulatus*). Herpetologica 51:487–496.

Tinkle, D. W., and R. E. Ballinger. 1972. *Sceloporus undulatus:* a study of the intraspecific comparative demography of a lizard. Ecology 53:570–584.

POLYCHROTIDAE

***Anolis carolinensis*, Green Anole**

Jenssen, T. A., M. B. Lovern, and J. D. Congdon. 2001. Field-testing the protandry-based mating system for the lizard, *Anolis carolinensis:* does the model organism have the right model? Behavioral Ecology and Sociobiology 50:162–172.

Jenssen, T. A., and S. C. Nunez. 1998. Spatial and breeding relationships of the lizard, *Anolis carolinensis:* evidence of intrasexual selection. Behaviour 135:603–634.

Lovern, M. B. 2000. Behavioral ontogeny in free-ranging juvenile male and female green anoles, *Anolis carolinensis,* in relation to sexual selection. Journal of Herpetology 34:274–281.

Lovern, M. B., and K. M. Passek. 2002. Sequential alternation of offspring sex from successive eggs by female green anoles, *Anolis carolinensis.* Canadian Journal of Zoology 80:77–82.

***Anolis sagrei*, Brown Anole**

Campbell, T. S. 1996. Northern range expansion of the brown anole (*Anolis sagrei*) in Florida and Georgia. Herpetological Review 27:155–157.

Campbell, T. S., and A. C. Echternacht. 2003. Introduced species as moving targets: body sizes of introduced lizards following experimental introductions and historical invasions. Biological Invasions 5:193–212.

Kolbe, J. J., R. E. Glor, L. Rodriguez-Schettino, A. C. Lara, A. Larsen, and J. B. Losos. 2004. Genetic variation increases during biological invasion by a Cuban lizard. Nature 431:177–181.

Parmley, D. 2002. Northernmost record of the brown anole (*Anolis sagrei*) in Georgia. Georgia Journal of Science 60:191–193.

SCINCIDAE

***Eumeces anthracinus*, Coal Skink**

Hotchkin, P. E., C. D. Camp, and J. L. Marshall. 2001. Aspects of the life history and ecology of the coal skink, *Eumeces anthracinus,* in Georgia. Journal of Herpetology 35:145–148.

Pyron, R. A., and C. D. Camp. 2006. Courtship and mating behaviours of two syntopic species of skink (*Plestiodon anthracinus* and *Plestiodon fasciatus*). Amphibia-Reptilia 28:263–268.

Trauth, S. E. 1994. Reproductive cycles in two Arkansas skinks in the genus *Eumeces* (Sauria: Scincidae). Proceedings Arkansas Academy of Science 48:210–218.

Walley, H. D. 1998. *Eumeces anthracinus.* Catalogue of American Amphibians and Reptiles 658:1–6.

***Eumeces egregius*, Mole Skink**

Branch, L. C., A. M. Clark, P. E. Moler, and B. W. Bowen. 2003. Fragmented landscapes, habitat specificity, and conservation genetics of three lizards in Florida scrub. Conservation Genetics 4:199–212.

Hamilton, W. J. Jr., and J. A. Pollack. 1958. Notes on the life history of the red-tailed skink. Herpetologica 14:25–28.

Mount, R. H. 1963. The natural history of the red-tailed skink, *Eumeces egregius* Baird. American Midland Naturalist 70:356–385.

Mount, R. H. 1965. Variation and systematics of the scincoid lizard, *Eumeces egregius* (Baird). Bulletin of the Florida State Museum, Biological Sciences 9:183–213.

***Eumeces fasciatus*, Five-lined Skink**

Fitch, H. S. 1954. Life History and Ecology of the Five-lined Skink, *Eumeces fasciatus.* University of Kansas Publications, Museum of Natural History 8:1–156.

Hecnar, S. J., and R. T. M'Closkey. 1998. Effects of human disturbance on five-lined skink, *Eumeces fasciatus,* abundance and distribution. Biological Conservation 85:213–222.

Pyron, R. A., and C. D. Camp. 2006. Courtship and mating behaviours of two syntopic species of skink (*Plestiodon anthracinus* and *Plestiodon fasciatus*). Amphibia-Reptilia 28:263–268.

Smith, H. M. 2005. *Plestiodon:* a replacement name for most members of the genus *Eumeces* in North America. Journal of Kansas Herpetology 14:15–16.

Vitt, L. J., and W. E. Cooper. 1986. Skink reproduction and sexual dimorphism: *Eumeces fasciatus* in the southeastern United States, with notes on *Eumeces inexpectatus.* Journal of Herpetology 20:65–76.

***Eumeces inexpectatus*, Southeastern Five-lined Skink**

Platt, S. G., and T. R. Rainwater. 2000. Aspects of reproduction in the southeastern five-lined skink (*Eumeces inexpectatus*): new data and a review. Proceedings of the Louisiana Academy of Sciences 2000:47–51.

Vitt, L. J., and W. E. Cooper Jr. 1986. Skink

reproduction and sexual dimorphism: *Eumeces fasciatus* in the southeastern United States, with notes on *Eumeces inexpectatus.* Journal of Herpetology 20:65–76.

***Eumeces laticeps,* Broadhead Skink**

Cooper, W. E., and L. J. Vitt. 1993. Female mate choice of large male broad-headed skinks. Animal Behaviour 45:683–693.

Vitt, L. J., and W. E. Cooper. 1985. The evolution of sexual dimorphism in the skink *Eumeces laticeps:* an example of sexual selection. Canadian Journal of Zoology 63:995–1002.

Vitt, L. J., and W. E. Cooper. 1985. The relationship between reproduction and lipid cycling in the skink *Eumeces laticeps* with comments on brooding ecology. Herpetologica 41:419–432.

***Scincella lateralis,* Ground Skink**

Brooks, G. R. 1964. Food habits of the ground skink. Quarterly Journal of the Florida Academy of Sciences 26:361–367.

Greenberg, C. H., D. G. Neary, and L. D. Harris. 1994. Effect of high-intensity wildfire and silvicultural treatments on reptile communities in sand-pine scrub. Conservation Biology 8:1047–1057.

TEIIDAE

***Cnemidophorus sexlineatus,* Six-lined Racerunner**

Fitch, H. S. 1958. Natural history of the six-lined racerunner (*Cnemidophorus sexlineatus*). University of Kansas Publications, Museum of Natural History 11:11–62.

Paulissen, M. A. 1988. Ontogenetic and seasonal comparisons of daily activity patterns of the six-lined racerunner, *Cnemidophorus sexlineatus* (Sauria: Teiidae). American Midland Naturalist 120:355–361.

Paulissen, M. A. 1988. Ontogenetic and seasonal shifts in microhabitat use by the lizard *Cnemidophorus sexlineatus.* Copeia 1988:1021–1029.

Reeder, T. W., C. J. Cole, and H. C. Dessauer. 2002. Phylogenetic Relationships of Whiptail Lizards of the Genus *Cnemidophorus* (Squamata: Teiidae): A Test of Monophyly, Reevaluation of Karyotypic Evolution, and Review of Hybrid Origins. American Museum Novitates 3365:1–64.

Trauth, S. E. 1983. Nesting habitat and reproductive characteristics of the lizard *Cnemidophorus sexlineatus* (Lacertilia: Teiidae). American Midland Naturalist 109:289–299.

Snakes

Gibbons, W., and M. E. Dorcas. 2005. Snakes of the Southeast. Athens: University of Georgia Press.

Seigel, R. A., and J. T. Collins, eds. 1993. Snakes: Ecology and Behavior. New York: McGraw-Hill.

Wright, A. H., and A. A. Wright. 1949. Handbook of Snakes. 3 vols. Ithaca, N.Y.: Comstock.

COLUBRIDAE

***Carphophis amoenus,* Eastern Worm Snake**

Barbour, R. W., M. J. Harvey, and J. W. Hardin. 1969. Home range, movements, and activity of the eastern worm snake (*Carphophis amoenus amoenus*). Ecology 50:470–476.

Russell, K. R., and H. G. Hanlin. 1999. Aspects of the ecology of worm snakes (*Carphophis amoenus*) associated with small isolated wetlands in South Carolina. Journal of Herpetology 33:339–344.

Willson, J. D., and M. E. Dorcas. 2004. Aspects of the ecology of small fossorial snakes in the western Piedmont of North Carolina. Southeastern Naturalist 3:1–12.

***Cemophora coccinea,* Scarlet Snake**

Enge, K. M., and J. D. Sullivan. 2000. Seasonal activity of the scarlet snake, *Cemophora coccinea,* in Florida. Herpetological Review 31:82–84.

Nelson, D. H., and J. W. Gibbons. 1972. Ecology, abundance and seasonal activity of the scarlet snake, *Cemophora coccinea.* Copeia 1972:582–584.

***Coluber constrictor,* Black Racer**

Fitch, H. S. 1963. Natural History of the Racer *Coluber constrictor.* University of Kansas Publications, Museum of Natural History 15:351–468.

Plummer, M. V., and J. D. Congdon. 1994. Radiotelemetric study of activity and movement of racers (*Coluber constrictor*) associated with a Carolina bay in South Carolina. Copeia 1994:20–26.

***Diadophis punctatus,* Ringneck Snake**

Fitch, H. S. 1975. A Demographic Study of the Ringneck Snake (*Diadophis punctatus*) in Kansas. Miscellaneous Publications of the University of Kansas, Museum of Natural History 62:1–53.

Fitch, H. S. 1999. A Kansas Snake Community:

Composition and Changes over 50 Years. Melbourne, Fla.: Krieger Publishing Company.
Myers, C. 1965. Biology of the ringneck snake, *Diadophis punctatus,* in Florida. Bulletin of the Florida State Museum, Biological Sciences, 10:43–90.

Drymarchon couperi, Eastern Indigo Snake

Diemer, J. E., and D. W. Speake. 1983. The distribution of the eastern indigo snake, *Drymarchon corais couperi,* in Georgia. Journal of Herpetology 17:256–264.
Moler, P. E. 1992. Eastern indigo snake: *Drymarchon corais couperi* (Holbrook). Pp. 181–186 *in* P. E. Moler, ed., Rare and Endangered Biota of Florida, vol. 3: Amphibians and Reptiles. Gainesville: University Press of Florida.
Moulis, R. 1976. Autecology of the eastern indigo snake, *Drymarchon corais couperi.* Bulletin of New York Herpetological Society 12:14–23.
Stevenson, D. J, K. J. Dyer, and B. A. Willis-Stevenson. 2003. Survey and monitoring of the eastern indigo snake in Georgia. Southeastern Naturalist 2:393–408.

Elaphe guttata, Corn Snake

Bechtel, H. B., and E. Bechtel. 1989. Color mutations in the corn snake (*Elaphe guttata guttata*): review and additional breeding data. Journal of Heredity 80:272–276.
Mitchell, J. C. 1977. Geographic variation of *Elaphe guttata* (Reptilia: Serpentes) in the Atlantic Coastal Plain. Copeia 1977:33–41.

Elaphe obsoleta, Rat Snake

Burbrink, F. T. 2001. Systematics of the Eastern Ratsnake Complex (*Elaphe obsoleta*). Herpetological Monographs 15:1–53.
Burbink, F. T., R. Lawson, and J. B. Slowinski. 2000. Mitochondrial DNA phylogeography of the polytypic North American rat snake (*Elaphe obsoleta*): a critique of the subspecies concept. Evolution 54:2107–2118.
Fitch, H. S. 1963. Natural history of the black rat snake (*Elaphe o. obsoleta*) in Kansas. Copeia 1963:649–658.

Farancia abacura, Mud Snake

Hall, P. M., and A. J. Meier. 1993. Reproduction and behavior of western mud snakes (*Farancia abacura reinwardtii*) in American alligator nests. Copeia 1993:219–222.
Mitchell, J. C. 1982. *Farancia* Gray. Mud and rainbow snakes. Catalogue of American Amphibians and Reptiles 292:1–3.
Riemer, W. J. 1957. The snake *Farancia abacura:* an attended nest. Herpetologica 13:31–32.
Semlitsch, R. D., J. H. K. Pechmann, and J. W. Gibbons. 1988. Annual emergence of juvenile mud snakes (*Farancia abacura*) at aquatic habitats. Copeia 1988:243–245.

Farancia erytrogramma, Rainbow Snake

Gibbons, J. W., J. W. Coker, and T. M. Murphy Jr. 1977. Selected aspects of the life history of the rainbow snake (*Farancia erytrogramma*). Herpetologica 33:276–281.
Neill, W. T. 1964. Taxonomy, natural history, and zoogeography of the rainbow snake, *Farancia erytrogramma* (Palisot de Beauvois). American Midland Naturalist 71:257–295.

Heterodon platirhinos, Eastern Hognose Snake

Edgren, R. A. 1955. The natural history of the hog-nosed snakes, genus *Heterodon:* a review. Herpetologica 11:105–117.
Platt, D. R. 1969. Natural History of the Hognose Snakes *Heterodon platyrhinos* and *Heterodon nasicus.* University of Kansas Publications, Museum of Natural History 18:253–420.

Heterodon simus, Southern Hognose Snake

Edgren, R. A. 1955. The natural history of the hog-nosed snakes, genus *Heterodon:* a review. Herpetologica 11:105–117.
Enge, K. M., and K. N. Wood. 2003. A pedestrian survey of the southern hognose snake (*Heterodon simus*) in Hernando County, Florida. Florida Scientist 66:189–203.
Jensen, J. B. 2004. Southern hognose snake: *Heterodon simus.* Pp. 42–43 *in* R. E. Mirarchi, M. A. Bailey, T. M. Haggerty, and T. L. Best, eds., Alabama Wildlife, vol. 3: Imperiled Amphibians, Reptiles, Birds, and Mammals. Tuscaloosa: University of Alabama Press.
Tuberville, T. D., J. R. Bodie, J. B. Jensen, L. LaClaire, and J. W. Gibbons. 2000. Apparent decline of the southern hog-nosed snake, *Heterodon simus.* Journal of the Elisha Mitchell Scientific Society 116:19–40.

***Lampropeltis calligaster,* Mole Kingsnake**

Ernst, C. H., S. W. Gotte, and J. E. Lovich. 1985. Reproduction in the mole kingsnake, *Lampropeltis calligaster rhombomaculata.* Bulletin of the Maryland Herpetological Society 21:16–22.

Krysko, K. L., L. E. Krysko, and C. Hurt. 2000. Reproduction and distribution of the South Florida mole kingsnake (*Lampropeltis calligaster occipitolineata*) from central peninsular Florida. Journal of the Elisha Mitchell Scientific Society 116:344–347.

Means, D. B. 1992. Mole snake: *Lampropeltis calligaster rhombomaculata* (Holbrook). Pp. 227–231 *in* P. E. Moler, ed., Rare and Endangered Biota of Florida, vol. 3: Amphibians and Reptiles. Gainesville: University Presses of Florida.

***Lampropeltis getula,* Common Kingsnake**

Gibbons, J. W., and R. D. Semlitsch. 1987. Activity patterns. Pp. 396–421 *in* R. A. Seigel, J. T. Collins, and S. S. Novak, eds., Snakes: Ecology and Evolutionary Biology. New York: Macmillan.

Krysko, K. L. 2002. Seasonal activity of the Florida kingsnake, *Lampropeltis getula floridana* (Serpentes: Colubridae), in southern Florida. American Midland Naturalist 148:102–114.

Krysko, K. L., and D. J. Smith. 2005. The decline and extirpation of kingsnakes, *Lampropeltis getula,* in Florida. Pp. 132–141 *in* W. E. Meshaka Jr. and K. J. Babbitt, eds., Status and Conservation of Florida Amphibians and Reptiles. Malabar, Fla.: Krieger Press.

***Lampropeltis triangulum elapsoides,* Scarlet Kingsnake**

Armstrong, M. P., D. Frymire, and E. J. Zimmerer. 2001. Analysis of sympatric populations of *Lampropeltis triangulum syspila* and *Lampropeltis triangulum elapsoides,* in western Kentucky and adjacent Tennessee with relation to the taxonomic status of the scarlet kingsnake. Journal of Herpetology 35:688–693.

Groves, J. D., and P. S. Sachs. 1973. Eggs and young of the scarlet kingsnake, *Lampropeltis triangulum elapsoides.* Journal of Herpetology 7:389–390.

Williams, K. L. 1988. Systematics and Natural History of the American Milk Snake, *Lampropeltis triangulum.* Milwaukee: Milwaukee Public Museum.

Williams, K. L. 1994. *Lampropeltis triangulum.* Catalogue of American Amphibians and Reptiles 594:1–9.

***Lampropeltis triangulum triangulum,* Eastern Milk Snake**

Neill, W. T. 1949. The distribution of milk snakes in Georgia. Herpetologica 5:8.

Williams, K. L. 1988. Systematics and Natural History of the American Milk Snake, *Lampropeltis triangulum.* Milaukee: Milwaukee Public Museum.

Williams, K. L. 1994. *Lampropeltis triangulum.* Catalogue of American Amphibians and Reptiles 594:1–9.

***Masticophis flagellum,* Coachwhip**

Wilson, L. D. 1970. The coachwhip snake, *Masticophis flagellum* (Shaw): taxonomy and distribution. Tulane Studies in Zoology and Botany 16:31–99.

Wilson, L. D. 1973. *Masticophis flagellum.* Catalogue of American Amphibians and Reptiles 145:1–4.

***Nerodia erythrogaster,* Plain-bellied Watersnake**

Gibbons, J. W., and M. E. Dorcas. 2004. North American Watersnakes: A Natural History. Norman: University of Oklahoma Press.

Mushinsky, H. R., J. J. Hebrard, and M. G. Walley. 1980. The role of temperature on the behavioral and ecological associations of sympatric water snakes. Copeia 1980:744–754.

Roe, J. H., B. A. Kingsbury, and N. R. Herbert. 2003. Wetland and upland use patterns in semi-aquatic snakes: implications for wetland conservation. Wetlands 23:1003–1014.

Roe, J. H., B. A. Kingsbury, and N. R. Herbert. 2004. Comparative water snake ecology: conservation of mobile animals that use temporally dynamic resources. Biological Conservation 118:79–89.

***Nerodia fasciata,* Banded Watersnake**

Gibbons, J. W., and M. E. Dorcas. 2004. North American Watersnakes: A Natural History. Norman: University of Oklahoma Press.

Lawson, R., A. J. Meier, P. G. Frank, and P. E. Moler. 1991. An allozyme study of the *Nerodia fasciata–Nerodia clarkii* complex of water snakes (Serpentes: Colubridae). Copeia 1991:638–659.

Mushinsky, H. R., J. J. Hebrard, and D. S. Vodopich. 1982. Ontogeny of water snake foraging ecology. Ecology 63:1624–1629.

Schwaner, T. D., and R. H. Mount. 1976. Systematic and ecological relationships of the water snakes *Natrix sipedon* and *N. fasciata* in Alabama and the Florida panhandle. Occasional Papers of the

University of Kansas Museum of Natural History 45:1–44.

***Nerodia floridana*, Eastern Green Watersnake**

Gibbons, J. W., and M. E. Dorcas. 2004. North American Watersnakes: A Natural History. Norman: University of Oklahoma Press.

Lawson, L. 1987. Molecular studies of thamnophiine snakes: I. The phylogeny of the genus *Nerodia*. Journal of Herpetology 21:140–157.

Thompson, J. S., and B. I. Crother. 1998. Allozyme variation among disjunct populations of the Florida green watersnake (*Nerodia floridana*). Copeia 1998:715–719.

***Nerodia sipedon*, Northern Watersnake**

Bauman, M. A., and D. E. Metter. 1977. Reproductive cycle of the northern water snake, *Natrix s. sipedon* (Reptilia, Serpentes, Colubridae). Journal of Herpetology 11:51–59.

Blaney, R. M., and P. K. Blaney. 1979. The *Nerodia sipedon* complex of water snakes in Mississippi and southeastern Louisiana. Herpetologica 35:350–359.

Brown, E. E. 1958. Feeding habits of the northern water snake, *Natrix sipedon sipedon* Linnaeus. Zoologica 43:55–71.

Gibbons, J. W., and M. E. Dorcas. 2004. North American Watersnakes: A Natural History. Norman: University of Oklahoma Press.

Weatherhead, P. J., F. E. Barry, G. P. Brown, and M. R. L. Forbes. 1995. Sex ratios, mating behavior and sexual size dimorphism of the northern water snake, *Nerodia sipedon*. Behavioral Ecology and Sociobiology 36:301–311.

***Nerodia taxispilota*, Brown Watersnake**

Camp, C. D., W. D. Sprewell, and V. N. Powders. 1980. Feeding habits of *Nerodia taxispilota* with comparative notes on the foods of sympatric congeners in Georgia. Journal of Herpetology 14:301–304.

Herrington, R. E. 1989. Reproductive biology of the brown water snake, *Nerodia taxispilota*, in central Georgia. Brimleyana 15:103–110.

Mills, M. S. 2004. The brown watersnake, *Nerodia taxispilota*. Pp. 197–211 *in* J. W. Gibbons and M. E. Dorcas, North American Watersnakes: A Natural History. Norman: University of Oklahoma Press.

Mills, M. S., C. J. Hudson, and H. J. Berna. 1995. Spatial ecology and movements of the brown water snake (*Nerodia taxispilota*). Herpetologica 51:412–423.

***Opheodrys aestivus*, Rough Green Snake**

Aldridge, R. D., J. J. Greenhaw, and M. V. Plummer. 1990. The male reproductive cycle of the rough green snake (*Opheodrys aestivus*). Amphibia-Reptilia 11:165–172.

Plummer, M. V. 1981. Habitat utilization, diet and movements of a temperate arboreal snake (*Opheodrys aestivus*). Journal of Herpetology 15:425–432.

Plummer, M. V. 1984. Female reproduction in an Arkansas population of rough green snakes (*Opheodrys aestivus*). Vertebrate Ecology and Systematics Special Publication 10:105–115.

Plummer, M. V. 1990. Nesting movements, nesting behavior, and nest sites of green snakes (*Opheodrys aestivus*) revealed by radiotelemetry. Herpetologica 46:190–195.

Plummer, M. V. 1997. Population ecology of green snakes (*Opheodrys aestivus*) revisited. Herpetological Monographs 11:102–123.

***Pituophis melanoleucus*, Pine Snake**

Burger, J., and R. T. Zappalorti. 1986. Nest site selection by pine snakes, *Pituophis melanoleucus*, in the New Jersey Pine Barrens. Copeia 1986:116–121.

Burger, J., and R. T. Zappalorti. 1991. Nesting behaviour of pine snakes (*Pituophis melanoleucus*) in the New Jersey Pine Barrens. Journal of Herpetology 25:152–160.

Burger, J., R. T. Zappalorti, M. Gochfeld, W. I. Boarman, M. Caffrey, V. Doig, S. D. Garber, B. Lauro, M. Mikovsky, C. Safina, and J. Saliva. 1988. Hibernacula and summer den sites of pine snakes (*Pituophis melanoleucus*) in the New Jersey Pine Barrens. Journal of Herpetology 22:425–433.

Rodriques-Robles, J. A., and J. M. de Jesus-Escobar. 2000. Molecular systematics of New World gopher, bull, and pine snakes (*Pituophis:* Colubridae), a transcontinental species complex. Molecular Phylogenetics and Evolution 14:35–50.

***Regina alleni*, Striped Crayfish Snake**

Franz, R. 1977. Observations on the food, feeding behavior, and parasites of the striped swamp snake, *Regina alleni*. Herpetologica 33:91–94.

Gibbons, J. W., and M. E. Dorcas. 2004. North American Watersnakes: A Natural History. Norman: University of Oklahoma Press.

Godley, J. S. 1980. Foraging ecology of the striped

swamp snake, *Regina alleni,* in southern Florida. Ecological Monographs 50:411–436.

Godley, J. S. 1982. Predation and defensive behavior of the striped swamp snake (*Regina alleni*). Florida Field Naturalist 10:31–36.

Regina rigida, Glossy Crayfish Snake

Gibbons, J. W., and M. E. Dorcas. 2004. North American Watersnakes: A Natural History. Norman: University of Oklahoma Press.

Huheey, J. E. 1959. Distribution and variation in the glossy water snake, *Natrix rigida* (Say). Copeia 1959:303–311.

Kofron, C. P. 1978. Foods and habitats of aquatic snakes (Reptilia, Serpentes) in a Louisiana swamp. Journal of Herpetology 12:543–554.

Regina septemvittata, Queen Snake

Branson, B. A., and E. C. Baker. 1974. An ecological study of the queen snake, *Regina septemvittata* (Say) in Kentucky. Tulane Studies in Zoology and Botany 18:153–171.

Gibbons, J. W., and M. E. Dorcas. 2004. North American Watersnakes: A Natural History. Norman: University of Oklahoma Press.

Rhadinaea flavilata, Pine Woods Snake

Myers, C. W. 1967. The pine woods snake, *Rhadinaea flavilata* (Cope). Bulletin of the Florida State Museum 11:47–97.

Seminatrix pygaea, Black Swamp Snake

Dorcas, M. E., J. W. Gibbons, and H. G. Dowling. 1998. *Seminatrix* Cope. Catalogue of American Amphibians and Reptiles 679:1–5.

Gibbons, J. W., and M. E. Dorcas. 2004. North American Watersnakes: A Natural History. Norman: University of Oklahoma Press.

Winne, C. T., M. E. Dorcas, S. Poppy, and S. M. Poppy. 2005. Population structure, body size, and seasonal activity of black swamp snakes (*Seminatrix pygaea*). Southeastern Naturalist 4:1–14.

Storeria dekayi, Brown Snake

Christman, S. P. 1982. *Storeria dekayi.* Catalogue of American Amphibians and Reptiles 306:1–4.

King, R. B. 1993. Determinants of offspring number and size in the brown snake, *Storeria dekayi.* Journal of Herpetology 27:175–185.

Meshaka, W. E. Jr. 1994. Clutch parameters of *Storeria dekayi* Holbrook (Serpentes: Colubridae) from south-central Florida. Brimleyana 21:73–76.

Storeria occipitomaculata, Red-bellied Snake

Ernst, C. H. 2002. *Storeria occipitomaculata* (Storer). Catalogue of American Amphibians and Reptiles 759:1–8.

Rossman, D. A., and R. L. Erwin. 1980. Geographic variation in the snake *Storeria occipitomaculata* (Storer) (Serpentes: Colubridae) in southeastern United States. Brimleyana 4:95–102.

Semlitsch, R. D., and G. B. Moran. 1984. Ecology of the redbelly snake (*Storeria occipitomaculata*) using mesic habitats in South Carolina. American Midland Naturalist 111:33–40.

Tantilla coronata, Southeastern Crowned Snake

Aldridge, R. D., and R. D. Semlitsch. 1992. Female reproductive biology of the southeastern crowned snake (*Tantilla coronata*). Amphibia-Reptilia 13:209–218.

Aldridge, R. D., and R. D. Semlitsch. 1992. Male reproductive biology of the southeastern crowned snake (*Tantilla coronata*). Amphibia-Reptilia 13:219–225.

Semlitsch, R. D., K. L. Brown, and J. P. Caldwell. 1981. Habitat utilization, seasonal activity, and population size structure of the southeastern crowned snake (*Tantilla coronata*). Herpetologica 37:40–46.

Tantilla relicta, Florida Crowned Snake

Mushinsky, H. R., and B. W. Witz. 1993. Notes on the peninsula crowned snake, *Tantilla relicta,* in periodically burned habitat. Journal of Herpetology 27:468–470.

Telford, S. R. 1966. Variation among the southeastern crowned snakes, genus *Tantilla.* Bulletin of the Florida State Museum 10:261–304.

Telford, S. R. 1980. *Tantilla relicta.* Catalogue of American Amphibians and Reptiles 257:1–2.

Thamnophis sauritus, Eastern Ribbon Snake

Bernardino, F. S. Jr., and G. H. Dalrymple. 1992. Seasonal activity and road mortality of the snakes of the Pa-hay-okee wetlands of Everglades National Park, USA. Biological Conservation 61:71–75.

Bowers, B. B., A. E. Bledsoe, and G. M. Burghardt. 1993. Responses to escalating predatory threat in garter and ribbonsnakes (*Thamnophis*). Journal of Comparative Psychology 107:25–33.

Carpenter, C. C. 1952. Comparative ecology of the common gartersnake (*Thamnophis s. sirtalis*), the ribbonsnake (*Thamnophis s. sauritus*), and Butler's gartersnake (*Thamnophis butleri*) in mixed populations. Ecological Monographs 22:235–258.

Rossman, D. A. 1963. The colubrid snake genus *Thamnophis:* a revision of the *sauritus* group. Bulletin of the Florida State Museum 7:99–178.

***Thamnophis sirtalis,* Common Garter Snake**

Rossman, D. A., N. B. Ford, and R. A. Seigel. 1996. The Garter Snakes: Evolution and Ecology. Norman: University of Oklahoma Press.

***Virginia striatula,* Rough Earth Snake**

Clark, D. R. Jr., and R. R. Fleet. 1976. The rough earth snake (*Virginia striatula*): ecology of a Texas population. Southwestern Naturalist 20:467–478.

Powell, R., J. T. Collins, and L. D. Fish. 1994. *Virginia striatula.* Catalogue of American Amphibians and Reptiles 599:1–6.

***Virginia valeriae,* Smooth Earth Snake**

Blem, C. R., and L. B. Blem. 1985. Notes on *Virginia* (Reptilia: Colubridae) in Virginia. Brimleyana 11:87–95.

Willson, J. D., and M. E. Dorcas. 2004. Aspects of the ecology of small fossorial snakes in the western Piedmont of North Carolina. Southeastern Naturalist 3:1–12.

ELAPIDAE

***Micrurus fulvius,* Eastern Coral Snake**

Greene, H. W. 1984. Feeding behavior and diet of the eastern coral snake, *Micrurus fulvius.* Pp. 147–162 *in* R. A. Seigel, L. E. Hunt, J. L. Knight, L. Malaret, and N. L. Zuschlag, eds., Vertebrate Ecology and Systematics: A Tribute to Henry S. Fitch. University of Kansas Museum of Natural History Special Publication 10:1–278.

Jackson, D. R., and R. Franz. 1981. Ecology of the eastern coral snake (*Micrurus fulvius*) in northern peninsular Florida. Herpetologica 37:213–228.

Neill, W. T. 1957. Some misconceptions regarding the eastern coral snake, *Micrurus fulvius.* Herpetologica 13:111–118.

VIPERIDAE

***Agkistrodon contortrix,* Copperhead**

Fitch, H. S. 1960. Autecology of the Copperhead. University of Kansas Publications, Museum of Natural History 13:85–288.

Gloyd, H. K., and R. Conant. 1990. Snakes of the *Agkistrodon* Complex. Oxford, Ohio: Society for the Study of Reptiles and Amphibians.

Thorson, A., E. J. Lavonas, A. M. Rouse, and W. P. Kerns. 2003. Copperhead envenomations in the Carolinas. Journal of Toxicology–Clinical Toxicology 41:29–35.

***Agkistrodon piscivorus,* Cottonmouth**

Gibbons, J. W., and M. E. Dorcas. 2002. Defensive behavior of cottonmouths (*Agkistrodon piscivorus*) towards humans. Copeia 2002:195–198.

Glaudas, X. 2004. Do cottonmouths (*Agkistrodon piscivorus*) habituate to human confrontations? Southeastern Naturalist 3:129–138.

Scott, D. E., R. U. Fischer, J. D. Congdon, and S. A. Busa. 1995. Whole body lipid dynamics and reproduction in the eastern cottonmouth, *Agkistrodon piscivorus.* Herpetologica 51:472–487.

***Crotalus adamanteus,* Eastern Diamondback Rattlesnake**

Martin, W. H., and D. B. Means. 2000. Distribution and habitat relationships of the eastern diamondback rattlesnake (*Crotalus adamanteus*). Herpetological Natural History 7:9–34.

Means, D. B. 2005. The value of dead tree bases and stumpholes as habitat for wildlife. Pp. 74–78 *in* W. Meshaka Jr. and K. Babbitt, eds., Amphibians and Reptiles: Status and Conservation in Florida. Melbourne, Fla.: Krieger Press.

Speake, D. W., and R. H. Mount. 1973. Some possible ecological effects of "rattlesnake roundups" on the southeastern Coastal Plain. Proceedings of the Annual Conference of the Southeastern Association of Game and Fish Commissions 24:267–271.

Timmerman, W. W., and W. H. Martin. 2003. Conservation Guide to the Eastern Diamondback Rattlesnake, *Crotalus adamanteus.* Society for the Study of Amphibians and Reptiles, Herpetological Circular 32:1–64.

***Crotalus horridus,* Timber Rattlesnake**

Brown, W. S. 1993. Biology, Status, and Management of the Timber Rattlesnake (*Crotalus horridus*): A Guide for Conservation. Society for the Study of Amphibians and Reptiles, Herpetological Circular 22:1–84.

Diemer-Berish, J. E. 1998. Characterization of rattlesnake harvest in Florida. Journal of Herpetology 32:551–557.

Gibbons, J. W. 1972. Reproduction, growth, and sexual dimorphism in the canebrake rattlesnake (*Crotalus horridus atricaudatus*). Copeia 1972:222–226.

Hamilton, W. J. Jr., and J. A. Pollack. 1955. The food of some crotalid snakes from Fort Benning, Georgia. Natural History Miscellany 140:1–4.

Parmley, D., and A. M. Parmley. 2001. Food habits of the canebrake rattlesnake (*Crotalus horridus atricaudatus*) in central Georgia. Georgia Journal of Science 59:172–178.

***Sistrurus miliarius*, Pigmy Rattlesnake**

Farrell, T. M., P. G. May, and M. A. Pilgrim. 1995. Reproduction in the dusky pigmy rattlesnake, *Sistrurus miliarius barbouri*, in central Florida. Journal of Herpetology 29:21–27.

Glaudas, X., T. M. Farrell, and P. G. May. 2005. Defensive behavior of free-ranging pygmy rattlesnakes (*Sistrurus miliarius*). Copeia 2005:196–200.

May, P. G., T. M. Farrell, S. T. Heulett, M. A. Pilgrim, L. A. Bishop, D. J. Spence, A. M. Rabatsky, M. G. Campbell, A. D. Aycrigg, and W. E. Richardson II. 1996. Seasonal abundance and activity of a rattlesnake (*Sistrurus miliarius barbouri*) in central Florida. Copeia 1996:389–401.

Turtles

Carr, A. 1952. Handbook of Turtles: The Turtles of the United States, Canada, and Baja California. Ithaca, N.Y.: Cornell University Press.

Ernst, C. H., J. E. Lovich, and R. W. Barbour. 1994. Turtles of the United States and Canada. Washington, D.C.: Smithsonian Institution Press.

CHELONIIDAE

***Caretta caretta*, Loggerhead Sea Turtle**

Carr, A. F. Jr. 1986. Rips, FADS and little loggerheads. BioScience 36:92–100.

Frick, M. G., K. L. Williams, and M. Robinson. 1998. Epibionts associated with nesting loggerhead sea turtles (*Caretta caretta*) in Georgia, USA. Herpetological Review 29:211–214.

Ruckdeschel, C., C. R. Shoop, and G. R. Zug. 2000. Sea Turtles of the Georgia Coast. Occasional Publications of the Cumberland Island Museum 1:1–100.

Winn, B. 1999. Loggerhead sea turtle: *Caretta caretta*. Pp. 65–66 *in* T. W. Johnson, J. C. Ozier, J. L. Bohannon, J. B. Jensen, and C. Skelton, eds., Protected Animals of Georgia. Social Circle, Ga.: Georgia Department of Natural Resources, Wildlife Resources Division.

***Chelonia mydas*, Green Sea Turtle**

Carr, A. 1967. So Excellent a Fishe: A Natural History of Sea Turtles. New York: Scribner's.

Mortimer, J. A., and A. Carr. 1987. Reproduction and migrations of the Ascension Island green turtle (*Chelonia mydas*). Copeia 1987:103–113.

Ruckdeschel, C., C. R. Shoop, and G. R. Zug. 2000. Sea Turtles of the Georgia Coast. Occasional Publications of the Cumberland Island Museum 1:1–100.

Winn, B. 1999. Green sea turtle: *Chelonia mydas*. Pp. 67–68 *in* T. W. Johnson, J. C. Ozier, J. L. Bohannon, J. B. Jensen, and C. Skelton, eds., Protected Animals of Georgia. Social Circle, Ga.: Georgia Department of Natural Resources, Wildlife Resources Division.

***Eretmochelys imbricata*, Hawksbill Sea Turtle**

Diez, C. E., and R. P. van Dam. 2002. Habitat effect on hawksbill turtle growth rates on feeding grounds at Mona and Monito islands, Puerto Rico. Marine Ecology Progress Series 234:301–309.

Mrosovsky, N. 2000. Sustainable Use of Hawksbill Turtles: Contemporary Issues in Conservation. Darwin, Australia: Key Centre for Tropical Wildlife Management, Northern Territory University.

Richardson, J. I., R. Bell, and T. H. Richardson. 1999. Population ecology and demographic implications drawn from an 11-year study of nesting hawksbill turtles, *Eretmochelys imbricata*, at Jumby Bay, Long Island, Antigua, West Indies. Chelonian Conservation and Biology 3:244–250.

Ruckdeschel, C., C. R. Shoop, and G. R. Zug. 2000. Sea Turtles of the Georgia Coast. Occasional Publications of the Cumberland Island Museum 1:1–100.

Winn, B. 1999. Hawksbill sea turtle: *Eretmochelys imbricata*. Pp. 77–78 *in* T. W. Johnson, J. C. Ozier, J. L. Bohannon, J. B. Jensen, and C. Skelton, eds., Protected Animals of Georgia. Social Circle, Ga.: Georgia Department of Natural Resources, Wildlife Resources Division.

***Lepidochelys kempii*, Kemp's Ridley Sea Turtle**

Winn, B. 1999. Kemp's ridley sea turtle: *Lepidochelys kempii*. Pp. 87–88 *in* T. W. Johnson, J. C. Ozier, J. L. Bohannon, J. B. Jensen, and C. Skelton, eds., Protected Animals of Georgia. Social Circle, Ga.: Georgia Department of Natural Resources, Wildlife Resources Division.

Zug, G. R., H. J. Kalb, and S. J. Luzar. 1997. Age and growth in wild Kemp's ridley seaturtles *Lepido-*

chelys kempii from skeletochronological data. Biological Conservation 80:261–268.

CHELYDRIDAE

Chelydra serpentina, Common Snapping Turtle

Aresco, M. J., M. A. Ewert, M. S. Gunzburger, P. A. Meylan, and G. W. Heinrich. 2006. Snapping turtle: *Chelydra serpentina.* Pp. 44–57 *in* P. A. Meylan, A. G. J. Rhodin, and P. C. H. Pritchard, eds., Conservation Biology of Florida Turtles. Lunenburg, Mass.: Chelonian Research Foundation.

Galbraith, D. A., R. J. Brooks, and M. E. Obbard. 1989. The influence of growth rate on age and body size at maturity in female snapping turtles (*Chelydra serpentina*). Copeia 4:896–904.

Iverson, J. B., H. Higgins, A. Sirurulnik, and C. Griffiths. 1997. Local and geographic variation in the reproductive biology of the snapping turtle (*Chelydra serpentina*). Herpetologica 53:96–117.

Packard, G. C., M. J. Packard, K. Miller, and T. J. Boardman. 1987. Influence of moisture, temperature, and substrate on snapping turtle eggs and embryos. Ecology 68:983–993.

Macrochelys temminckii, Alligator Snapping Turtle

Dobie, J. L. 1971. Reproduction and growth in the alligator snapping turtle, *Macrochlemys temminckii* (Troost). Copeia 1971:645–658.

Jensen, J. B. 1999. Alligator snapping turtle: *Macroclemys temminckii.* Pp. 89–90 *in* T. W. Johnson, J. C. Ozier, J. L. Bohannon, J. B. Jensen, and C. Skelton, eds., Protected Animals of Georgia. Social Circle, Ga.: Georgia Department of Natural Resources, Wildlife Resources Division.

Jensen, J. B., and W. S. Birkhead. 2003. Distribution and status of the alligator snapping turtle (*Macrochelys temminckii*) in Georgia. Southeastern Naturalist 2:25–34.

Powders, V. N. 1978. Observations of oviposition and natural incubation of eggs of the alligator snapping turtle, *Macroclemmys temminckii,* in Georgia. Copeia 1978:154–156.

Pritchard, P. C. H. 1989. The Alligator Snapping Turtle: Biology and Conservation. Milwaukee: Milwaukee Public Museum.

DERMOCHELYIDAE

Dermochelys coriacea, Leatherback Sea Turtle

Dutton, P. H., B. W. Bowen, D. W. Owens, A. Barragan, and S. K. Davis. 1999. Global phylogeography of the leatherback turtle (*Dermochelys coriacea*). Journal of Zoology 248:397–409.

Keinath, J. A., and J. A. Musick. 1993. Movements and diving behavior of a leatherback turtle, *Dermochelys coriacea.* Copeia 1993:1010–1017.

Rabon, D. R., S. A. Johnson, R. Boettcher, M. Dodd, M. Lyons, S. Murphy, S. Ramsey, S. Roff, and K. Stewart. 2003. Confirmed leatherback turtle (*Dermochelys coriacea*) nests from North Carolina, with a summary of leatherback nesting activities north of Florida. Marine Turtle Newsletter 101:4–8.

Ruckdeschel, C., C. R. Shoop, and G. R. Zug. 2000. Sea Turtles of the Georgia Coast. Occasional Publications of the Cumberland Island Museum 1:1–100.

Winn, B. 1999. Leatherback sea turtle: *Dermochelys coriacea.* Pp. 73–74 *in* T. W. Johnson, J. C. Ozier, J. L. Bohannon, J. B. Jensen, and C. Skelton, eds., Protected Animals of Georgia. Social Circle, Ga.: Georgia Department of Natural Resources, Wildlife Resources Division.

EMYDIDAE

Chrysemys picta, Painted Turtle

Gibbons, J. W. 1968. Population structure and survivorship in the painted turtle, *Chrysemys picta.* Copeia 1968:260–268.

Grayson, K. L., and M. E. Dorcas. 2004. Seasonal temperature variation in the painted turtle (*Chrysemys picta*). Herpetologica 60:325–336.

Mitchell, J. C. 1988. Population ecology and life histories of the freshwater turtles *Chrysemys picta* and *Sternotherus odoratus* in an urban lake. Herpetological Monographs 2:40–61.

Clemmys guttata, Spotted Turtle

Barnwell, M. E., P. A. Meylan, and T. Walsh. 1997. The spotted turtle (*Clemmys guttata*) in central Florida. Chelonian Conservation and Biology 2:405–408.

Jensen, J. B. 1999. Spotted turtle: *Clemmys guttata.* Pp. 69–70 *in* T. W. Johnson, J. C. Ozier, J. L. Bohannon, J. B. Jensen, and C. Skelton, eds., Protected Animals of Georgia. Social Circle, Ga.: Georgia Department of Natural Resources, Wildlife Resources Division.

Litzgus, J. D., and T. A. Mousseau. 2003. Multiple

clutching in southern spotted turtles. Journal of Herpetology 37:17–23.

***Clemmys muhlenbergii,* Bog Turtle**

Jensen, J. B. 1999. Bog turtle: *Clemmys muhlenbergii.* Pp. 71–72 *in* T. W. Johnson, J. C. Ozier, J. L. Bohannon, J. B. Jensen, and C. Skelton, eds., Protected Animals of Georgia. Social Circle, Ga.: Georgia Department of Natural Resources, Wildlife Resources Division.

Lovich, J. E., D. W. Herman, and K. M. Fahey. 1992. Seasonal activity and movements of bog turtles (*Clemmys muhlenbergii*) in North Carolina. Copeia 1992:1107–1111.

***Deirochelys reticularia,* Chicken Turtle**

Buhlmann, K. A. 1995. Habitat use, terrestrial movements, and conservation of the turtle, *Deirochelys reticularia,* in Virginia. Journal of Herpetology 29:173–181.

Buhlmann, K. A., and J. W. Gibbons. 2001. Terrestrial habitat use by aquatic turtles from a seasonally fluctuating wetland: implications for wetland conservation boundaries. Chelonian Conservation and Biology 4:115–127.

Demuth, J. P., and K. A. Buhlmann. 1997. Diet of the turtle *Deirochelys reticularia* on the Savannah River Site, South Carolina. Journal of Herpetology 31:450–453.

Gibbons, J. W. 1969. Ecology and population dynamics of the chicken turtle, *Deirochelys reticularia.* Copeia 1969:669–676.

***Graptemys barbouri,* Barbour's Map Turtle**

Crenshaw, J. W. Jr., and G. B. Rabb. 1949. Occurrence of the turtle *Graptemys barbouri* in Georgia. Copeia 1949:226.

Jensen, J. B. 1999. Barbour's map turtle: *Graptemys barbouri.* Pp. 81–82 *in* T. W. Johnson, J. C. Ozier, J. L. Bohannon, J. B. Jensen, and C. Skelton, eds., Protected Animals of Georgia. Social Circle, Ga.: Georgia Department of Natural Resources, Wildlife Resources Division.

Lee, D. S., R. Franz, and R. A. Sanderson. 1975. A note on the feeding habits of male Barbour's map turtles. Florida Field Naturalist 3:45–46.

Wahlquist, H., and G. W. Folkerts. 1973. Eggs and hatchlings of Barbour's map turtle, *Graptemys barbouri* Carr and Marchand. Herpetologica 29:236–237.

***Graptemys geographica,* Common Map Turtle**

Jensen, J. B. 1999. Common map turtle: *Graptemys geographica.* Pp. 83–84 *in* T. W. Johnson, J. C. Ozier, J. L. Bohannon, J. B. Jensen, and C. Skelton, eds., Protected Animals of Georgia. Social Circle, Ga.: Georgia Department of Natural Resources, Wildlife Resources Division.

Pluto, T. G., and E. D. Bellis. 1986. Habitat utilization by the turtle, *Graptemys geographica,* along a river. Journal of Herpetology 20:22–31.

Pluto, T. G., and E. D. Bellis. 1988. Seasonal and annual movements of riverine map turtles, *Graptemys geographica.* Journal of Herpetology 22:152–158.

Vogt, R. C. 1981. Food partitioning in three sympatric species of map turtle, genus *Graptemys* (Testudinata, Emydidae). American Midland Naturalist 105:102–111.

White, D. Jr., and D. Moll. 1991. Clutch size and annual reproductive potential of the turtle *Graptemys geographica* in a Missouri stream. Journal of Herpetology 25:493–494.

***Graptemys pulchra,* Alabama Map Turtle**

Cagle, F. R. 1952. The status of the turtles *Graptemys pulchra* Baur and *Graptemys barbouri* Carr and Marchand, with notes on their natural history. Copeia 1952:223–234.

Jensen, J. B. 1999. Alabama map turtle: *Graptemys pulchra.* Pp. 85–86 *in* T. W. Johnson, J. C. Ozier, J. L. Bohannon, J. B. Jensen, and C. Skelton, eds., Protected Animals of Georgia. Social Circle, Ga.: Georgia Department of Natural Resources, Wildlife Resources Division.

Lovich, J. E., and C. J. McCoy. 1992. Review of the *Graptemys pulchra* group (Reptilia: Testudines: Emydidae), with descriptions of two new species. Annals of the Carnegie Museum 61:293–315.

***Malaclemys terrapin,* Diamondback Terrapin**

Gibbons, J. W., J. E. Lovich, A. D. Tucker, N. N. FitzSimmons, and J. L. Greene. 2001. Demographic and ecological factors affecting conservation and management of the diamondback terrapin (*Malaclemys terrapin*) in South Carolina. Chelonian Conservation and Biology 4:66–74.

Tucker, A. D., J. W. Gibbons, and J. L. Greene. 2001. Estimates of adult survival and migration for diamondback terrapins: conservation insight from local extirpation within a metapopulation. Canadian Journal of Zoology 79:2199–2209.

Tucker, A. D., S. R. Yeomans, and J. W. Gibbons. 1997. Shell strength of mud snails (*Ilyanassa obsoleta*) may deter foraging by diamondback terrapins (*Malaclemys terrapin*). American Midland Naturalist 138:224–229.

Pseudemys concinna, River Cooter

Buhlmann, K. A., and M. R. Vaughan. 1991. Ecology of the turtle *Pseudemys concinna* in the New River, West Virginia. Journal of Herpetology 25:72–78.

Jackson, D. R. 1995. Systematics of the *Pseudemys concinna-floridana* complex (Testudines: Emydidae): an alternative interpretation. Chelonian Conservation and Biology 1:329–333.

Jackson, D. R., and R. N. Walker. 1997. Reproduction in the Suwannee cooter, *Pseudemys concinna suwanniensis*. 1997. Bulletin of the Florida Museum of Natural History 41:69–167.

Seidel, M. E. 1994. Morphometric analysis and taxonomy of cooter and red-bellied turtles in the North American genus *Pseudemys* (Emydidae). Chelonian Conservation and Biology 1:117–130.

Pseudemys floridana, Florida Cooter

Gibbons, J. W., and J. W. Coker. 1977. Ecological and life history aspects of the cooter, *Chrysemys floridana* (LeConte). Herpetologica 33:29–33.

Jackson, D. R. 1995. Systematics of the *Pseudemys concinna-floridana* complex (Testudines: Emydidae): an alternative interpretation. Chelonian Conservation and Biology 1:329–333.

Seidel, M. E. 1994. Morphometric analysis and taxonomy of cooter and red-bellied turtles in the North American genus *Pseudemys* (Emydidae). Chelonian Conservation and Biology 1:117–130.

Pseudemys nelsoni, Florida Red-bellied Cooter

Jackson, D. R. 1978. *Chrysemys nelsoni*. Catalogue of American Amphibians and Reptiles 210:1–2.

Jackson, D. R. 1988. Reproductive strategies of sympatric freshwater emydid turtles in northern peninsular Florida. Bulletin of the Florida State Museum of Biological Sciences 33:113–158.

Lardie, R. I. 1973. Notes on courtship, eggs, and young of the Florida red-bellied turtle, *Chrysemys nelsoni*. HISS News Journal 1:183–184.

Terrapene carolina, Eastern Box Turtle

Congdon, J. D., R. E. Gatten Jr., and S. J. Morreale. 1989. Overwintering activity of box turtles (*Terrapene carolina*) in South Carolina. Journal of Herpetology 23:179–181.

Dodd, C. K. Jr. 2001. North American Box Turtles: A Natural History. Norman: University of Oklahoma Press.

Trachemys scripta, Pond Slider

Burke, V. J., J. L. Greene, and J. W. Gibbons. 1995. The effect of sample size and study duration on metapopulation estimates for slider turtles (*Trachemys scripta*). Herpetologica 51:451–456.

Garstka, W. R., W. E. Cooper Jr., K. W. Wasmund, and J. E. Lovich. 1991. Male sex steroids and hormonal control of male courtship behavior in the yellow-bellied slider turtle, *Trachemys scripta*. Comparative Biochemistry and Physiology 98A:271–280.

Gibbons, J. W. 1990. Life History and Ecology of the Slider Turtle. Washington, D.C.: Smithsonian Institution Press.

KINOSTERNIDAE

Kinosternon baurii, Striped Mud Turtle

Ewert, M. A., and D. S. Wilson. 1996. Seasonal variation of embryonic diapause in the striped mud turtle (*Kinosternon baurii*) and general considerations for conservation planning. Chelonian Conservation and Biology 2:43–54.

Iverson, J. B. 1978. Variation in striped mud turtles, *Kinosternon baurii* (Reptilia, Testudines, Kinosternidae). Journal of Herpetology 12:135–142.

Iverson, J. B. 1979. The female reproductive cycle in north Florida *Kinosternon baurii* (Testudines: Kinosternidae). Brimleyana 1:37–46.

Lamb, T. 1983. On the problematic identification of *Kinosternon* (Testudines: Kinosternidae) in Georgia, with new state localities for *Kinosternon baurii*. Georgia Journal of Science 41:115–120.

Lamb, T. 1983. The striped mud turtle (*Kinosternon bauri*) in South Carolina: a confirmation through multivariate character analysis. Herpetologica 39:383–390.

Kinosternon subrubrum, Eastern Mud Turtle

Burke, V. J., J. W. Gibbons, and J. L. Greene. 1993. Prolonged nesting forays by common mud turtles (*Kinosternon subrubrum*). American Midland Naturalist 131:190–195.

Gibbons, J. W. 1983. Reproductive characteristics and

ecology of the mud turtle, *Kinosternon subrubrum* (Lacépède). Herpetologica 39:254–271.

Sternotherus minor, Loggerhead Musk Turtle

Bels, V. L., and Y. J. M. Crama. 1994. Quantitative analysis of the courtship and mating behavior in loggerhead musk turtle *Sternotherus minor* (Reptilia: Kinosternidae) with comments on courtship behavior in turtles. Copeia 1994:676–684.

Etchberger, C. R., and L. M. Ehrhart. 1987. The reproductive biology of the female loggerhead musk turtle, *Sternotherus minor minor*, from the southern part of its range in central Florida. Herpetologica 43:66–73.

Etchberger, C. R., and R. H. Stovall. 1990. The seasonal variation in the testicular cycle of the loggerhead musk turtle, *Sternotherus minor minor*, from central Florida. Canadian Journal of Zoology 68:1071–1074.

Iverson, J. B. 1977. *Sternotherus minor* (Agassiz). Loggerhead musk turtle. Catalogue of American Amphibians and Reptiles 195:1–2.

Onorato, D. 1996. The growth rate and age distribution of *Sternotherus minor* at Rainbow Run, Florida. Journal of Herpetology 30:301–306.

Sternotherus odoratus, Common Musk Turtle

Iverson, J. B., and W. E. Meshaka. 2006. Common musk turtle (*Sternotherus odoratus*). Pp. 207–223 *in* P. A. Meylan and G. Heinrich, eds., The Conservation of the Turtles of Florida. Chelonian Conservation and Biology Monographs 3.

Gibbons, J. W. 1970. Reproductive characteristics of a Florida population of musk turtles (*Sternotherus odoratus*). Herpetologica 26:268–270.

Gibbons, J. W., J. L. Greene, and K. K. Patterson. 1982. Variation in reproductive characteristics of aquatic turtles. Copeia 1982:776–784.

TESTUDINIDAE

Gopherus polyphemus, Gopher Tortoise

Birkhead, R. D., C. Guyer, S. M. Hermann, and W. K. Michener. 2005. Species composition and seasonal abundance of seeds ingested by gopher tortoises (*Gopherus polyphemus*) in a southeastern pine savanna. American Midland Naturalist 154:143–151.

Boglioli, M. D., W. K. Michener, and C. Guyer. 2000. Habitat selection and modification by the gopher tortoise, *Gopherus polyphemus*, in Georgia longleaf pine forest. Chelonian Conservation and Biology 3:699–705.

Eubanks, J. O., W. K. Michener, and C. Guyer. 2003. Patterns of movement and burrow use in a population of gopher tortoises (*Gopherus polyphemus*). Herpetologica 59:311–321.

Landers, J. L., J. A. Garner, and W. A. McRae. 1980. Reproduction of gopher tortoises (*Gopherus polyphemus*) in southwestern Georgia. Herpetologica 36:351–361.

Tuberville, T. D., E. E. Clark, K. A. Buhlmann, and J. W. Gibbons. 2005. Translocation as a conservation tool: site fidelity and movements of repatriated gopher tortoises (*Gopherus polyphemus*). Animal Conservation 8:349–358.

TRIONYCHIDAE

Apalone ferox, Florida Softshell

Aresco, M. J. 2005. Mitigation measures to reduce highway mortality of turtles and other herpetofauna at a north Florida lake. Journal of Wildlife Management 69:549–560.

Meylan, P. A., R. Schuler, and P. Moler. 2002. Spermatogenic cycle of the Florida softshell turtle, *Apalone ferox*. Copeia 2002:779–786.

Apalone spinifera, Spiny Softshell

Plummer, M. V., and C. C. Burnley. 1997. Behavior, hibernacula, and thermal relations of softshell turtles (*Trionyx spiniferus*) overwintering in a small stream. Chelonian Conservation and Biology 2:489–493.

Plummer, M. V., N. E. Mills, and S. L. Allen. 1997. Activity, habitat, and movement patterns of softshell turtles (*Trionyx spiniferus*) in a small stream. Chelonian Conservation and Biology 2:514–520.

Stone, P. A., J. L. Dobie, and R. P. Henry. 1992. Cutaneous surface area and bimodal respiration in soft-shelled (*Trionyx spiniferus*), stinkpot (*Sternotherus odoratus*), and mud turtles (*Kinosternon subrubrum*). Physiological Zoology 65:311–330.

Stone, P. A., J. L. Dobie, and R. P. Henry. 1992. The effect of aquatic oxygen levels on diving and ventilatory behavior in soft-shelled (*Trionyx spiniferus*), stinkpot (*Sternotherus odoratus*), and mud turtles (*Kinosternon subrubrum*). Physiological Zoology 65:331–345.

Webb, R. G. 1973. *Trionyx spiniferus*. Catalogue of American Amphibians and Reptiles 140:1–4.

Checklist of Georgia's Amphibians and Reptiles

Amphibians

Frogs

BUFONIDAE

- ❏ *Bufo americanus* American toad
- ❏ *Bufo fowleri* Fowler's toad
- ❏ *Bufo quercicus* oak toad
- ❏ *Bufo terrestris* southern toad

HYLIDAE

- ❏ *Acris crepitans* northern cricket frog
- ❏ *Acris gryllus* southern cricket frog
- ❏ *Hyla avivoca* bird-voiced treefrog
- ❏ *Hyla chrysoscelis* Cope's gray treefrog
- ❏ *Hyla cinerea* green treefrog
- ❏ *Hyla femoralis* pine woods treefrog
- ❏ *Hyla gratiosa* barking treefrog
- ❏ *Hyla squirella* squirrel treefrog
- ❏ *Pseudacris brachyphona* mountain chorus frog
- ❏ *Pseudacris brimleyi* Brimley's chorus frog
- ❏ *Pseudacris crucifer* spring peeper
- ❏ *Pseudacris feriarum* upland chorus frog
- ❏ *Pseudacris nigrita* southern chorus frog
- ❏ *Pseudacris ocularis* little grass frog
- ❏ *Pseudacris ornata* ornate chorus frog

LEPTODACTYLIDAE

- ❏ *Eleutherodactylus planirostris* greenhouse frog (introduced)

MICROHYLIDAE

- ❏ *Gastrophryne carolinensis* eastern narrow-mouthed toad

PELOBATIDAE

- ❏ *Scaphiopus holbrookii* eastern spadefoot

RANIDAE

- ❏ *Rana capito* gopher frog
- ❏ *Rana catesbeiana* bullfrog
- ❏ *Rana clamitans* green frog
- ❏ *Rana grylio* pig frog
- ❏ *Rana heckscheri* river frog
- ❏ *Rana palustris* pickerel frog
- ❏ *Rana sphenocephala* southern leopard frog
- ❏ *Rana sylvatica* wood frog
- ❏ *Rana virgatipes* carpenter frog

Salamanders

AMBYSTOMATIDAE

- ❏ *Ambystoma cingulatum* flatwoods salamander
- ❏ *Ambystoma maculatum* spotted salamander
- ❏ *Ambystoma opacum* marbled salamander
- ❏ *Ambystoma talpoideum* mole salamander
- ❏ *Ambystoma tigrinum* tiger salamander

AMPHIUMIDAE

- ❏ *Amphiuma means* two-toed amphiuma
- ❏ *Amphiuma pholeter* one-toed amphiuma

CRYPTOBRANCHIDAE

- ❏ *Cryptobranchus alleganiensis* hellbender

PLETHODONTIDAE

- ❏ *Aneides aeneus* green salamander
- ❏ *Desmognathus aeneus* seepage salamander
- ❏ *Desmognathus apalachicolae* Apalachicola dusky salamander
- ❏ *Desmognathus auriculatus* southern dusky salamander
- ❏ *Desmognathus conanti* spotted dusky salamander
- ❏ *Desmognathus folkertsi* dwarf black-bellied salamander
- ❏ *Desmognathus marmoratus* shovel-nosed salamander
- ❏ *Desmognathus monticola* seal salamander
- ❏ *Desmognathus ocoee* Ocoee salamander
- ❏ *Desmognathus quadramaculatus* black-bellied salamander
- ❏ *Eurycea chamberlaini* Chamberlain's dwarf salamander

- ❑ *Eurycea cirrigera* southern two-lined salamander
- ❑ *Eurycea guttolineata* three-lined salamander
- ❑ *Eurycea longicauda* long-tailed salamander
- ❑ *Eurycea lucifuga* cave salamander
- ❑ *Eurycea quadridigitata* dwarf salamander
- ❑ *Eurycea wilderae* Blue Ridge two-lined salamander
- ❑ *Gyrinophilus palleucus* Tennessee cave salamander
- ❑ *Gyrinophilus porphyriticus* spring salamander
- ❑ *Haideotriton wallacei* Georgia blind salamander
- ❑ *Hemidactylium scutatum* four-toed salamander
- ❑ *Plethodon chattahoochee* Chattahoochee slimy salamander
- ❑ *Plethodon chlorobryonis* Atlantic Coast slimy salamander
- ❑ *Plethodon glutinosus* northern slimy salamander
- ❑ *Plethodon grobmani* southeastern slimy salamander
- ❑ *Plethodon metcalfi* southern gray-cheeked salamander
- ❑ *Plethodon ocmulgee* Ocmulgee slimy salamander
- ❑ *Plethodon petraeus* Pigeon Mountain salamander
- ❑ *Plethodon savannah* Savannah slimy salamander
- ❑ *Plethodon serratus* southern red-backed salamander
- ❑ *Plethodon shermani* red-legged salamander
- ❑ *Plethodon teyahalee* southern Appalachian salamander
- ❑ *Plethodon variolatus* South Carolina slimy salamander
- ❑ *Plethodon ventralis* southern zigzag salamander
- ❑ *Plethodon websteri* Webster's salamander
- ❑ *Pseudotriton montanus* mud salamander
- ❑ *Pseudotriton ruber* red salamander
- ❑ *Stereochilus marginatus* many-lined salamander

PROTEIDAE

- ❑ *Necturus* cf. *beyeri* Alabama waterdog
- ❑ *Necturus maculosus* mudpuppy
- ❑ *Necturus punctatus* dwarf waterdog

SALAMANDRIDAE

- ❑ *Notophthalmus perstriatus* striped newt
- ❑ *Notophthalmus viridescens* eastern newt

SIRENIDAE

- ❑ *Pseudobranchus striatus* northern dwarf siren
- ❑ *Siren intermedia* lesser siren
- ❑ *Siren lacertina* greater siren

Reptiles

Crocodilians

ALLIGATORIDAE

- ❑ *Alligator mississippiensis* American alligator

Worm Lizards

RHINEURIDAE

- ❑ *Rhineura floridana* Florida worm lizard

Lizards

ANGUIDAE

- ❑ *Ophisaurus attenuatus* slender glass lizard
- ❑ *Ophisaurus compressus* island glass lizard
- ❑ *Ophisaurus mimicus* mimic glass lizard
- ❑ *Ophisaurus ventralis* eastern glass lizard

GEKKONIDAE

- ❑ *Hemidactylus turcicus* Mediterranean gecko (introduced)

PHRYNOSOMATIDAE

- ❑ *Sceloporus undulatus* eastern fence lizard

POLYCHROTIDAE

- ❑ *Anolis carolinensis* green anole
- ❑ *Anolis sagrei* brown anole (introduced)

SCINCIDAE

- ❑ *Eumeces anthracinus* coal skink
- ❑ *Eumeces egregius* mole skink
- ❑ *Eumeces fasciatus* five-lined skink
- ❑ *Eumeces inexpectatus* southeastern five-lined skink
- ❑ *Eumeces laticeps* broadhead skink
- ❑ *Scincella lateralis* ground skink

TEIIDAE

- ❑ *Cnemidophorus sexlineatus* six-lined racerunner

Snakes

COLUBRIDAE

- ❑ *Carphophis amoenus* eastern worm snake
- ❑ *Cemophora coccinea* scarlet snake
- ❑ *Coluber constrictor* black racer
- ❑ *Diadophis punctatus* ringneck snake
- ❑ *Drymarchon couperi* eastern indigo snake

- ❑ *Elaphe guttata* corn snake
- ❑ *Elaphe obsoleta* rat snake
- ❑ *Farancia abacura* mud snake
- ❑ *Farancia erytrogramma* rainbow snake
- ❑ *Heterodon platirhinos* eastern hognose snake
- ❑ *Heterodon simus* southern hognose snake
- ❑ *Lampropeltis calligaster* mole kingsnake
- ❑ *Lampropeltis getula* common kingsnake
- ❑ *Lampropeltis triangulum elapsoides* scarlet kingsnake
- ❑ *Lampropeltis triangulum triangulum* eastern milk snake
- ❑ *Masticophis flagellum* coachwhip
- ❑ *Nerodia erythrogaster* plain-bellied watersnake
- ❑ *Nerodia fasciata* banded watersnake
- ❑ *Nerodia floridana* eastern green watersnake
- ❑ *Nerodia sipedon* northern watersnake
- ❑ *Nerodia taxispilota* brown watersnake
- ❑ *Opheodrys aestivus* rough green snake
- ❑ *Pituophis melanoleucus* pine snake
- ❑ *Regina alleni* striped crayfish snake
- ❑ *Regina rigida* glossy crayfish snake
- ❑ *Regina septemvittata* queen snake
- ❑ *Rhadinaea flavilata* pine woods snake
- ❑ *Seminatrix pygaea* black swamp snake
- ❑ *Storeria dekayi* brown snake
- ❑ *Storeria occipitomaculata* red-bellied snake
- ❑ *Tantilla coronata* southeastern crowned snake
- ❑ *Tantilla relicta* Florida crowned snake
- ❑ *Thamnophis sauritus* eastern ribbon snake
- ❑ *Thamnophis sirtalis* common garter snake
- ❑ *Virginia striatula* rough earth snake
- ❑ *Virginia valeriae* smooth earth snake

ELAPIDAE

- ❑ *Micrurus fulvius* eastern coral snake

VIPERIDAE

- ❑ *Agkistrodon contortrix* copperhead
- ❑ *Agkistrodon piscivorus* cottonmouth
- ❑ *Crotalus adamanteus* eastern diamondback rattlesnake
- ❑ *Crotalus horridus* timber rattlesnake
- ❑ *Sistrurus miliarius* pigmy rattlesnake

Turtles

CHELONIIDAE

- ❑ *Caretta caretta* loggerhead sea turtle
- ❑ *Chelonia mydas* green sea turtle
- ❑ *Eretmochelys imbricata* hawksbill sea turtle
- ❑ *Lepidochelys kempii* Kemp's ridley sea turtle

CHELYDRIDAE

- ❑ *Chelydra serpentina* common snapping turtle
- ❑ *Macrochelys temminckii* alligator snapping turtle

DERMOCHELYIDAE

- ❑ *Dermochelys coriacea* leatherback sea turtle

EMYDIDAE

- ❑ *Chrysemys picta* painted turtle
- ❑ *Clemmys guttata* spotted turtle
- ❑ *Clemmys muhlenbergii* bog turtle
- ❑ *Deirochelys reticularia* chicken turtle
- ❑ *Graptemys barbouri* Barbour's map turtle
- ❑ *Graptemys geographica* common map turtle
- ❑ *Graptemys pulchra* Alabama map turtle
- ❑ *Malaclemys terrapin* diamondback terrapin
- ❑ *Pseudemys concinna* river cooter
- ❑ *Pseudemys floridana* Florida cooter
- ❑ *Pseudemys nelsoni* Florida red-bellied cooter
- ❑ *Terrapene carolina* eastern box turtle
- ❑ *Trachemys scripta* pond slider

KINOSTERNIDAE

- ❑ *Kinosternon baurii* striped mud turtle
- ❑ *Kinosternon subrubrum* eastern mud turtle
- ❑ *Sternotherus minor* loggerhead musk turtle
- ❑ *Sternotherus odoratus* common musk turtle

TESTUDINIDAE

- ❑ *Gopherus polyphemus* gopher tortoise

TRIONYCHIDAE

- ❑ *Apalone ferox* Florida softshell
- ❑ *Apalone spinifera* spiny softshell

Index of Scientific Names

Boldface page numbers refer to species accounts.

Index of Common Names

Boldface page numbers refer to species accounts.

DADE
CATOOSA
WHITFIELD
FANNIN
TOWNS
UNION
RABUN
WALKER
MURRAY
GILMER
WHITE
CHATTOOGA
LUMPKIN
HABERSHAM
STEPHENS
GORDON
PICKENS
DAWSON
HALL
BANKS
FRANKLIN
HART
FLOYD
BARTOW
CHEROKEE
FORSYTH
JACKSON
MADISON
ELBERT
POLK
COBB
GWINNETT
BARROW
CLARKE
OGLETHORPE
PAULDING
HARALSON
OCONEE
WALTON
WILKES
LINCOLN
DOUGLAS
FULTON
DEKALB
ROCKDALE
CARROLL
CLAYTON
MORGAN
GREENE
NEWTON
TALIAFERRO
COLUMBIA
HENRY
WARREN
MCDUFFIE
FAYETTE
COWETA
HEARD
JASPER
PUTNAM
RICHMOND
SPALDING
BUTTS
HANCOCK
GLASCOCK
MERIWETHER
PIKE
TROUP
LAMAR
MONROE
JONES
BALDWIN
JEFFERSON
BURKE
WASHINGTON
UPSON
HARRIS
BIBB
WILKINSON
JENKINS
TALBOT
CRAWFORD
TWIGGS
JOHNSON
SCREVEN
EMANUEL
MUSCOGEE
TAYLOR
PEACH
LAURENS
CHATTA-
HOOCHEE
MARION
MACON
HOUSTON
BLECKLEY
TREUTLEN
CANDLER
BULLOCH
EFFINGHAM
SCHLEY
STEWART
DOOLY
PULASKI
DODGE
WHEELER
MONTGOMERY
TOOMBS
EVANS
BRYAN
CHATHAM
WEBSTER
SUMTER
QUITMAN
CRISP
WILCOX
TELFAIR
TATTNALL
TERRELL
JEFF DAVIS
LIBERTY
RANDOLPH
LEE
BEN HILL
APPLING
TURNER
LONG
CLAY
CALHOUN
DOUGHERTY
WORTH
IRWIN
COFFEE
BACON
WAYNE
MCINTOSH
TIFT
EARLY
BAKER
BERRIEN
ATKINSON
PIERCE
MITCHELL
COLQUITT
GLYNN
MILLER
COOK
BRANTLEY
WARE
LANIER
SEMINOLE
DECATUR
GRADY
THOMAS
BROOKS
LOWNDES
CLINCH
CHARLTON
CAMDEN
ECHOLS